AF880317

Landscape Series

Volume 16

Series Editors

Henri Décamps, Centre National de la Recherche Scientifique, Toulouse, France
Bärbel Tress, TRESS & TRESS GbR, Munich, Germany
Gunther Tress, TRESS & TRESS GbR, Munich, Germany

For further volumes:
http://www.springer.com/series/6211

Aims and Scope

Springer's innovative Landscape Series is committed to publishing high-quality manuscripts that approach the concept of landscape from a broad range of perspectives. Encouraging contributions on theory development, as well as more applied studies, the series attracts outstanding research from the natural and social sciences, and from the humanities and the arts. It also provides a leading forum for publications from interdisciplinary and transdisciplinary teams.

Drawing on, and synthesising, this integrative approach the Springer Landscape Series aims to add new and innovative insights into the multidimensional nature of landscapes. Landscapes provide homes and livelihoods to diverse peoples; they house historic—and prehistoric—artefacts; and they comprise complex physical, chemical and biological systems. They are also shaped and governed by human societies who base their existence on the use of the natural resources; people enjoy the aesthetic qualities and recreational facilities of landscapes, and people design new landscapes.

As interested in identifying best practice as it is in progressing landscape theory, the Landscape Series particularly welcomes problem-solving approaches and contributions to landscape management and planning. The ultimate goal is to facilitate both the application of landscape research to practice, and the feedback from practice into research.

Pablo Campos · Lynn Huntsinger
José L. Oviedo · Paul F. Starrs
Mario Díaz · Richard B. Standiford
Gregorio Montero
Editors

Mediterranean Oak Woodland Working Landscapes

Dehesas of Spain and Ranchlands of California

Springer

Editors
Pablo Campos
José L. Oviedo
Institute of Public Goods and Policies
Spanish National Research Council
Madrid
Spain

Lynn Huntsinger
Richard B. Standiford
Department of Environmental Science,
Policy, and Management
University of California
Berkeley, CA
USA

Mario Díaz
Museo Nacional de Ciencias Naturales
Spanish National Research Council
Madrid
Spain

Paul F. Starrs
Department of Geography
University of Nevada
Reno, NV
USA

Gregorio Montero
Forest Research Centre
National Institute for Agriculture
and Food Research and Technology
Madrid
Spain

ISSN 1572-7742 ISSN 1875-1210 (electronic)
ISBN 978-94-007-6706-5 ISBN 978-94-007-6707-2 (eBook)
DOI 10.1007/978-94-007-6707-2
Springer Dordrecht Heidelberg New York London

Library of Congress Control Number: 2013939347

© Springer Science+Business Media Dordrecht 2013
This work is subject to copyright. All rights are reserved by the Publisher, whether the whole or part of the material is concerned, specifically the rights of translation, reprinting, reuse of illustrations, recitation, broadcasting, reproduction on microfilms or in any other physical way, and transmission or information storage and retrieval, electronic adaptation, computer software, or by similar or dissimilar methodology now known or hereafter developed. Exempted from this legal reservation are brief excerpts in connection with reviews or scholarly analysis or material supplied specifically for the purpose of being entered and executed on a computer system, for exclusive use by the purchaser of the work. Duplication of this publication or parts thereof is permitted only under the provisions of the Copyright Law of the Publisher's location, in its current version, and permission for use must always be obtained from Springer. Permissions for use may be obtained through RightsLink at the Copyright Clearance Center. Violations are liable to prosecution under the respective Copyright Law.
The use of general descriptive names, registered names, trademarks, service marks, etc. in this publication does not imply, even in the absence of a specific statement, that such names are exempt from the relevant protective laws and regulations and therefore free for general use.
While the advice and information in this book are believed to be true and accurate at the date of publication, neither the authors nor the editors nor the publisher can accept any legal responsibility for any errors or omissions that may be made. The publisher makes no warranty, express or implied, with respect to the material contained herein.

Printed on acid-free paper

Springer is part of Springer Science+Business Media (www.springer.com)

Foreword

How exciting and remarkable that this book, long in the making, is now done. Within these pages economy meets natural resources and ecology, in a union that honors both science and the practice of management. Compared within are two geographically set-apart agroforestry ecosystems that are nonetheless near neighbors in terms of climate, ecology, and cultural-historical linkages: Spain's extensive *dehesas* and the oak woodland ranches of California. This study of working woodlands in areas of Mediterranean-type climate sets aside proprietary approaches, laying out instead a body of knowledge and field-gathered data for use by professionals, managers, and policymakers. Those of us who have long sought to globalize studies of natural resource management, recognizing that economies and ecosystems are today wholly internationalized, see in this work author skills and interests that demolish all those conventional disciplinary limitations that typically restrain—and hamstring—scientific research.

The scope of this undertaking is commensurate with the complexity of the ecosystems and economies studied. Interdisciplinary collaboration demands breaking down a traditional aloofness among specialties and countries, and, with that, overcoming technical terminology. It is nearly without precedent for authors to have forged such commonalities in language, methodology, and focus. Overcoming a looming Tower of Babel of arcane specialized subfields, approaches, and language is difficult and irksome. Then, of course, when outlines were done, findings had to be rendered in the scientific vernacular of today, which is English. To do that, the entire working group necessarily grew comfortable with the Spanish of Castile and the English of California, accepting a vernacular with words like woodlands, *dehesa*, *monte*, and shrublands. At hand was a living lesson: an evolving process of mutual exchange and linguistic enrichment. Reliable data was drawn into support arguments and observations, and often was laboriously gleaned from places where information seemed initially unavailable.

It is therefore satisfying that the analyses in this volume ultimately derive from a huge collection of data, obtained for the most part directly by the researchers who wrote and illustrated each chapter. This offers a fertile synergy where the analysis in a chapter includes concrete data on motivation and behavior, income and production, historical process or ecosystem function—or all of the above. The discussion of the origins and evolution of land claims and the law of property in Spain and California, for example, lays out practices that historically shaped dehesas and ranches, making this book a sizable step forward in comparative studies that will edify and clarify. It is neither possible nor desirable for me to go here into questions of detail, but I would add that, to my way of thinking, this book stands as a before-and-after benchmark; it compares the facts, fancies, and function of dehesa properties and woodland ranches, which is significant not just because of what is said in these pages, but also because a firm and unmistakeable foundation is laid for any future investigation.

Coordinating a large team of researcher-authors is, at the best of times, arduous. Let me stress the importance of the experience and expertise that bound together this group of authors, and constructed the vision of the editors, who are united by an untallyable count of meetings, field visits, and exchanges where they knotted together friendships and cemented an ongoing collaboration. The book itself, with its extensive photographic material, reveals a fusion of intellect and shared affection that shows how human exchange encourages creativity, enthusiasm, and exuberance to the mutual improvement of researcher and results. I think it is also notable that this book has gone ahead with authors who gave freely of their time. Authors toiled on this because they believe in working landscapes and the people who work them, they enjoy learning about residents on the land, and ultimately in gaining understanding of the human role in ecosystem conservation. This compels me to note a paradox: How often does the richest learning and result come from studies that issue primarily from interest and affection?

The book poses philosophical reflections that go well beyond agroforestry ecosystems. In-depth study of complex systems such as dehesas and oak woodland ranches suggests the limitations posed by conventional sources of academic knowledge. That division starts with a specious separation between the natural or earth sciences and the social or human sciences. Barriers purportedly loom like redwoods or chestnut trees, separating humans from the natural world, dividing economy and environment, sundering quality of life considerations from environmental quality. Yet dehesas and ranches produce both sellable goods and "environmental services," which put the lie to standard sequestering of such services into spaces, parks, or ecosystems that are described as "natural," where they are supposedly incompatible with any form of extractive economic activity. Paradoxically, when these book authors write about ecosystem services, they show that an oak woodland agroforestry ecosystem not only makes sellable goods, it also produces an ecosystem that generates a rich range of "environmental services." Humans relish these services.

With so many amenities to offer, the much-managed dehesa landscape is appreciated as much or more than a forest where humans as stewards and producers are absent. In fact, such a forest is quite unnatural, given thousands of years of human occupation and use in California and Spain. An enjoyment and love of time spent in the built landscape of dehesas and ranches guides managers, owners, and visitors to oak woodland properties, which makes them a product of human choice as much as pecuniary goals. Let us, as a result, consider as ancient prejudice any argument whatsoever that insists on separating economy and ecology. Nor does nature's economy function without humans: agroecology, industrial ecology, and urban ecology are part of the same fundamental economy of our time on Earth. In general, I would argue that this ambitious work demonstrates that investigations uniting systematic study, including processes (economics, history, ecology, geography) pushes authors to transcend reductive borders. The result, here, is a model for understanding not just the dehesa and the oak woodland ranch, but for undertaking economic analysis in general.

In sum, this book exemplifies the salutary advantages of transdisciplinary research in the widening terrain of studies formed by an open economy. Not only are oak woodland ranches in California and the Spanish dehesa illuminated with a fine touch, so too are studies of working landscapes and economic processes. Sometimes what is laid bare are landscape deficiencies and economic problems; in other cases, what is suggested are improvements and benefits. Humankind may as a result be able to make saner, safer, sounder use of resources. We may learn from centuries of traditional agriculture, the institutions that build social capital, and the curious yet elegant vernacular architecture that results from this.

Madrid, November 20, 2012

José Manuel Naredo
Economist
Ad honorem Lecturer of Madrid School of Architecture,
Spanish National Award on Environment (2000),
Geocritica International Award (2008),
WWF Award for Natural Environment Conservation (2011)

Contents

Part I Setting

1 **Working Landscapes of the Spanish Dehesa and the California Oak Woodlands: An Introduction** 3
Lynn Huntsinger, Pablo Campos, Paul F. Starrs, José L. Oviedo, Mario Díaz, Richard B. Standiford and Gregorio Montero

2 **History and Recent Trends** 25
Peter S. Alagona, Antonio Linares, Pablo Campos and Lynn Huntsinger

Part II Vegetation

3 **Climatic Influence on Oak Landscape Distributions** 61
Sonia Roig, Rand R. Evett, Guillermo Gea-Izquierdo, Isabel Cañellas and Otilio Sánchez-Palomares

4 **Soil and Water Dynamics** 91
Susanne Schnabel, Randy A. Dahlgren and Gerardo Moreno-Marcos

5 **Oak Regeneration: Ecological Dynamics and Restoration Techniques** 123
Fernando Pulido, Doug McCreary, Isabel Cañellas, Mitchel McClaran and Tobias Plieninger

6 **Overstory–Understory Relationships** 145
Gerardo Moreno, James W. Bartolome, Guillermo Gea-Izquierdo and Isabel Cañellas

7 **Acorn Production Patterns** 181
Walter D. Koenig, Mario Díaz, Fernando Pulido, Reyes Alejano, Elena Beamonte and Johannes M. H. Knops

Part III Management, Uses, and Ecosystem Response

8 **Effects of Management on Biological Diversity and Endangered Species** 213
Mario Díaz, William D. Tietje and Reginald H. Barrett

9 **Models of Oak Woodland Silvopastoral Management** 245
Richard B. Standiford, Paola Ovando, Pablo Campos and Gregorio Montero

10 **Raising Livestock in Oak Woodlands** 273
Juan de Dios Vargas, Lynn Huntsinger and Paul F. Starrs

11 **Hunting in Managed Oak Woodlands: Contrasts Among Similarities** 311
Luke T. Macaulay, Paul F. Starrs and Juan Carranza

Part IV Economics

12 **Economics of Ecosystem Services** 353
Alejandro Caparrós, Lynn Huntsinger, José L. Oviedo, Tobias Plieninger and Pablo Campos

13 **The Private Economy of Dehesas and Ranches: Case Studies** 389
José L. Oviedo, Paola Ovando, Larry Forero, Lynn Huntsinger, Alejandro Álvarez, Bruno Mesa and Pablo Campos

Part V Landscape

14 **Recent Oak Woodland Dynamics: A Comparative Ecological Study at the Landscape Scale** 427
Ramón Elena-Rosselló, Maggi Kelly, Sergio González-Avila, Alexandra Martín, David Sánchez de Ron and José M. García del Barrio

Part VI Conclusions

15 **Whither Working Oak Woodlands?** 463
Paul F. Starrs, José L. Oviedo, Pablo Campos, Lynn Huntsinger, Mario Díaz, Richard B. Standiford and Gregorio Montero

Index 499

Part I
Setting

Chapter 1
Working Landscapes of the Spanish Dehesa and the California Oak Woodlands: An Introduction

Lynn Huntsinger, Pablo Campos, Paul F. Starrs, José L. Oviedo, Mario Díaz, Richard B. Standiford and Gregorio Montero

Frontispiece Chapter 1. Gateway to a Californian oak woodland cattle ranch. California and Spain share an economic, ecosystemic, and cultural tradition of extensive properties that produce diverse goods and services. (Photograph by L. Huntsinger)

L. Huntsinger (✉) · R. B. Standiford
Department of Environmental Science, Policy, and Management, University of California, Berkeley, 130 Mulford Hall , Berkeley, CA, MC 3110, 94720 USA
e-mail: huntsinger@berkeley.edu

P. Campos et al. (eds.), *Mediterranean Oak Woodland Working Landscapes*, Landscape Series 16, DOI: 10.1007/978-94-007-6707-2_1,
© Springer Science+Business Media Dordrecht 2013

Abstract Oak woodlands have offered a welcoming environment for human activities for tens of thousands of years, but how that history has unfolded has many variations. The long-time collaboration that led to this book ran into complications arising from the different meanings attached to many a term, including struggles over the most appropriate title, settling on common units of measurement and area, quantifying the woodland's extent in Spain and California, and even in deciding how many oaks constitute a woodland. Defining with anything approaching international precision such terms as oak woodlands, oak woodland ranches, and wooded dehesas is nuanced, and is compounded by distinctions in culture and language. But our efforts to dovetail one inscrutable system with another may offer insight into the relationship of humans with environments long occupied and modified, as further shaped by location, history, and opportunity. In 15 chapters we offer a comparison of conservation and management on California oak woodland ranches and in the dehesas of Spain, including economic, institutional, ecological, spatial, and geographical aspects, from how to raise an Iberian pig to what we can learn about oak woodlands with remote sensing.

Keywords Translations · Comparative study · International exchange · Multi-functional · Landscapes

R. B. Standiford
e-mail: standifo@berkeley.edu

P. Campos · J. L. Oviedo
Institute of Public Goods and Policies (IPP), Spanish National Research Council (CSIC), Albasanz 26-28, 28037 Madrid, Spain
e-mail: pablo.campos@csic.es

J. L. Oviedo
e-mail: jose.oviedo@csic.es

P. F. Starrs
Department of Geography, University of Nevada, Reno, MS 0154, Reno, NV 89557–0048, USA
e-mail: starrs@unr.edu

M. Díaz
Department of Biogeography and Global Change, Museo Nacional de Ciencias Naturales (BGC-MNCN), Spanish National Research Council (CSIC), Serrano 115bis 28006 Madrid, Spain
e-mail: Mario.Diaz@ccma.csic.es

G. Montero
Forest Research Centre, National Institute for Agriculture and Food Research and Technology, Ctra. de la Coruña km. 7,5 28040 Madrid, Spain
e-mail: montero@inia.es

1.1 Origins, Language, and Expectations

When we began writing a book in 2002 comparing what might on the surface appear to be similar oak woodland landscapes in California and Spain, the proposition hardly seemed difficult. Over the years, a group of scholars from Europe and the United States interested in oaks and who study the ways that humans occupy oak landscapes developed close connections and a shared interest in what a comparison of landscapes could offer. Every chapter in this book is written by locally based authors well versed in the oak woodlands of Spain, of California, or both. There is added work from colleagues resident in Germany, Portugal, France, other parts of Europe, and from across the United States. Our goal is to compare the history, economics, ecology, and management of the oak woodland ranches of California and the dehesas of Spain (Fig. 1.1). But as the work progressed over coffee breaks, on joint field studies, and when sifting through the many-languaged and various-disciplined contributions, we found our efforts to navigate comparisons washing up regularly on savage shoals of awkward translation.

Problems arose from the different dimensions of meaning attached to many a terms that permeate this book, starting with struggles over the most appropriate title, but extending to working with different units of measurement and area, defining the woodland's extent in Spain and California, and even deciding how many oaks are needed to constitute a woodland. But we hope these efforts to dovetail the inscrutability of one system to another may offer insight into the

Fig. 1.1 The Iberian pigs historically characteristic of the dehesa were often accompanied by swineherds, now a rarity, but featured with long cloak and shepherd's staff in this 1960 view captured by the Berkeley geographer James J. Parsons. Herds of black-hued pigs such as these are still common users and grazers on the oak woodlands of Spain, although escorts are less common now than 50 years ago. (Photograph from the collection of P.F. Starrs)

Fig. 1.2 A gateway, with ornate tiles of painting by the famed illustrator Mariano Aguayo showing dogs assembled for a hunt, offers an entry into a dehesa near Cazalla de la Sierra, north of Seville. Properties in the oak woodlands of Spain and California are reflections of the aspirations and pleasures of their owners—whether absentee or resident on the land. (Photograph by S. García)

distinct ways that the relationship of humans with the environments they have occupied and modified plays out, as shaped by location, history, and opportunity.

This is very much an effort to forge an understanding of landscapes grounded in economies, geographies, histories, and ecologies that are distinct yet allied by increasingly common outwash from the global economy including the elusive—yet findable—human preference. The main difference in studying a "working" landscape, as compared to another kind of landscape, is that the human dimension is at least as important as the ecological one (Fig. 1.2).

1.2 Complexities in Translation and Definition

Translating a word from one language to another might seem a straightforward process, but when it comes to oak woodland dehesas and ranches, problems of translation reflect the need to translate one world to another: old to new, Iberian Peninsula to North America, Spain to California. As an opening example there is no word in American English that does justice to the term "dehesa." For the 47 million residents of Spain, the California term "oak woodland" may seem vague, ill defined, and even banal. Certainly lacking are the savory connotations of a southern European vocabulary describing the remarkably varied and humanized woodlands of the dehesa. Those have been appreciated and exploited for—literally—millennia, since well before Roman and Arab occupiers began spreading everything from culture, hunting, seeds, livestock, economies, and ambitions around the Mediterranean basin. The French historian-geographer Fernand Braudel pondered this while imprisoned during World War II and came to the conclusion that occupation of the Mediterranean realm involved one of the great transformations of human society (Braudel 1975). We do nevertheless in this book attempt to explain one continent to the other, to share, synthesize, and compare what is known, loaned, and retrieved from each society.

Words do not always add up to worlds, and instead require context and explanation. To offer one example, an oak woodland in California generally refers to an oak-dominated area with 10 % or more canopy cover of oaks and a canopy that is open enough to allow a grassland and occasionally a shrub understory (Gaman and Firman 2006). Landscapes of lower canopy cover but still with oaks as a prominent feature are often referred to as oak savanna. The oaks may be of more than 10 different species or their readily-formed hybrids; they may be deciduous or evergreen, a monoculture or of mixed species. The oak woodland may be owned by government—either local or national—or it may be the property of private individuals, families, corporations, or non-governmental organizations (NGOs). A nature preserve may include oak woodland—and in fact often does. Landscape ecologists define the beginning and end of oak woodland by oak tree cover and density over a given area. When oaks are set far apart, and the understory is grass, it becomes an oak savanna (another term argued over). When the oaks are close together, and the closed canopy puts the understory in shade all the time, it becomes an oak forest. Sometimes, if a specific species or oak type predominates, the term is modified to specify this, as in "blue oak woodland" (*Quercus douglasii*) or "live oak woodland" (evergreen oaks). The name oak woodland, however, does not necessarily carry with it any implication of a particular use or form of management.

A ranch, on the other hand, is an enterprise traditionally grounded in the raising of livestock, though the term can be used for one that focuses on wildlife or recreation, especially if modified, as in "wildlife ranch" or "dude ranch." A ranch implies a place in the western United States of extensive acreage—the term has been borrowed for many types of enterprises, including "chicken ranches" and housing developments, and the always-popular "mobile home ranch." It is loosely used. It does not imply any particular vegetation type, other than one in the American West, and a ranch is generally relatively dry in prevailing climate. When you put "oak woodland" and "ranch" together you get an "oak woodland ranch," which moves the terminology closer to dehesa, but nonetheless, can mean a chicken ranch or a wildlife ranch, or just an expansive property located in the oak woodlands. An "oak woodland cattle ranch" would at least mean some form of livestock production, but it says nothing about the complex oak management and multifunctional agriculture that is embodied within the simple term "dehesa." Throughout this book, when we use the term "oak woodland ranch" we mean

Fig. 1.3 A California cowboy is preparing to rope the back legs of a calf to bring it to a ground crew, where the animals will be branded to identify ownership. While the "cowboy" may seem characteristically American, the reality is that the chaps, the bit in the horse's mouth, the rope, and even the techniques of branding are all borrowed or transfers from Spain that came with the Spanish–Mexican presence into Alta California in the late eighteenth century. The rope, for example, is a lariat in English—from *la reata,* or alternatively, a lasso—from *lazoga,* both long-ago Spanish terms. (Photograph by L. Huntsinger)

"oak woodland livestock ranch" as the closest approximation that we can get to dehesa (Fig. 1.3).

The dehesa is an enterprise and a kind of vegetation. The two are inseparable. Dehesa by government definition must meet specific parameters, but "dehesa" is also a form of agro-sylvo-pastoral economy with oaks managed deliberately for a well-developed grass or crop understory, as part of a multifunctional agricultural unit that often includes the grazing of more than one type of livestock and vegetation type and other enterprises such as cork production, cereal and grain croping, hunting, mushroom harvesting, and beekeeping. There are a number of species of oaks that can be managed as a dehesa—but by far the most common are holm oaks (*Q. ilex*) and cork oaks (*Q. suber*)—although there are longstanding disputes, about the exact cladistics of holm oak. Most dehesa is owned by individuals and families, but in all dehesa regions except Andalucía, for which there is no available data, 17 % of dehesa is in collective ownership. This includes properties shared by a community or municipality; generally, a *dehesa boyal*. In Andalucía, collective ownership is less common than in other dehesa regions (MARM 2008: 34 and 40).

In Spain, if we want to talk about oak woodlands, there is a term for each type. *Alcornoque* is a cork oak, *alcornocal* a woodland of such trees (Fig. 1.4). *Encinar* is largely comprised of holm oaks, known as *encina*. *Quejigal* is a woodland largely made up of *quejigo*, the semi-deciduous Lusitanian oak or Algerian oak

Fig. 1.4 Learning from the land, in this case a cork oak woodland or *alcornocal*. (Photograph by L. Huntsinger)

(*Q. faginea or Q. canariensis*). *Melojo* is the deciduous Pyrenean oak (*Q. pyrenaica*), found at higher latitudes and elevations, in a woodland referred to as *melojar* (or *melojal*). And so on. The terms roll off the tongue in a way evocative of the environment. There are similar syntactic and definitional problems with *monte*, a Spanish term sometimes translated as montane, forest, or wildland, that also refers to vegetation, with *monte abierto* or *hueco* specifying an open woodland without identifying a particular kind of tree.

We don't know of a generic word for oak woodland in Spain that is lacking in species specificity—except *dehesa*, which also means a particular kind of eco-social enterprise that includes a mosaic of oak woodland, grassland, shrubs and cropped areas. Part of this is a result of the fact that dehesa disappears without regular human intervention and the California oak woodland, though no doubt shaped by the management of indigenous Californians over millennia, persists for an as-yet unknown length of time without human intervention. Unmanaged oak woodland in Spain is most often what would be referred to in California as chaparral or shrubland.

The dehesa derives from a history that goes back more than 2000 years, part of a deliberate effort to maximize the production of multiple goods and services from the ecosystem. The question is, has the culture and the practice changed so much that the dehesa is being abandoned by the people and the practices needed to sustain it? The oak woodlands of California are usually viewed as the creation of a previous era, flourishing in open stands when Native Californians managed with fire. Later many woodlands were cut down for mining, and where irrigable and reasonably flat, cleared for farming by colonists. The non-arable hilly remnants are today grazed by livestock, providing a large part of the resource base for the range livestock industry in California. No one is sure how the future woodlands will develop in an environment that is so much changed. The long lifespan of oaks means the woodlands retain today evidence of earlier management and use-regimes that reach into the future: What imprint are we today making on the landscape? What management (and supported by whose funds) will persevere and prevail? Is the motive force personal profit, societal benefit and social capital, or biodiversity—or a heady mix of all of these?

1.3 What this Book is About

Californian oak woodlands and Spanish dehesas are beautiful Mediterranean-type landscapes. Oaks share space with annual grasses and shrubs. Both woodlands are vulnerable to demographic, economic, and climatic change. Each environment is rich in biodiversity, and important historically and culturally.

Most important to understand is that today these are landscapes at risk. Scientists, academics, managers, and policy-makers are working on both sides of the Atlantic Ocean to understand the dynamics and drivers of these ecosystems. An overriding goal is to sustain their value as economic and ecological systems, and to preserve the oak woodlands themselves, trying to adjust to current climate change effects and changes in societal preferences.

In California, scientists and policymakers are beginning to learn how to foster the conservation and stewardship of oak woodland ranches. The term "working landscape" has come to embody the goal of joining agricultural commodity production to a flow of diverse ecosystem services like carbon sequestration, sight lines and view shed, watershed, and wildlife habitat. Spain's ancient dehesa reflects dozens of generations—over several millennia—of stewardship and efforts to enhance production of multiple goods and services from the ecosystem (Chap. 2). A dehesa does not exist without human care and maintenance—it is truly a working landscape created in large part by human labor, livestock, tending of cork and acorn-bearing trees, and steady use. Our rapidly changing society and economic base have vast implications for each of these landscapes.

The oak woodlands known as a dehesa in Spain (and in Portugal, *montado*), are prevalent in the south–west portion of Spain. The government definition of dehesas is that they are livestock producing properties, including the grasslands and shrublands that typically form a mosaic with dehesa oak woodlands, with at least 20 % of their area occupied by oak woodland with a canopy cover of between 5 and 60 % (MARM 2008, 7). The dehesa area in Spain according to this definition totals 3.6 million hectares in 5 Autonomous Regions (known as *Comunidades*

Table 1.1 Dehesa in the Spanish autonomous regions according to the Ministry of the environment's definition (MARM 2008, 7)

Autonomous Region (Spain)	Dehesa area (ha) including croplands, shrublands, grasslands, and woodlands.	Percentage of dehesa that is at least 20 % oak woodland with a canopy cover of 5–60 %.
Extremadura	1,065,188	77.8
Castilla-La Mancha	1,048,713	46.4
Andalucía	743,774	62.1
Castilla-León	687,407	57.1
Madrid	61,069	54.2
Total	3,606,151	61.1

Fig. 1.5 Defining the dehesa is no simple matter, as the main text and Table 1.1 reveal. To establish with precision just how much of an area is "dehesa" requires accurate estimates of canopy cover and knowledge of whether or not the area is used for livestock production. As this aerial view of a dehesa region in the Sierra Norte de Sevilla (Andalucía) suggests, oak density can be remarkably variable even across a small area. (Photograph by P.F. Starrs)

Autonomas), which are Andalucía, Extremadura, Castilla-La Mancha, Castilla-León and Madrid (MARM 2008, 8). This area includes 2.2 million hectares of oak woodland with 5–60 % cover (MARM 2008, 43). Although holm oak is the dominant oak species, and present in 84 % of the woodlands (MARM 2008, 34), cork oaks dominate in a few areas (e.g.: Alcornocales Natural Park in Cádiz province) and are commonly interspersed with holm oaks. Table 1.1 shows the distribution of dehesa and the percentage of oak woodland within it for the five Autonomous regions that have dehesa in Spain (Fig. 1.5).

Fig. 1.6 Areas defined as "oak woodland" in California are less intensively managed and may or may not be grazed by livestock. Stands may be dense and nearly closed, or have only a few isolated trees. Those with low canopy cover are often called oak savanna. This view in Shasta County, California, illustrates the irregular canopy cover throughout the woodlands. (Photograph by R.B. Standiford)

Extremadura is the most representative dehesa region, with its high proportion of oaks to grassland, and from there came many Spanish colonists, explorers, and missionaries who went to Mexico and eventually California to establish religious and secular range livestock enterprises starting in the eighteenth century. Spanish officials of early California often came from noble families who owned dehesas. While the lower reaches of the Guadalquivir River provided the origins of Mexican-Spanish livestock ranching culture that transferred Spanish practices to the Americas, today many of the traditions common to modern-day Californian range culture derive from those early migrations from Andalucía and Extremadura (Doolittle 1987; Jordan 1989; Butzer 1988; Starrs 1997; Starrs and Huntsinger 1998; Sluyter 1996).

The closest equivalent to Spain's wooded dehesa is described as oak woodland in California, and covers 3.4 million ha (Gaman and Firman 2006), about two-thirds of which is grazed by livestock as part of ranching activity (Huntsinger et al. 2010). Five of the state's oak species—blue oaks, coast live oaks (*Q. agrifolia*), interior live oaks (*Q. wislizenii*), valley oaks (*Q. lobata*) and Englemann oak (*Q. engelmannii*)—are the dominant overstory oaks across most of the state's grazed woodlands (Pavlik et al. 1991). Tree canopy and density vary throughout the region (Fig. 1.6). California's oak-dominated landscapes occur mainly in Mediterranean climate zones in the Coast Ranges, Transverse Ranges, and western foothills of the Sierra-Cascade Range (CDF-FRAP 2003). More than 350 vertebrate species inhabit them (CIWTG 2005), and they provide some of California's richest wildlife habitat (Chap. 8).

Table 1.2 provides a general comparison of dehesa and ranch characteristics. About 85 % of the dehesa regions are private properties. They are frequently larger than 350 ha and rearing livestock and the periodic harvest of cork are the primary commercial activities (Parsons 1962a, b; Campos 1984).

More than 80 % of California's oak woodlands are in private ownership (CDF-FRAP 2003), and despite the rapid land use and demographic change of recent decades, most of those areas are still managed as oak woodland ranches (Huntsinger et al. 2010). The quantity and quality of understory grazing forage varies seasonally with the climate and life cycles of hundreds of plant species, including several dozen varieties of native and introduced grasses (Stromberg et al. 2007) (Chap. 6).

1.4 Broader Themes: Chapters in this Volume

The chapters included in this volume are on topics as specific as acorn crop fluctuations linked to climate, and as overarching as a comparative history of landownership and use. Because this book attempts to address a broad spectrum of woodland uses and incorporates diverse analytic approaches, numerous authors and professional specialties are involved. With the goal of enabling an in-depth appreciation of the two systems, we have focused on California and Spain (Figs. 1.7, 1.8 and 1.9), although other Mediterranean oak woodlands are scattered about the world. Research has been conducted on Portuguese montados, cork oak woodlands in Tunisia, oak woodlands in Morocco, and Mediterranean forests in France and Italy. However, the vast amount of research devoted to Spanish dehesas and California oak woodland ranches is unique, and makes possible a detailed comparison between these ecologically significant working landscapes.

1.4.1 History and Recent Trends

Appropriately, this volume begins with a story: a comparative history of the woodlands. Contemporary ranchlands set in oak woodlands and the dehesas of Spain result from dissimilar histories involving centuries of human use. What are now recognized as dehesas began forming during Roman rule, developed in Arab-dominated Iberia, and by the fifteenth century at the time of the Christian reconquest were subject to diversified management involving grazing, hunting, farming, and non-timber forest products such as firewood, charcoal, and even the harvest of palm fronds from stock driveways that cut through the dehesa (Fig. 1.10). California woodlands were modified by thousands of years of Native

Table 1.2 Characteristics of Spanish dehesas and Californian oak woodland ranches landowners

Characteristics	Oak ranches in California	Spanish dehesa
Extent	1.9 million ha owned by ranchers out of 3.4 million ha total (Gaman and Firman 2006)	3.6 million ha (MARM 2008, 43)
Typical range of property sizes	600–1,000 ha (Huntsinger et al. 2010) (see also Chap. 10)	100–1,000 ha (MARM 2008); 465 ha on average in Andalucía (RECAMAN project,[a] unpublished data)
Most common oak	Blue oak (*Q. douglasii*)	Holm oak (*Q. ilex*)
Land use	66 % grazed by livestock (Huntsinger et al. 2010)	82 % grazed by livestock (RECAMAN project, unpublished data)
Commodity products	Beef, lamb, wool, firewood, game/hunting, grazing	Beef, Iberian pigs, lamb, acorns, firewood, charcoal, hay, cereals, grazing, wool, goat meat and milk, game, truffles, cheese, fodder, honey, cork.
Ownership	80 % in private ownership; mean ownership 39 years, 3 % corporate; 17 % in trust. (Huntsinger et al. 2010)	85 % in private ownership; mean ownership 25 years in 2010; 79 % in family ownership, 8 % corporate; 13 % in other private ownership (RECAMAN project, unpublished data)
Management	80 % are resident managers/owners; caretakers may manage larger properties. (Huntsinger et al. 2010)	83 % of landowners are involved in dehesa management (RECAMAN project, unpublished data). 9 % are resident owners. 71 % have a residential house in the dehesa for weekends and vacation.
Age of principal landowner	62 years (Huntsinger et al. 2010)	58 years (RECAMAN project, unpublished data)
Education	60 % with a university degree in 2004; for 1985, 50 % (Huntsinger et al. 2010)	41 % have some university education in 2010
Contribution to household economy	14 % earn majority of income from woodlands; more on larger properties. (Huntsinger et al. 2010)	Dehesa management is the main job for one-third of the landowners (MARM 2008).
Labor	Mostly resident landowners; some hired labor.	20 % employ family; 60 % hire $\geq$ 1 non-family member (MARM 2008); 10–15 h of labor per ha are required to manage property.

[a] The RECAMAN project (*Valoración de la Renta y el Capital de los Montes de Andalucía*) of the Junta de Andalucía is ongoing and applies the Agroforestry Accounting System at the regional scale to measure total income and capital from the montes of Andalucía in Spain.

Fig. 1.7 A tour of Spanish researchers to California helped kick off the collaboration. Here a group of Spanish and Californian researchers pose beneath an old cork oak at Mission San Juan Bautista in the central coast of California in 2004. (Photograph by P. Gil)

Fig. 1.8 A visit by Californian researchers to Spain sealed the deal, in the Montes de Jerez in the Sierra de Cádiz (Andalucía) in 2003. (Photograph unattributed)

Fig. 1.9 Learning from a pair of landowners in Spain in 2011. (Photograph by A. Caparrós)

Fig. 1.10 The diverse uses of the dehesa, and a complicated landscape history, is reflected in this view from the Montes de Toledo (Castilla-La Mancha). An abandoned and unroofed building, with chimneys still evident, is surrounded by repopulating oaks, and adjoining the ruins is a field recently harvested for grain production—something less common now than it once was in the dehesa. (Photograph by M. Díaz)

American use, including widespread burning. In the eighteenth century Spanish settlers brought livestock into California, along with new plants that replaced the oak woodland understory. Chapter 2 takes us through periods of over-exploitation that, in some forms, are still ongoing. There is in Spain, for example, deep concern about a lack of oak regeneration, causing some to refer to cork production as the mining of "brown gold" from a putatively renewable resource that is failing to be renewed. In California, the woodlands are considered prime real estate for exurban development, and are being fragmented, converted, and developed, although this is currently slowed by the economic recession that began in 2008. The chapter moves us to the present, with an embedding of oak ranchlands and dehesas in the global economy, and to the shared concern of both countries and hemispheric powers for the future of the woodlands.

1.4.2 Environmental Setting

We then move into the environmental setting, exploring first the climate in Chap. 3 and then soil and water dynamics in Chap. 4. Climate constrains the presence and specific characteristics of California oak woodlands and Spanish dehesas. The authors summarize studies conducted in the two regions, using different methodologies to investigate the influence of climatic factors on the distribution of oak species (Fig. 1.11). Climate strongly influences oak distribution in California. Soil characteristics and socioeconomic issues are more important factors than climate for the creation and maintenance of dehesa in Spain.

Climate conditions, terrain morphology and parent material, but also land use and management, play a crucial role in the functioning of oak woodland ranches and dehesas. The authors review research results to gain understanding of human influences on soil and water through land-use and management practices. Soils in the Spanish dehesa have been subject to many centuries of agricultural use. Erosion by runoff and rivers resulting in the reduction of organic matter and physical degradation are the most important phenomena. For California, the authors present results from studies on water quality and the effects of vegetation conversion on water yield, soil stability, and erosion.

1.4.3 Vegetation

Vegetation is the focus of the next several chapters, examining the critical question of whether or not the oaks are reproducing adequately. Oak woodland area in both regions was greatly reduced in the twentieth century. Scientists and the public are deeply concerned about the sustainability of the remaining woodlands, and a baseline requirement for that is whether or not there is enough seedling survival to replace aging trees. Chapter 5 is about oak regeneration, examining both what we

Fig. 1.11 The massive and often solitary valley oak (*Q. lobata*) is a long-time fixture of the fertile bottomlands and alluvial soils of the valleys and riparian areas in the Central Valley and the coast ranges. However, today most of its range has been converted to field crops. (Photograph by F. Bruno Navarro)

know about the ecology of oak reproduction, and how to restore oak woodlands that have lost oaks. In Chap. 6, the relationship between oaks and their understory is discussed. Although California oak woodlands and the Spanish dehesa may often look very much alike, the dynamics of the understory vegetation are quite different. Shrubs are swift invaders into dehesa and are excluded vigorously by managers (Fig. 1.12), yet they facilitate oak regeneration by protecting seedlings from summer drought and by maintaining populations of acorn dispersers that move acorns outside of oak canopies. In California woodlands, shrub invasion happens more slowly if at all, but with fire suppression is becoming more common.

Acorns, once the staff of life and still of cultural significance for Native Californians, are important livestock feed in Spain and offer wildlife forage in both places. Acorn production is highly variable from year to year, and researchers are working to explore what factors explain this variability, including ongoing—and changing—dehesa management practices. Chapter 7 explores this body of research, and the potential differences in dynamics between Spain and California and between dehesas and nearby oak forests in Spain.

Fig. 1.12 When shrubs and brush are cleared from hillsides, left in the open are often holm oaks, which by Spanish law are under moderate protection regimes. The dehesa is more readily invaded by shrubs than California's oak ranchlands, and requires regular maintenance. Nonetheless, there is high biodiversity and productivity in the mosaic of vegetation patterns seen in both environments, and a great deal of habitat for game and non-game species as well as livestock enterprises can be sustained. (Photograph by M. Díaz)

1.4.4 Management, Uses, and Ecosystem Response

Chapters 8, 9, 10, and 11 examine the interaction of economic enterprises and the ecosystems of working landscapes. We begin with a look at biological diversity in dehesa and oak woodland ranches, and how management benefits from it and influences it, in Chap. 8. Intensive land use, long-term abandonment of livestock enterprises and active management, and development into housing certainly threatens habitat mosaics that foster both high biodiversity and oak woodland functioning at multiple spatial and temporal scales. The chapter reviews how different management practices can affect the provision of this biodiversity.

In Chap. 9, silvopastoral management models are used to analyze how dehesa and Californian oak woodlands support the production of multiple goods and services. Management scenarios for supporting oak regeneration in dehesa are reviewed and compared to outcomes without such management. Silvopastoral models for California woodlands illustrate the importance of reflecting actual landowner behavior in policy analysis to accurately represent the trajectory of future oak woodland status, whereas Spanish models emphasize the need for public short-term support to landowners to achieve higher longer-term economic and environmental benefits.

Fig. 1.13 With a rough mixture of oaks behind them, including the pointed and sharp leaves of *Q. coccifera*, which in its shrub form is a particularly difficult form of oak to travel through, these hunters are working their way toward assigned posts, part of a *montería* in the Sierra Norte de Sevilla (Andalucía). Such activities, which used to attract mainly wealthy landowners, are now accessible (for a fee) to hunting enthusiasts. The leather chaps are trappings carried over from earlier times when hunters derived as much enjoyment from pushing dogs after game in the oak understory as they did from shooting; a rarity in this day and age when dog handlers are mostly hired and travel from hunt to hunt with their packs of dogs. (Photograph by P.F. Starrs)

Extensive livestock production in dehesa and oak woodland is examined in Chap. 10. In both countries, cattle, sheep, and goats are all found in the woodlands, though cattle are overwhelmingly the most common in California. In Spain, the Iberian pig is fattened on acorns in the oak woodland to produce high quality *jamón* (air-dried ham). In California, acorns are mostly used by wildlife, including wild pigs, an import to California from Europe.

Chapter 11 presents hunting as a source of income for landowners in the woodlands, but also as a product enjoyed by the owner and shared with friends (Fig. 1.13). Distinct cultural and legal histories governing property rights over wildlife and land tenure have created dissimilar hunting systems in Spain and California with differences that are manifest in the methods of hunting, the economic return to landowners, the actions taken to manage game species, and the accompanying environmental effects.

1.4.5 Oak Woodland Economics

The term ecosystem services was coined in the 1980s to describe the valuation of a full range of human benefits from ecosystems, including provisioning services, regulation and maintenance functions, and cultural services. In Chap. 12, authors explore efforts to identify oak woodland ecosystem services that may be difficult to quantify and value economically, and which therefore are often undervalued in policy decision-making processes that draw on such analysis. The chapter reviews several studies that attempt to incorporate non-market values of ecosystem services into economic assessments.

In Chap. 13, the authors use case studies of oak woodland ranches and dehesas to reveal how landowners on the ground use their woodlands and profit from them, with tabulations that include the value owners derive from enjoying the amenities that come from owning and managing the land (Fig. 1.14). A complete Agroforestry Accounting System is applied to enumerate operating income and capital gains or losses. The studies reveal important vulnerabilities in the economic functioning of the enterprises that support the management of these woodlands. Oak woodlands in ranches and dehesas provide multiple public goods, and society and landowners alike need to work together and make compromises in order to pass the natural capital of the woodlands on to future generations (Fig. 1.15).

Fig. 1.14 Field research investigating dehesas in Spain and oak woodland ranches in California invariably requires ongoing contact with landowners who rightly see themselves as linchpins in the management and maintenance of a complex system of products and benefits that derive from a healthy and thriving ecosystem. In both environments, owners make decisions that reflect their goals and motivations. (Photograph by S. García)

Fig. 1.15 An oak woodland ranch in the Gold Rush country of the Sierra Nevada foothills in California. (Photograph by Lynn Huntsinger)

1.4.6 Dehesa and Oak Woodland Ranch Landscapes

Chapter 14 extends a Geographic Information System (GIS) and remote sensing based model for looking at landscape change in Spain and California, exploring our ability to use the techniques of landscape ecology to understand the drivers of pattern and change in oak woodlands.

We conclude with recommendations for conservation in California and Spain, and our findings of issues that need to be addressed, in Chap. 15. There, the takeaway points from this study are laid out, along with a discussion of the salient advantages of a comparative analytical approach to working landscapes.

References

Braudel F (1975) The Mediterranean and the Mediterranean World in the Age of Philip II, vol 1. Harper and Row, New York (orig. published 1949 in French)

Butzer KW (1988) Cattle and Sheep from old to New Spain: historical antecedents. Annals Assoc Am Geog 78:29–56

Campos P (1984) Economía y energía en la dehesa extremeña. MAPA, Madrid

CDF-FRAP [California Department of Forestry and Fire Protection-Fire and Resource Assessment Program, CalFire] (2003) Changing California: forest and range 2003 assessment. Sacramento, CA: State of California Resources Agency. http://www.frap.cdf.ca.gov/assessment2003. Accessed June 2012

CIWTG [California Interagency Wildlife Task Group] (2005) California WILDLIFE HABITAT RELATIONSHIPS (CWHR) System version 8.1, personal computer program. California Department of Fish and Game, Sacramento. http://www.dfg.ca.gov/biogeodata/cwhr/. Accessed Aug 2012

Doolittle WE (1987) Las Marismas to Pánuco to Texas: the transfer of open range cattle ranching from Iberia through Northeastern Mexico. Yearbook, Conf Lat Americanist Geog 23:3–11

Gaman T, Firman J (2006) Oaks 2040: The status and future of oaks in California. California oak foundation, Oakland http://www.forestdata.com/Oaks2040_long_version_web.pdf. Accessed Sept 2012

Huntsinger L, Johnson M, Stafford M, Fried J (2010) California hardwood rangeland landowners 1985 to 2004: ecosystem services, production, and permanence. Rangel Ecol Mgmt 63:325–334

Jordan TG (1989) An Iberian lowland/highland model for Latin American cattle ranching. J Hist Geog 15:111–125

MARM [Ministerio de Medio Ambiente y Medio Rural y Marino] (2008) Diagnóstico de las Dehesa Ibéricas Mediterráneas. Tomo I. Available at: http://www.magrama.gob.es/es/biodiversidad/temas/montes-y-politica-forestal/anexo_3_4_coruche_2010_tcm7-23749.pdf. Accessed Aug 2012

Parsons JJ (1962a) The Acorn-Hog Economy of the oak-woodlands of Southwestern Spain. Geogr Rev 52:211–235

Parsons JJ (1962b) The cork oak forests and the evolution of the cork industry in Southern Spain and Portugal. Econ Geog 38:195–214

Pavlik BM, Muick PC, Johnson SG, Marjorie Popper (1991) Oaks of California. Cachuma Press, Los Olivos

Sluyter A (1996) Ecological origins and consequences of cattle ranching in sixteenth-century New Spain. Geogr Rev 86:161–177

Starrs PF (1997) Let the cowboy ride: cattle ranching in the American West. Johns Hopkins University Press, Baltimore

Starrs PF, Huntsinger L (1998) The cowboy & buckaroo in American ranch hand styles. Rangel 20:36–40

Stromberg MR, Corbin JD, D'Antonio CM (eds) (2007) California grasslands: ecology and management. University of California Press, Berkeley and Los Angeles

Chapter 2
History and Recent Trends

Peter S. Alagona, Antonio Linares, Pablo Campos and Lynn Huntsinger

Frontispiece Chapter 2. A characteristically multihued livestock herd grazes in the Sierra de Cádiz, a dehesa area in southern Spain. (Photograph by J.L. Oviedo)

P. S. Alagona (✉)
Department of History and Environmental Studies Program, University of California, Humanities and Social Sciences Building, 4231, Santa Barbara, CA 93106-9410, USA
e-mail: alagona@history.ucsb.edu

A. Linares
Department of Economy, University of Extremadura, Avda de Elvas S/N 06071 Badajoz, Spain
e-mail: alinares@unex.es

P. Campos et al. (eds.), *Mediterranean Oak Woodland Working Landscapes*,
Landscape Series 16, DOI: 10.1007/978-94-007-6707-2_2,

© Springer Science+Business Media Dordrecht 2013

Abstract Contemporary ranches and dehesas are layered onto centuries of human use. The Spanish dehesas began forming during Roman rule, and by the time of the Christian Reconquest were managed for grazing, hunting, farming, foraging, and forestry. California's oak woodlands were shaped by thousands of years of Native American management, including widespread burning that was eventually suppressed after European settlement. With the first settlers from Spain came livestock and crops from the Old World, as well as grasses and other species that have since naturalized across the state. California woodlands have undergone periods of expropriation, scientific management, conservation, and integrated management. Spanish dehesas, meanwhile, have experienced periods of consolidation, development, decay, and resurgence. California oak woodland ranches have not been managed as intensively as the Spanish dehesa, but since World War II both landscapes have experienced pressures associated with development, technology, demographics, and globalization, leading to profound social and ecological change.

Keywords Environmental history · California Indians · Spanish colonialism · Mesta · Missions · Roman period · Menhirs

2.1 Setting

California and Spain are nearly 6,000 miles apart. Separated by a great continent and a vast ocean, they share only a handful of native species. Before 1542, when Juan Rodríguez Cabrillo sailed a 200 ton galleon, the *San Salvador*, and a pair of accompanying ships up the Pacific Coast of western North America, these two places might have existed on different planets. Today, however, a visitor from California who travels to the countryside of south-west Spain will find a landscape of oak woodlands, rangelands, pastures, and farms, with grassy rolling hills and distant arid mountains that is unmistakably—even eerily—familiar. The oak-dominated rural landscapes of California and Spain appear alike in part due to their physical geographies and climate. But their similarities are also the result of transformative human action. California's oak woodland ranches and the dehesas of Spain have different social, cultural, political, and economic histories. Yet, over

P. Campos
Institute of Public Goods and Policies (IPP), Spanish National Research Council (CSIC), Albasanz 26-28 28037 Madrid, Spain
e-mail: pablo.campos@csic.es

L. Huntsinger
Department of Environmental Science, Policy, and Management, University of California, Mulford Hall , Berkeley, CA, MC 3110 94720, USA
e-mail: huntsinger@berkeley.edu

the past 250 years, these histories have increasingly converged, producing similar landscapes, with similar attributes and similar problems, even while some aspects (hunting, cork planting, government policy) have diverged in notable ways. We begin with a history of California's oak woodlands, before turning to the history of the Spanish dehesa. The story of Portugal's montado is, of course, significant in its own right, but its history of management and use is quite distinct, and so is reserved for another venue.

2.2 History and Recent Trends in California's Oak Dominated Landscapes

Oaks have waxed and waned in abundance in California throughout recent geologic history. Oaks declined during major glacial cooling cycles, reaching a low point by 20,000–40,000 years BP as evidenced in the pollen record (Fig. 15.10 in Millar and Brubaker 2006). The oak woodlands of today can be traced back to the retreat of the glaciers after the last ice age, when a warming climate fostered the spread of grasslands and members of the genus *Quercus* or oaks. That advance accelerated even further around the time of the arrival of Spanish and Mexican immigrants, likely as a result of reduced Native American burning and changes in grazing and woodcutting regimes (Byrne et al. 1991; Mensing 2005, 2006). Today, California's oak woodland ranches form a bucolic countryside that is perhaps the state's most attractive and familiar rural landscape (Fig. 2.1).

2.2.1 Early Historical Use

Human settlement in California dates back more than 13,000 years. Prior to European contact, California was home to a diverse indigenous population comprised of at least 300,000 people, divided into six language families and more than 300 dialects (Fagan 2004). Native Californians describe a pre-contact system of access to lands and natural resources derived from common and usufructuary rights, distributed spatially and with access varying through the seasons, though the form these systems took varied among the many groups. Native Californians took part in vast trade networks extending far into the continent, transferring plant and animal materials. Oak-dominated landscapes were among the region's most densely populated and heavily used environments, and supporting one of the highest population densities in native North America were the abundant, nutrient rich, acorn crops, which offered a significant food source (Fig. 2.2). Thousands of years of human occupation created a distinctly cultural landscape long before European colonization.

Native Californians engaged in a range of activities including game hunting, acorn gathering, seasonal burning, and planting that shaped the region's ecology

Fig. 2.1 Oak woodlands are predominantly on private land in California, and may be used for ranching, wildlife habitat, viewshed, and, as here in the Sutter Buttes of the Sacramento Valley, kept as a space where visitors may on occasional go for a hike—with permission of the landowner. (Photograph by P. F. Starrs)

Fig. 2.2 Acorns were often ground using rock outcrops. Generation after generation of Native Americans used a pestle to create acorn flour, a staple food that could be stored, creating these depressions in many of the rocks of the oak woodlands. (Photograph by L. Huntsinger)

and environment (Anderson et al. 1997). Lightning frequency records, oral histories, and tree ring data confirm that Native Californians in some areas shortened wildfire intervals from natural cycles of about once a century to a frequency of a decade or less (Keeley et al. 2003; Syphard et al. 2007). Indigenous Californians used fire on a broad scale to enhance the production and collection of acorns and to improve habitat for wild game, and such burning continues in small areas today as permitted by law, sometimes as part of efforts to restore native vegetation.

2.2.2 Spanish and Mexican Era Woodland Use

Beginning in 1769, Spanish colonists settled along California's coast in a system of missions, presidios, pueblos, and large land grants—called *ranchos* by their proprietors—which they used for livestock production. The Spanish Crown granted about 30 ranchos, of several thousand hectares each, usually to retired soldiers or officials as a reward for their service. The early Spanish colonists, known as *Californios*, became a landed gentry of the New World. Blending Spanish tradition with New World imperatives, they established unique cattle handling practices, some of which persist to this day (Starrs and Huntsinger 1998). The *Californios* introduced not only the tools of the trade, like the lariat (*la reata* in Spanish) and the branding iron, but major livestock management institutions as well. The Judges of the Plains (*Jueces del Campo*) presided at regular round-ups where livestock were sorted to their correct owners and disputes among owners resolved, an institution that was eventually codified in California's constitution at statehood and can be argued to persist in the form of brand inspectors. The Judges evolved from the similar *Alcaldes de la Mesta* in Spain, recognized as a valid institution by the Crown as early as 1273, and the Judge of the Plains survived as a public office in California until the early 1950s (Stanford 1969).

Most ranchos were located in oak-dominated coastal or valley landscapes suited to supporting a colonial economy based on livestock grazing. These areas housed many of the region's largest populations of Native Americans, whom the padres hoped to convert to Christianity. The missionary project was a disaster. Between 1769 and 1834, disease, violence, and displacement ravaged California's indigenous population, and the number of Native Americans living along the coast between San Diego and Sonoma declined by 75 % (Hackel 2005). The ranching project was more successful. By the 1820s, California's 21 missions acquired a vast pastoral empire of some 17 million acres (nearly 7 million ha), grazing around 300,000 sheep and 400,000 head of cattle (Fig. 2.3; Burcham 1981).

Indigenous lifeways were woven into the fabric of the pre-Columbian landscape, but generations of European colonizers undermined, ignored, and even sought to erase this historical legacy. Spanish newcomers set changes in motion that transformed California's hardwood rangelands. Cattle compacted the soil and sheep denuded the slopes. Plant seeds brought in ship ballast, crop seeds,

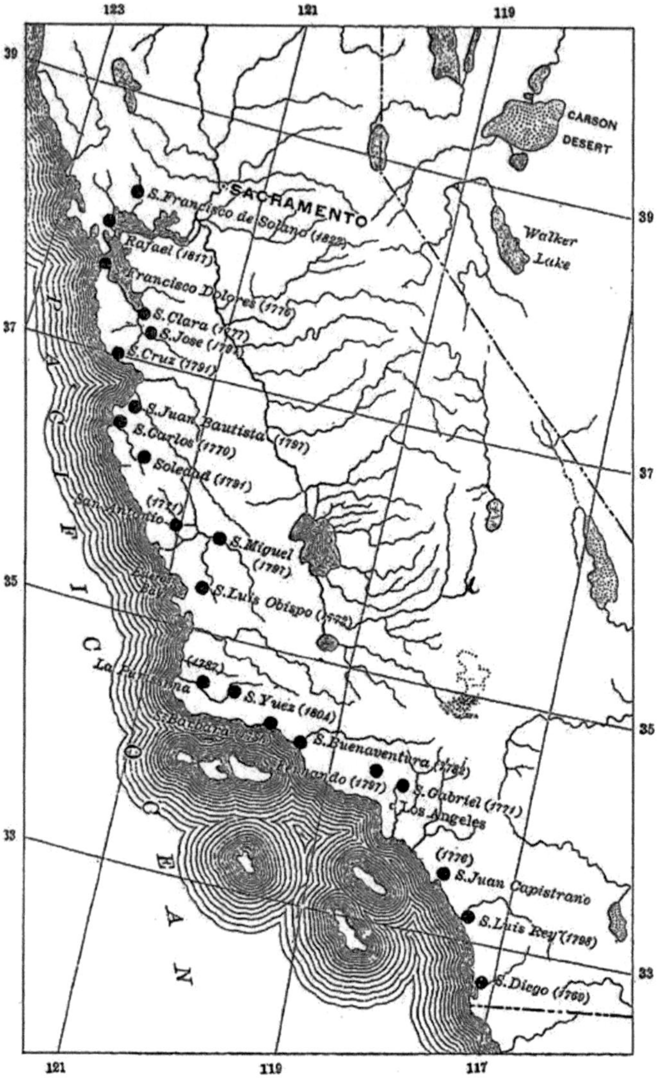

Fig. 2.3 A total of 21 missions were established in California, with San Diego de Alcala (1769), San Gabriel (Los Angeles) (1771), and San Francisco Dolores (1776) among the earliest. The even spacing is widely attributed to a desire of the padres to have mission sites no farther than a long days trek apart. Later, these missions would be secularized. But their sway was great, acting as a nucleus for ranching and farming operations that extended well into the Sacramento and San Joaquin valleys. (Map modified from original in files of the California Missions Foundation)

and livestock feed spread widely and colonized the understory while feral goats and pigs tilled the soil and devoured the acorns. Ploughing land created opportunities for the introduction and spread of exotic grasses and weeds. Some wild game populations suffered due to disease or competition from introduced livestock, but reports from the first half of the nineteenth century suggest that others benefited from reductions in Native American hunting and other ecological changes, and even from an abundance of livestock carcasses. Populations of large predators, particularly the California grizzly, appear to have increased as livestock became more available for consumption (Preston 2002).

In 1821 California became part of a newly independent Mexico. The Mexican government accelerated the Spanish policy of distributing lands, with more than 770 grants to individuals throughout California's southern and coastal regions (Pérez 1982, 1996). This process accelerated after the Mexican Congress passed the Secularization Act of 1833, which enabled the confiscation and sale of mission lands, or in some cases their conversion into pueblos. After secularization, private citizens assumed control of the ranchos, which they maintained as hacienda-style livestock operations.

At the end of the Mexican–American War in 1848, California became a territory of the United States, achieving statehood in 1850. In the decades that followed, Anglo-American settlers used the courts to dispossess most of the Mexican grantees of their lands, and the ranchos were often broken into smaller parcels

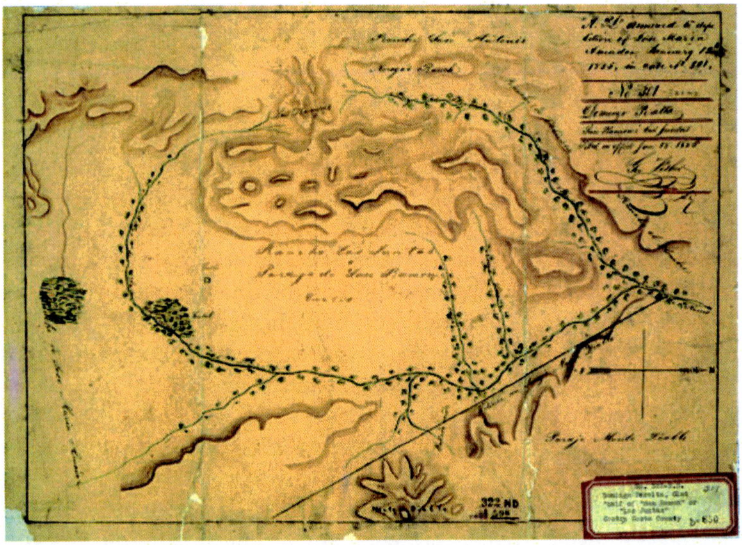

Fig. 2.4 Diseños, or property maps, were required by the Board of California Land Commissioners, established in 1851. The Board required Mexican and Spanish landowners to prove a legal right to land granted them by Mexican and Spanish authorities. The maps often showed more artistry than accuracy, which proved a problem in establishing the validity of claims, many of which were overturned over twenty years of legal cases. This oak-dotted diseño, dating from 1855, is a product of the Domingo Peralta claim to the land of Rancho San Ramon, Land Case 322-ND, in Contra Costa County. (Becker 1964) (Map from federal Land Case archives, in The Bancroft Library, University of California, Berkeley)

(Hornbeck 1983) (Fig. 2.4). Despite the subdivision of these original land grants, California's ranch properties remain relatively large. The few original ranchos in existence today cover thousands of hectares, and California oak woodland ranches still average 800–1,000 hectares in size (Huntsinger et al. 2010) .

2.2.3 *Conflicting Claims and a Rising Federal Role*

The Gold Rush of 1849 resulted in a huge population increase in California as gold seekers from the eastern United States, Latin America, Asia, Europe, and elsewhere flooded into the territory. Because the ore drew miners to the foothills and mountains, tribal groups that previously avoided the direct impacts of Spanish and Mexican settlement were deeply and suddenly affected. Already reduced by around three-quarters under Spanish and Mexican governance, native populations resumed a precipitous decline. When large-scale gold mining ended in the western foothills of the Sierra Nevada, miners returned home or moved to San Francisco, Oakland, Sacramento, Stockton, and other cities, leaving behind ecological devastation, crumbling infrastructure, and deserted settlements (Isenberg 2005).

By the 1860s, a few industrial livestock corporations, based in San Francisco, began to acquire vast landholdings in the Sacramento-San Joaquin valleys and adjacent foothills (Igler 2001). A system of transhumance developed in which ranchers moved their cattle and sheep from lowland winter pastures into the state's sparsely populated mountains. Yet by the late 1870s, barbed wire and enclosure laws began to restrict wintertime access of livestock in fertile valleys. The development in the twentieth century of large-scale industrial agriculture, supported by irrigation subsidies from federal and state government and employing costly new farm equipment, would complete this process. Montane summer range helped compensate for the loss of valley pastures to crop production. Indeed, grazing management was an important goal in the establishment of federal forest reserves, later called national forests, beginning in the 1890s (Miller 2011). This concentration of authority over land use in a single federal agency proved controversial among the local land users and owners (Fig. 2.5).

Montane cattle grazing continued after 1906 under permit from the U.S. Forest Service, but ranchers who did not own adjacent foothill properties were routinely excluded, and itinerant sheepherders, including many of Basque heritage, were usually the first to go. The establishment of national parks, such as Yosemite and Sequoia-Kings Canyon, further restricted high country grazing access. Since World War II, transhumance has continued to decline due to fire suppression efforts that reduce livestock forage, government environmental regulations, reductions of grazing permits on public lands, and land development patterns that interfere with traditional stock routes.

Fig. 2.5 Considerable controversy surrounded the Forest Reserve Act (1891) and Organic Act (1897), which established federal control over much of California's higher-elevation forested areas. Gifford Pinchot, the first Chief of the U.S. Forest Service was vaunted but also vilified, as in this cartoon from around 1908. (Archives of the U.S. Forest Service)

2.2.4 Current-Day Uses

The contemporary geography of California's oak woodland ranches is a product of this history. Spanish and Mexican rule fostered the establishment of unusually large properties, compared to other parts of the American West where the U.S. government disbursed lands to private holders in smaller parcels. These large holdings had, and have, distinct advantages in terms of wildlife habitat and extensive management practices that provide much higher levels of environmental protection. The advent of large-scale agriculture led to an exclusion of most livestock from lowland pastures and the conversion of irrigated pasture to crops, and the establishment of national parks and forests resulted in the gradual loss of access to summer higher-elevation ranges (Starrs and Goin 2010). Today, grazing is concentrated in the oak woodlands that occupy the narrow elevation band between its lower valleys and higher mountains. Unlike California's deserts and montane forests, most of which are on public land, 82 % of the state's oak woodland rangelands remain in private ownership (CDF-FRAP 2003).

2.3 California's Oak Woodland History: Four Eras

Within a broader environmental history, California's woodlands have undergone changes through at least four major historical periods since 1850, each defined by a shift in management practices and institutional arrangements (Fig. 2.6).

Fig. 2.6 Reminiscent of eighteenth century paintings of European landed gentry, this work by James Walker (1819–1889), depicts a "Patron" with silver conchos along the vest-front and pants, ornate silver-mounted spurs and bit, and the high-stepping horse of Arabian heritage, which testifies to the ranch-owner's prosperity. (Courtesy of The Bancroft Library, University of California, Berkeley)

2.3.1 Expropriation and Ranch Enlargement

A first phase, lasting from 1850 to around 1920, began when Anglo-American settlers started usurping lands owned by Mexican ranchers and developed a more efficient, market-oriented approach to livestock production (Fig. 2.7). Demand for livestock products was initially high, with population growth in mining and trading towns spurring high prices. Livestock were imported from other territories, and new herds established with the animals that did not go immediately to market. The newcomers believed that although California's environment was dynamic and unpredictable, its resources would be limitless if it could be properly subdued and transformed into a capitalist wealth-producing machine.

The fantasy of controlling nature and the myth of nature's inexhaustibility shaped the use of oak woodland ranches during this period. These ideas helped produce a series of booms and busts in livestock markets that, in some areas, resulted in severe rangeland degradation (Burcham 1981; Cleland 1941; Igler 2001). By the beginning of the twentieth century, the false faith in nature's infinite productivity gave way to an equally unfounded sense of inevitable decline.

Fig. 2.7 The influence of early Spanish and Mexican heritage left a deep imprint in California's place names, as with this small town in the eastern San Joaquin Valley. *Bellota* is Spanish for acorn, and oaks, acorns, and the mast that would accumulate below oak trees were important feed sources from Native American times onward in California. (Photograph by P. F. Starrs)

2.3.2 Efforts Toward Scientific Management

A second phase began in the early 1920s with the advent of scientific range management. Range management as a scientific, rather than a vernacular activity, came to California's oak woodland ranches around 1922 when Arthur Sampson, who had studied under Frederick Clements at the University of Nebraska and worked under Gifford Pinchot at the U.S. Forest Service, accepted the University of California's first professorship in this new field. Sampson believed that productivity was an intrinsic quality of the landscape and that the range manager's task was to restore and maintain sustainable levels of natural productivity (Sampson 1914, 1923). To achieve this, he mobilized the state's agricultural extension program, in partnership with local cattlemen's groups, to provide ranchers with useable scientific knowledge and organize ranching communities to engage in coordinated efforts. These included grass seeding, livestock management, and seasonal burning. By the end of World War II, Sampson and his colleagues had enrolled most of the state's oak woodland ranches in cooperative conservation programs to inhibit shrub growth and encourage forage productivity (George 1987).

With range science in its infancy, efforts to develop a hardwood forest products industry were also underway. Attempts by industrial scientists and Extension foresters at the University of California to establish cork oaks in California dated to 1858 (Metcalf 1947). By the early twentieth century, gaskets and effective

Fig. 2.8 The arboretum at California State University, Chico, includes an oak grove established in 1904 to evaluate prospects for growing cork in California. Traces of harvests attempted from 1940 to 1947 appear on the trunks. The ground is covered with cork oak (*Q. suber*) seedlings, and over time, an unmanaged thicket of trees has evolved that is much appreciated as a walking path. (Photograph by L. Huntsinger)

sealants were in short supply, and cork was a crucial insulating material in the years between World Wars. While cork oaks were planted on a variety of California sites including Chico, Davis, and Napa, an absence of skilled cork harvesters and the development of alternative fireproofing, insulating, and noise-reduction technologies reduced demand and left cork oak stands isolated and neglected (Ryan 1948) (Fig. 2.8).

2.3.3 A Concern for Conservation

The third phase, which lasted from 1950 to 1985, comprised an era of "big conservation." Beginning around 1950, the farming and ranching industries in California expanded to supply commodities for growing markets. Over the next 25 years, California's cattle population rose by 280 %, reaching a peak of about 3.2 million head in 1976 (Burcham 1981). California's ranchers benefitted from financial and technical support provided by the state and federal governments, corporations, private donors, and a new generation of range managers who launched ambitious research, education, and outreach programs.

Unlike the range managers of Arthur Sampson's day, post-war range managers spoke out for large-scale landscape manipulation that could fundamentally alter the productivity of the landscape. They were advocates for the use of herbicides, heavy machinery, and other tools to reduce tree density, which they believed would increase forage availability, improve stream flow, and raise livestock carrying capacity. Using state annual range improvement reports, Bolsinger (1988) found that from 1945 to 1974 about 0.8 million ha of hardwoods and chaparral were cleared in the name of rangeland improvement.

By 1980 a cohort of scientists and managers began criticizing what was in essence an industrial conservation approach to hardwood range management. This new group was allied more closely with 1970s environmental activists than with the big program conservationists of an earlier generation. They argued that clearing too many oak trees was counterproductive in many cases because it impaired important ecological processes, and their studies suggested that short-term gains in productivity would soon be followed by long-term declines (Holland 1976). And significantly, they spotted a threat that 1950s and 1960s managers had not addressed: Oak woodland ranch subdivision for residential and agricultural development.

2.3.4 Integrated Management

A fourth period in the management history of California's hardwood rangelands began in 1986, when the University of California launched a new cooperative program to address escalating conflicts, and conservation concerns, over privately owned oak woodlands, referred to in the program as "hardwood rangelands." The Integrated Hardwood Range Management Program (IHRMP) helped avert the controversial prospect of state regulation of oak use and management with increased support for voluntary research and education. The IHRMP provided a central clearinghouse for statewide programs in hardwood rangeland science, conservation, and restoration (Standiford and Bartolome 1997). This included projects in ecology, natural resource economics, rural sociology, and public policy, and specific initiatives to deal with the spectacular growth of wine grape cultivation in the 1990s and the spread of sudden oak death in the 2000s. In 2009 the University terminated IHRMP funding during a time of state-level budget cuts to higher education and environmental programs. The IHRMP's work continues today through a coalition of scientists, extension specialists, and private ranchers, and through the University of California's Oak Woodland Conservation Workgroup and other programs (CA-OWCW 2012).

The burst of research that followed the formation of the IHMRP provided a rich trove of information about the status of California's oak woodland ranches. California livestock ranching on oak properties remains a family business, but not often a lucrative one from a commercial point of view. More than 80 % of ranchers live on their ranch with their families and manage the enterprise themselves, with few, if any, employees. Yet less than 15 % of ranchers make the majority of their income from ranching (Huntsinger et al. 2010). The romance of a rural lifestyle has attracted large numbers of exurbanites who have sought the amenities of ranch living, but who are unaccustomed to the sights, sounds, and smells of commodity production in working landscapes. Local political conflicts have often surrounded these clashes of urban and rural cultures (Walker and Fortmann 2003).

2.3.5 At the Moment

Today, California's oak woodland ranches exist in a complex public policy environment. Subsidies for tree clearing gradually ended, at both the state and federal levels, but other subsidies for agricultural and real estate development remain, and markets for biomass harvesting for cogeneration creates local incentives to remove trees. Agricultural and open space easement programs, including tax rebate opportunities such as the California Land Conservation Act of 1965—universally referred to as the Williamson Act, though threatened with elimination in California's currently troubled economy—can entice landowners to make conservation commitments. Yet ranchers often view these as weak incentives when compared to the pressure of rising land market value and formidable estate taxes, and recent austerity cuts in the state budget have put such programs at risk.

The State of California has chosen not to regulate oaks under the Forest Practice Act of 1974, which gives the state the authority to oversee harvesting or clearing projects for marketable timber species, and it has successfully defended this policy in court. As a result, regulation of oaks has devolved to local governments (Doak et al. 1988). By 1990, over 100 city and county governments in California had laws to protect native oaks. Today, many more such regulations exist. These local oak tree ordinances provide guidelines and regulatory frameworks for community oak management, but often focus on maintaining individual oak trees rather than functional ecosystems, lack adequate enforcement mechanisms, have weak or poorly enforced mitigation requirements, and create an uneven regulatory landscape across the state's many complex governance structures and jurisdictions.

2.4 History and Recent Trends in the Spanish Dehesa

The long history of oak woodland management in the dehesa region dates back some 6,000 years, according to pollen core evidence of vegetation change. Pollen studies suggest that early forms of management involved livestock grazing and human-wielded fire. These activities created the oak dotted savannas that characterize the landscape today, although past dehesas probably also included cultivated chestnuts, olives, and grapes (Stevenson and Harrison 1992). Work since on Neolithic cave sites in Cáceres corroborates the early human transformation of oak woodland into a managed dehesa system (López Sáez et al. 2007a, b).

Contemporary dehesas are dominated by human activities that prevent shrub encroachment and maintain an open understory of grasslands and patchy farmlands, systems that require constant maintenance (Chaps. 6, 8; Martín Bolaños 1943; Parsons 1962a, b; Gade 2010). After clearing, shrubs recolonize the understory within a few years, absent intensive grazing, clearing, or crop cultivation (Díaz et al. 1997).

The dehesa appeared in a recognizable form in south-west Spain in the first millennium AD when the region's lands were divided among retired Roman legionnaires from Extremadura (Fig. 2.9) (Cerrillo 1984; Díaz et al. 1997). Dehesa became more permanent and widespread during the Reconquest, which lasted from the eleventh through the fifteenth centuries (Linares and Zapata 2003; Stevenson and Harrison 1992). During this period, the kingdoms of Castile, León, and Aragón captured areas previously under Muslim control with the help of the northern nobility and military orders, including knights from the Orders of Alcántara, Santiago, and Calatrava.

Several factors beyond the region's soil and climate encouraged emergence of the dehesa as a dominant land-use system (Linares 2012). From the Islamic

Fig. 2.9 A detail from a mosaic, restored at the Museo Romano in Mérida (Emerita Augusta) and provisionally dated 324 AD, shows a well-garbed Roman hunter spearing a wild boar in a managed woodland. The taller oak shows the effects of pruning, and in the foreground are spiny oaks that would key out as either *Q. ilex* or *Q. coccifera*. (Photograph by P. F. Starrs)

conquest in the eighth century until at least the twelfth century, the south and west of Spain comprised a military frontier with weak institutions and fragmented local communities. Settlement patterns shaped the arrival of humanized rural landscapes including the present-day dehesa.

Extremadura was notably less fertile than areas to its south (Andalucía) and east (Murcia), and unattractive to settlers. Population growth lagged until the second half of the thirteenth century, when monarchs encouraged settlement with concessions of large properties to nobility and the creation of new municipalities, which received their own landholdings to use, manage, and exploit. Growing demand from the textile industry encouraged the development of new political and economic alliances that brought the nobility, military, and local governments together with northern woolgrowers to make Extremadura a prime lowland seasonal range for Merino sheep trailed from northern Spain as part of a transhumance between the north and the south. The woolgrowers became a guild known as the *Mesta*, whose tax payments to the Crown earned them considerable royal support, much to the chagrin of local farmers and stock producers seeking to protect their lands from migrant flocks that would graze their way across Spain. Nearly 600 years later, by the mid-nineteenth century, the Mesta had dissolved, and private landowners—including gentry, Church orders, and municipalities—were vying for grazing land. Despite vast changes leading up to the late 1800s, the livestock industry continued to flourish in the Iberian Peninsula.

As is the norm in many a traditional society, everyday life in historical Spain gave a great deal of attention to hunting (Chap. 11). The taking of big game (wild boars and deer) and small game that ranged from hares and rabbits and doves to starlings and larger birds led to a complicated combination of hunting for food and sport (Parsons 1960; CdV 1986). On private dehesa estates hunting was an activity pursued by nobles, usually for their own pleasure or with friends. For the more prosperous municipalities (*municipios*), where a village or community owned a dehesa that included hunting rights, residents might be allowed to hunt (Owens 1977; López Ontiveros et al. 1988; López Ontiveros and Valle Buenestado 1989). And poachers roamed, with the combined rewards that could include alleviating boredom and getting men and boys out of the house. Game taken would either be eaten at home or sold as provender to local bars and roadside inns. Of course, significant penalties could come with being caught—penalties that continued well into the 1970s, as the Guardia Civil wielded an iron authority over rural Spain and protected landowner interests (López Ontiveros 1986, 1994).

Rights to draw on woodland and forest resources in the dehesa existed in a complicated scheme of access, penalties, rights, and traditions, with significant variations from landholding, municipality, region, and kingdom (Gómez Mendoza 1989; Chap. 11). In terms of hunting, no region is so well studied, in historical context and management, as Andalucía's Córdoba, with researchers working on hunting history and on aspects of current hunting interest, examining each from the perspectives of social life, veterinary health, and management of game (Buenestado 1978; de Urquijo 1988; López Ontiveros and Valle Buenestado 1989). Hunting of

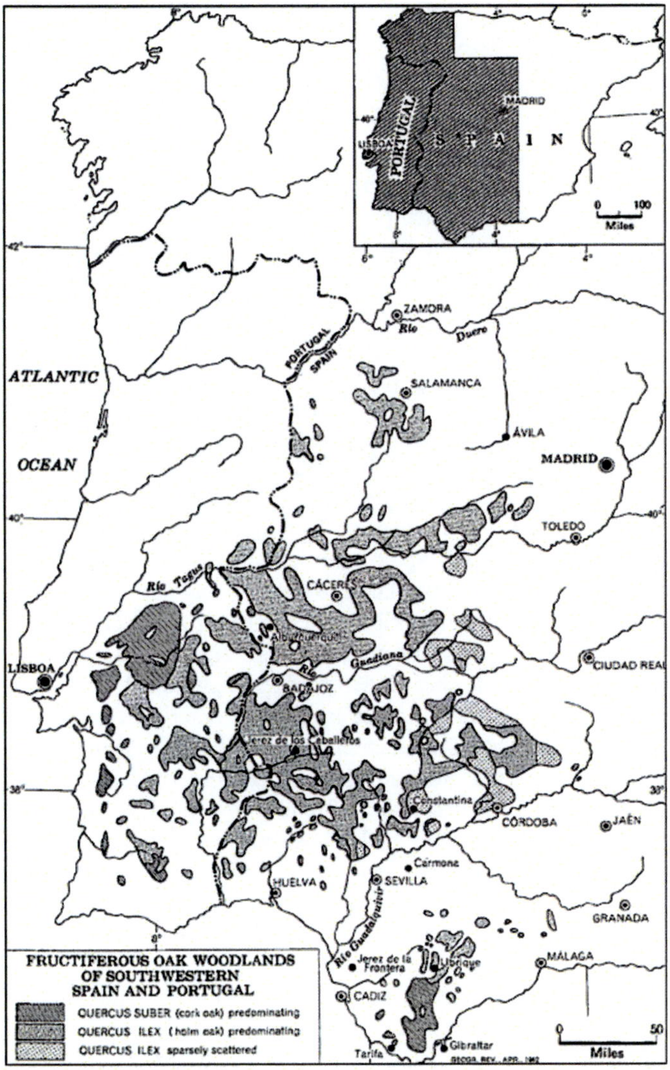

Fig. 2.10 While delimiting the exact extent of the dehesa remains a matter of some cartographic uncertainty, this effort supervised by the geographer James J. Parsons offers a reasonable idea of the extent of the dehesa and its characteristic cork and holm oaks in the southern and western Iberian Peninsula. The definitive maps were produced for some, but not all of, the dehesa area by the Spanish Ministry of Agriculture in the 1970s and 1980s. (Modified from an original map in Parsons 1962a)

rural dehesa lands was a sought-after activity as early as the years of Roman and Arab control of Spain, and the full picture of its history is slowly being revealed in historical and geographical sources (Fig. 2.10).

2.5 Four Historical Eras in the Spanish Dehesa

The historical roots of today's dehesa landscape derive from the time of the Christian repopulation. Consolidation of power and authority in Church and gentry grew in a region of weak urban networks, low population density, large landholdings, and a livestock-based economy. Since medieval times, the dehesa has undergone countless smaller changes. The remainder of this section focuses on changes that relate to the ownership, management, and use of the dehesa, dividing it into four periods: consolidation, development, decay, and present-day trends.

2.5.1 Consolidation of the Dehesa

A period of consolidation, lasting from the mid-thirteenth-century to the mid-eighteenth-century, was a time of great change in dehesa lands. Place names and data collected in the *Book of Hunting* by Alfonso XI document the Christian colonization of south-western Spain and the accompanying shift from Mediterranean hardwood forests and shrublands to pastures and farms (Bernal 1998). Yet a lack of definitive data makes it difficult to characterize the vegetation that existed immediately before the period of consolidation, and only very recently has the precise spatial extent of the area in dehesa been defined (MARM 2008).

The exact process by which dehesa landscapes emerged during the consolidation period remains unclear. Resettlement charters describe land use practices that could have resulted in the creation and maintenance of dehesa systems. These included cutting timber for use as farm implements, building materials, firewood, and charcoal, stripping cork to make beehives, and the cultivation of crops, hunting of game, and gathering of mushrooms, wild herbs, and medicinal plants. Extensive grazing and acorn gathering appear to be the most important and widespread land use practices during this period on both communal and privately owned lands (Linares 2002; Clemente 2007).

By the mid-fifteenth century, when more detailed information began to emerge, extensive dehesa systems were already fully formed (Linares 2001). Within a fringing ring around the populated areas—usually on the margins between municipalities and farmlands—noblemen, military orders, municipalities, and neighboring communities maintained open stands of oak. Over time, towns and farmlands began to encroach on these woodlands, while seasonal grazing expanded the dehesa along its outer edges (Linares and Zapata 2003). Similar patterns of encroachment and expansion have continued in more recent centuries.

Sources available for the period of consolidation distinguish between two types of dehesa properties: private dehesa owned by the nobility, clergy, or agrarian oligarchy, and public dehesa (*dehesas boyales*) controlled by municipalities or communities. Private lands were managed by administrators, or, in some areas, as around Cáceres, by associations of multiple administrators who shared in the ownership of a single property and divided the revenue from its uses (Melón 1989). For public dehesa, governing boards from each municipality assumed the authority to manage the use of their communally held properties—an approach still common today (Linares 2002).

On private and public lands, grazing and browsing of livestock was the most important commercial use, especially on fresh pastures during the fall and winter. Except in the commons, grazing rights were usually leased to members of the Honorable Council of the Mesta, an arrangement that guaranteed the seasonal presence of Merino sheep in south-west Spain (Klein 1920; García-Sanz 1985b).

The second most important use of these lands was for acorn foraging by domestic pigs. Pig foraging permits were leased to the highest bidder on the private estates and at an appraised rate on the public lands. Livestock use permits were leased at lower rates for summer pastures, and any additional agricultural by-products, including forage and hunting for small game on fallowed fields and post-harvest stubble, remained free for use by local residents (Linares 2006).

Crop cultivation was the third most important commercial activity on the dehesas during the period of consolidation. The predominant crop was cereal for human consumption, grown in biennial or triennial rotation systems with intervening fallow years. On private woodlands, this practice was often subleased to third parties by the northern stockbreeders. On public dehesas, it was offered to community members free of charge or sold to the highest bidder (Linares 2002).

Forestry uses had only minor commercial value, but they were important for land management. Tree pruning was thought to increase the production of acorns (see, however, Chap. 7), and the few existing contemporary sources suggest local residents were permitted to cut branches for use as lumber, firewood, and charcoal at no cost. The information available about cork harvest is less clear. In the eighteenth century, cork harvesting was linked to uses such as tanning and beehive construction, although even today Extremadura has villages where the skill of its specialized cork-harvesters-for-hire is famed. Sources say little about landowner cork oak management. The same goes for other dehesa uses, such as hunting, fishing, stone working, beekeeping, and the harvesting of mushrooms, plants, truffles, and aromatic or healing herbs. All of these practices were underway, but little is known about their extent, application, or management (Linares 2008).

2.5.2 The Developing Dehesa

A second phase—the period of development—lasted from the mid-eighteenth century to the mid-twentieth century (Fig. 2.11). Reports compiled in the *Catastro of Ensenada* (1750–1754) indicate that as this period began dehesa covered more than 30 % of all useable land in the former provinces of Salamanca, Toledo, La Mancha, Extremadura, Sevilla, Córdoba, and Jaén. In Extremadura, dehesa covered 55 % of usable land (Grupo'75 1977). Multiple activities continued on the dehesa, but Merino sheep predominated on the region's autumnal pastures.

Fig. 2.11 The Plaza de España of Seville, built in 1928 for the Ibero-American Exposition, is regarded as a Renaissance Revival architectural marvel. Alongside the Alcoves of the Provinces in the Plaza appears this map showing part of the Sierra Morena, north of Seville, in southwestern Spain, showcasing the region's long history of herders and their livestock, along with villages and features of local renown. Livestock are a reflection of the dehesa landscape, with goats, sheep, cattle, dairy animals, pigs, and donkeys. (Photograph by P. F. Starrs)

Local residents called for change in response to the continued presence of the northern Merino herds. After 1808, tumultuous events—including the Spanish War of Independence (1808–1814), the enactment of the first Spanish Constitution of 1812, and a crisis in the Merino industry brought on by increased international competition—diminished the influence of northern wool growers and began a new era on the dehesa (Llopis 1985). Transhumance declined as Mesta authority faded, which meant landowners could elect to take land out of pasture and turn to raising grain crops and other livestock (Llopis 1989) (Fig. 2.12 and Fig 2.13).

During the early nineteenth century, liberal reforms amplified shifts in Spanish political and economic life (García-Sanz 1985a). Privatization of church properties and the General Disentitlement Act of 1855 accelerated a process of land privatization begun during the War of Independence (Linares 2001). During the sixty-nine years covered by the General Disentitlement Act (*desamortización*, 1855–1924), thousands of municipal and communal estates throughout southwestern Spain once managed as public lands became privately owned properties.

The reforms of the nineteenth century resulted in an almost total disappearance of public land in the Spanish south and west. By the 1860s, few municipal or communal estates had escaped privatization. In 1863 national government officials launched a planning process they hoped would foster more efficient forest management on the remaining public lands. By 1924, efforts to impose rational scientific management on a complex system shaped by centuries of social relationships, cultural practices, and deeply held traditional local knowledge had

Fig. 2.12 The Iberian black-footed pig was traditionally placed on the dehesa in October to consume acorns that would in turn flavor meat in ways characteristic of Spanish ham (Chap. 10). While long respected within the regional diet of south-west Spain, the Iberian pig has in the last several decades become a gastronomic superstar, and added considerable value to dehesa livestock production (Chaps. 10, 13). Over the long term, only cork and, locally, income from commercial hunting competes with hog production. (Photograph by S. García)

proven futile (Linares 2007). Since then, the municipalities have once again assumed authority over management of most public lands.

Privatization contributed to dramatically new landscape patterns. The owners and leasers of land invested significant sums toward dehesa improvement. Efforts included thinning forests to create canopy openings, pruning trees to stimulate acorn production, and constructing buildings to provide spaces for living and working—in each case, hired labor and the skills of itinerant specialists were required. These projects were meant to increase dehesa productivity (Zapata 1986), but they also increased the region's ecological uniformity (Fig. 2.14).

The spread of the moldboard plow and the use of the first synthetic fertilizers allowed dehesa users to expand the region's cultivated area. Instead of leading to a reduction in livestock, however, the increase in farming only furthered shifts in animal husbandry that had begun in the early nineteenth century (García-Sanz 1994). Grain production enabled ranchers to increase their herds. Sheep continued to graze in the pastures, but transhumance declined as the flocks became more integrated into local farming systems as producers of manure, wool, leather, and, increasingly, meat. Rising demand for animal protein, which paralleled an increase in per capita income, encouraged a rise in cattle and pig populations. The changing nature of dehesa agriculture also led to the gradual replacement of oxen by mules, donkeys, and horses in many agricultural tasks (Linares 2007).

Fig. 2.13 The pruning of holm oaks can go to extremes, as here in the Sierra Morena, leaving oaks a spidery wisp of their former selves. Longstanding traditional knowledge holds that pruning increases the fall of acorns; scientific measurements over the last two decades argue that there is no increase in production from pruning (Chaps. 5, 6, 7, 10). Nonetheless, an area of *encinas recién podadas* (recently pruned oaks) is a distinctive sight. (Photograph by P. F. Starrs)

Fig. 2.14 In central Badajoz province the full force of dehesa management is readily apparent: the holm oaks show distinct browse lines, where livestock consume oak leaves from the ground; the effects of pigs on the ground cover are apparent both in the oak understory, and in their nighttime enclosure, and the park-like expanse of trees stretch into the distance, seen from Monasterio de Rocamador, near Almendralejo. (Photograph by P. F. Starrs)

Fig. 2.15 Slabs of cork, steamed and pressed flat, await transport and conversion into cork bottle stoppers, flooring, or other uses. Once considered an essential strategic product, and used as an insulator and as a flexible gasket-sealant, cork remains a high-value product of the dehesa (and the Portuguese *montado*), but faces competition from the plastic "stopper" industry in wine-producing countries that are not able to produce cork and must buy their supply. (Photograph by P. F. Starrs)

Traditional forestry practices continued to evolve. Wood harvested through felling and pruning operations had low commercial value in forest management, but was considered significant in terms of woodland health (Linares 2007). An important change involved cork (Fig. 2.15). During the nineteenth century, the status of cork production made a transition from a marginal resource into the single most important industrial woody forest product of the dehesa (Parsons 1962b). This shift had little to do with regional changes in land ownership or property rights. Rather, it resulted from the emergence of a new bottle and barrel stopper industry, the rise of international demand for cork as a versatile building material, and the role of cork as a sound insulator in strategic defense industries, especially for warships (Metcalf 1947; Zapata 2002).

Additional research remains to be done on the social, economic, and ecological aspects of the dehesa during this period. For example, although government foresters of the 1920s noted an increase in hunting and poaching, confirming news reports from the first decades of the Franco era, little is known about specific trends or events (Linares 2012). What is clear is that in the mid-twentieth-century the dehesa remained a region of diverse and complementary land uses (Linares and Zapata 2003). In Extremadura the region's working landscapes offered a source of some social and demographic stability in rural areas, although in the 1960s and 1970s more than half a million residents left Extremadura to go abroad or to other parts of Spain.

2.5.3 The Mid- and Late Twentieth-Century: A Dehesa in Decay

The third phase—the period of decay—began in the mid-twentieth-century with influences continuing to the present day. Grain production rose during the 1950s when the Franco government required that all arable land be tilled under penalty of law, but declined by the mid-1960s. Since the mid-twentieth-century, the total area covered by dehesa has declined, coinciding with a crisis of traditional agriculture in Spain (Plieninger 2006a). This crisis was, in part, a consequence of the substitution of capital for labor in the farming sector related to a decline in relative commodity prices caused by the Green Revolution. The rapid adoption of new technologies—including tractors, harvesters, fertilizers, herbicides, and pesticides—has increased productivity, but it led to a growing dependence on agricultural inputs purchased from outside the dehesa region.

In recent decades, agricultural mechanization and the deposition of large amounts of soil fertilizer have enabled yet another increase in the total area of croplands in the dehesa region, and a shift from untilled or rested fallow to seeded fallow, where nitrogen-rich cover crops would be plowed under to provide so-called green manure. This intensification of farming practices and an expansion of arable land enabled production of more varied agricultural commodities. Grain for human, and increasingly animal, consumption remains the region's predominant tilled crop, but other species—such as bean, chickpea, sunflower, and vetch—have also proliferated on once-fallow fields (Campos 1984).

Increases in forage crops and external inputs, such as supplemental feeds, along with incentives in the Common Agricultural Policy, have encouraged the continued growth of livestock populations (Plieninger 2006b). As in previous eras, this growth has been accompanied by changes in the composition of herds. Sheep still maintain a slight numerical advantage over other species, but sheep and pigs are both being displaced by cattle and deer, raised commercially on fenced dehesa. The relative simplicity of these latter two species' management, together with the growth of per capita income and demand for red meat facilitated this livestock substitution process (Campos et al. 2003). The adoption of Green Revolution mechanized technologies has reduced the importance of horses as draft power—even while horses have gained ground as pets (Chap. 10; Linares 2012).

Traditional forestry practices face considerable challenges. A decline in demand for biomass energy, including firewood and charcoal, has reduced oak pruning. Many residents believe that this has reduced the quality of acorns and diminished trees' capacity for regeneration. Cork has substantial commercial value, but its days, too, may be numbered. A lack of active management, amounting to neglect by some cork oak owners, along with the emergence of substitute wine stoppers made of plastic or synthetic resin, continues to place downward pressure on cork prices (Zapata 2010).

Changes in forest and range management practices have accompanied a general decline in oak tree prevalence and health on the dehesa. A massive government-driven reduction in Iberian pig population in the 1960s, as a response to an infestation of African swine fever, slashed an important market for acorn-bearing trees, although pig numbers have recovered in the decades since. A second pervasive problem is the high cost of the traditional practices required to promote oak regeneration. A third issue is the lack of public programs to encourage resource management by private landowners, many of whom do not believe that conservation is their responsibility (Campos 2008). Public sector reforms, such as incentive programs, could help promote sustainable and diversified dehesa management. However, the public sector no longer has the leverage it once did, and European Union subsidies and initiatives have met with varying degrees of enthusiasm, and sometimes even outright rejection. Two centuries of economic liberalism have left private landowners largely in control. Today, public ownership is residual, while private property is dominant.

Not surprisingly, these economic, technological, and administrative changes have had serious repercussions for the social fabric of the dehesa region. Once an area of relative demographic stability, the region's agricultural industrialization has become a significant push factor promoting out-migration to urban areas. In recent decades there has been an unprecedented exodus from the countryside to the city.

There is, however, a positive side to these changes. During the past couple of decades, increases in income have fostered a revival of traditional crafts and land use practices, such as stone working and hunting, and increased the value of newer recreational uses, including horseback riding, hiking, and nature tourism. The dehesa is now recognized for its potential to maintain biodiversity and ecosystem services, in particular by the European Union. The growing appreciation of these new values will only promote conservation, however, if it can be converted into profits for people who live and work in the dehesa (Campos 2008), and who depend on the land for their livelihoods.

2.5.4 A Post-millennial Dehesa, Resurgent?

Today, livestock production remains the dominant commercial activity in the dehesa. Most livestock income depends to a degree on subsidies under European Union agricultural policies. With surplus production in many EU countries, policies tend to favor products that are locally sought-after and use this to encourage low-intensity agriculture linked to biodiversity hot spots (Chap. 8). This fits the dehesa well (López-López et al. 2011). The current livestock census of the dehesa also shows a significant dependence on supplements from outside sources. Yet such reliance on subsidies and inputs increases the vulnerability of commercial operations that are faced with the dual uncertainties of changing subsidy policies and input prices. In some cases, landowners have responded by increasing the size of their herds, with consequent improvements in labor productivity, but this can increase their dependence on outside inputs even more. In other cases, landowners have switched the species they cultivate from swine and sheep to cattle and wild game—and even the semi-domesticated game that are fed grain and other supplements to add their size and increase their desirability as trophy animals.

Recreational hunting is increasing in the dehesa, but subsistence and social hunting remains important for private landowners. Some landowners see hunting as an alternative to livestock production that can contribute to the conservation of ecological services in the dehesa. Many newer, wealthier landowners value hunting, but usually more for recreation. Wildlife is not the only source of biodiversity worth conserving in the region. The dehesa contains several unique breeds of domestic livestock, some of which may be in danger of extinction (Chap. 10). Concern about has led to the establishment of public compensation programs that support the maintenance of some of these native or heritage breeds.

The value of dehesa forest products—including grass, cork, and firewood—has continued to decline as an income source for most landowners. Decreases in tree pruning and brush trimming permits shrubs to reinvade the oak woodlands, and pastures increasingly show signs of trampling by pigs, cattle, sheep, and goats—sometimes in sequence—searching out palatable forage. The decline of oak and cork trees, only partially offset by recent reforestation efforts, reduces acorn production for swine fodder, known as *montanera*. Agricultural crops with several year rotations traditionally encouraged soil recuperation and prevented the livestock compaction typical of shrub lands. Today, however, the decline of farming has an unintended effect of diminishing soil quality and productivity.

Increasingly, income to landowners does not justify the private investments required to produce livestock, wild game, forest products, and crops in an integrated system. To make dehesa ranching worthwhile, residents must accept that non-commercial environmental services have value. Landowners may accept lower incomes and higher expenses for the opportunity to live in a dehesa setting that provides them with non-commercial environmental values (Fig. 2.16). All that some landowners need is a modest financial return to justify the considerable additional benefits of their dehesa ranch life. In this way, the dehesa is moving toward a management system that may be both commercially remunerative and ecologically healthful.

Forest and grass management are essential for maintaining dehesa environmental services. Yet given the scope of the problems and the vast area involved, effective public policy will require the prioritization of management objectives for woodlands, pastures, habitats, and species. There may be increases in public investments and incentive programs in the coming years with financial instruments of the EU Common Agricultural Policy. Such support can contribute to maintaining the economic viability of the dehesa with its associated high biodiversity. However, incentives should also finance timely solutions to complex biodiversity issues. For example, moderate shrub cover can encourage oak recruitment, although it may reduce local biodiversity and forage productivity. Management choices will need to be made, some with considerable social and ecological trade-offs.

Although dehesa landownership is largely privatized (Fig. 2.17), a public support system provides resources for fire protection, visitor services, biodiversity conservation, and environmental management programs. Future public programs

Fig. 2.16 Most Spanish fighting bulls are and have been raised on dehesa lands, and perhaps unsurprisingly, bullfighters and their managers were among the prominent purchasers of dehesa properties in the late twentieth century. On a dehesa north of Córdoba, a bullfighter-owned property offers a testimony to the source of wealth that allowed the new property owner to acquire his land. (Photograph by A. Caparrós)

must be linked to credible priorities, with implementation of territorial agreements that provide adequate compensation to landowners for the provision of environmental services. The implementation of such agreements requires better economic data on the costs and benefits of various approaches to dehesa management.

2.6 Synchronicity and Divergence: A Conclusion

California oak woodland ranches and the dehesa of south-west Spain have many common attributes. With similar climates and physiographic features, they are defined by graceful oak trees, often arranged in park-like stands (Fig. 2.18); they each have annual grass understories that change from bright green during the winter growing season to golden brown during long, dry summers. In some areas, ranches and dehesas appear so similar that landscape photographs taken in the two places can be virtually indistinguishable, even for educated naturalists and long-time residents. It is no wonder that the Spanish padres who arrived in California during the eighteenth century felt so at home compared to Catholic missionaries in other parts of the New World.

Despite the similarities of their climates, terrains, ecologies, and appearances, the ranchos and dehesas have experienced markedly different histories. California's ranchos emerged in landscapes occupied for more than 13,000 years by Native

Fig. 2.17 Hunting was once an activity largely pursued by village men during non-agricultural seasons, and by the gentry who would seek out game on their own land or as invited guests on the property of neighbors. Hunting today is highly commercialized, especially on dehesa lands where hunting is reserved, as indicated by the sign declaring a hunting preserve near Guadalcanal, north of Seville. Fees to hunt on attractive properties where hunters can take trophy-quality animals may rise to $10,000 or more, and fees above $1,000 are routine. (Photograph by P. F. Starrs)

Californians whose primary tool for landscape manipulation was fire. Some exotic plants probably made it to California well before the Spanish colonial era, but the introduction and proliferation of European livestock and cultivation occurred in the span of just a few decades. The emergence of a pastoral empire in California, during the periods of Spanish and Mexican control, came with a host of ecological changes that scholars are even now still struggling to understand (Fig. 2.19). It is astonishing to recognize that by the time California became part of the United States, it had supported large-scale livestock grazing for less than eight decades—no more than the length of a single human lifetime. The contemporary rural geography of California's oak woodlands emerged during the late nineteenth century, as the state's fertile valleys shifted to intensive agriculture and federal officials began to regulate seasonal grazing on public lands in the higher elevation mountains.

The Spanish dehesa landscape began to acquire many of its contemporary characteristics during the Catholic repopulation of the thirteenth century, and has remained under Spanish and Portuguese control. Their cultural lineage supported the gradual development of an intricate agro-forestry system built on seasonal grazing, foraging, cultivation of tree- and understory crops, and relatively intensive forest management practices designed to promote these primary uses. Until the nineteenth century, the church, military orders, municipal councils, and northern nobility—with their rangy flocks of Merino sheep—controlled the

Fig. 2.18 Massive holm oaks can be sizable acorn producers, as here in southern Salamanca province. Oaks can live for several hundred years, but during that time must eventually regenerate through seedlings that have to survive the hunger of livestock and wildlife that are well accustomed to eating sweet acorns and shrubs in an oak understory. The long-term future is uncertain. (Photograph by P. F. Starrs)

Fig. 2.19 The oak woodland landscape of California, much of it in private ownership, reflects a 250 year-old presence of grasses, weeds, and other exotic species introduced by livestock that came with Spanish and Mexican colonizers who arrived even before the first permanent Spanish settlement at Mission San Diego de Alcalá in 1769. Wild oats (*Avena fatua),* pictured here, is a grass common to both places. As a result, the look of the land is remarkably familiar to almost any visitor from the southern Mediterranean, and especially from Spain. (Photograph by P. F. Starrs)

dehesas. A series of liberal economic and legal reforms, beginning with the Spanish War of Independence, resulted in the transfer of most dehesas to private control. Changes in commodity markets, such as increased demand for meat and cork, shaped the use of the dehesa and contributed to modern landscape patterns.

At the beginning of the twentieth century, California oak woodland ranches and Spanish dehesas were mostly in private ownership. By the early 1920s, officials in Spain had largely given up on efforts to institute scientific management. In California, however, such efforts were just getting underway, as ranchers voluntarily joined cooperative, university-sponsored programs to restore and improve the productivity of their lands. California ranches have never been as intensively managed or used as the Spanish dehesas. Yet both landscapes experienced dramatic escalations in the scope and intensity of agricultural mechanization following World War II. Growth in demand for meat was one reason, but public subsidies, private investments, and complex changes in global agricultural markets also fostered important shifts in land use programs and patterns.

Local residents and conservationists increasingly recognize both landscapes as valuable, for their amenities as for their commodities, and for their ability to support sustainable agriculture while promoting biodiversity conservation and the maintenance of ecosystem services (Chap. 12). But a range of economic factors—

Fig. 2.20 The huge black, long-legged and fast-traveling Negra Avileña cow is a feature of the dehesa, once common but still in evidence in an annual transhumant movement from Extremadura up the old Roman roads to the *Sierra de Gredos* and, eventually, Avila. Native (autocthonous) breeds such as these are a prized part of dehesa life, and currently encouraged under EU policies, but the duration of support remains an uncertain affair. (Photograph by L. Macaulay)

including international competition and increases in property values, rents, and other development pressures—continue to promote land use change. This makes it more important now than ever to articulate the value of the diverse social, cultural, and ecological goods and services these systems provide (Fig. 2.20).

California ranches and the dehesas of Spain have dramatically different histories. Yet, during the past 250 years their stories have converged. They look now more similar than ever before, with more plants and animals in common than at any previous point in time. They have experienced similar privatization and modernization efforts. Each region has been shaped by global agricultural markets. Today, both landscapes are the subjects of intensive study and conservation efforts, and face similar social and ecological challenges. The remaining chapters in this book examine these congruencies and departures in greater detail.

References

Anderson KM, Barbour MG, Whitworth V (1997) A world of balance and plenty: land, plants, animals, and humans in a pre-European California. Ca Hist 76:12–47

Becker RH (1964) Diseños of California ranchos: maps of thirty-seven land grants, 1822–1846 from the records of the United States district court. San Francisco Book Club, San Francisco

Bernal Estévez Á (1998) Poblamiento, transformación y organización social el espacio extremeño (siglos XIII al XV). Editorial Regional de Extremadura, Mérida

Bolsinger CL (1988) The hardwoods of California's timberlands, woodlands, and savannas. Res Bull PNW-RB-148. U.S. Department of Agriculture, Forest Service, Pacific Northwest Research Station, Portland

Burcham LT (1981) California rangelands in historical perspective. Rangel 3:95–104

Byrne R, Edlund E, Mensing S (1991) Holocene changes in the distribution and abundance of oaks in California. Symposium on oak woodlands and hardwood rangeland management, Oct 31–Nov 2, Davis, California, pp 182–188. USDA For Serv Gen Tech Rep PSW-126. U.S. Department of Agriculture, Forest Service, Pacific Southwest Research Station, Albany, CA

CA-OWCW [California Oak Woodland Conservation Working Group] (2012). Website, http://ucanr.org/sites/anrstaff/Divisionwide_Programs/Workgroups/Workgroup_Directory/?thiswg=66. Accessed 20 Aug 2012

Campos P (1984) Economía y energía en la dehesa extremeña. MAPA, Madrid. Campos P, Casado, JM, Azuqueta, D (2004) Cuentas Ambientales y Actividad Económica. Colegio de Economismas, Madrid

Campos P (2008) La economía de la dehesa. Biodiversidad, efecto invernadero y cambio climático. In: Agricultura Familiar en España, 103-108. Unión de Pequeños Agricultores y Ganaderos, Madrid

Campos P, Cañellas, I, Montero G (2003) Evolución y situación actual del monte adehesado. In: Pulido F, Campos P, Montero G (eds) La gestión forestal de las dehesas. Hist, Ecolog, Selvicul, Econ, 27–37. IPROCOR, Cáceres

CDF-FRAP [California Department of Forestry and Fire Protection-Fire and Resource Assessment Program, CalFire] (2003) Changing California: forest and range 2003 assessment. Sacramento, State of California Resources Agency, CA. http://www.frap.cdf.ca.gov/assessment2003. Accessed 01 June 2012

CdV [Equipo Pluridiciplinar de la Casa de Velázquez] (1986) El Paper de las Actividades Cinegéticas: Los Cotos de Caza. In: Superviviencia de la Sierra Norte: Evolución de los Paisajes y Ordenación del Territorio en Andalucía Occidental, 299–335. Casa de Velázquez and Ministerio de Agricultura, Pesca, y Alimentación, Madrid

Cerrillo E (1984) La vida rural romana en Extremadura. Servicio de Publicaciones, Universidad de Extremadura, Cáceres

Cleland RG (1941) The cattle on a thousand hills: Southern California, 1850–1880. The Huntington Library, San Marino

Clemente J (2007) La Tierra de Medellín (1234–c.1450). Dehesas, Ganadería y Oligarquía. Diputación de Badajoz

de Urquijo A (1988) Los Serreños: retazos cinegéticos y camperos de Sierra Morena. Aldaba Ediciónes, Editorial Olivo, Córdoba

Díaz M, Campos P, Pulido FJ (1997) The Spanish dehesa: A diversity in land use and wildlife. In: Pain DJ, Pienkowski MW (eds) Farming and birds in Europe: the common agricultural policy and its implications of bird conservation. Academic Press, London, pp 178–209

Doak S, Green K, Fairfax SK, Johnson S (1988) The legal environment for hardwood land ownerships in California. California Department of Forestry and Fire Protection, Sacramento

Fagan BM (2004) Before California: an archaeologist looks at our earliest inhabitants. Rowman Altamira, New York

Gade DW (2010) Parsons on pigs and acorns. Geogr Rev 100:598–606

García-Sanz Á (1985a) Crisis de la agricultura tradicional y revolución liberal (1800-1850). In: García-Sanz Á and Garrabou R (eds) Historia agraria de la España contemporánea. I. Cambio social y nuevas formas de propiedad (1800–1850), 7–99. Crítica Ed, Barcelona

García-Sanz Á (1985b) La agonía de la Mesta y el hundimiento de las exportaciones laneras: un capítulo de la crisis económica del antiguo régimen en España. In: García-Sanz Á and Garrabou R (eds) Historia agraria de la España contemporánea. I. Cambio social y nuevas formas de propiedad (1800–1850), 174–216

García-Sanz Á (1994) La ganadería española entre 1750 y 1865: los efectos de la reforma agraria liberal. Agric y Soc 72:81–119

George M (1987) Management of hardwood range: a historical review. University of California. Davis Range Sci Rep 12:3–4

Gómez Mendoza J (1989) El Entendimiento del Monte en la Genesis de la Politica Forestal Español. Seminario Sobre el Paisaje, pp 64–78. Obras Publicas, Junta de Andalucìa, Sevilla

Grupo'75 (1977) La economía del Antiguo Régimen. La 'Renta Nacional' de la Corona de Castilla. Universidad Autónoma de Madrid, Madrid

Hackel SW (2005) Children of coyote, missionaries of Saint Francis: Indian-Spanish relations in colonial California, 1769–1850. University of North Carolina Press, Durham

Holland VL (1976) In defense of blue oaks. Fremontia 4:3–8

Hornbeck D (1983) California patterns: a geographical and historical atlas, with Kane, P, Fuller DL. Mayfield Publishing Company, Palo Alto

Huntsinger L, Johnson M, Stafford M, Fried J (2010) California hardwood rangeland landowners 1985 – 2004: ecosystem services, production, and permanence. Rangel Ecol Man 63:325–334

Igler D (2001) Industrial cowboys: miller and lux and the transformation of the far west, 1850–1920. University of California Press, Berkeley and Los Angeles

Isenberg AD (2005) Mining California: an ecological history. Hill and Wang, New York

Keeley JE, Lubin D, Fotheringham CJ (2003) Fire and grazing impacts on plant diversity and alien plant invasions in the Southern Sierra Nevada. Ecol Appl 13: 1355–1374. http://www.esajournals.org/doi/abs/10.1890/02-5002. Accessed 15 July 2012

Klein J (1920) The Mesta: a study in Spanish economic history, 1273–1836. Cambridge UP, Cambridge

Linares A (2001) Estado, comunidad y mercado en los montes municipales extremeños (1855–1924). Revis Hist Econ 19:17–52

Linares A (2002) El proceso de privatización de los patrimonios de titularidad pública en Extremadura, 1750–1936. Universitat de Barcelona

Linares A (2006) Tapando grietas. Hacienda local y reforma tributaria en Extremadura (1750–1936). Invest Hist Econ 5:71–103

Linares A (2007) Forest planning and traditional knowledge in collective woodlands of Spain: the dehesa system. For Ecol Manage 249:71–79

Linares A (2008) The commons's versatility in the southwest of Spain. Twelfth biennial conference of the international association for the study of commons, 14–18 July, international association for the study of commons, Cheltelham. http://dlc.dlib.indiana.edu/dlc/bitstream/handle/10535/2379/Linares_201602.pdf?sequence=1. Accessed 15 July 2012

Linares A (2012) La evolución de la dehesa: entre la persistencia y el cambio. In: Linares A, Llopis E, Pedraja F (eds) Santiago Zapata Blanco: Economía e Historia Económica. Fundación Caja de Extremadura, Cáceres, pp 11–35

Linares A, Zapata S (2003) La dehesa: una visión panorámica de ocho siglos. In: Pulido F, Campos P, Montero G (eds) La gestión forestal de las dehesas. Historia, Ecología, Selvicultura y Economía, pp 13-25, IPROCOR, Cáceres

Llopis E (1985) Algunas consideraciones acerca de la producción agraria en los veinticinco últimos años del Antiguo Régimen. In García-Sanz, Ángel, and Garrabou, Ramón (eds.), Historia agraria de la España contemporánea. I. Cambio social y nuevas formas de propiedad (1800–1850), pp 267–290. Crítica Ed, Barcelona

Llopis E (1989) El agro extremeño en el Setecientos: crecimiento demográfico, 'invasión mesteña' y conflictos sociales. In: MAPA (eds) Estructuras agrarias y reformismo ilustrado en la España del siglo XVIII, pp 267-290, MAPA, Madrid

López-López P, Maiorano L, Falucci A, Barba E (2011) Hotspots of species richness, threat, and endemism for terrestrial vertebrates in SW Europe. Acta Oecologica 37:399–412

López Ontiveros A (1986) Caza y actividad agraria en España y Andalucìa y su evolución reciente. Agricult y Soc 40:67–98

López Ontiveros A (1994) Caza, actividad agraria, y geografìa en España. Documents D'Análisi Geográfica 24:111–130

López Ontiveros A, Valle Buenestado B (1989) Caza y explotación cinegética en las provincias de Córdoba y Jaén. Junta de Andalucía, Instituto Andaluz de Reforma Agraria, Córdoba

López Ontiveros A, Valle Buenestado B, García Verdugo FR (1988) Caza y paisaje geografico en las Tierras Beticas segun el *Libro de la Montería.* Andalucía entre Oriente y Occidente (1236–1492), pp 281–307, Tipografía Catolica, Córdoba

López Sáez JA, González Cordero A, Cerillo Cuenca E (2007a) Paleoambiente y paleoeconomía durante el neolítico antiguo y al calcolítico en Extremadura: análisis arqueopalinológico del yaciemiento del cerro de la Horca (Pasenzuela, Cáceres, España). Zephyrus: Rev Prehist y Arqueol 60:145–153

López Sáez JA, López Garcia P, López-Merino L, Cerillo Cuenca E, González Cordero A, Prada Gallardo A (2007b) Origen prehistórico de la dehesa en Extremadura: una perspectiva paleoambiental. Revi Estud Extremeños 63:493–510

MARM (2008) Diagnóstico de la Dehesa MARM (2008) Diagnóstico de las dehesas ibéricas mediterráneas. 766 En línea: http://www.mma.es/portal/secciones/biodiversidad/montes_politica_forestal/sanidad_forestal/pdf/anexo_3_4_coruche_2010.pdf Accessed March 2013

Martín Bolaños M (1943) Consideraciones sobre los encinares de España. Año 14, Núm 27, Instituto Forestal de Investigaciones y Experiencias, Madrid

Melón MA (1989) Extremadura en el Antiguo Régimen. Economía y Sociedad en tierras de Cáceres, 1700-1814. Universidad de Extremadura, Cáceres

Mensing SA (2005) The history of oaks in California, part I: the paleoecologic record. Calif Geog 45:1–38

Mensing SA (2006) The history of oaks in California, Part II: the native American and historic period. Calif Geog 46:1–31

Metcalf W (1947) The cork tree in California. Econ Bot 1:26–46

Millar CI, Brubaker LB (2006) Climate change and paleoecology: new contexts for restoration ecology. In: Falk DA, Palmer MA, Zedler JB (eds) Foundations of restoration ecology. Society for Ecological Restoration International, Island Press, Covelo, pp 315–340

Miller C (2011) How counting sheep saved the U.S. Forest Service. Peeling back the bark: exploring the collections, acquisitions, and treasures of the forest history society [blog site]. http://fhsarchives.wordpress.com/2011/05/03/how-counting-sheep-saved-the-u-s-forest-service/. Accessed 01 May 2012

Owens JB (1977) Diana at the bar: hunting, aristocrats, and the law in renaissance castile. Sixteenth Cent J 8:17–36
Parsons JJ (1960) Starlings for Seville. Landscape 10:28–31
Parsons JJ (1962a) The acorn-hog economy of the oak-woodlands of Southwestern Spain. Geogr Rev 52:211–235
Parsons JJ (1962b) The cork oaks forests and the evolution of the cork industry in Southern Spain and Portugal. Econ Geog 38:195–214
Pérez CN (1982) Grants of land in California made by Spanish or Mexican authorities. Boundary determination office, state lands commission, boundary investigation unit
Pérez CN (1996) Land grants in Alta California. Landmark Enterprises, Rancho Cordova
Plieninger T (2006a) Habitat loss, fragmentation, and alteration: quantifying the impact of land-use changes on a Spanish dehesa landscape by use of aerial photography and GIS. Landsc Ecol 21:91–105
Plieninger T (2006b) Las dehesas de la penillanura cacereña: Origen y evolución de un paisaje cultural. Universidad de Extremadura, Cáceres
Preston W (2002) Post-Columbian wildlife irruptions in California: implications for cultural and environmental change. In: Kay CE, Simmons RT (eds) Wilderness and Political Ecology: Aboriginal Influences and the Original State of Nature. University of Utah Press, Salt Lake City, pp 111–140
Ryan VA (1948) Some geographic and economics aspects of the cork oak. Crown Cork and Seal, Inc., Baltimore
Sampson AW (1914) Natural revegetation of range lands based upon growth requirements and life history of the vegetation. J Ag Res 3:93–147
Sampson AW (1923) Pasture and range management. Wiley, Hoboken
Standiford RB, Bartolome J (1997) The integrated hardwood range management program: education and research as a conservation strategy. USDA Forest Service Gen Tech Rep PSW-GTR-160, pp 569-581. PSW, Albany
Stanford L (1969) San Diego's judges of the plains. J San Diego Hist 15: 27–32. http://www.sandiegohistory.org/journal/69fall/judges.htm. Accessed 05 July 2012
Starrs PF, Huntsinger L (1998) The Cowboy and Buckaroo in American ranch hand styles. Rangel 20:36–40
Starrs PF, Goin PJ (2010) A field guide to California agriculture. University of California Press, Berkeley and Los Angeles
Stevenson AC, Harrison RJ (1992) Ancient forests in Spain: a model for land-use and dry forest management in South-west Spain from 4000 BC to 1900 AD. Proc Prehist Soc 58:227–447
Syphard AD, Radeloff VC, Keeley JE, Hawbaker TJ, Clayton MK, Stewart SL, Hammer RB (2007) Human influence on California fire regimes. Ecol Appl 17: 1388–1402. http://www.esajournals.org/doi/abs/10.1890/06-1128.1. Accessed 15 July 2012
Valle Buenestado B (1978) Los Cotos de caza mayor en la provincia de Córdoba: notas para su estudio geografico. In Medio Fisica, Desarollo Regional, Y Geografía, Granada Secretaria de Publicaciones de la Villa de Granada
Walker P, Fortmann LP (2003) Whose landscape? a political ecology of the "Exurban" Sierra. Cult Geog 10:469–491
Zapata S (1986) La producción agraria de Extremadura y Andalucía occidental, 1875–1935. Universidad Complutense de Madrid, Madrid
Zapata S (2002) Del suro a la cortiça. El ascenso de Portugal a primera potencia corchera del mundo. Rev Hist Indus 22:109–137
Zapata S (2010) La "revolución vitivinícola" y sus efectos sobre el negocio corchero. Documentos de Trabajo de la Asociación Española de Historia Económica: numero DT-AEHE 1002. http://szapata.com/investigacion/publicaciones/otras-publicaciones. Accessed 13 July 2012

Part II
Vegetation

Chapter 3
Climatic Influence on Oak Landscape Distributions

Sonia Roig, Rand R. Evett, Guillermo Gea-Izquierdo, Isabel Cañellas and Otilio Sánchez-Palomares

Frontispiece Chapter 3. Summer drought is typical of both ranches and dehesas. As a result, the herbaceous understory is green and flowering in spring (above) and brown and dry in summer (below) as in these photos from Spain (Photographs by S. Roig)

P. Campos et al. (eds.), *Mediterranean Oak Woodland Working Landscapes*, Landscape Series 16, DOI: 10.1007/978-94-007-6707-2_3,

© Springer Science+Business Media Dordrecht 2013

Abstract Climate is one determinant of the distribution and structure of California oak woodlands and Spanish dehesas. We summarize studies conducted in the two regions that use different methodologies to investigate the influence of climate on the distribution of oak species in California and on the development of dehesa silvopastoral systems in Spain. Results show some common climatic characteristics, mainly strong summer drought, medium to low overall year rainfall, and high summer temperatures. However, the influence of climate on oak distribution and management differs sharply. Climate strongly influences oak distribution in California. However, it has little effect on whether or not oak forests are managed as dehesa in Spain. Soil characteristics and socioeconomic issues are more important factors than climate for the creation and maintenance of dehesas in Spain.

Keywords Dehesa · Oak woodlands · Climatic change · *Quercus* · Summer drought

3.1 Introduction

Climate is a main factor determining the presence, distribution, development, dynamics, and function of California oak woodlands and Spanish dehesa ecosystems. Bioclimatic studies are typically used to quantify species distribution at different scales. However, humans have modeled Spanish dehesa landscapes and developed silvopastoral systems adapted to the climate intensively and consistently for a long time, making it clear that other factors can significantly interact with climate. This chapter examines the results of research that describe and compares the influence of climate on the presence and distribution of the major oak species found in California oak savanna and oak woodland landscapes (Sect. 3.2) with the combined influence of climatic variation and human management on the Spanish dehesa (Sect. 3.3).

S. Roig (✉)
Department of Silviculture and Pastures, Technical University of Madrid ETSI Montes, Ciudad Universitaria s/n 28040 Madrid, Spain

R. R. Evett
Department of Environmental Science, Policy, and Management, University of California, 137 Mulford Hall, Berkeley, CA 94720-3114, USA
e-mail: revett@sonic.net

G. Gea-Izquierdo · I. Cañellas · O. Sánchez-Palomares
Forest Research Centre, National Institute for Agriculture and Food Research and Technology, Ctra A Coruña km 7.5 28040 Madrid, Spain
e-mail: gea.guillermo@inia.es

I. Cañellas
e-mail: canellas@inia.es

3.2 The Climate of Californian Oak Woodlands

California has tremendous geographical diversity, leading to a wide range in regional climates. These climates are largely controlled by four factors: latitude, zonal atmospheric circulation, topography (elevation and aspect), and proximity to the Pacific Ocean. California spans nearly 10° of latitude; at similar elevations, mean annual temperature increases substantially with decreasing latitude. Precipitation is largely controlled by the location of the Pacific subtropical high, which normally lies in the Pacific Ocean off Oregon in the summer, blocking unstable air from the mid-latitude westerlies. It migrates to lower latitudes in the winter, allowing the passage of moisture-laden frontal storms with frequency depending on latitude, creating a north–south precipitation gradient. The Coast Ranges and the Sierra Nevada, mountains located perpendicular to the westerlies, strongly influence mean annual precipitation regimes through orographic effects, and there is an adiabatic decrease in mean temperature with elevation. The Pacific Ocean moderates temperatures in coastal California, particularly in the summer when the cold water of the California Current cools descending summer air masses, causing substantial coastal fog that leads to strong coastal-inland temperature gradients.

The dominant climatic feature unifying California is the seasonality of precipitation, with wet winters and dry summers, typical of Mediterranean climates throughout the world. Plant communities are forced to adapt to an environment where moisture availability is mostly out of phase with the higher temperatures that define the growing season (Major 1988). This is particularly true for the approximately 3.5 million ha oak woodland region where the majority of ranching activity in California occurs. Oak woodland plant communities ring California's Central Valley, from 100 to 700 m in northern California and up to 1,500 m in the Coast Ranges and Sierra Nevada foothills. Several native species of oak, well-adapted to seasonal drought, are dominant in the overstory. The understory is currently dominated by exotic annual species, principally of Mediterranean origin, that escape the summer drought by producing seeds and senescing when available soil moisture is exhausted. The species composition of the understory prior to the exotic annual invasion is unknown but was likely much more spatially variable than seen today; historical, relict, and fossil evidence suggests native annual forbs were abundant and native perennial grasses were not dominant (Minnich 2008; Evett and Bartolome, manuscript in preparation).

Here we focus on six of the eight major species of oak found in California's open oak woodlands. Characteristics of these species are described in Table 3.1. The geographical ranges of many of the species overlap considerably, and hybridization is very common. The goal is to gain understanding of the distribution of oak woodland in California by answering the following question: why do these six oak species occur where they do? We describe three approaches that have been used to build climatic niche models for California oak species by combining

Table 3.1 Characteristics of six out of the eight major species of oak found in California's open oak woodlands (compiled from, Allen et al. 1990, Griffin and Critchfield 1972, Jepson 1910, Pavlik et al. 1991 and Plumb 1980)

	Coast live oak	Interior live oak	Canyon live oak	Valley oak	Blue oak	California black oak
Scientific name	*Quercus agrifolia*	*Quercus wislizenii*	*Quercus chrysolepis*	*Quercus lobata*	*Quercus douglasii*	*Quercus kelloggii*
Average height (m)	10	10–20	<15	20–30	7–20	10–25
Canopy form	Broad, dense	Broad, round	Highly variable	Broad	Rounded	Rounded, open
Northern boundary	Mendocino Co.	Siskiyou Co.	South-west Oregon	Shasta Lake	Shasta Co.	Central Oregon
Southern boundary	Baja California	Baja California	Baja California	San Fernando Valley	Los Angeles Co.	Baja California
Western boundary	Pacific Ocean	Outer Coast Ranges	Pacific Ocean	Near Pacific Ocean	Outer Coast Ranges	Outer Coast Ranges
Eastern boundary	Central Valley	Sierra foothills	Central Arizona	Sierra foothills	Sierra foothills	Sierra foothills
Topographic position	Valley floors Steep hillsides	Foothill slopes Valley floor in CR	Steep, rocky canyons Valley floors	Valley floors Hill slopes	Foothills	
Elevation (m)	<1,600	<1,600	90–2,740	<1,300	<1,300	<2,000
Growth preferences	Warm winter Moderate summers Well-drained soils	Cool, wet winter Hot, dry summer Well-drained soil	Wide range of precip Wide range of temp. Wide range of soils	Cool, moist winter Hot, dry summer Roots tap water table	Cool, wet winters Hot, dry summer Shallow, dry soil	Winter cool w/snow Dry summer Well-drained soil
Fire resistance	High	Low	Intermediate	High	Low	Low
Special adaptations	Fire resistant bark Sprouting bole buds Sprouting root buds	Sprouting root buds	Sprouting root buds	Fire resistant bark Sprouting trunk base	Waxy leaves Summer deciduous Fast growing roots Weak sprouters	Sprouting trunk base

Engelmann (*Q. engelmannii*) and Oregon (*Q. garryana*) oaks were not considered due to their restricted distribution in California current climate data with species distribution data. We also examine how these models have been used to make predictions of changes in species distributions based on projected changes in future climate.

3.2.1 The Bioclimatic Niche of California Oaks

3.2.1.1 Bioclimatic Analyses

The simplest approach to describing the climatic niche is bioclimatic analysis, defining the climate envelope of a species by estimating the mean and range for each climatic parameter (Nix 1986; Lindenmayer et al. 1991). This approach is based on the principle of homoclime matching, which states that if a species is known to occur at a site with a particular set of bioclimatic characteristics–its bioclimatic envelope—it is also likely to occur at other climatically similar sites, and has been used to make predictions of present, past, and future distribution of a species based on limited data. The main drawback of the approach for niche studies is that the relative importance of climatic parameters cannot be ascertained because quantitative multivariate analysis is not possible (Santika and Hutchinson 2009).

Bioclimatic analysis was applied to California oaks (Evett 1994) using presence-absence data from the vegetation type map (VTM) survey of California that mapped the vegetation of nearly half of California based on species cover data for 13,000 plots (0.081 ha/plot) from 1928 to 1940 (Wieslander 1935). As part of an oak woodland classification project (Allen et al. 1990), 4,315 plots containing at least one of the six oak species were identified. Climatic data, based on average climatic data from California meteorological stations for the period 1930–1960, were digitized from isoline maps of 11 climate parameters for each county drawn by the California State Climatologist (Elford 1970). Isolines for these 11 climate parameters, as well as three additional derived climate parameters (Table 3.2), were digitized into IDRISI (Eastman 1992), a Geographic Information System (details in Evett 1994). Species cover data for each oak plot were digitized into the GIS, enabling estimation of climate parameters for each plot by interpolation.

The bioclimatic envelope for each species was estimated using climate data (interpolated using IDRISI) for each VTM site where that species was present (Table 3.3). While the complete range for each climate parameter for each species was probably not spanned because VTM data did not cover the entire state, mean values may be more accurate.

Comparison of bioclimatic envelopes for the six species suggests there are many similarities in climatic requirements but also some important differences (Table 3.3). All of the oak species were found on sites receiving <320 mm/yr rainfall, but only black oak and canyon live oak were found on sites with precipitation >1,525 mm/yr. The species means arrayed along the mean annual precipitation gradient were (lowest to highest): valley oak, coast live oak, blue oak,

Table 3.2 Climatic variables used in the California oak niche study (Evett 1994)

Variable	Description
MAT	Mean annual temperature
MAP	Mean annual precipitation
JUMA	July mean maximum temperature
JUMI	July mean minimum temperature
JAMA	January mean maximum temperature
JAMI	January mean minimum temperature
HDD	Heating degree-days (based on 18.33 °C)
GS32	Length of 0 °C growing season
GS28	Length of −2.22 °C growing season
PE	Potential evapotranspiration
ET	Actual evapotranspiration
JAMEAN	Mean January temperature = (JAMA + JAMI)/2
JUMEAN	Mean July temperature = (JUMA + JUMI)/2
TEMPR	Mean annual temperature range = JUMEAN–JAMEAN

interior live oak, canyon live oak, and black oak. The xeric end of this array is surprising because blue oak is considered more drought-tolerant than valley oak or coast live oak. This suggests that valley oak taps into groundwater in areas with reduced precipitation, while moderate summer temperatures and coastal fog lead to a more favorable moisture balance for coast live oak. The only species found on sites with mean annual temperature <10 °C were black oak and canyon live oak. The species means arrayed along the mean annual temperature gradient were, from lowest to highest: black oak, canyon live oak, coast live oak, valley oak, interior live oak, and blue oak. The range between mean July temperature and mean January temperature (TEMPR) was considerably less for coast live oak than for other species, probably reflecting its largely coastal distribution.

3.2.1.2 Bioclimatic Modeling Results: Logistic Regression

One of the most frequently used approaches to modeling species distributions based on correlation with climatic variables is logistic regression, a special form of generalized linear modeling useful for dealing with presence/absence data (Rushton et al. 2004; Austin 2007; Santika and Hutchinson 2009; Bedia et al. 2011). Climate parameters, ideally reflecting physiological requirements of a species (such as absolute minimum temperature, moisture availability, pH) are introduced in statistical order of importance into the regression in a stepwise procedure. If relevant climatic data are not available, proxy data that combine several climatic parameters, such as elevation, have been used.

Evett (1994) used logistic regression, based largely on climatic parameters, to model the realized environmental niche as defined by Austin et al. (1990) for the six oak species in the VTM database discussed above. In addition to the variables used in the bioclimatic analysis (Table 3.2), three non-climatic variables were

Table 3.3 Summary of bioclimatic parameters for six Californian oak species from the VTM database (Evett 1994)

Variable		Coast live oak	Interior live oak	Valley oak	Black oak	Blue oak	Canyon live oak
MAP (mm)	Mean	551	771	544	1011	557	970
	Range	203–1554	274–1524	275–1439	317–2032	203–1439	254–1863
MAT (°C)	Mean	14.46	14.81	14.50	13.10	14.95	13.13
	Range	12.22–17.11	10.37–17.06	11.5–16.67	5.55–16.24	11.6717.14	6.39–16.17
JUMA (°C)	Mean	28.58	33.70	30.59	32.04	33.43	31.30
	Range	18.00–36.11	17.96–37.78	18.3–37.54	18.14–37.78	18.75–37.78	18.06–38.33
JUMI (°C)	Mean	11.43	14.37	12.28	13.23	13.75	13.16
	Range	8.77–18.89	8.33–20.03	8.89–19.48	5.83–20.00	8.89–20.03	6.19–21.72
JAMA (°C)	Mean	14.79	12.03	13.72	10.60	13.04	11.35
	Range	10.00–18.47	7.92–16.53	8.09–17.78	2.78–17.22	8.52–17.22	5.56–18.39
JAMI (°C)	Mean	2.62	1.39	1.86	−0.32	1.32	0.00
	Range	1.69–7.13	−5.00 to 6.70	−3.33 to 6.11	−7.98 to 5.56	−3.33 to 6.67	−6.67 to 6.71
PE (mm)	Mean	628	687	646	585	677	585
	Range	406–818	368–914	444–902	176–914	444–940	279–902
ET (mm)	Mean	229	246	213	221	212	209
	Range	51–356	51–406	76–381	51–406	70–406	51–406
JUMEAN (°C)	Mean	20.01	24.04	21.43	22.63	23.59	22.23
	Range	14.11–27.41	14.11–28.65	14.70–28.44	14.07–28.57	14.44–28.65	14.14–30.00
JAMEAN (°C)	Mean	8.70	6.71	7.79	5.14	7.18	5.67
	Range	5.72–12.33	2.01–10.84	2.38–11.24	−2.50 to 10.46	2.96–11.16	0.56–10.83
TEMPR (°C)	Mean	11.30	17.33	13.64	17.50	16.41	16.56
	Range	3.21–20.28	3.28–23.08	3.91–21.77	3.61–21.98	3.61–23.08	3.32–21.43
JULYRAD (cal)	Mean	667.1	680.1	679.5	680.1	678.9	667.1
	Range	511–733	512–740	532–743	493–750	519–736	493–755
GS32 (days)	Mean	254.9	236.3	236.2	195.3	224.8	201.8
	Range	150–352	106–351	150–350	50–350	150–344	65–351
GS28 (days)	Mean	307.3	283.5	292.1	252.2	277.4	254.7
	Range	200–360	175–352	194–360	125–350	175–351	150–352

tested: JULYRAD (average monthly solar radiation in July, in calories per cm^2 and day) at a site, based on latitude, slope, elevation, and aspect); PM (soil parent material from categories in VTM data); and TEX (six soil texture categories, from rock or gravel through clay). At each step in the regression, each variable was fitted into the model; the variable that reduced deviance (a statistical measure of error in logistic regression, roughly analogous to variance) by the greatest amount was added to the model. Variables were added until no additional variable was statistically significant. Linear terms for each variable were tested first, followed by second order terms (if significant, indicating a Gaussian shape of the response curve along the environmental gradient) and third order terms (if significant, indicating a skewed response curve). Interactions between each pair of variables were also tested.

The results of Evett's (1994) logistic regression (Table 3.4), which estimated the probability of presence for an oak species at each point within the environmental hyperspace defined by predictor variable gradients, are summarized as follows:

Coast live oak—This niche model was the most successful of all species models analyzed, accounting for more than 60 % of the total deviance. Although there were five variables included, the final model was largely defined by a single gradient, mean annual temperature range, whose probability of presence had a Gaussian (i.e., hump-shaped) distribution peaking at 11 °C, confirming the observation that coast live oak distribution is dependent on a temperate climate, with relatively warm winters and cool summers found only in coastal regions in California.

Interior live oak—Two variables, mean annual precipitation, with a Gaussian response curve peaking at 171 cm, and July mean temperature, with a skewed response curve peaking at 19 °C, accounted for most of the 32 % deviance reduction. However, these two variables are significantly correlated, so that part of the effect of each one can be due to the effect of the other on it. Joint effects are thus lower that shown by the model.

Valley oak—Seven variables were included in the final model, accounting for only 24 % of total deviance. The model suggested valley oak has complex responses to the mean annual temperature range (two peaks, probably reflecting coastal versus Central Valley populations), mean annual precipitation (peaking at 54 cm; Fig. 3.1), and July maximum temperature (also with two peaks). Probability of presence was greater on finer textured soils.

Black oak—Mean annual precipitation accounted for almost two-thirds of the 47 % total deviance reduction in the model. The response curve was weakly Gaussian, but over the range of precipitation sampled, the probability of presence assumed an S-shape, showing increased probability of presence with increased precipitation (Fig. 3.1).

Blue oak—The niche model, accounting for 43 % of total deviance, was the most complex of all, containing eight variables. The model indicated there is increased probability of blue oak presence at sites with lower winter temperatures

Table 3.4 Summary of forward stepwise logistic regression climatic niche models for six Californian oak species (Evett 1994)

	Coast live oak		Interior live oak		Valley oak		Black oak		Blue oak		Canyon live oak	
Step #	Variable added	Deviance change	Variable added	Deviance change	Variable added	Deviance change	Variable Added	Deviance change	Variable added	Deviance change	Variable added	Deviance change
0	MEAN	(4745)	MEAN	(3850)	MEAN	(3092)	MEAN	(5345)	MEAN	(5473)	MEAN	(2901)
1	TEMPR	2293	MAP	10	TEMPR	151	MAP	1667	JAMEAN	106	MAP	219
	$TEMPR^2$	310	MAP^2	530	$TEMPR^2$	189	MAP^2	56	$JAMEAN^2$	938		
					$TEMPR^3$	19						
2	MAP	50	JUMEAN	364	MAP	76	JAMEAN	306	MAP	503	MAT	30
	MAP^2	37	$JUMEAN^2$	64	MAP^2	33					MAT^2	60
	MAP^3	45	$JUMEAN^3$	33	MAP^3	8						
3	JAMEAN	4	JAMEAN	27	JUMA	7	PM	181	MAT	327	TEX	82
	$JAMEAN^2$	0	$JAMEAN^2$	76	$JUMA^2$	39						
	$JAMEAN^3$	69			$JUMA^3$	38						
4	JUMA	4	JUMEAN x MAP	79	TEX	65	JUMA	6	PM	137	PM	104
	$JUMA^2$	23					$JUMA^2$	109				
5	PM	13	JUMA	30	PM	39	JULYRAD	42	JULYRAD	41	JULYRAD	63
							$JULYRAD^2$	0	$JULYRAD^2$	16		
							$JULYRAD^3$	50	$JULYRAD^3$	43		
6					MAT	20	JUMA x MAP	73	TEX	56	MAT x MAP	29
					MAT^2	16						
7					JULYRAD	27	JUMEAN	47	$(JAMEAN \times MAT)^2$	72		
8									JUMA	61		

See Table 3.2 for definitions of variables. Note that for some variables square values were included in models

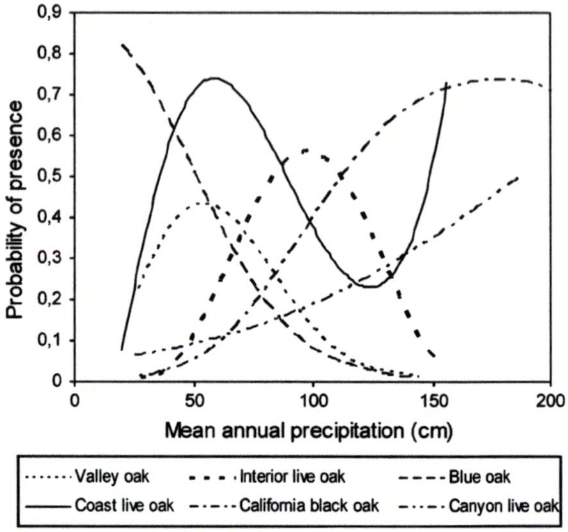

Fig. 3.1 Response curves for six Californian oak species along the mean annual precipitation gradient based on niche models derived from VTM data (Evett 1994)

(Gaussian curve peaking at 5 °C), lower precipitation, and higher mean annual temperature.

Canyon live oak—The model only accounted for 20 % of the total deviance, lowest of all models. Probability of presence increased linearly with mean annual precipitation, had a Gaussian response to mean annual temperature (peaking at 8 °C), increased on rocky soils, and decreased on finer textured soils.

The regression models were used to predict the distribution of each species. Predicted distribution was then compared with the actual distribution according to maps in Griffin and Critchfield (1972) using several measures of goodness of fit (Evett 1994). Coast live oak, California black oak, interior live oak, and blue oak models were judged adequate; canyon live oak and valley oak models were judged inadequate. Canyon live oak, tolerant of a wide range of climate, apparently responds more to local habitat conditions (including soil moisture, light, and biotic factors) than to regional climate factors (Myatt 1975). Valley oak, generally found on deep alluvial soils near watercourses, has low xylem sap tension, even during drought (Griffin 1973), indicating the species probably taps into the water table. Depth to water table for each VTM site was not available, so the poor fit of the valley oak model is not surprising. However, despite the poor fit, the model still provides valuable information on climatic requirements for valley oak.

To visualize the results of these complex regression models, probability of presence for each species was plotted along a single gradient while holding all other variables constant at their mean values from Table 3.3 (Evett 1994). The mean annual precipitation gradient, a significant variable in each niche model, showed blue oak dominating the xeric end, followed by extensively overlapping valley oak and coast live oak, interior live oak, and overlapping black oak and canyon live oak dominating the mesic end (Fig. 3.1). The mean January

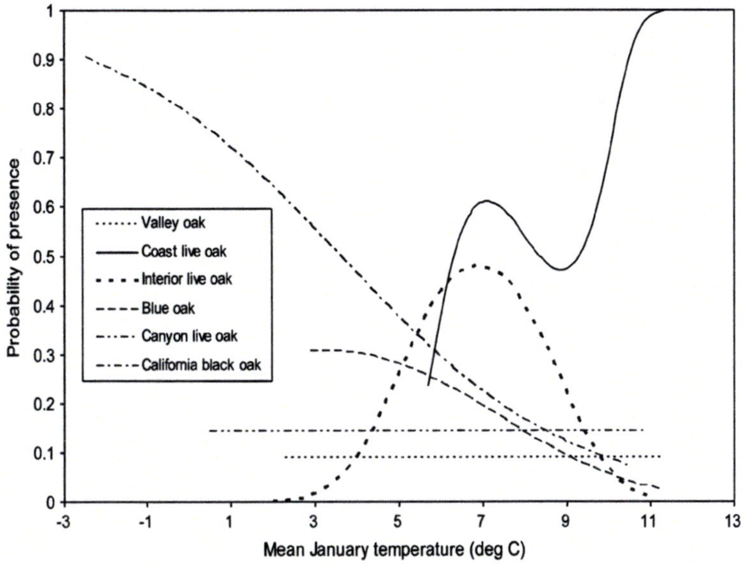

Fig. 3.2 Response curves for six California oak species along the mean January temperature gradient based on niche models derived from VTM data (Evett 1994)

temperature gradient, significant in four niche models, showed California black oak dominating the colder end, followed by blue oak, interior live oak, and coast live oak dominating the warmer end; valley oak and canyon live oak did not significantly respond to this gradient (Fig. 3.2). However, while providing a perspective based on one interpretation, each figure is only one slice of a complex multi-dimensional space and niche relationships do not necessarily hold given different values for the variables held constant. For example, the figure suggests little overlap of probability of presence along the precipitation gradient for blue oak and interior live oak, even though the two species commonly occur together. Range overlaps are however, evident when considering multivariate spaces, although these spaces cannot be plotted on only two–three dimensions.

Regression model results were also visualized by plotting niche space diagrams defined by the two most significant climatic variables for each species (Fig. 3.3) while holding other significant variables constant (Evett 1994). Each diagram only includes the portion of each gradient where the species was found according to VTM data; mean annual precipitation is shown as the x-axis for each diagram because it is one of the two most significant variables for each model. Blue oak and California black oak diagrams are relatively simple and comparable because they have the same variables, showing increased probability of presence at colder January temperatures but opposite responses to increased precipitation (Fig. 3.3). Diagrams for coast live oak, valley oak and interior live oak are more complex. Probability of presence never reaches 0.50 for canyon live oak or 0.25 for valley oak, another indication that these models fit the data poorly.

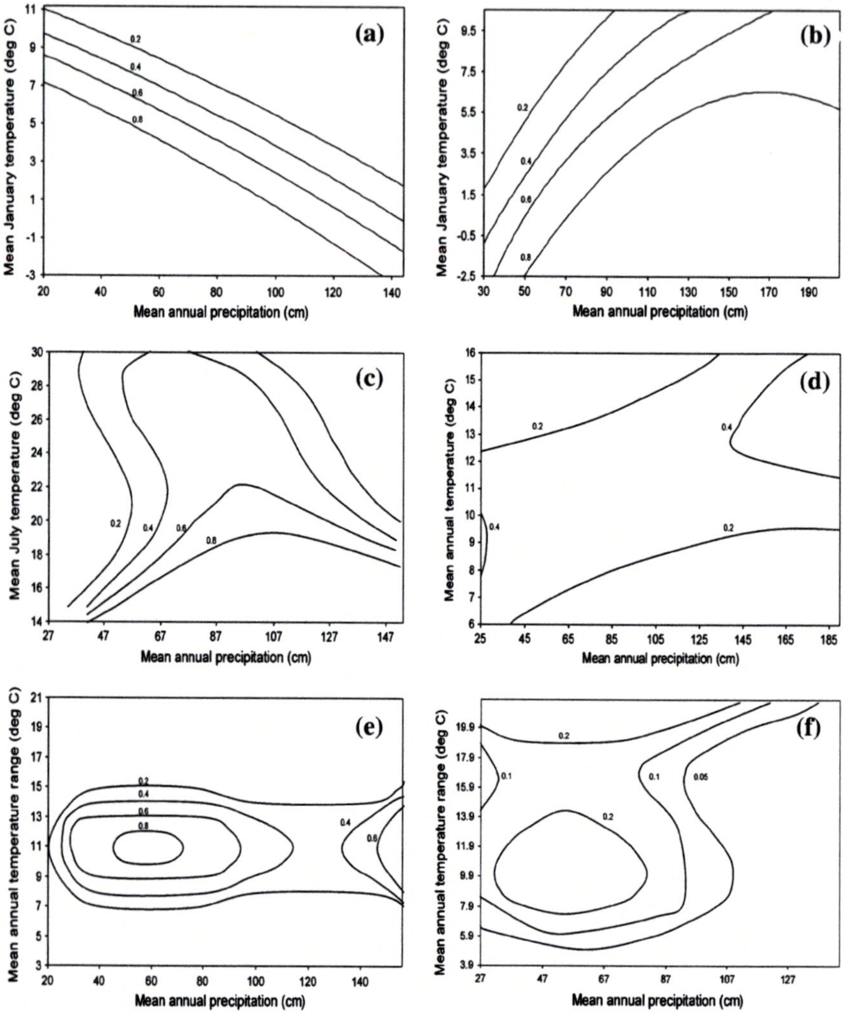

Fig. 3.3 Niche space diagrams describing probability of presence for six California oak species given the two most significant climatic parameters from logistic regression niche models (Evett 1994): **a** blue oak; **b** California black oak; **c** interior live oak; **d** canyon live oak; **e** coast live oak; **f** valley oak

3.2.1.3 Bioclimatic Modeling Results: Classification and Regression Trees

Vayssieres et al. (2000) used a classification and regression tree (CART) approach to determine the bioclimatic niche for three Californian oak species using the VTM dataset. CART is a non-parametric method based on developing a decision tree composed of simple data-based rules to predict species occurrence given climatic and other environmental data for a site. Vayssieres et al. (2000) approach largely confirmed and clarified the results of Evett (1994). They found that mean annual temperature range was a very good predictor of coast live oak distribution, with presence more likely if the temperature range is below 15.3 °C. Mean annual temperature range was also the most important CART variable for valley oak distribution, with presence more likely below 15.65 °C, but they found that the overall predictive power of the model including all available environmental variables was mediocre, probably because data on the most important variable, access to the soil water table, were not available. The blue oak model was the most complex, with many variables included in the decision tree. Mean annual precipitation was the most important predictor of species distribution, with presence more likely below 91 cm/yr, but contrary to Evett (1994) mean January temperature was not an important variable.

3.2.2 *Effects of Projected Climate Change on California Oak Distribution*

Global and regional climate models predict substantial changes in climate in most of California over the next 100 years (mean annual temperature increasing 2–3 °C statewide; up to 40 mm decreased or increased mean annual precipitation, depending on region) as a result of the anthropogenic increase in greenhouse gases in the atmosphere (Kueppers et al. 2005). California oaks are sensitive to their temperature and moisture environment at all stages of their life history, but particularly during the seedling stage (Tyler et al. 2006; Zavaleta et al. 2007); climate change may substantially affect oak distribution.

Several bioclimatic models have been used to predict changes in California oak distribution resulting from projected climate change. Evett (1994), using the same methods for the same species described above for VTM data, built logistic regression models to predict species presence or absence based on a Geographic Information System (GIS) approach, overlaying statewide maps of oak distribution (from Griffin and Critchfield 1972) with maps of climatic variables drawn from the Atlas of California (Donley et al. 1979). After calibration and validation, these GIS models, very similar to the models based on VTM data, provided good predictions of current distribution for all species except valley oak. Conservative values for an altered climate scenario (all temperature variables increased 1.5 °C and mean annual precipitation increased 10 %) were plugged into the regression models and species projected distribution maps were generated (Fig. 3.4). The results suggested there will be widespread migration of species. California black oak (Fig. 3.4b) and canyon live oak (Fig. 3.4d), species favoring mesic, cooler conditions, were predicted to respond to warmer climate by migrating upslope. Little change in coast live oak distribution was predicted (Fig. 3.4e), probably because global and regional climate models anticipate little change in the variable most important for determining current distribution, mean annual temperature range (Kueppers et al. 2005). Interior live oak was predicted to benefit the most from global warming, expanding its range considerably without relinquishing any of its current distribution (Fig. 3.4c).

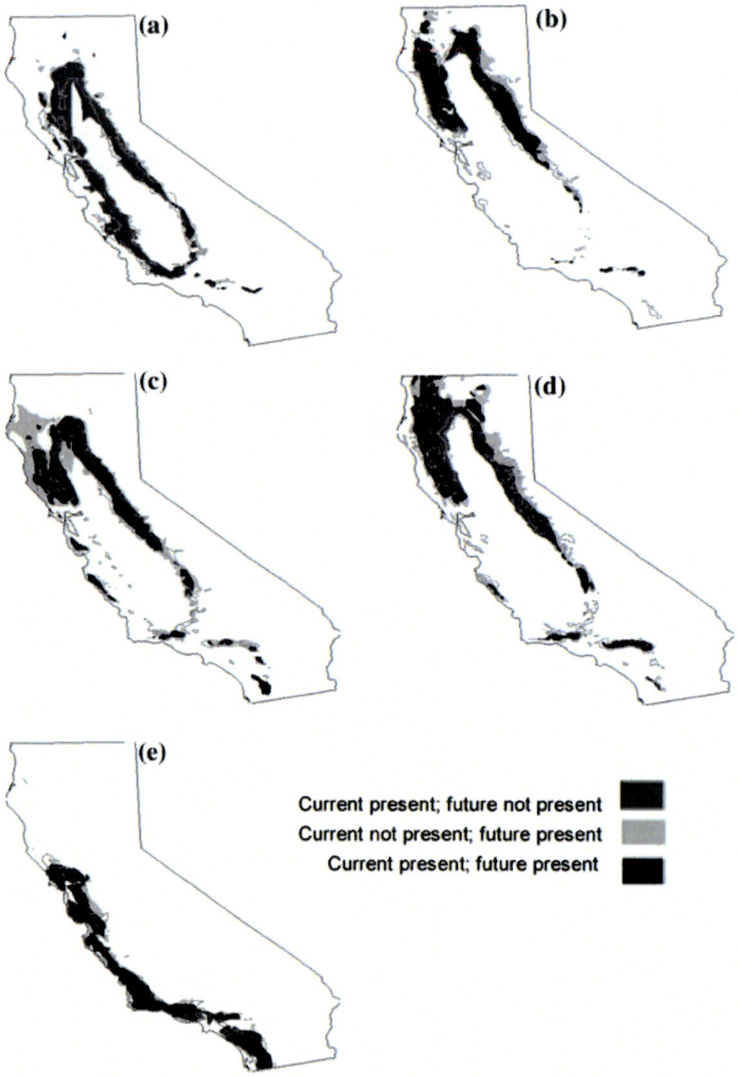

Fig. 3.4 Predicted distribution from niche models (Evett 1994) of five California oak species under an altered climate scenario (temperature variables increased 1.5 °C; mean annual precipitation increased 10 %): **a** blue oak; **b** California black oak; **c** interior live oak; **d** canyon live oak; **e** coast live oak

Blue oak showed the most dramatic response, a range contraction of 95 %, to the altered climate scenario (Fig. 3.4a). Evett (1994) suggested that blue oak, whose distribution is (according to the niche model) most correlated with mean January temperature, may currently be near the upper limits for this climatic variable throughout much of its range. Although upward migration to sites with cooler January temperatures might be the expected response, increased precipitation at higher elevations may lead to overwhelming competition from species adapted to higher precipitation regimes. The observation that blue oak recruitment is poor throughout much of its range (Bartolome et al. 1987), commonly attributed to browsing by livestock and wildlife and competition with exotic grasses (Tyler et al. 2006), may indicate that range contraction due to climate change is already underway.

Kueppers et al. (2005) employed discriminant analysis to build statistical models to predict the current distribution of blue oak and valley oak in California using four climate and three soil variables. They plugged temperature and precipitation values based on predictions of future climate scenarios drawn from general and regional climate models into their distribution models. Based on regional climate model data, they predicted that the range of blue oak would decrease 59 % and valley oak would decrease 54 % compared to the present, with considerable migration to new sites, particularly in northwest California, for each species.

Shafer et al. (2001) employed response surface modeling methods (similar to Evett 1994) with three bioclimatic variables to predict changes in tree species distributions using altered climate scenario data drawn from three different general circulation models. They concluded that valley oak distribution would be greatly diminished in California, but the range may expand dramatically northward along the Pacific coast and eastward into Nevada, Arizona, New Mexico, Utah and Colorado. Similar contraction of the current range with eastward expansion was predicted for Oregon white oak (*Q. garryana*).

Studies of predicted changes in broad vegetation types under altered climate scenarios in California support the results from studies of individual oak species. Lenihan et al. (2003) using a dynamic vegetation model coupled with two general circulation models, predicted oak woodland cover would decrease ~40 % statewide, while grassland cover would increase ~70 %. Similarly, Hayhoe et al. (2004) predicted oak woodland would decrease ~20 % and grassland would increase ~25 %.

3.3 The Climate of Spanish Dehesas

Spain is also a good example of geographical diversity that supports a wide variety of regional climates. The majority of the country has a Mediterranean climate, characterized by long, usually hot and dry summers with rainy periods concentrated in spring, autumn and winter, though these are notably erratic in character. Variability is the key factor in Spanish Mediterranean climate zones and habitats.

The dehesa is an agrosilvopastoral system originating from a combination of thinning, seeding, and coppicing where trees, native grasses, croplands and livestock interact under specific management regimes (Chap. 1). They are among the most prominent and widespread agroforestry systems in Europe (Papanastasis 2004). Although climate is a key factor defining species and ecosystem components, the long history of human management is the main force that has molded the dehesa ecosystem. However, knowing the characteristics of the dehesa climate is essential for management planning to assure the persistence and conservation of the system within the framework of global change. There are several types of dehesa in Spain because many different tree species have been locally thinned and managed as silvopastoral systems (San Miguel et al. 2002). Typical dehesas are populated by holm oak (*Quercus ilex ballota*), cork oak (*Q. suber*), and/or other oaks such as Lusitanian oak (*Q. faginea broteroi*) in areas of higher soil moisture. However, there are also dehesas dominated by the eastern subspecies of Lusitanian oak (*Quercus faginea faginea*), by Pyrenean oaks (*Quercus pyrenaica*), and even by ash trees (*Fraxinus angustifolia*) or stone pines (*Pinus pinea*). Other accompanying tree species, which may be locally important, especially in eastern and southeastern Iberian thermophyllous dehesas, include wild olive (*Olea europaea*) and junipers (*Juniperus oxycedrus* and *J. thurifera*). Due to the extensive area they occupy, we focus here on the typical dehesa of holm and/or cork oaks.

3.3.1 Climate Studies on Tree Species in Dehesa

There are few studies on the relationship between plant species and climate in Spanish dehesa. The most important works on the ecology of Spanish tree species, including in-depth climatic analysis, are parametric autoecological studies (Gandullo and Sánchez Palomares (1994) for pines in Spain; Rubio et al. (1999) for chestnut (*Castanea sativa*); Sánchez et al. (2003) for beech (*Fagus sylvatica*); or Díaz-Maroto et al. (2006) for Pyrenean oaks). Autecological studies of holm oak are now being developed in Spain, and a cork oak autecology study has just been finished (Sánchez Palomares et al. 2007), in both cases analyzing the habitat of the two species within diverse forest types and management regimes (Fig. 3.5). Dehesas dominated by holm or cork oak are just one of the specific ecosystems where these species can be found in Spain, as particularly holm oak occupies a great variety of soil substrates within the Mediterranean climate zone (Barbero et al. 1992; Fig. 3.6). It is important to remember that these studies are based on the current distribution of a species, reflecting not only current habitat conditions, but also the history of its distribution after glacial migrations, human selection and propagation, and the dynamics and structure of forest stands: in the Mediterranean region, these are very complex phenomena.

The most common climatic model used in autecological studies of the main Spanish forest species was developed by Sanchez Palomares et al. (1999) using a set of climatic parameters associated to the spatial X and Y UTM coordinates and

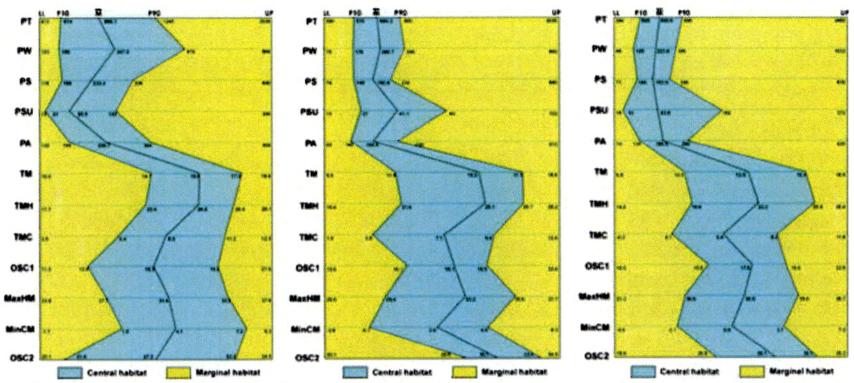

Fig. 3.5 Climatic habitat of cork oak (left), dehesas of cork and holm oaks (center), and holm oak (right) from data of the National Forest Inventory (NFI) and climatic models of Sanchez Palomares et al. (Sánchez Palomares et al. 1999)

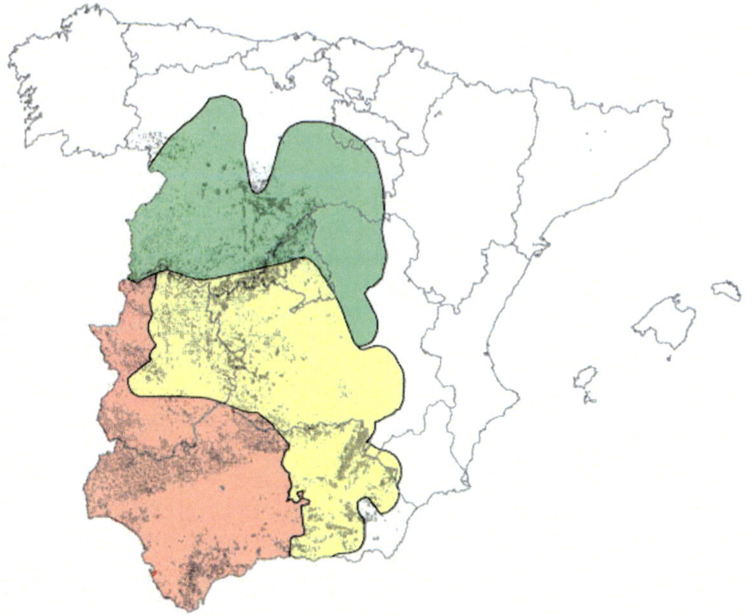

Fig. 3.6 The three climatic zones of dehesa in Spain (green, yellow and pink patches) overlying the location of the National Forest Inventory plots classified as low density stands of holm or cork oak (dark patches). Grey boundary lines in the background indicate the Spanish autonomous regions. *Source* Sánchez et al. (2007) and Third NFI

the altitude above sea level. During the last decade, the National Forest Inventory (NFI) has been the main database for this kind of ecological and silvicultural sampling in Spain.

Table 3.5 Dasonomic characterization of the dehesa system of holm or cork oak in Spain (n = 5,144 plots of the third Spanish NFI)

	Mean	Median	Amplitude (Max–Min)	Std. Dev.
N (No. stems/ha)	59.6	47.5	199.9	43.75
DG (diameter mm)	410.8	386.1	1526.0	155.76
BA (Basal area, m^2/ha)	6.5	5.5	49.8	4.05
Fcc (Canopy cover, %)	69.9	85.0	100.0	30.33
Mean height (m)	7.6	7.4	21.2	1.88
Mean diameter (mm)	431.8	418.8	1526.0	147.26
Ho (maximum height, m)	8.9	8.5	28.0	2.43
Do (maximum diameter, mm)	555.2	530.5	1526.0	188.53

The NFI is a systematic sampling of Spanish forestlands, repeated every 10 years. Plots are located on a 1 km^2 grid where dasymetric structure (diameters, heights and locations of trees measured in four concentric radial plots) and other ecological variables are measured (Bravo et al. 2002). The information obtained from the NFI is very valuable because the frequency of the measurements (every ten years) allows monitoring changes in the state of forest stands. We use here the NFI database to analyze climatic correlates of dehesa systems, assuming that dehesas can be identified from NFI parameters (Roig et al. 2007). From a dasonomic and silvicultural point of view, and including dehesa stands dominated by Pyrenean and Lusitanian oaks and junipers, we consider dehesas in this study to be forest systems with low tree density (<200 stems per hectare), dominated by holm and cork oak (measured by density or basal area), that allow sporadic cultivation (slope <20 %) and with large trees (quadratic diameter >20 cm). The spatial distribution of NFI plots with these characteristics match the distribution of dehesas as defined in Chap. 1 (Roig et al. 2007).

Combining the NFI database with a digital model of the entire country, we found that the third Spanish NFI includes 5,144 plots that meet the dehesa selection criteria (a dasynomic synthesis of the selected group is shown in Table 3.5). For every selected plot, estimates of the main climatic variables (mean monthly and annual precipitation and temperature, maximum and minimum temperature; see Tables 3.6, 3.7, 3.8) were calculated from X, Y coordinates and altitude using the database of Sánchez Palomares et al. (1999). The quality and accuracy of the selection criteria were validated by a study of the NFI-selected plot locations compared with the distribution of the dehesa systems on Spanish vegetation and landscape maps (Roig et al. 2007).

3.3.2 Three Climatic Zones of the Spanish Dehesa

Previous studies, using REDPARES (Spanish Rural Landscapes Network or *Red de Paisajes Rurales Españoles*), Netplot, and the landscape concept of dehesa, have identified three climatic zones of the Spanish dehesa dominated by holm and

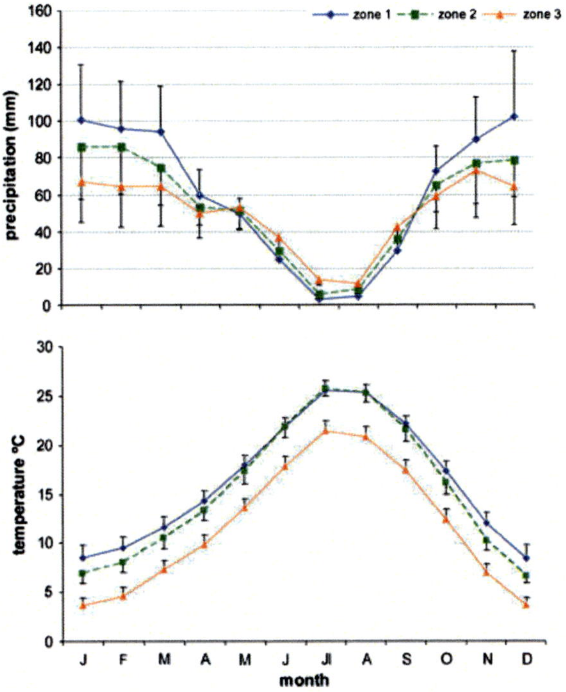

Fig. 3.7 Distribution of precipitation and mean temperature during the year at the three climatic zones of the Spanish dehesa. Bars show the standard deviation of the mean

cork oaks (García del Barrio et al. 2004; Fig. 3.6; see also Chap. 14). According to Sánchez de Ron et al. (2007), dehesa distribution is mainly explained by temperature, drought, and summer precipitation. The variance vectors divided all dehesas plots along two main axes: warm-cold and mesic-xeric. We overlaid the three previously defined climatic zones of the Spanish dehesa with climate parameters from our NFI dehesa plots to describe in a general way the climate of the dehesa system in Spain.

The main and most extensive group (zone 1; Table 3.6), where dehesas are very frequent, is in the south-west of the Iberian Peninsula, including most of Andalucía, Badajoz and western Cáceres, coinciding with the evergreen oak open woodland that we identified as typical dehesa. The second group (zone 2; Table 3.7) is mainly in Castilla-La Mancha, eastern Andalucía, and Cáceres, Badajoz and Madrid provinces. The third and most heterogeneous climatic group (Table 3.8) is located in Castilla-León, north of Madrid, Guadalajara and Cuenca. Zone 3 is separated from the other two by its lower annual temperature regime, while zones 1 and 2 have significant differences in precipitation (zone 2 has longer drought). Zone 3 is likely to include species such as Spanish junipers *Juniperus thurifera* and a number of other tree species in mixed stands with holm and cork oaks.

Cork oak is frequently dominant in zone 1, although holm oak-dominated dehesas are also very important. Zone 2 includes colder, more xeric dehesas that

Table 3.6 Climatic characterization of zone 1 of the Spanish dehesa

	Monthly precipitation						Monthly temperatures				
	Mean	Median	Min	Max	Std. Dev.		Mean	Median	Min	Max	Std. Dev.
J	101.0	95.0	53.0	249.0	30.1	J	8.5	8.3	5.9	12.3	1.3
F	95.7	90.0	56.0	249.0	26.3	F	9.5	9.3	6.9	12.8	1.1
M	94.0	89.0	49.0	230.0	25.2	M	11.6	11.6	8.4	14.5	1.1
A	60.0	58.0	35.0	188.0	13.6	A	14.4	14.4	10.4	16.4	0.9
M	49.3	48.0	28.0	146.0	8.8	M	18.0	18.0	13.3	20.7	1.0
J	24.9	25.0	10.0	48.0	4.3	J	21.9	21.9	17.8	24.0	0.9
Jl	3.0	3.0	0.0	7.0	1.5	Jl	25.6	25.9	22.2	27.9	1.0
A	5.0	5.0	1.0	12.0	1.4	A	25.4	25.6	21.6	27.3	0.8
S	29.6	28.0	14.0	76.0	5.4	S	22.2	22.2	18.7	24.4	0.8
O	72.6	71.0	47.0	153.0	13.7	O	17.3	17.3	13.7	19.8	1.1
N	90.2	85.0	49.0	187.0	22.8	N	12.0	11.9	9.4	15.4	1.2
D	102.1	95.0	55.0	259.0	35.8	D	8.5	8.1	6.0	12.8	1.4
Total	727.4	689.0	466.0	1781.0	170.8	*Annual mean*	16.2	16.3	13.1	18.4	0.9
						TMH	25.7	25.9	22.2	27.9	0.9
						TMC	8.3	8.0	5.9	12.3	1.3
						Max HM	33.6	34.1	27.1	37.6	1.9
						Min CM	3.9	3.6	1.4	8.3	1.4
						OSC1	17.4	17.9	12.5	20.0	1.6
						OSC2	29.7	30.2	20.1	34.4	2.9

have significantly longer drought (Table 3.7), are correlated with soil properties (higher pH and presence of calcium in soil), and altered species composition (Lusitanian oak replaces cork oak). Zone 3 shows great variability of species composition and habitats.

3.3.3 The Climate of the Spanish Dehesas

The definition of the main climatic variables obtained from the NFI-based distribution of the dehesa and the climatic model of Sánchez Palomares et al. (1999) for the three established zones in Sánchez Ron et al. (2007) are shown in Tables 3.6, 3.7 and 3.8. The monthly rainfall and temperature regimes for the three zones show that summer is the hottest season, accompanied by drought, making it the harshest period for vegetation in the dehesa, similar to other Mediterranean ecosystems (Fig. 3.7). Topography and human management can ameliorate conditions for production by increasing nutrient and moisture availability for plants and increasing plant nutrient availability through livestock management (Puerto and Rico 1992; Moreno and Pulido 2009). July and August are the worst months

Table 3.7 Climatic characterization of zone 2 of the Spanish dehesa

	Monthly precipitation						Monthly temperatures				
	Mean	Median	Min	Max	Std. Dev.		Mean	Median	Min	Max	Std. Dev.
J	86.4	81.0	13.0	181.0	28.7	J	6.9	7.1	3.1	9.0	1.1
F	86.1	82.0	11.0	174.0	25.6	F	8.1	8.3	3.6	10.2	1.0
M	74.8	72.0	17.0	149.0	20.2	M	10.5	10.6	5.8	12.9	1.0
A	53.4	50.0	27.0	100.0	10.0	A	13.4	13.5	8.2	15.6	1.0
M	51.7	49.0	27.0	83.0	11.1	M	17.3	17.7	7.8	19.7	1.3
J	29.8	29.0	18.0	47.0	4.2	J	21.9	22.2	15.6	24.1	1.1
Jl	6.0	5.0	0.0	21.0	3.1	Jl	25.8	26.0	20.4	28.1	0.9
A	8.5	8.0	3.0	22.0	2.5	A	25.3	25.5	17.8	27.7	0.9
S	36.3	35.0	21.0	71.0	6.9	S	21.5	21.9	15.0	24.0	1.2
O	65.1	61.0	31.0	146.0	14.6	O	16.2	16.5	10.5	18.3	1.3
N	76.9	73.0	15.0	207.0	21.9	N	10.2	10.3	6.6	12.4	1.0
D	78.7	74.0	22.0	138.0	20.1	D	6.7	6.8	3.3	8.9	0.7
Total	653.7	617.0	237.0	1281.0	148.2	*Annual mean*	15.3	15.6	10.3	17.5	1.0
						TMH	25.8	26.0	20.4	28.1	0.9
						TMC	6.6	6.8	3.1	8.7	0.8
						Max HM	34.4	34.6	26.6	37.0	1.4
						Min CM	1.9	1.9	-0.8	3.8	0.9
						OSC1	19.2	19.2	15.5	21.0	0.5
						OSC2	32.5	32.8	24.3	34.5	1.2

for production and vegetation survival (Figs. 3.8a and 3.8b). Winter climatic variables are also different for the three zones (Figs. 3.9 and 3.10). The third zone, including the northern limit of Spanish dehesa distribution, has the lowest winter temperatures, while the first zone shows the mildest temperature regime.

This climatic study is based on two tools that could be improved. The low tree densities that characterize the dehesa system (Fig. 3.11) and large heterogeneity of tree distribution means there are a small number of NFI dehesa plots that do not completely cover the dehesa area estimated from vegetation maps. Also, the models we used to characterize the dehesa climate are based on mean values of the main climatic variables. The huge inter-annual climatic heterogeneity typical of the region is not well analysed with this methodology and this variation should be incorporated in future models. Nevertheless, and in spite of large regional variation in climate throughout the dehesa range, climatic factors do not seems to influence whether oak forests have been managed to form dehesas.

Tree species distribution in the Iberian Peninsula has been greatly influenced by human management during the last millennia, particularly since the Middle Ages (Chap. 2). Humans have increased the distribution of those species they consider beneficial, a trend particularly evident in agroforestry systems. Among other uses, holm oaks have long been used as food sources for both livestock and humans.

Table 3.8 Climatic characterization of zone 3 of the Spanish dehesa

	Monthly precipitation						Monthly temperatures				
	Mean	Median	Min	Max	Std. Dev.		Mean	Median	Min	Max	Std. Dev.
J	67.5	63.0	27.0	211.0	22.2	J	3.7	3.8	1.6	6.9	0.8
F	64.8	61.0	32.0	221.0	22.3	F	4.6	4.8	2.2	7.8	0.8
M	64.8	61.0	22.0	199.0	21.9	M	7.4	7.5	4.7	10.1	0.9
A	49.9	47.0	25.0	143.0	13.2	A	9.8	10.0	6.8	12.9	1.0
M	53.0	50.0	28.0	128.0	11.7	M	13.6	13.8	10.7	17.1	1.0
J	36.9	35.0	29.0	60.0	5.7	J	17.8	18.0	14.8	21.3	1.1
Jl	14.0	13.0	6.0	25.0	3.4	Jl	21.4	21.5	18.5	24.5	1.0
A	12.0	11.0	6.0	23.0	2.8	A	20.9	20.9	18.1	24.0	1.0
S	42.6	41.0	26.0	101.0	8.6	S	17.5	17.6	14.7	21.3	1.0
O	59.4	56.0	29.0	183.0	17.9	O	12.4	12.5	9.6	16.1	1.0
N	73.4	69.0	34.0	264.0	26.1	N	7.0	7.1	4.8	9.8	0.8
D	64.8	61.0	19.0	216.0	21.5	D	3.7	3.7	1.8	6.8	0.8
Total	603.0	570.0	293.0	1758.0	165.8	*Annual mean*	11.7	11.8	9.0	14.9	0.9
						TMH	21.4	21.5	18.5	24.5	1.0
						TMC	3.6	3.7	1.6	6.8	0.8
						Max HM	29.7	29.8	26.5	34.0	1.1
						Min CM	-1.0	-0.9	-2.8	2.1	0.8
						OSC1	17.8	17.8	16.9	19.7	0.4
						OSC2	30.6	30.7	27.7	33.5	0.5

They undoubtedly occupy a larger area today than they occupied before humans shaped the ecosystem, complicating ecological models of species distribution, particularly climatic models. Human management may also affect tree mortality, resulting in range contraction. Intensive grazing reduces seedling survival. A pathogen, *Phytophthora cinnamomii*, linked to the *seca* disease that results in rapid oak mortality, is currently affecting oaks in large areas of the dehesa in southwestern Spain (Brasier 1992). Although the specific factors explaining oak mortality are not totally clear, such mortality seems to be preferentially affecting the most aggressively managed holm oaks in the most water-stressed dehesas, those subjected to root-killing damage caused by inter-tree deep plowing.

It is essential to monitor species distribution and response to changing climate. Recently, Gea-Izquierdo et al. (2009) produced one of the first holm oak dendrochronological studies. They observed that at the northern limit of dehesa (where oak mortality from disease is lower than in more southern locations), holm oaks have increased sensitivity to climate. They are probably expanding their growing season as a consequence of warming, but summer is becoming more stressful, influencing growth. More research on the relationship between other dehesa tree species and climate is necessary.

The observed lack of tree regeneration that today threatens dehesas probably has several causes, including climate change (Pulido and Díaz 2005). Among

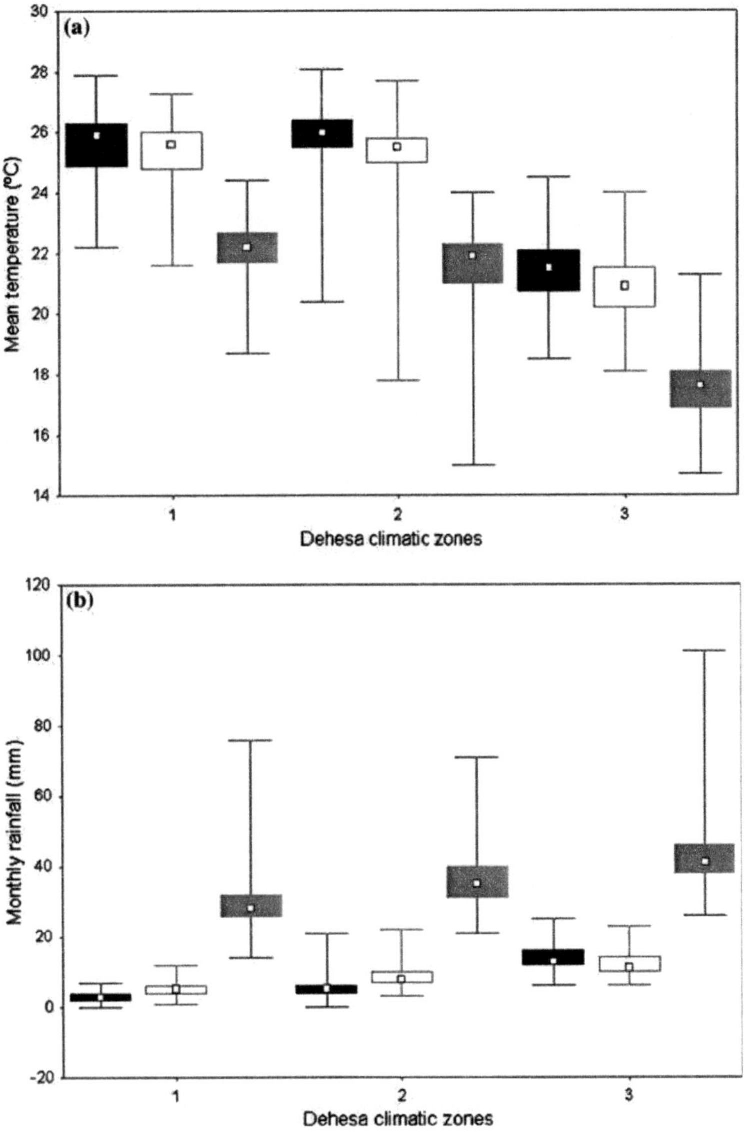

Fig. 3.8 Median (squares), interquartile range (boxes) and range (wiskers) of the mean temperature (**a**) and monthly rainfall (**b**) during the summer months (July: black; August: white; September: grey) for the three climatic zones of the Spanish dehesa

them, summer moisture stress is limiting seedling survival in stands with the lowest tree densities, probably in most sandy soils, and in locations where summer evapotranspiration is most limiting.

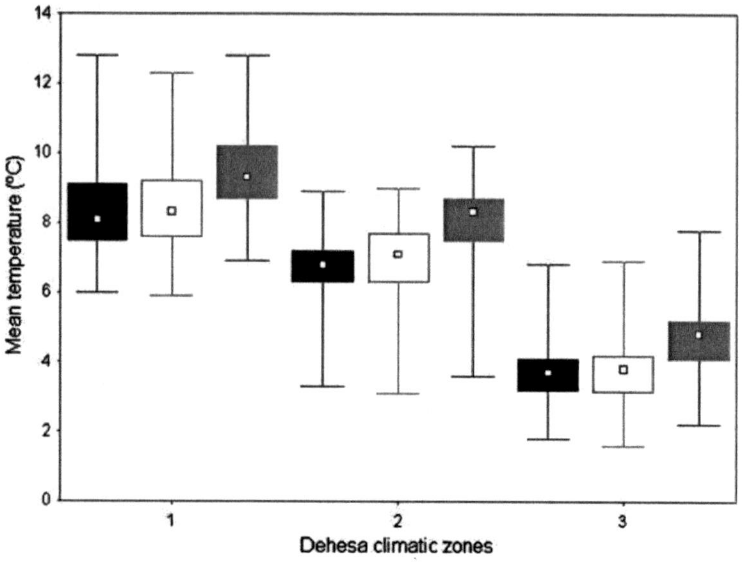

Fig. 3.9 Median (squares), interquartile range (boxes) and range (wiskers) of the mean temperature during the winter months (December: black; January: white; February: grey) for the three climatic zones of the Spanish dehesa

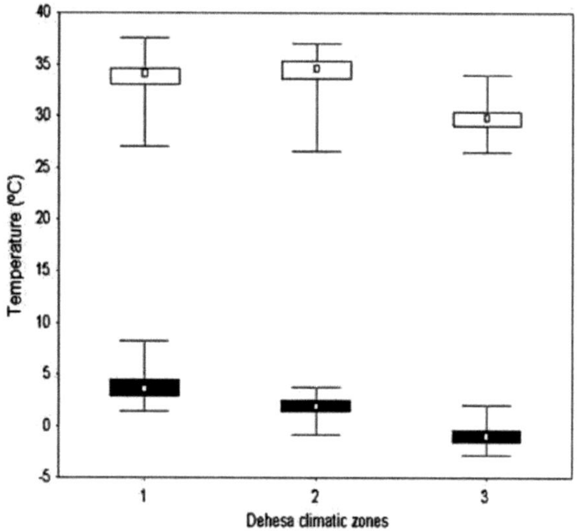

Fig. 3.10 Median (squares), interquartile range (boxes) and range (wiskers) of the maximum temperature of the hottest month (open boxes) and minimum temperature of the coldest month (closed boxes) for the three climatic zones of the Spanish dehesa

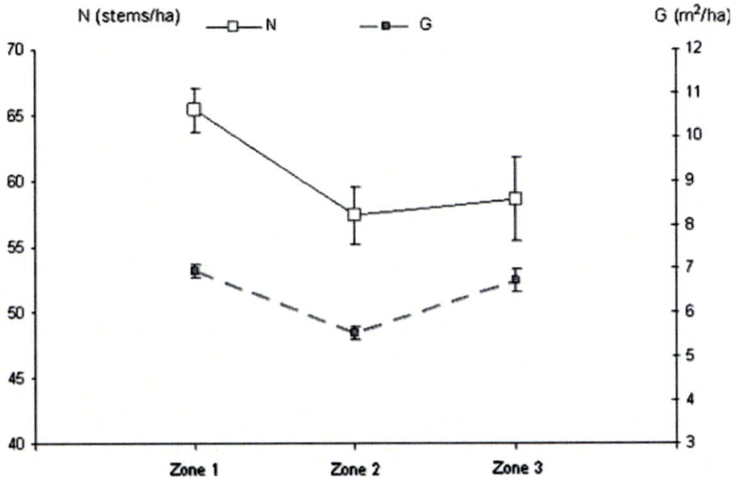

Fig. 3.11 Weighted means (vertical bars denote 0.95 confidence intervals) of the density (N, No stems/ha) and basal area (G, square meters per ha) of stands within the three climatic zones of the Spanish dehesa

3.4 Comparisons and Conclusions

Understanding the climatic determinants of oak species distribution in California and Spain is essential for current management, particularly conservation planning to maintain these important ecosystems in the face of global climate change (Santos and Thorne 2010). The distribution of major oak species in California is largely determined by climatic factors. While human management has been a locally important determinant of oak distribution in California for thousands of years, initially through regular burning by Native Americans (Anderson 2006) and for the past two centuries following European settlement through activities such as firewood cutting and oak clearing to increase grass cover for grazing (Chap. 2), human management has been much more important for the establishment and maintenance of dehesa ecosystems in Spain. Not surprisingly for plants in water-limited Mediterranean ecosystems, the distribution of each oak species examined in California and Spain responds to mean annual precipitation, which is probably a crude proxy for a physiologically relevant variable: available soil moisture during the growing season. Regression models suggest the sequence along the precipitation gradient in California is blue oak at the xeric end, followed by valley oak, coast live oak, interior live oak, black oak, and canyon live oak at the mesic end, while in the Spanish dehesa, holm oak is at the xeric end and cork oak at the mesic end. Each species also responds to temperature: in California, the most important temperature parameter is mean annual temperature range for coast live oak and valley oak, mean July temperature for interior live oak, mean January temperature for blue oak and black oak, and mean annual temperature for canyon live oak.

In Spain, mean annual temperature is also an important parameter, with dehesa zone 1 sites located at the warm end and zone 2 sites at the cold end of the gradient.

Various researchers have used climate data to build models that predict the current presence of oak species on a site with reasonable accuracy and used them to predict future distribution given altered climate scenarios. All available models indicate substantial local extinction and migration is likely for most oak species in response to projected global climate change. Blue oak and valley oak woodlands and savannas, where most ranching activity in California occurs, may be particularly adversely affected.

The distribution of dehesa in Spain is determined not only by climatic factors but also by soil characteristics and human management. The two main species comprising the tree layer of the dehesa are at different ends of the bioclimatic spectrum based on climatic characteristics typical of Mediterranean regions, including summer drought, medium to low precipitation and very high temperatures during summer. In sum, there are three main climate zones in Spanish dehesas: the most widespread is in the southern and southwestern part of Spain, under the influence of humid winds from the Atlantic Ocean, and is the one focused on in this volume; the second is a more xeric and cold zone, in the northern part of the distribution of the dehesa; and the third zone is in between the two others, with intermediate climate characteristics.

3.5 Future Research Directions

The California and Spanish climate studies rely on relatively coarse-scale climate data. This means that despite substantial fine-scale heterogeneity in topography and plant communities, the climatic niche for important trees in each region is described broadly. Increased resolution of climatic and topographic data will allow more fine-scale resolution of species distributions. In California, we quantitatively calculated the probability of presence for six oak species on a site using equations based on climatic parameters, while in Spain we provided qualitative climatic descriptions for three classes of dehesa sites rather than individual tree species. Future work in Spain will focus on building more quantitative bioclimatic models of tree species and vegetation type distributions that will allow more precise estimates of dehesa response to climate change. Future California research will focus on developing quantitative bioclimatic models of vegetation type distributions because there is considerable overlap in individual oak species distributions.

Acknowledgments The authors are very grateful to David Sanchez de Ron for contributing photos, and for his valuable comments on this work. Comments made by referees were also very helpful during revision.

References

Allen BH, Holzman BA, Evett RR (1990) A classification system for California hardwood rangeland. Hilgardia 59:1–45

Anderson MK (2006) Tending the wild: native american knowledge and the management of California's natural resources. University of California Press, Berkeley

Austin MP (2007) Species distribution models and ecological theory: a critical assessment and some possible new approaches. Ecol Model 200:1–19

Austin MP, Nicholls AO, Margules CR (1990) Measurement of the realized qualitative niche: environmental niches of five *Eucalyptus* species. Ecol Monogr 60:161–177

Barbero M, Loisel R, Quézel P (1992) Biogeography, ecology and history of Mediterranean *Quercus ilex* ecosystems. Vegetatio 99–100:19–34

Bartolome JW, Muick PC, McClaran MP (1987) Natural regeneration of Californian hardwoods. In: Plumb TR, Pillsbury NH (tech. coords.) Proceedings of the symposium on multiple-use management of California's hardwood resources. General technical report PSW–100, USDA forest service PSW forest and range experiment station, Berkeley

Bedia J, Busque J, Gutierrez JM (2011) Predicting plant species distribution across an alpine rangeland in Northern Spain. A comparison of probabilistic methods. Appl Veget Sci 14:415–432

Brasier CM (1992) Oak tree mortality in Iberia. Nature 360:539

Bravo F, Rivas JC, Monreal JA, Ordóñez C (2002) BASIFOR 2.0: Aplicación informática para el manejo de las Bases de datos del Inventario Forestal Nacional. Departamento de Producción Vegetal y Silvopascicultura. Universidad de Valladolid

Díaz-Maroto IJ, Fernández-Parajes J, Vila-Lameiro PJ (2006) Autoécologie du chêne tauzin (*Quercus pyrenaica* Willd.) en Galice (Espagne). Ann For Sci 63:157–167

Donley MW, Allan S, Caro P, Patton C (1979) Atlas of California. Pacific Book Center, Culver City

Eastman JR (1992) IDRISI, Version 4.0. Worcester, Massachusetts, Clark University, Graduate School of Geography

Elford CR (1970) Climates of California Counties. Published by various state agencies. Available in the water resources library on the U.C. Berkeley campus, Vol 1–44

Evett RR (1994) Determining environmental realized niches for six oak species in California through direct gradient analysis and ecological response surface modeling, Ph.D. dissertation. University of California, California

Gandullo JM, Sánchez Palomares O (1994) Estaciones ecológicas de los pinares españoles. MAPA–ICONA. Colección Técnica, Madrid (Spain)

García del Barrio JM, Bolaños F, Ortega M, Elena-Roselló R (2004) Dynamics of land use and land cover change in dehesa landscapes of the REDPARES network between 1956 and 1988. Adv Geoecol 37:47–54

Gea-Izquierdo G, Martín-Benito D, Cherubini P, Cañellas I (2009) Climate-growth variability in *Quercus ilex* West Iberian open woodlands of different stand density. Ann For Sci 66:802

Griffin JR (1973) Valley oaks—the end of an era? Fremontia 1:5–9

Griffin JR, Critchfield WB (19729 The distribution of forest trees in California. USDA Forest Service, Pacific southwest forest and range experiment station research paper PSW–82

Hayhoe K, Cayan D, Field CB, Frumhoff PC, Maurer EP, Miller NL, Moser SC, Schneider SH, Nicholas Cahill K, Cleland EE, Dale L, Drapek R, Hanemann RM, Kalkstein LS, Lenihan J, Lunch CK, Neilson RP, Sheridan SC, Verville JH (2004) Emissions pathways, climate change, and impacts on California. PNAS 101:12422–12427

Jepson WL (1910) The silva of California. In: Memoirs of the University of California, Vol 2 University of California, Berkeley

Kueppers LM, Snyder MA, Sloan LC, Zavaleta ES, Fulfrost B (2005) Modeled regional climate change and California endemic oak ranges. PNAS 102:16281–16286

Lenihan JM, Drapek R, Bachelet D, Neilson RP (2003) Climate change effects on vegetation distribution, carbon, and fire in California. Ecol Appl 13:1667–1681
Lindenmayer DB, Nix HA, McMahon JP, Hutchinson MF, Tanton MT (1991) The conservation of Leadbeater's possum, *Gymnobelideus leadbeateri* (McCoy): a case study of the use of bioclimatic modelling. J Biogeo 18:371–383
Major J (1988) California climate in relation to vegetation. In: Barbour MG, Major J (eds) Terrestrial vegetation of California, 2nd edn. California Native Plant Society, Sacramento
Minnich RA (2008) California's fading wildflowers: lost legacy and biological invasions. University of California Press, Berkeley
Moreno G, Pulido F (2009) Dehesa functioning, management, and persistence. In: Rigueiro A, Mosquera MR, McAdam J (eds) Agroforestry in Europe. Springer, Berlin, pp 127–160
Myatt RG (1975) Geographic and ecological variation in *Quercus chrysolepis*, Liebm Ph.D. Dissertation, University of California, Davis
Nix HA (1986) A biogeographic analysis of the Australian elapid snakes. In: Longmore R (ed) Atlas of elapid snakes. Australian Flora and Fauna Series 7:4–15
Papanastasis VP (2004) Vegetation degradation and land use changes in agrosilvopastoral systems. In: Schnabel S, Ferreira A (eds) Advances in Geoecology 37: sustainability of agrosilvopastoral systems. Catena Verlag, Reiskirchen, pp 1–12
Pavlik BM, Muick PM, Johnson SG, Popper M (1991) Oaks of California. Cachuma Press, Los Olivos
Plumb TR (1980) Response of oaks to fire. In: Plumb TR (tech. coor.) Proceedings of the symposium on the ecology, management, and utilization of California oaks. General technical report PSW–44, USDA Forest service PSW forest and range experiment station, Berkeley
Puerto A, Rico M (1992) Spatial variability on slopes of Mediterranean grasslands: structural discontinuities in strongly contrasting topographic gradients. Vegetatio 98:23–31
Pulido FJ, Díaz M (2005) Regeneration of a Mediterranean oak: a whole-cycle approach. Ecoscience 12:92–102
Roig S, Alonso-Ponce R, Sánchez-González MO, García del Barrio JM, Cañellas I (2007) Caracterización de la dehesa española de encina y alcornoque a partir del Inventario Forestal Nacional. Reunión GT Sistemas Agroforestales de la SECF. Plasencia. Abril Cuader Soc Ciens Forest 22:163–170
Rubio A, Elena R, Sanchez O, Blanco A, Sanchez F, Gómez V (1999) Autoecología de los castañares catalanes. Invest Agrar: Sist Recur For 8:387–405
Rushton SP, Ormerod SJ, Kerby G (2004) New paradigms for modelling species distributions? J Appl Ecol 41:193–200
San Miguel A, Roig S, Cañellas I (2002) Las prácticas agroforestales en la Península Ibérica. Cuadernos de la SECF 14:33–38
Sánchez de Ron D, Elena-Roselló R, Roig S, García del Barrio JM (2007) Los paisajes de dehesa en España y su relación con el ambiente geoclimático. Cuadernos de la SECF. Reunión GR Sistemas Agroforestales 22:171–176
Sánchez Palomares O, Sánchez Serrano F, Carretero Carrero MP (1999) Modelos y cartografía de estimaciones climáticas para la España peninsular, MAPA
Sánchez Palomares O, Rubio A, Blanco A, Elena R, Gómez V (2003) Autoecología paramétrica de los hayedos de Castilla y León. Invest Agrar: Sist Recur For 12:87–110
Sánchez Palomares O, Jovellar LC, Sarmiento LA, Rubio A, Gandullo JM (2007) Las estaciones ecológicas de los alcornocales españoles. Monografías INIA: Serie forestal n° 14. INIA. Madrid, Spain
Santika T, Hutchinson MF (2009) The effect of species response form on species distribution model prediction and inference. Ecol Model 220:2365–2379
Santos MJ, Thorne JH (2010) Comparing culture and ecology: conservation planning of oak woodlands in Mediterranean landscapes of Portugal and California. Environ Conserv 37:155–168
Shafer SL, Bartlein PJ, Thompson RS (2001) Potential changes in the distributions of western North America tree and shrub taxa under future climate scenarios. Ecosystems 4:200–215

Tyler CM, Kuhn B, Davis FW (2006) Demography and recruitment limitations of three oak species in California. Q Rev Biol 81:127–152
Vayssieres MP, Plant RE, Allen-Diaz BW (2000) Classification trees: an alternative non-parametric approach for predicting species distributions. J Veget Sci 11:679–694
Wieslander AE (1935) A vegetation type map of California. Madroño 3:140–144
Zavaleta ES, Hulvey KB, Fulfrost B (2007) Regional patterns of recruitment success and failure in two endemic California oaks. Divers Distrib 13:735–745

Chapter 4
Soil and Water Dynamics

Susanne Schnabel, Randy A. Dahlgren and Gerardo Moreno-Marcos

Frontispiece Chapter 4. Dehesas found on an erosion surface dissected by the drainage system (Almonte River belonging to the Tagus watershed in the Cáceres province). (Photograph by S. Schnabel)

S. Schnabel (✉)
GeoEnvironmental Research Group, Universidad de Extremadura, Avda. de la Universidad, 10071 Cáceres, Spain
e-mail: schnabel@unex.es

R. A. Dahlgren
Department of Land, Air and Water Resources, University of California, Davis, CA 95616, USA
e-mail: radahlgren@ucdavis.edu

G. Moreno-Marcos
Grupo de Investigación Forestal, Universidad de Extremadura, Plasencia 10600, Spain
e-mail: gmoreno@unex.es

P. Campos et al. (eds.), *Mediterranean Oak Woodland Working Landscapes*, Landscape Series 16, DOI: 10.1007/978-94-007-6707-2_4,
© Springer Science+Business Media Dordrecht 2013

Abstract Soil properties and water dynamics play a crucial role in the function of oak woodland ranches and dehesas. They are largely controlled by climate conditions, terrain morphology and parent material, but also by land use and management. We review results obtained from research carried out in California and Spain on topics related to soil quality, soil degradation, and water dynamics. Of particular interest is gaining understanding of the influence of land-use and management practices. The distribution of vegetation produces spatial and temporal variation in soil properties that are described in detail. The influence of trees on soil water content is discussed and the dynamics of catchment hydrology is presented, for both California and Spanish cases. An important characteristic is high variability in precipitation, with the occurrence of prolonged dry periods (droughts) that affect water availability for plants. On ranches the effects are two-fold, influencing pasture productivity and water resources for livestock rearing. Soils in the Spanish dehesas have been subject to degradational processes as a consequence of centuries of agricultural use. Water erosion resulting in the reduction of organic matter and physical degradation is the most important phenomena. For California, with a much shorter history of plowing and livestock grazing, we present results from studies on water quality and the effects of vegetation conversion on water yield, soil stability and erosion.

Keywords Dehesas • Oak woodlands • Land degradation • Soil properties • Water dynamics • Islands of fertility

4.1 Introduction

Soil properties and water dynamics play a fundamental role in the function of oak woodland ranches and dehesas. Soils provide plants with nutrients and physical support and with water, thanks to the storage capacity of soil. Soils are important as a crucial element in the hydrological cycle, are the habitat of soil organisms, and store large amounts of carbon, exceeding that of living biomass. Apart from nutrients, plant growth in Mediterranean climate is largely controlled by water availability. Water is a limiting factor and plants in this type of climate are adapted to the seasonality of rainfall, with markedly dry conditions in summer when temperatures are high and more humid conditions from autumn through spring. An important characteristic is a high variability of precipitation with the occurrence of prolonged dry periods (droughts) that affect water availability for plants. In rangelands the effects are two-fold, influencing pasture productivity and the water resources for livestock rearing. The latter is related to runoff production and aquifer recharge.

Soil properties and water dynamics are largely controlled by climate conditions, terrain morphology, rock type, and by human activities. Human interventions in dehesa often cause changes to vegetation cover that can in turn influence soil characteristics and the hydrological cycle, changes that from an environmental point of view are not necessarily negative. The spatially diverse management practices of the dehesa have changed through time. As one example, cultivation of cereals in rotation was common in the past. Since the beginning of the 1980s agricultural cropping activities were abandoned on many dehesas, and the land used only for livestock grazing, a change that inevitably affects soils and the water cycle.

In California, oak communities are the landscape where interactions at the urban-wildland-agricultural interface are most pronounced. These ecosystems are used extensively for livestock grazing, firewood production, wildlife habitat, watersheds, recreation, and increasingly for the development of small ranchettes and rural communities that have been subdivided from formerly larger land parcels. Land-use practices associated with these activities, which may include grazing, wild or controlled burning, tree removal, and creation of impervious surfaces, result in ecosystem disturbances that affect soil and water quality. Changes in soil quality may in turn affect oak regeneration success, plant community structure, and ecosystem services.

4.2 Dehesa Soils

Most dehesas in Spain and Portugal are located on flat to gently sloping land, with elevations ranging from 300 to 800 m above sea level. The dominant landforms are peneplains, well-aged erosion surfaces formed in ancient rocks that are predominantly schist, greywacke, and granite. Gently undulating surfaces intersected by valleys have increasingly steep slopes as they approach main rivers (Frontispiece). Another element interrupting the peneplains are localized ridges (*sierras*), composed mainly of alternating quartzite and schist, where open woodlands are established on moderately steep slopes. Flat areas, when found, are related to either tectonic depressions or pediments at the base of major mountain ridges. They are formed of soft sandstones and unconsolidated sediments. All of these rocks are of siliceous nature, which explains a predominantly acid reaction of soils. Calcareous rocks such as limestone are noticeably limited. Owing to the large size of dehesa properties, a diversity of landforms may be found on any given site, which gives rise to varying slope gradients and differences in soils.

Common soil types include Cambisols and Leptosols (FAO 2006). Cambisols often manifest a shallow Ah horizon and a weakly developed Bw horizon. These will be underlain by parent material that is either slightly weathered or nearly unweathered. Dorronsoro Fernández (1992) reports depths on the order of 60 cm for Cambisols in dehesas of the province of Salamanca. Studies carried out in 54 farms in the region of Extremadura revealed shallower soils, with most Cambisols showing a thickness of less than 50 cm (Schnabel et al. 1996). In many areas, unweathered parent material is found at shallow depths of less than 25 cm, giving rise to Leptosols. Soil depths can vary strongly in accordance with the irregular surface of the underlying rocks. More developed soils with a clay-enriched subsoil, such as Acrisols and Luvisols are on occasion reported (García Navarro and López

Piñeiro 2002), but are far less frequent. A common feature of dehesa soils is a thin A horizon with a thickness ranging from 2 to 8 cm and a sharp lower limit, where organic matter is clearly higher than in the underlying horizon. Fine roots of herbaceous vegetation are abundant in this uppermost part of the soil. Given the advanced age of the terrain, soils are expected to be more developed on slightly sloping land, an indication of erosion.

Soil texture is predominantly silty to sandy loam, with variations depending on lithology (CSIC 1970; Dorronsoro Fernández 1992). To characterize the properties of soils in dehesas, results of a study carried out in 54 farms distributed throughout the region of Extremadura are presented, representing the most important types of rangelands in SW Iberian Peninsula. Sampling was carried out in layers of 0–5 and 5–10 cm, and at greater depths in a selection of soil profiles.

Table 4.1 presents properties of the 5–10 cm layer that can be considered representative of the soils. Sand is the dominant particle size fraction with a mean value of 50.2 %, followed by silt with an average of approximately 39 %. Clay content amounts to 10.8 %, though variation is high with a standard deviation of 4.9 %. Soils are fairly stony with an average content of rock fragments of 20 %. Soil organic carbon content (SOC) is generally low, with a mean value of 11.6 g kg^{-1}. Cultivated areas have even less SOC. Soils are acid, with 80 % of samples strongly to moderately acid (pH = 5.0–5.9). They are poor in nutrients, as indicated by the low contents of exchangeable cations and available phosphorus, the latter having a median value of 2.0 mg kg^{-1} (Table 4.1). Bulk density is fairly high, with an average of 1.52 g cm^{-3}, corresponding to a total porosity of 43 %.

Soils of the surface horizons have a poorly developed crumb structure and aggregates are of low stability, mainly related to low content of organic matter and clay (Dorronsoro Fernández 1992; Lagar Timón et al. 2006). Soil structure

Table 4.1 Soil properties at 5–10 cm depth

Soil property	Mean	Median	Percentile 0.1	Percentile 0.9	Standard deviation
Clay (%)[a]	10.8	10.1	5.3	18.0	4.9
Silt (%)[a]	38.9	40.0	18.5	53.2	12.5
Sand (%)[a]	50.2	49.4	35.1	68.1	12.8
Rock fragments (%)[b]	20.0	18.5	8.1	32.6	12.5
BD (g cm^{-3})	1.52	1.52	1.42	1.63	0.09
pH	5.43	5.40	4.99	5.87	0.46
CEC (cmol kg^{-1})	8.3	8.0	4.1	11.9	3.3
Ca (cmol kg^{-1})	3.3	3.2	1.2	5.6	2.4
Mg (cmol kg^{-1})	1.0	0.7	0.2	2.0	1.1
K (cmol kg^{-1})	0.2	0.2	0.1	0.4	0.2
Na (cmol kg^{-1})	0.7	0.7	0.1	1.6	0.4
Base saturation (%)	66.5	63.0	36.4	95.2	35.8
N (g kg^{-1})	1.0	0.9	0.4	1.7	0.6
P (mg kg^{-1})	5.8	2.0	0.4	16.9	9.4
SOC (g kg^{-1})	11.6	11.0	6.3	17.4	4.6

[a] clay, silt and sand expressed as percentage weight of the fine fraction
[b] rock fragments present the percentage weight of the bulk sample

influences water retention capacity, and the stability of aggregates defines the soil's resistance to erosion. Dorronsoro Fernández (1992) reports the volumetric water content of dehesa soils at field capacity range 20–35 %, and available water 11–21 %. A disadvantage for dehesa plant growth is the shallow soil depth. During periods of dry and sunny weather, with high evapotranspiration, plant-available water is quickly exhausted. Water deficit during average years runs from June until September. During the wet season, rainless periods are particularly unfavorable to development of herbaceous plants, especially where soils are shallow. Trees, with an extensive and deep root system, are less affected.

Soil properties of the dominant rock types are as follows (only considering samples taken in open areas, uncovered by shrubs or by tree canopies): Soils on granite have the highest sand content (72 %) and soils on schist the highest silt content. Schist, however, has a high content of fine sand, as compared to granite. The highest bulk density mean value is observed for soils that develop on sediments (1.60 g cm^{-3}), and there is no significant difference between granites and schist, with mean values of $\sim$1.45 g cm^{-3}. No significant differences in pH are observable between the rock types. For exchangeable cations, granites and sediments show the lowest values with 6.1 and 7.7 cmol kg^{-1}, respectively. A comparison of soil organic carbon, nitrogen, and phosphorus content between the two most frequent rock types, granite and schist, shows fairly similar values.

An interesting feature of dehesa soils is the vertical distribution of their properties. Soil organic carbon content, as related to depth, is shown in Fig. 4.1, which includes data for soils below the canopy of trees, shrubs, and open areas. SOC decreases strongly with depth. Differences in organic carbon content are particularly large in the surface layer (0–5 cm), and variable at 5–10 cm depth. Below 20 cm SOC content is low ($<$0.5 g kg^{-1}). Main nutrients, such as nitrogen, phosphorus, and potassium, behave in a similar way, with high variations in the uppermost soil layer, a strong decrease with depth, and low values below 20 cm. Jobbágy and Jackson (2001) in a global study on the distribution of soil nutrients with depth showed that the topsoil concentrations of nutrients in the soil profiles are particularly pronounced where the elements were more scarce. Their study supports the idea that plant cycling exerts a dominant control on the vertical distribution of the most limiting elements for plants.

4.2.1 Nutrient Cycling and Soil Fertility

In dryland ecosystems isolated trees have an important effect on the spatial and temporal heterogeneity of soils, which can determine the structure and function of the herbaceous and animal communities in the soil (Gallardo et al. 2000). The role of trees on nutrient dynamics is critical because dehesa has a mostly internal nutrient cycle. Both nutrient inputs via atmospheric deposition and via animal harvesting are low as compared to internal fluxes (Escudero 1992; Moreno and Gallardo 2003).

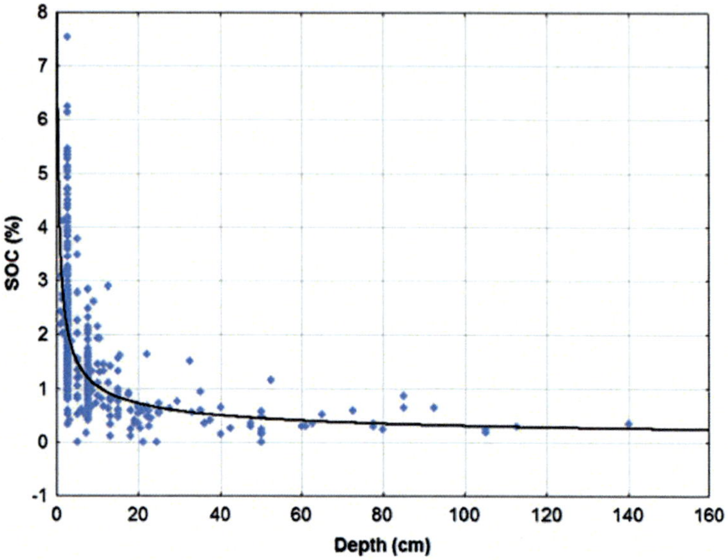

Fig. 4.1 Soil organic carbon distribution with depth. The samples were obtained from 54 farms distributed in extremadura ($n = 573$, variance accounted for is 0.3448, $R = 0.587$, $p < 0.001$, $SOC = [3.54651]\ Depth^{[-0.53566]}$)

Trees immobilize a great amount of nutrients in their living and dead structures, which can be seen both as having a detrimental immediate effect on the understory and as future nutrient storage (Escudero 1992). Trees bring up nutrients from lower soil layers, inaccessible to herbaceous vegetation, and pump nutrients laterally from areas beyond the canopy (Scholes and Archer 1997; McPherson 1997). As a result, more than 50 % of the nutrients are annually recycled beneath the canopy in dehesas with a canopy cover of only 20 % of the dehesa surface (Escudero 1992).

Litterfall in dehesas is unusually high, with 1,900 kg ha^{-1} as compared to 1,600 kg ha^{-1} in dense holm oak sites (Escudero 1992). This comprises an annual input to soil of 0.3–1.43 % of the soil pool of nitrogen beneath the canopy, 21–59 % of available phosphorus, 1.8–9.5 % of exchangeable potassium and 1.1–9.9 % of exchangeable calcium (Escudero et al. 1985).

Additionally, the turnover rate on the soil surface of dehesa ecosystems is unusually high (Escudero et al. 1985). Dehesa litterfall decomposes up to 24 times faster than that in dense forest. The amount of litterfall accumulated on the soil surface is estimated at, respectively, 400 and 8,000 kg ha^{-1} in dehesa and dense forest (Escudero et al. 1985). This rapid decomposition is explained by the action of herbivores, which can consume and recycle up to 85 % of the plant mass, and also by the rapid alternation of dry and wet periods, common in Mediterranean climates (Gallardo et al. 2000). Trees play an additional prominent role in the process, since net mineralization is higher beneath than beyond the canopy cover, as Gallardo et al. (2000) reported for nitrogen dynamics.

As a result of the nutrient dynamics in the dehesa, soils beneath tree canopies are richer in soil organic matter and nutrients than soils beyond the canopy, as widely demonstrated (González-Bernáldez et al. 1969; Escudero 1985; Puerto 1992; Gallardo 2003; Obrador-Olán et al. 2004). Although the effect of trees is reported by Joffre and Rambal (1988) to be noticeable in the whole soil profile, other authors show that significant differences on soil properties beneath and beyond canopy are usually only found for the uppermost soil layer (0 to 20–30 cm) (Escudero 1985; Moreno et al. 2007).

Beside trees, shrubby vegetation may cause a significant modification of soil fertility, although the information available is scarce. Moro et al. (1997) document a positive effect of Mediterranean shrubs on soil fertility. In ungrazed dehesa plots encroached upon by shrubs, Rodríguez et al. (1987), Obrador et al. (2004) and Moreno et al. (2007) each reported an increase of organic matter, total nitrogen, and exchangeable calcium and potassium, but a decrease of available phosphorus and mineral nitrogen, compared to values in grazed dehesa.

4.2.2 Soil Degradation

Soil degradation is a deterioration that reduces biological potential and soil productive capacity, and includes a variety of processes that affect physical, chemical, and biological soil properties (Imeson 1988). Degradation of soils is commonly associated with other phenomena of land degradation and is caused by human activities including deforestation, agricultural practices, and overgrazing (GLASOD 1990; Blum 1998). Increased awareness and concern exists about the degradation of dehesa systems and their long-term sustainability (Shakesby et al. 2002; Moreno and Pulido 2009). The lack of tree regeneration, which threatens the future of the woodlands, is recognized as among the biggest problems facing dehesas (Montero et al. 1998; Chaps. 5, 6, 8, 9, 10). A reduction of tree cover will presumably affect soil properties given the linkages explained above.

Soil degradation constitutes an additional problem in dehesas, mainly through soil erosion and increased runoff (Schnabel 1997). Few studies exist on soil degradation in these ecosystems, and information that can be applied for planning and management purposes is scarce because of the extreme diversity of dehesas, which differ with respect to natural conditions and more intensive land-use and management. A common feature is coexisting extensification and intensification, causing distinct forms of degradation. For example, payments to landowners for total numbers of animals grazed has led to an unchecked rise in stock numbers, increasing the risk of degradation of pasture and soils. On the other hand, abandonment of livestock grazing produces marked vegetation changes, leading to shrub encroachment and increased risk of wildfires that, in turn, degrade soil.

The most important potential soil degradation processes in dehesa include water erosion, physical degradation (compaction, loss of structure, reduction of infiltration capacity and water retention capacity), and biological degradation

Fig. 4.2 Signs of sheet erosion as a result of a rainstorm falling at the beginning of autumn in Spanish dehesa. (Photograph by S. Schnabel)

(impoverishment of biological activity and soil organic matter content). Most of these processes are interrelated and strongly related to plant cover.

Erosion processes found in dehesa fall into three classes. First is sheetwash, also termed interrill erosion, which is the dominant process on hillslopes. It is caused by the impact of raindrops (splash erosion) and by the erosional effect of water running off the hillslopes (Schnabel 1997) (Fig. 4.2). Second, gully erosion produces grooves with depths and widths in excess of 0.5 m. Gullies typically form in small valley bottoms of slightly-to-moderately undulating land (Fig. 4.3). Most of the small valleys only register periodic water flow and are filled with an alluvial sediment layer 1–2 m thick. Finally, rill erosion is less common in dehesa, except where land was plowed for cultivation or shrub clearing. Soil erosion research was carried out at different spatial scales since 1990 in the region of Extremadura (Schnabel 1997; Schnabel and Gómez Amelia 1993; Gómez Gutiérrez et al. 2009a; Gómez Gutiérrez et al. 2012).

The relationship between vegetation cover and soil loss for different rainfall intensities is illustrated in Fig. 4.4. With vegetation covering more than 60 % of the ground surface, even exceptionally high intensity storms (I-30 > 40 mm h^{-1}) produced soil losses below 0.3 Mg ha^{-1}. On the other hand, with a ground cover of less than 20 %, moderately intense storms, which are not uncommon, may produce significant soil losses. Changes in soil cover are influenced by seasonal and interannual variations in precipitation, and by livestock density. In the study catchment a great reduction of surface cover was observed with a moderate stocking density during a prolonged drought. The same effect was noted in periods with more abundant rainfall as a consequence of increased livestock numbers (Schnabel et al. 2009).

Mean soil loss amounted to 0.63 Mg ha^{-1} yr^{-1}, a value higher than that obtained in a Mediterranean holm oak forest in Cataluña (Sala 1988), which can be considered a less human-modified system than oak woodlands in Extremadura and Andalucía. Although the erosion rate is not much higher than a tolerance value of 0.20 mg ha^{-1} yr^{-1} (Kirkby 1980), soils are shallow with poor productivity and present-day erosion contributes to further soil degradation.

Fig. 4.3 Active gully erosion in the valley bottom of the Guadal catchment of Spain. (Photograph by S. Schnabel)

Fig. 4.4 Relationship between soil loss and vegetation cover for different 30 min maximum rainfall intensities (mm/hour). Based on mean values of 12 erosion plots and 88 events in Spain. Variance accounted for is 0.901, $p < 0.001$) (Schnabel et al. 2009)

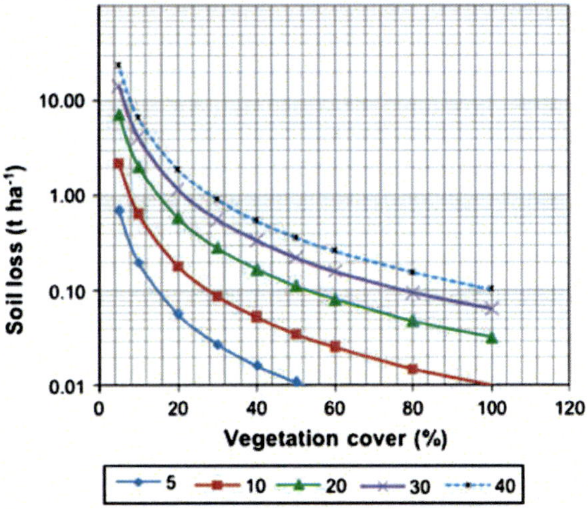

In addition, soil losses showed a great temporal variation, related to vegetation cover. Table 4.2 shows sheet erosion rates for two groups of data: VEG representing periods with moderate to high ground cover, and DEG periods with a strongly reduced cover. Reduction of the vegetation cover was produced by a drought during the first part of the study period and by an increase in livestock density at the end of the monitoring period. As a result, mean erosion rates increased more than ten-fold (0.12 and 1.34 Mg ha^{-1} yr^{-1}). With high stocking densities soil erosion may increase strongly, but even with moderate animal numbers the risk of erosion may be enhanced during unfavorable rainfall conditions (droughts) when there is a reduction of grazing capacity due to low pasture production. Soil erosion related to excessive stocking rates is reported by Coelho et al. (2004) and Shakesby et al. (2001) for montados in Portugal and Morocco, respectively.

Rill erosion is often seen where land is plowed for cultivation or for clearing understory vegetation. High soil losses are produced in particular when tilling immediately precedes an exceptional rainstorm. Soil losses produced by rill erosion of approximately 100 Mg ha^{-1} were registered as a consequence of an extreme rainfall event in a recently-plowed field (Schnabel et al. 2001). An increase of overland flow and erosion due to plowing was likewise reported by Coelho et al. (2004) in Portuguese montado.

Gullies represent only a small fraction of the total land and are more frequent on schist and greywacke than on granites. Individual gullies may present high soil losses, constitute an obstacle for traffic and enhance drainage of subsurface flow. Sediment losses produced by gullying vary strongly as evidenced in the Guadal catchment, where the average rate was 1.55 Mg ha^{-1} yr^{-1}. However, mean soil loss during the first six years of observation amounted to only 0.12 Mg ha^{-1} yr^{-1} and during the last two years of the study period increased to 5.57 Mg ha^{-1} yr^{-1} (Table 4.2). Gully erosion is related to the amount of overland flow generated on hillslopes that are characterized by shallow soils with low infiltration capacity. High intensity rainstorms generate rapid runoff response in a channel, even under dry soil conditions. During wet periods, when continuous rainfall saturates the valley bottom, large amounts of runoff are generated producing high sediment losses. In addition, extreme rainfall events may provoke exceptionally high gully erosion. In fact, the high soil loss rate observed during 1997 (Table 4.2) was mainly the result of a rainstorm with a return period of 200 years. The lower rate of gully erosion registered in the Parap catchment is explained by the rainfall conditions predominating during the study period.

The influence of land use and management on gully erosion was studied for the period from 1947 to 2002 (Gómez Gutiérrez et al. 2009b). Gully activity was highest in the 1950s when nearly half of the catchment was cultivated, in accordance with Spanish government edicts that all cultivable land be put to use. Abandonment of cultivation from roughly 1970 onwards reduced gullying in the system. During the last decade a new increase in gullying has been observed that is

Table 4.2 Comparison of soil loss rates for different erosion processes and their variations in Spain (Schnabel et al. 2009)

Erosion process	Mean loss (Mg ha^{-1} yr^{-1})	Variation of loss (Mg ha^{-1} yr^{-1})
Sheetwash	0.63	Periods VEG 0.12
Guadal catchment (Sept 1990—Dec 1996)		Periods DEG 1.34
Gully erosion		
Guadal catchment (Sept 1990—Nov 1997)	1.55	1990–1995 0.12
Parap catchment (Dec 2001—Jun 2007)	0.07	1995–1997 5.57
Rill erosion, plowed slope, extreme event	100.50 (Mg ha^{-1})	4/11/97

related to growing livestock numbers, and has resulted in the appearance of lateral bank headcuts caused by animal trampling in the vicinity and along gully margins (Gómez Gutiérrez et al. 2009b). Figure 4.5 shows a heavily degraded area with abundant animal paths, a lot of bare soil, and signs of water erosion.

A survey carried out in a large number of dehesa properties showed that only 13.2 % of those were reported to suffer heavy sheet erosion and 26.4 % experienced moderate erosion (Schnabel et al. 2006). In the rest of the dehesas sheet erosion was low or undetectable. Other soil degradation processes important in dehesas are soil compaction, reduction of organic matter content, and decrease of plant available water. The few quantitative data available to date show that in large parts of the region soils are degraded: they are shallow, have low organic matter content, and high bulk density. The number of farms classified as strongly degraded was 22.6 % and a further 37 % were moderately degraded (Schnabel et al. 2006). For montados in Portugal and Morocco, Coelho et al. (2004) demonstrated that heavy grazing increases soil compaction, overland flow, and erosion.

4.3 Water Dynamics in Dehesas

Across their geographic range dehesas exhibit low precipitation and high evaporative demand. Rainfall is highly variable, with low amounts registered in summer coinciding with high potential evapotranspiration. Most dehesas are found in areas

Fig. 4.5 The effect of animal trampling on gully erosion in Spanish dehesa. Note headcuts and animal paths crossing a gully. The photo was taken at the end of summer when vegetation was greatly reduced. (Photograph by S. Schnabel)

with climates ranging from semi-arid to dry sub-humid (Chap. 3). Rainfall year-to-year is highly variable, giving rise to droughts, which alternate with wetter periods. A disadvantage is that shallow soils result in low water-holding capacity. A major ecological influence on dehesas is water availability (Infante et al. 2003).

4.3.1 Catchment Hydrology

Channel flow in small upland catchments in dehesas depends highly on the antecedent moisture conditions and particularly on the water content of sediment-filled valley bottoms (Ceballos and Schnabel 1998). When these accumulation zones are saturated, with infiltration capacity markedly reduced, high amounts of runoff are produced during rainfall events. During a high-intensity rainstorm overland flow is rapidly generated on hillslopes producing a rapid response of runoff in the channel. However, water flow is of short duration and the amount of discharge small. In contrast, during periods with above-average rainfall soil water content is high and valley bottom sediments are saturated (Ceballos and Schnabel 1998). As a consequence, discharge amounts are then high, and water flow in a channel may last several weeks.

The effect of rainfall variability on the generation of surface water is highlighted in Fig. 4.6, which shows the relationship between annual rainfall and discharge in two small headwater catchments. Mean precipitation during the 12 years of record was 517 mm, generating on average 57 mm ($l\ m^{-2}\ yr^{-1}$) of runoff, which was approximately 12 % of rainfall. A rainfall amount at the average value was only observed once (Fig. 4.6). The remaining years registered either substantially lower or higher values. The amount of water generated in dry years was small, constituting only 3.3 % of precipitation which means that more than 96 % of rainfall is evaporated by plants and soils (assuming low deep drainage due

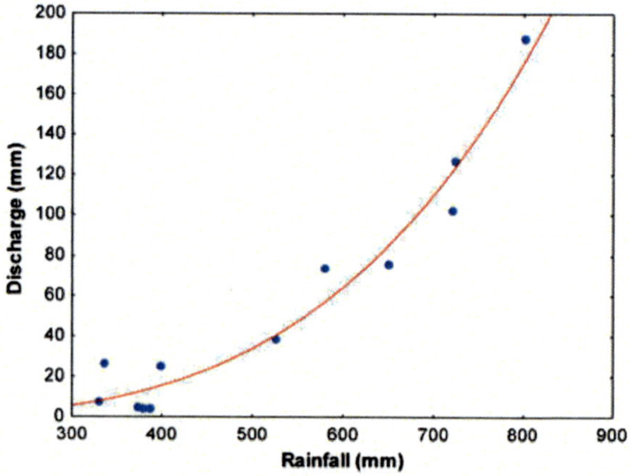

Fig. 4.6 Rainfall-discharge relationship of two small catchments in the Spanish province of Cáceres for a 12-year period

to the nearly impervious rocks). Only during years of higher precipitation did the catchments generate higher volumes of runoff with values in excess of 100 mm. This behaviour is typical for semi-arid areas, where most of the precipitation is lost to the atmosphere by evapotranspiration (Rodier 1975).

These data illustrate the possible effects of climate change on water resources. Climate change models project a decrease in precipitation and an increase of rainfall variability for the western part of the Mediterranean area (IPCC 2007). The consequence of a rainfall decline on catchment hydrology would be a reduction of water resources in general terms and increased rainfall variability would provoke more extreme droughts with little channel flow, alternating with humid periods of water flow of short duration.

4.3.2 Soil Water Dynamics

Puerto (1992) and Joffre and Rambal (1993) reported that soil water content was always higher beneath than beyond the tree canopy in northern and southern sub-humid dehesas, respectively. Joffre and Rambal (1988) estimated that maximum soil water storage was 40–110 mm higher beneath than beyond trees. This increased soil moisture occurred in spite of the soil beneath the canopy receiving significantly less water than the interstitial area between trees as a consequence of rainfall interception. Luis-Calabuig (1992) and Mateos and Schnabel (2002) reported values of 36.7 and 26.8 % of interception expressed as percentage of annual rainfall, respectively.

The presence of trees affects water balance. Trees significantly increase water consumption by transpiration, while water is lost by deep drainage and/or surface runoff beyond the tree canopies. Joffre and Rambal suggested in 1993 that a water yield of 200 mm of annual rainfall is produced away from trees in contrast to 570 mm under tree cover. Their water balance calculations mistakenly assume the amount of rainfall reaching the soil surface below the canopy cover is the same as in open spaces, considering interception losses negligible. But on the order of 27 % of rainfall is deflected or absorbed by foliage, leaving the amount of water reaching the ground below the tree canopy (throughfall and stemflow) at 73 % of total rainfall (Mateos and Schnabel 2002). The quantity of water available for vegetation consumption, runoff, and deep drainage is significantly less than gauged amounts, measured in the open. In the case of drier sites with an average annual rainfall of 500 mm, only 365 mm would be available below tree canopies.

These results are best understood in terms of improved microclimate and soil physical properties beneath the tree cover. Apart from increasing soil organic matter content, trees modify soil texture by increasing the abundance of fine particles (Joffre and Rambal 1988; Puerto 1992). However, Escudero (1985) found no significant difference in soil texture beneath versus beyond the canopy. A positive effect of trees on dry bulk density (1.51 vs. 1.58 g cm^{-3}), infiltration rate, and available soil water (243 vs. 155 mm) was found in dehesas. As a result, the drought period in the soil showed a delay of 1 and 1.5 months, as reported by Joffre and Rambal (1988) and Puerto (1992), respectively.

Nevertheless, a recent study in semi-arid dehesas (annual rainfall around 500 mm) shows that soils dry almost simultaneously in time and intensity beneath and beyond the tree cover (Cubera and Moreno 2007). A similar pattern is reported by Nunes et al. (2005) in an area with annual rainfall of 666 mm yr^{-1}. The widely accepted idea that trees improve the soil water status of a dehesa is uncertain at best, especially in the driest dehesas. The volume of water extracted by tree roots affects spatial and temporal changes in soil moisture. Trees reach water located beyond the canopy cover (Joffre and Rambal 1993), and tap subterranean water located up to 20 m away from the canopy edge (Cubera and Moreno 2007).

In accordance with this, surface runoff on hillslopes is lower below tree canopies than in open areas (Schnabel 1997). This is attributable to reduced water reaching the ground below the canopy, as a result of rainfall interception, rather than to higher infiltration capacity, as measured in runoff experiments (Cerdà et al. 1998; Ceballos et al. 2002). These were carried out with simulated high intensity rainfall, at equivalent rates below trees and in open areas. With both wet and dry soils, similar runoff was observed in the open and in areas below tree canopy. Runoff may even be enhanced below the tree canopy thanks to soil hydrophobicity (Cerdà et al. 1998). Reduced runoff production below the tree canopy is more likely the effect of lower rainfall reaching the ground as compared to open areas.

4.4 Soils in California Oak Communities

4.4.1 Soil Properties

Soils in California oak communities are found at an elevation between 100 and 700 m in northern California and up to 1,500 m in southern California. These soils form in a xeric soil moisture regime (dry summers/wet winters) and generally have a thermic soil temperature regime (mean annual soil temperature is 15–22 °C).

They have formed on a variety of bedrock types (granite, andesite, basalt, greenstone, shale, sandstone), and on sediments originating from these bedrocks. The primary factors affecting soil characteristics are type of parent material, climate, and topography. Mean annual rainfall within oak communities of central California ranges from approximately 400–800 mm, with nearly all precipitation occurring as rainfall between October and March (Chap. 3). Soils formed on bedrock in this zone are typically shallow, ranging from $\sim$50 cm on erosional surfaces (steeper topography) to $\sim$1.5 m on stable and depositional surfaces.

In general, soils within the oak communities are slightly-to-moderately acidic (pH = 5.5–7) with higher pH values occurring in drier and more base-rich parent materials. Soils are dominated by Inceptisols (weakly developed) in the lower elevations and transform to predominately Alfisols (formation of clay-rich B horizons) in the higher elevations as increasing precipitation leads to greater chemical weathering, clay formation and clay translocation (Dahlgren et al. 1997; Rasmussen et al. 2007, 2010). The amount of clay varies as a function of parent rock type with more basic rocks (basalt) and some sedimentary/metasedimentary rocks (shale, phyllite) that have greater maximum clay contents (30–50 %) than granitic rocks (15–20 %). Mollisols (and mollic epipedons in Alfisols) are often found in these more clay-rich soils owing to preservation of organic matter by the higher clay concentrations. Iron-rich, basic igneous rocks tend to develop a strong red color due to accumulation of iron oxides, which bind with phosphate often resulting in low phosphorus availability (Fig. 4.7).

Fig. 4.7 A typical California oak woodland soil profile formed in greenstone in the Sierra Nevada foothills. Annual grass roots are limited to the upper 40 cm of the profile while oak roots are found throughout the soil profile (typically 1.5 m). (Photograph by R. Dahlgren)

4.4.2 Spatial and Temporal Variability

Oak canopy coverage ranges from a few scattered trees (savanna) through oak woodlands, to nearly complete coverage (oak forest), with increasing elevation and amount of precipitation (Fig. 4.8 and Chap. 1). In oak savannas, scattered blue oak (*Q. douglasii)* trees create a mosaic of open grasslands and oak/understory plant communities having a profound influence on soil quality and fertility. Beneath blue oak canopies, several physical (bulk density, infiltration rates), chemical (pH, organic matter, nutrient concentrations), and biological (microbial biomass, soil respiration) properties are significantly enhanced compared to the adjacent grassland plant communities (Dahlgren et al. 1997) (Table 4.3). Enhanced soil quality and fertility are dominantly displayed in the upper ~30 cm of the soil profile with the largest enrichments occurring in the surface horizon (~12 cm). While the enhanced soil properties are most prominent directly beneath the canopy, the enhancement of soil properties extends to approximately two canopy radii from the tree trunk (Figs. 4.9, 4.10) (Table 4.3).

These "islands" of enhanced soil quality and fertility are apparent beneath the oak canopy for both grazed and non-grazed sites indicating that grazing is neither responsible for formation of these islands nor does it destroy them (Camping et al. 2002). Soils beneath the oak canopy generally had thicker A horizons, suggesting that oak trees promote the development of thicker topsoil horizons through

Fig. 4.8 Scattered oak trees create a mosaic of open grasslands and oak-understory plant communities on the California landscape. Oak tree density declines with decreasing precipitation. (Photograph by S. Bledsoe)

Table 4.3 Mean (±standard deviation) for selected soil properties for non-grazed soils beneath the oak canopy compared to soils in open grasslands on basic metavolcanic (greenstone) bedrock in the Sierra Nevada foothills of northern California (Dahlgren et al. 1997)

Soil property	Oak canopy soil	Grassland soil
A horizon thickness (cm)	12.1 (2.4)	8.4 (2.1)
Bulk density (g cm^{-3})	0.92 (0.08)	1.12 (0.04)
Infiltration rate (cm hr^{-1})	10.4 (2.9)	6.9 (1.2)
pH (H_2O)	7.16 (0.15)	6.44 (0.15)
Organic C (g kg^{-1})	66.0 (8.3)	40.9 (4.1)
Total N (g kg^{-1})	4.44 (0.88)	2.98 (0.45)
C/N ratio (atomic)	17.5 (1.3)	16.2 (1.6)
Microbial biomass C (g kg^{-1})	1.25 (0.21)	0.78 (0.26)
TP (mg kg^{-1})	718 (204)	406 (71)
Available P (Bray—mg kg^{-1})	39.8 (14.1)	11.8 (3.0)
Exchangeable Ca ($cmol_c kg^{-1}$)	16.8 (1.9)	7.9 (1.0)
Exchangeable Mg ($cmol_c kg^{-1}$)	3.0 (0.3)	2.1 (0.5)
Exchangeable K ($cmol_c kg^{-1}$)	0.91 (0.39)	0.44 (0.07)
Base saturation (%)	71.8 (8.4)	50.9 (9.2)

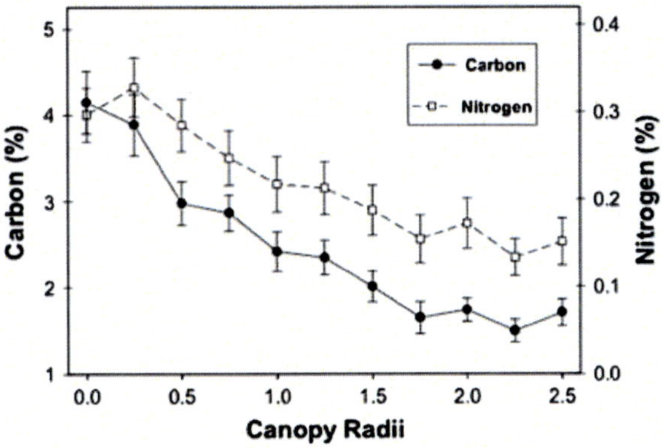

Fig. 4.9 Distribution of organic carbon and nitrogen (mean ± standard deviation) as a function of distance (canopy radii) from the tree bole in grazed soils formed on basic metavolcanic (greenstone) bedrock in the Sierra Nevada foothills of northern California (n = 16; 4 trees × north, east, south and west transects from bole for each tree)

enhanced organic matter production and cycling. Because of enhanced soil quality beneath the oak canopy, soil fauna (pocket gophers, ants) may preferentially inhabit the soils beneath the canopy, leading to greater physical mixing of organic matter into the soil profile. Increased mixing of soil by biota, coupled with enhanced soil structure due to increased organic matter and soil fauna/flora activity, contribute to lower bulk densities, improved water infiltration rates and improved gas exchange in the upper soil layers. The resultant water infiltration and gas exchange in turn promotes a more favorable environment for many soil organisms.

Enrichment of organic matter and nutrients beneath the canopy of oak trees results in large part from return of litterfall and its associated nutrients to the soil surface. The cycling of base cations (calcium, magnesium and potassium) by oaks

Fig. 4.10 Islands of enhanced soil fertility and moisture retention beneath the oak canopy are especially noticeable during the spring. Note the luxuriant growth of annual grasses beneath the oak canopy compared with the adjacent open grasslands. (Photograph by R. Dahlgren)

often leads to higher base saturation in the surface horizons beneath the oak canopy, leading to higher soil pH values. In a three year study of litterfall deposition in the Sierra Nevada foothills, blue oaks returned an average of 9,100 kg ha^{-1} yr^{-1} of litterfall to the soil surface, including associated nutrients (Dahlgren et al. 1997). Additionally, canopy throughfall (precipitation dripping from the canopy) contributes appreciable fluxes of nutrients (Dahlgren and Singer 1994). Nutrient fluxes in canopy throughfall originate from the capture of atmospheric aerosols and particulate matter (atmospheric deposition of nitrogen ranges from 3 to 10 kg ha^{-1} yr^{-1} in California oak communities), as well as from root uptake. Because oak roots are found at greater depths than those of annual grasses (generally less than 35 cm), nutrient uptake by oak roots lessens leaching losses of nutrients from the soil profile (Millikin and Bledsoe 1999). The extension of oak roots beyond the edge of the canopy may contribute to nutrient differences between soils beneath the oak canopy and open grasslands by concentrating nutrients beneath the oak canopy. Cattle seeking shade and preferentially defecating under trees may contribute to nutrient enrichment beneath the canopy and in the transition zone between the oak canopy and open grasslands.

A further beneficial effect of the oak canopy on enriching nutrients beneath the canopy occurs through reduced leaching and erosion, which results in more nutrients being retained in the upper soil layers. At a study site in the Sierra Nevada foothills evapotranspiration was estimated to be about 30 % greater beneath the oak canopy as compared with the open grasslands (Dahlgren and Singer 1994). This is because of greater extraction of water from the soil profile by deeply rooted oak trees and because precipitation is intercepted by the oak canopy (and then evaporates). Oak trees can extract some water from the soil profile throughout the dry summer months while the annual grasslands have no significant transpiration component during the summer as they are dead/dormant and not transpiring. The increased loss of water due to evapotranspiration by oaks reduces the leaching intensity beneath the oak canopy more than in grassland sites. In addition, higher organic matter concentrations reduce soil bulk density and increase infiltration rates, which reduces surface runoff and the loss of nutrients through erosion. There are, therefore, several biogeochemical processes by which oak trees concentrate nutrients and create islands of enhanced soil quality and fertility beneath their canopy.

Dahlgren et al. (2003) sought to determine the effects of parent material differences on formation of islands of soil fertility, oak woodlands with contrasting soil parent materials (granite, sandstone/shale, and greenstone [metamorphosed basalt]. The dominant oak species at all sites was blue oak, and the age of trees estimated at 75–120 years. Parent material differences resulted in large variations in soil texture. Granite parent material tends to result in sandy loam, sedimentary material in loam, and greenstone in silty clay loam. Clay mineralogy also has typical soil texture relationships with granites resulting in mica and kaolinite with minor vermiculite; sandstone/shale in interstratified vermiculite-chlorite, vermiculite, illite, smectite, kaolinite and gibbsite; greenstone in interstratified vermiculite-chlorite, and vermiculite, chlorite and kaolinite with small amounts of smectite. In addition, differences in climatic factors between sites may result in differences in net primary production. In spite of these differences, pools of organic carbon and total nitrogen in the 0–15 cm layer were similar among the sites (Fig. 4.11). Microbial biomass carbon and potentially mineralizable nitrogen tended to be higher in sedimentary parent material. Available phosphorus was lower in the greenstone parent material, probably due to the higher concentrations of iron oxides that strongly adsorb phosphate. Exchangeable potassium concentrations were generally similar across all sites, in spite of large differences in potassium concentrations in the parent materials. The differences imposed by parent materials among sites were expected to result in much larger differences between organic carbon and nutrient pools; however, for the most part differences between sites were small. One conclusion is that vegetation has a much stronger effect on soil organic matter and nutrient pools than do differences in soil parent material.

We were particularly interested in determining how long the islands of enhanced soil quality and fertility persist once oak trees are removed. Studies examining changes in soil solution chemistry revealed an immediate shift in soil solution nutrient concentrations toward that of the grassland soils in the year following tree removal (Dahlgren et al. 2003). This suggests that islands of soil fertility are quickly reverting to nutrient conditions similar to open grassland soils following tree removal.

Solid-phase soil properties from plots where oak trees were removed 5–34 years prior to sampling showed appreciable loss of soil organic matter (Dahlgren et al. 2003). Organic carbon concentrations showed a significant

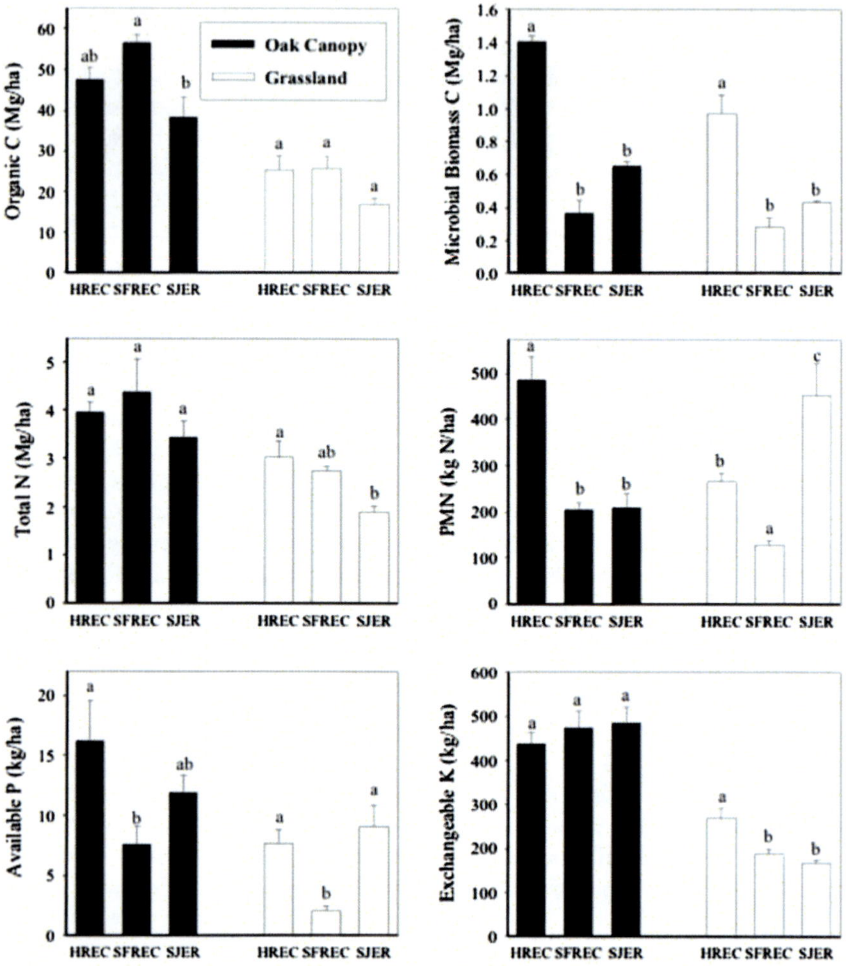

Fig. 4.11 Organic carbon and nutrient pools (mean ± standard error; n = 5) for the 0–15 cm depth increment of soils beneath the oak canopy and adjacent grasslands for sites on sandstone/shale (*HREC* Hopland research extension center), greenstone (*SFREC* Sierra foothill research extension center), and granite (*SJER* San Joaquin experimental range). Mean values with the same lower case letters within each vegetation type (oak vs. grassland) are not statistically different at $P = 0.05$

decrease after 10 years in the 0–5 cm layer and approached that of grassland soils after 30 years (Fig. 4.12). Organic carbon concentrations decreased in the 5–15 cm layer, but at a slower rate. Total nitrogen concentrations followed a pattern similar to organic carbon with a significant decrease after 10 years and a decline to levels similar to grassland soils after 10–30 years. After 30 years, the organic carbon pool in the 0–15 cm layer decreased by about 30 Mg ha^{-1} (44 % decrease) and 18 Mg ha^{-1} (34 % decrease) at the greenstone and sandstone/shale

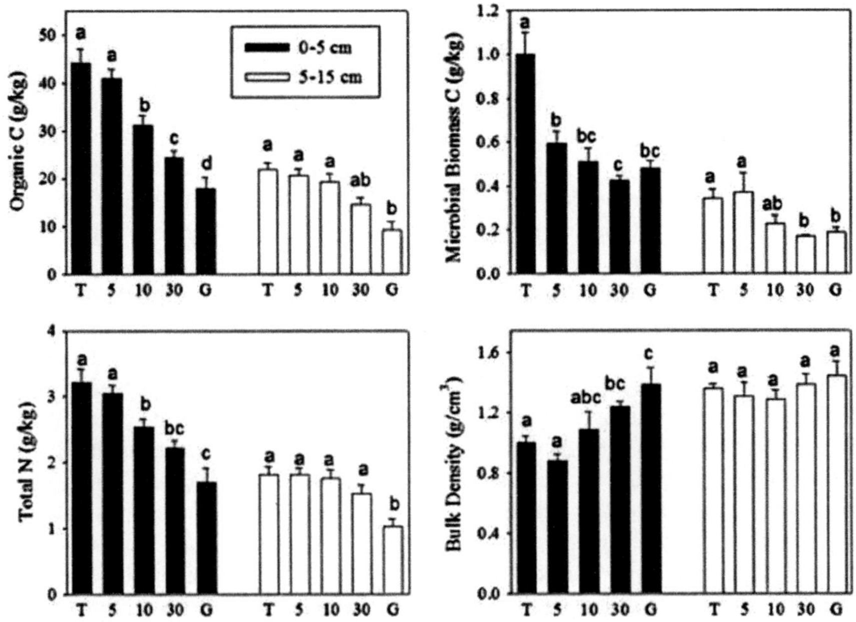

Fig. 4.12 Selected soil quality and fertility parameters (mean ± standard error; n = 5) for the 0–5 and 5–15 cm depth increments of soils beneath the oak canopy (T) and adjacent grasslands (G), and for soils where oak trees were removed 5, 10 and 30 years ago. Means with same lower case letters within each depth increment 0–5 and 5–15 cm are not statistically different at $P = 0.05$

sites, respectively. A similar comparison for the total nitrogen pool showed losses of about 1.4 Mg ha^{-1} (29 % decrease) and 0.7 Mg ha^{-1} (19 % decrease) at the greenstone and sandstone/shale sites, respectively. The majority of the organic carbon and nitrogen was lost within the first 10–20 years following tree removal. Bulk density in the 0–5 cm layer showed a significant increase over time and microbial biomass showed a rapid decrease following tree removal; both responses are probably associated with the loss of soil organic matter.

The rapid and large decreases in the organic carbon and nitrogen pools result, in part, from the immediate loss of litterfall from oak trees once a tree is removed. The return of about 4,500 kg ha^{-1} yr^{-1} of organic carbon and 84 kg ha^{-1} yr^{-1} of nitrogen in litterfall to the soil beneath the oak canopy provides a large annual input of organic matter and nutrients. A loss of litterfall inputs coupled with rapid decomposition in the tree removal sites result in a shift in soil organic matter concentrations until a new steady-state, less enriched with nutrients, is reached with respect to organic matter inputs from the annual grasses that dominate following oak removal. While planting oaks can sequester relatively large amounts of carbon into oak community soils, the sequestered carbon may be quickly released back to the atmosphere upon oak removal.

Understory grass productivity in California oak communities has been shown to range from <25 % of open grassland productivity to greater than 200 % that of surrounding grasslands (Callaway et al. 1991). In general, landscapes with lower densities of blue oak had enhanced forage yield while landscapes with higher canopy density had reduced forage production. Within a given landscape, differences in below-canopy forage production among trees are observed. Trees with low understory forage productivities had substantially higher amounts of oak fine roots in the upper 50 cm of soil than trees with higher understory forage productivities, suggesting greater competition for water and nutrients (Callaway et al. 1991). Oak tree removal has been suggested as a way to increase forage production by decreasing competition for light, water, and nutrients. Short-term increases in forage production were commonly observed following tree removal in relatively open stands (Chap. 9). However, this benefit lasts less than two decades before forage production returns to levels found in the adjacent grasslands. In our study, forage production beneath the oak canopy was ~70 % that of the open grasslands; however, forage production in the transition zone between the oak canopy and open grasslands (an area one canopy radius beyond the canopy edge) was elevated compared to the open grassland. For a typical tree with a 5 m radius canopy, forage production is decreased for a 79 m^2 area while it is enhanced within the transition zone having an area of 236 m^2 offsetting the forage loss beneath the canopy. These findings are consistent with others showing that scattered trees have a positive overall impact on forage quantity and quality, with a tree cover of 25–35 % being most profitable for rangelands (Walpole 1999; Barnes et al. 2011).

4.5 Water Dynamics in California Oak Communities

4.5.1 Watershed Studies

Watersheds dominated by oak communities play a critical role in California's water supply system providing runoff primarily from winter rainfall events and hosting two-thirds of the state's drinking water reservoirs. Understanding water storage and streamflow regulation by soils in these watersheds is essential for water resource planning under future climate change scenarios. More than 85 % of the annual precipitation in California oak woodlands occurs from October to March, in keeping with the area's Mediterranean-type climate. In the dry summer the flow in many first order streams ceases or is greatly diminished. During the fall wet-up period, stream flow generation (or hydrograph response to precipitation) does not occur until sufficient water infiltration brings the dry soil to field capacity. Once the water-holding capacity of the soil is filled, additional precipitation generates streamflow. While the amount of precipitation required to "prime" a watershed depends on soil water-holding capacity, it was shown to range from 150 to 250 mm for many watersheds in California oak woodlands (Dahlgren et al. 2001).

Vegetation type and density have an appreciable effect on watershed hydrology through their influence on evapotranspiration and canopy interception. Based on detailed measurements of soil profile water content beneath an oak canopy and adjacent open grassland, there was approximately twice as much evapotranspiration from soils beneath the canopy (660 mm) as compared to the grassland soil profile (320 mm) (Dahlgren and Singer 1994). The differences in evapotranspiration were attributed to canopy interception by the oak canopy (23 % of total precipitation) and greater transpiration ($\sim$7 %) due to greater water extraction by the deeply rooted oak trees.

Hortonian overland flow (the tendency of water to flow horizontally across land surfaces when rainfall has exceeded infiltration capacity and depression storage capacity) is not often observed on California oak woodlands due to relatively high infiltration rates; however, saturation overland flow is common in water-accumulating areas (concave landscape surfaces). Infiltration rates vary greatly due to grazing and vegetation (grassland vs. oak canopy) characteristics. Depending on the degree of grazing, infiltration rates can be greatly reduced due to compaction by cattle (8.0 cm hr^{-1} non-grazed to 1.5 cm hr^{-1} grazed grassland at a Sierra Nevada foothill site). In contrast, the accumulation of greater litter and soil organic matter beneath the oak canopy typically results in a three-fold increase in infiltrations rates beneath grazed oak canopies (5.3 cm hr^{-1}) as compared to the grazed open grasslands (1.5 cm hr^{-1}). On hillslopes, surface runoff from grassland areas may actually infiltrate as the runoff enters the more permeable surface soils beneath the oak canopy.

Previous studies in oak woodland watersheds have demonstrated large variability in annual runoff-to-rainfall ratios, which ranged from 0.19 to 0.76 during a 17 year record (Lewis et al. 2000). Other studies have highlighted the importance of complex interactions between soils and plants in regulating soil moisture storage during the year. For example, the water balance in open grasslands compared with under an oak tree canopy may differ by 50 % (Joffre and Rambal 1993; Dahlgren and Singer 1994); soil properties are different under oak than in open grasslands (Dahlgren et al. 1997, 2003) and soil water loss through evapotranspiration can be higher under oak than in grasslands (Jackson et al. 1990; Dahlgren and Singer 1994). The runoff-to-rainfall ratio was shown to be affected by both the total annual precipitation and distribution of rainfall throughout the year. Since significant runoff occurs during the winter-wet period when evapotranspiration is low, additional rainfall results in greater runoff (Swarowsky et al. 2011). In contrast, precipitation during fall wet-up and spring dry-down periods will largely be lost as evapotranspiration. Temporal shifts in precipitation as a result of possible climate change may have the greatest impact on future stream runoff patterns in California oak woodlands.

Topography has long been considered one of the main factors that affect runoff processes and the primary predictor of watershed-scale hydrologic flowpaths. However, recent research highlights the importance of soil stratigraphy in regulating streamflow. Integrated hydrologic measurements showed a close synchrony between streamflow and subsurface lateral flow in AB and Bt horizons overlying a hydraulically restrictive claypan, a feature common in California oak woodland soils (Swarowsky et al. 2012). The thickness of the perched water table controlled the magnitude of subsurface lateral flow, which was greatest when AB horizons became saturated. Stream recession characteristics were controlled by lateral flow in less permeable horizons (Bt) directly overlying the claypan. Over the course of the water year, subsurface lateral flow from near surface horizons (A and AB) increased as antecedent soil moisture and thickness of the perched water table increased catchment-wide. The dynamic nature of hydrologic flow paths in this system has implications for water quality as water is short-circuited through upper soil horizons.

4.5.2 Water Quality

Temporal variability in water quality occurs at the storm-event, seasonal and inter-annual time scales on California oak woodlands (Tate et al. 1999; Holloway and Dahlgren 2001; Ahearn et al. 2004; Dahlgren et al. 2004). It is common for a large portion (>80 %) of the total yearly stream discharge and nutrient and sediment loads to occur during storm events. The large runoff associated with storms allows these events to export large fluxes of non-point source constituents (sediments, nutrients, *Escherichia coli* bacteria) from the watersheds.

Interactions between hydrological and biological processes produce a distinct seasonal pattern in nutrient concentrations due to an asynchrony within nutrient cycling (Fig. 4.13) (Ahearn et al. 2004). Instead of a continuous nitrogen feedback among senescing plants and litter, their soils, and new growth (biotic uptake), nitrogen in oak woodlands is mineralized and accumulates in soils during dry summer and fall months. With the onset of winter rains, water begins to flow through the soil profile mobilizing the accumulated nitrate before new growth can uptake nutrients. Each storm progressively flushes this nitrogen pool so that by March there is little if any nitrogen found in streamflow. Similarly, annual grass

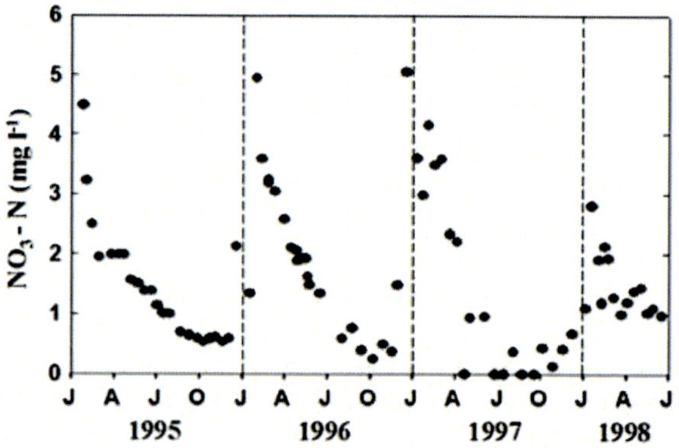

Fig. 4.13 Seasonal variability in nitrate-nitrogen concentrations resulting from temporal asynchrony between nutrient availability, biological uptake, and hydrologic flushing in California oak woodland/ annual grassland (adapted from Ahearn et al. 2004)

biomass accumulation is large in late-March to May and deciduous oaks break bud in April resulting in high biological demand for nutrient uptake. As such, oak woodlands are inherently leaky with respect to nutrients, and especially nitrate and potassium.

In contrast, coniferous forest vegetation in the higher elevation watersheds has the ability to uptake nutrients all year round, including autumn when soil moisture becomes available. Nutrient uptake and availability are more evenly synchronized in a coniferous forest than in the deciduous oak/annual grasslands found at lower elevations. Oak woodland annual rangelands are susceptible to nitrate leaching, even in the absence of management activities.

There is considerable variability in the magnitude of constituents exported from watersheds on an annual time step. A 21-year record from a Sierra Nevada foothill watershed showed that annual nitrate (0.2–4.0 kg N ha^{-1} yr^{-1}) and sediment (25–465 kg ha^{-1} yr^{-1}) fluxes varied by more than an order of magnitude over the period of record (Lewis et al. 2006).

4.5.3 Vegetation Conversion Effects on Water Yield

Efforts to increase water yield from oak woodlands have converted oak woodlands to grasslands through oak removal. Removal of woody vegetation with a deep rooting system reduces the amount of water extracted from the lower soil profile. Any residual soil water storage carried forward to the next water year reduces the amount of priming necessary to generate stream flow in the next water year resulting in greater water yields.

A case study in the northern Coast Ranges of California examined conversion of oak woodlands to grass vegetation and demonstrated decreased runoff during storms and nearly a doubling in the length of storm hydrographs (Burgy 1968). Peak runoff rates were reduced by about 25 % after conversion to grass. The longer hydrograph response indicates a prolongation of runoff and a greater contribution from subsurface flow. These changes result from a greater density of grass cover that retards overland flow and permits greater opportunity for infiltration. Reduced evapotranspiration from removal of deeply-rooted trees resulted in a long-term increase in runoff and extension of base flow through the dry season. Total runoff was increased due to decreased interception by grasses as compared to tree and brush vegetation. Interception losses by brush and oak trees were 10–25 % of the precipitation compared to negligible interception by grasses (Burgy and Pomeroy 1958). Average increase in stream discharge after vegetation conversion was about 60 %, with an increase in runoff on the order of 100 mm per year (Burgy and Papazifiriou 1971b).

Annual runoff was strongly correlated with annual precipitation, and the amount of runoff increased geometrically as precipitation increased (Burgy and Papazifiriou 1971b). While runoff was highly correlated with total precipitation, seasonal distribution of precipitation was an important factor affecting the total annual water yield.

4.5.4 Vegetation Conversion Effects on Soil Stability and Erosion

Vegetation type and amount has a tremendous influence on slope stability and erosion. Plant and litter cover protects soil aggregates from destruction by raindrop impact and slows overland flow providing greater opportunity for water infiltration. Removal of deeply rooted vegetation (oaks, shrubs, and brush) reduces the mechanical reinforcement within soil profiles provided by the root system. Fire has been shown to influence erosion by reducing infiltration rates, especially by removing surface cover and by contributing to soil hydrophobicity (water repellency). Generally, reduction of infiltration is considered to be proportional to the severity of burning.

Following removal of woody vegetation and fire, sedimentation in a headwater northern Coast Range watershed increased from about 4.3 to 43 Mg ha^{-1} yr^{-1} (Burgy and Papazifiriou 1971a). As grasses developed, erosion was reduced to rates ultimately below those of the original woody vegetation. Erosion rates increased again due to an increase in mass wasting events associated with the decay of the woody vegetation roots. While there were no massive soil movements prior to vegetation conversion, there were a total of 61 soil slips in the 10 years following removal of woody vegetation (Burgy and Papazifiriou 1971a). Two types of mass wasting were identified: slippage of slopes along a failure plane and mud flows due to liquefaction. The majority of the mass wasting events occurred in the vicinity of stream channels with stream scouring and bank cutting preceding the occurrence of the event. Minimum slope gradient where mass wasting events occurred was approximately 45 %, and the number of events per year directly proportional to the annual rainfall amount. Vegetation conversion from woody plants to grasses reduces slope stability by removing the soil reinforcement provided by root systems and increasing soil moisture content, causing a decrease in soil strength.

4.6 Conclusions

A comparison of soil and water dynamics between Californian oak woodlands and Spanish dehesas reveals notable similarities but also differences. Soils in California are generally of higher quality, with greater depth and fertility, especially those found on basic parent material and in areas with low slope gradients. The shallow and poorly developed soils found even on almost flat land in dehesas point to the influence of soil erosion, both past and present. Water erosion was reported to produce excessive soil losses under high grazing pressure due to a strong reduction of ground cover, an effect that was enhanced during prolonged periods of drought. Furthermore, large areas were cultivated in the last 50 years and well before, probably provoking high soil erosion rates. In the Californian case soil erosion is mainly observed on steep slopes, due to vegetation conversion from woody plants to grasses and after forest fires. Mass wasting was reported, especially of river banks. These phenomena are not common in dehesa because of low slope gradients and rock types that are not prone to mass movements. Soil degradation in dehesa varies greatly in space and depends largely on land use and management of the properties.

Common to open oak woodlands is an effect called "islands" of higher soil quality and fertility below the tree canopies, as compared to the open grasslands. This is related to biogeochemical processes, including deposition of tree litter, lower soil erosion, and accumulation of livestock excrement. Enhanced soil quality and fertility are largest in the uppermost layer of the soil profile and extend beyond the tree canopy. However, in California these islands of enhanced soil quality and fertility were apparent beneath the oak canopy for both grazed and non-grazed sites, indicating that grazing is neither responsible for formation of these islands nor does it destroy them. Although the same is described for dehesas, on heavily grazed dehesa sites soil degradation may lead to similar soil quality below and between the tree canopies. Studies in California suggest that tree removal provokes rapid change, reverting to nutrient conditions similar to open grassland soils. The effects of converting oak woodlands to grasslands in California were reduced soil quality and increased water yield.

Both study areas exhibit low precipitation and high evaporative demand, producing a water deficit, especially during summer. Rainfall is strongly variable between years giving rise to droughts alternating with more humid periods. Water availability is a major factor for plant growth, and for pasture production, but also is important for supplying water to livestock. Furthermore, watersheds dominated by oak communities play a critical role in California's water supply system providing runoff primarily from winter rainfall events. Likewise, in Spain and Portugal large reservoirs are found in river basins in grazed oak woodland and grassland landscapes. The water dynamics of small catchments in California and Spain are markedly different. In the case of California, Hortonian overland flow was rarely observed and runoff coefficients were higher than in the Spanish catchments. These differences may be explained by the annual rainfall total and soil properties. In California high amounts of runoff are generated due to saturation of the upper soil layer as a consequence of a nearly impervious clay horizon. In contrast, the Spanish catchments generate Hortonian overland flow during high intensity rainstorms, though part of the water reinfiltrates in the valley bottoms that are filled with sediments. Only during exceptionally humid years are considerable amounts of runoff produced.

An increase of rainfall variability and/or potential evapotranspiration as suggested by climate change models would increase the variability of runoff production with increased drought intensity. The effects of these changes on soils, water supply and the medium to long-term sustainability of oak woodlands need further investigation.

Acknowledgments The investigation carried out in dehesas was made possible through funding offered by the Spanish Ministry of Science and Technology (AMB92–0580, AMB95–0986–C02–02, HID98–1056–C02–02, CGL2004–04919–C02–02, CGL2008–01215, CGL2011–23361). Special thanks to all the colleagues and graduate students who offered valuable contributions to the dehesa research, especially Antonio Ceballos Barbancho, Marco Maneta López, Álvaro Gómez Gutiérrez, Manuel Pulido Fernández, Francisco Lavado Contador and Silvia Nadal Chillemi.

References

Ahearn DS, Sheibley RW, Dahlgren RA, Keller KE (2004) Temporal dynamics of stream water chemistry in the last free-flowing river draining the Sierra Nevada, California. J Hydrol 295:47–63

Barnes P, Wilson WP, Trotter MG, Lamb DW, Reid N, Koen T, Bayerlein L (2011) The patterns of grazed pasture associated with scattered trees across an Australian temperature landscape: an investigation of pasture quantity and quality. Rangel J 33:121–130

Blum WH (1998) Basic concepts: degradation, resilience and rehabilitation. In: Lal R, Blum WH, Valentine C, Stewart BA (eds) Methods for assessment of soil degradation. CRC Press, Boca Ratón, pp 1–16

Burgy RH (1968) Hydrological studies and watershed management on brushlands. Annual Report No 8 to California Department of Water Resources and UC Water Resources Ctr

Burgy RH, Papazifiriou ZG (1971a) Effects of vegetation management on slope stability, Hopland experimental watershed II at Hopland field station. Abstract for water resources center advance council meeting

Burgy RH, Papazifiriou ZG (1971b) Vegetative management and water yield relationships. In: Proceedings of 3rd international seminar for hydrology professors, Purdue University, pp 315–331

Burgy RH, Pomeroy CR (1958) Interception losses in grassy vegetation. Trans Am Geophys Union 39:1095–1100

Callaway RM, Nadkarni NM, Mahall BE (1991) Facilitation and interference of *Quercus Douglasii* on understory productivity in central California. Ecol 72:1484–1499

Camping TJ, Dahlgren RA, Tate KW, Horwath WR (2002) Changes in soil quality due to grazing and oak tree removal in California oak woodlands. In: Oaks in California's changing landscape. USDA Forest Service Gen. Tech. Rep. PSW–GTR–184, Berkeley, CA, pp 75–85

Ceballos A, Schnabel S (1998) Hydrological behaviour of a small catchment in the dehesa landuse system (extremadura, SW Spain). J Hydrol 210:146–160

Ceballos A, Cerdà A, Schnabel S (2002) Runoff production and erosion processes on a dehesa in Western Spain. Geogr Rev 92:333–353

Cerdà A, Schnabel S, Ceballos A, Gómez Amelia D (1998) Soil hydrological response under simulated rainfall in the dehesa land system (extremadura, SW Spain) under drought conditions. Earth Surf Proces 23:195–209

Coelho COA, Ferreira AJD, Laouina A, Hamza A, Chaker M, Naafa R, Regaya K, Boulet AK, Keizer JJ, Carvalho TMM (2004) Changes in land use and land management practices affecting land degradation within forest and grazing ecosystems in the Western Mediterranean. In: Schnabel S, Ferreira A (eds) Sustainability of agrosilvopastoral systems. Adv GeoEcology vol 37. Catena Verlag, Reiskirchen, Germany, pp 137–153

CSIC [Consejo Superior de Investigaciones Científicas] (1970) Suelos, estudio agrobiológico de la provincia de Cáceres. Centro de Edafología y Biología Aplicada de Salamanca, Salamanca
Cubera E, Moreno G (2007) Effect of single *Quercus ilex* trees upon spatial and seasonal changes in soil water content in dehesas of Central Western Spain. Ann For Sci 64:355–364
Dahlgren, RA, Singer MJ (1994) Nutrient cycling in managed and non-managed oak woodland-grass ecosystems. Final report: integrated hardwood range management program. Land, air and water resources paper–100028. UC Davis, Davis, CA
Dahlgren RA, Boettinger JL, Huntington GL, Amundson RG (1997) Soil development along an elevational transect in the western Sierra Nevada, California. Geoderma 78:207–236
Dahlgren RA, Tate KW, Lewis DJ, Atwill ER, Harper JM, Allen-Diaz BH (2001) Watershed research examines rangeland management effects on water quality. Cal Agric 55:64–71
Dahlgren RA, Horwath WR, Tate KW, Camping TJ (2003) Blue oak enhance soil quality in California oak woodlands. Cal Agric 57:42–47
Dahlgren RA, Tate KW, Ahearn DS (2004) Watershed scale, water quality monitoring—water sample collection. In: Down RD, Lehr JH (eds) Environmental instrumentation and analysis handbook. Wiley, New York, pp 547–564
Dorronsoro Fernández C (1992) Suelos. In: Gómez Gutiérrez JM, El libro de las dehesas salmantinas, Junta de Castilla y Leon, Salamanca, pp 71–121
Escudero A (1985) Efectos de árboles aislados sobre las propiedades químicas del suelo. Rev Ecol Biol Sol 22(2):149–159
Escudero A (1992) Intervención del arbolado en los ciclos de los nutrientes. In: Gómez Gutiérrez JM (ed): El libro de las dehesas salmantinas, Junta de Castilla y León. Salamanca, Spain, pp 241–257
Escudero A, García B, Luis E (1985) The nutrient cycling in *Quercus rotundifolia* and *Q. pyrenaica* ecosystems ("dehesas") of Spain. Oecol Plant 6:73–86
FAO (2006) World reference base for soil resources. World Soil Resources Reports No. 103. FAO, Rome
Gallardo A (2003) Effect of tree canopy on the spatial distribution of soil nutrients in a Mediterranean dehesa. Pedobiol 47:117–125
Gallardo A, Rodríguez-Saucedo JJ, Covelo F, Fernández-Alés R (2000) Soil nitrogen heterogeneity in a dehesa ecosystem. Plant Soil 222:71–82
García Navarro, A, López Piñeiro A (2002) Mapa de suelos de la provincia de Cáceres, escala 1:300.000, Universidad de Extremadura, Cáceres, Spain
GLASOD (1990) World map of the status of human-induced soil degradation. ISRIC/UNEP, Wageningen
Gómez Gutiérrez Á, Schnabel S, Lavado Contador F (2009a) Gully erosion, land use and topographical thresholds during the last 60 years in a small rangeland catchment in SW Spain. Land Degrad Dev 20:535–550
Gómez Gutiérrez Á, Schnabel S, Lavado Contador F (2009b) Modelling the occurrence of gullies in rangelands of SW Spain. Earth Surf Proces 34:1893–1902
Gómez Gutiérrez Á, Schnabel S, de Sanjosé JJ, Lavado Contador F (2012) Exploring the relationships between gully erosion and hydrology in rangelands of SW Spain. Z Geomorphol 56(suppl 1):27–44
González-Bernáldez F, Morey M, Velasco F (1969) Influences of *Quercus ilex rotundifolia* on the herb layer at El Pardo woodland. Bol Soc Esp Hist Nat 67:265–284
Holloway JM, Dahlgren RA (2001) Seasonal and event-scale variations in solute chemistry for four Sierra Nevada catchments. J Hydrol 250:106–121
Imeson AC (1988) Una vía de ataque eco-geomorfológica al problema de la degradación y erosión del suelo. In: MOPU (ed) Desertificación en Europa, MOPU, Madrid, pp 161–181
Infante JM, Domingo F, Fernández-Aléz R, Joffre R, Rambal S (2003) *Quercus ilex* transpiration as affected by a prolonged drought period. Biol Plant 46:49–55
IPCC (2007) Climate change 2007: the physical science basis. Contribution of working group I to the fourth assessment report of the intergovernmental panel on climate change. In: Solomon S,

Qin D, Manning M, Chen Z, Marquis M, Averyt KB, Tignor M, Miller HL (eds) Cambridge University Press, Cambridge
Jackson LE, Strauss RB, Firestone MK, Bartolome JW (1990) Influence of tree canopies on grassland productivity and nitrogen dynamics in deciduous oak savanna. Agr Ecosyst Environ 32:89–105
Jobbágy EG, Jackson RB (2001) The distribution of soil nutrients with depth: global patterns and the imprint of plants. Biogeochem 53:51–77
Joffre R, Rambal S (1988) Soil water improvement by trees in the rangelands of Southern Spain. Acta Oecol 9:405–422
Joffre R, Rambal S (1993) How tree cover influences the water balance of Mediterranean rangelands. Ecology 74:570–582
Kirkby MJ (1980) The problem. In: Kirkby MJ, Morgan RPC (eds) Soil erosion. Wiley, Chichester, pp 1–16
Lagar Timón D, Schnabel S, Gómez Gutiérrez A, Sánchez-Lorenzo A (2006) Efectos de los factores físicos y químicos del suelo sobre la estabilidad estructural en espacios adehesados de Extremadura. In Espejo Díaz M, Martín Bellido M, Matos C, Mesías Díaz (eds) Gestión ambiental y económica del ecosistema dehesa en la Península Ibérica, Junta de Extremadura, Mérida, pp 81–87
Lewis D, Singer MJ, Dahlgren RA, Tate KW (2000) Hydrology in a California oak woodland watershed: a 17-year study. J Hydrol 240:106–117
Lewis DJ, Singer MJ, Dahlgren RA, Tate KW (2006) Nutrient and sediment fluxes from a California rangeland watershed. J Environ Qual 35:2202–2211
Luis-Calabuig E (1992) Bioclima. In: Gómez-Gutiérrez JM (ed) El libro de las dehesas salmantinas, Junta de Castilla-León, Salamanaca, pp 241–260
Mateos B, Schnabel S (2002) Rainfall interception by holm oaks in mediterranean open woodland In: Garcia-Ruiz JM, Jones JAA, Arnaez J (eds) Environmental change and water sustainability, Consejo Superior de Investigaciones Científicas and University of La Rioja Press, La Rioja, Spain, pp 31–42
McPherson GR (1997) Ecology and management of North American savannas. University of Arizona Press, Tucson
Millikin CS, Bledsoe CS (1999) Biomass and distribution of fine and coarse roots from blue oak (*Quercus douglasii*) trees in northern Sierra Nevada foothills of California. Plant Soil 214:27–38
Montero G, San Miguel A, Cañellas I (1998) Sistemas de selvicultura mediterránea. La dehesa. In: Jiménez Diáz RM, Lamo de Espinosa J, Agricultura sostenible. Ediciones Mundi-Prensa, Madrid, pp 519–554
Moreno G, Gallardo JF (2003) Atmospheric deposition in oligotrophic *Quercus pyrenaica* forest: implications for forest nutrition. Forest Ecol Manag 171:17–29
Moreno G, Pulido (2009) The functioning, management, and persistence of dehesas. Adv Agroforest 6:127–160
Moreno G, Obrador JJ, Garcia-López E, Cubera E, Montero MJ, Pulido FJ, Dupraz C (2007) Competitive and facilitative interactions in dehesas of C-W Spain. Agrofor Syst 70:25–40
Moro MJ, Pugnaire FI, Haase P, Puigdefábregas J (1997) Effect of the canopy of *Retama sphaerocarpa* on its understorey in a semiarid environment. Funct Ecol 11:175–184
Nunes J, Madeira M, Gazarini L (2005) Some ecological impacts of *Quercus rotundifolia* trees on the understorey environment in the "montado" agrosilvopastoral system, Southern Portugal. In: Mosquera-Losada MR., Riguero-Rodriguez A, McAdam J (Eds.) Silvopastoralism and sustainable land management, CAB International, Oxfordshire, pp 275—277
Obrador-Olán JJ, García-López E, Moreno G (2004) Consequences of dehesa land use on nutritional status of vegetation in Central-Western Spain. In: Schnabel S, Ferreira A (eds) Sustainability of agrosilvopastoral systems, Adv GeoEcology vol 37. Catena Verlag, Reiskirchen, Germany pp 327–340
Puerto A (1992) Síntesis ecológica de los productores primarios. In Gómez-Gutiérrez JM (ed) El libro de las dehesas salmantinas, Junta de Castilla-León, Salamanaca, pp 583–632

Rasmussen C, Matsuyama N, Dahlgren RA, Southard RJ, Brauer N (2007) Soil genesis and mineral transformation across an environmental gradient on andesitic lahar in California. Soil Sci Soc Am J 71:225–237

Rasmussen C, Dahlgren RA, Southard RJ (2010) Basalt weathering and pedogenesis across an environmental gradient in the southern cascade range, California USA. Geoderma 154:473–485

Rodier J (1975) Evaluation of annual runoff in tropical African Sahel. ORSTOM Document, p 145

Rodriguez R, Puerto A, García JA, Saldaña A (1987) Algunas comunidades oligotróficas derivadas de la degradación de las dehesas. Pastos, pp 336–347

Sala M (1988) Slope runoff and sediment production in two mediterranean mountain environments. Catena Suppl 12:13–29

Schnabel S (1997) Soil erosion and runoff production in a small watershed under silvo-pastoral landuse (dehesas) in extremadura, Spain. Geoforma Ediciones, Logroño, Spain

Schnabel S, Gómez-Amelia D (1993) Variability of gully erosion in a small catchment in South-West Spain. Acta Geológica Hispánica 28:27–35

Schnabel S, González F, Murillo M, Moreno V (2001) Different techniques of pasture improvement and soil erosion in a wooded rangeland in SW Spain. Methodology and preliminary results. In: Conacher A (ed) Land Degradation. Kluwer Academic Publishers, The Netherlands, pp 241–256

Schnabel S, Lavado Contador, Gómez Gutiérrez A, Lagar Timón (2006) La degradación del suelo en las dehesas de Extremadura. In: Espejo Díaz M, Martín Bellido M, Matos C, Mesías Díaz (eds) Gestión ambiental y económica del ecosistema dehesa en la Península Ibérica, Junta de Extremadura, Mérida, pp 63–71

Schnabel S, Gómez Gutiérrez Á, Lavado Contador JF (2009) Grazing and soil erosion in dehesas of SW Spain. In: Romero Díaz A, Belmonte Serrato F, Alonso-Sarriá F, López Bermúdez F (eds) Advances in studies on desertification, Editum, Murcia, pp 725–728

Scholes RJ, Archer SR (1997) Tree-grass interactions in savannas. Ann Rev Ecol Syst 28:517–544

Shakesby RA, Coelho COA, Schnabel S, Keizer JJ, Clarke MA, Lavado Contador JF, Walsh RPD, Ferreira AJD, Doerr SH (2002) A ranking methodology for assessing relative erosion risk and its application to dehesas and montados in Spain and Portugal. Land Degrad Dev 13:129–140

Swarowsky A, Dahlgren RA, Tate KW, Hopmans J, O'Geen AT (2011) Catchment-scale soil water dynamics in a mediterranean oak woodland. Vadose Zone J 10:800–815

Swarowsky A, Dahlgren RA, O'Geen AT (2012) Linking subsurface lateral flowpath activity with streamflow characteristics in a mediterranean headwater catchment. Soil Sci Soc Am J 76:532–547

Tate KW, Dahlgren RA, Singer MJ, Allen-Diaz B, Atwill ER (1999) Timing, frequency of sampling affect accuracy of water-quality monitoring. Cal Agric 53:44–48

Walpole SC (1999) Assessment of the economic and ecological impacts of remnant vegetation on pasture productivity. Pac Conserv Biol 5:28–35

Chapter 5
Oak Regeneration: Ecological Dynamics and Restoration Techniques

Fernando Pulido, Doug McCreary, Isabel Cañellas, Mitchel McClaran and Tobias Plieninger

Frontispiece Chapter 5. After several decades of continuous grazing and regeneration failure, dehesas have a savanna-like aspect with low oak canopy cover. Grazing exclusion and/or planting is then clearly needed to ensure regeneration. (Photograph by T. Plieninger)

F. Pulido (✉)
Grupo de Investigación Forestal, Universidad de Extremadura, E-10600 Plasencia, Spain
e-mail: nando@unex.es

P. Campos et al. (eds.), *Mediterranean Oak Woodland Working Landscapes*, Landscape Series 16, DOI: 10.1007/978-94-007-6707-2_5,
© Springer Science+Business Media Dordrecht 2013

Abstract The acreage of oak woodlands has decreased in California and Spain, especially in the twentieth century. Currently, most surviving stands in Spain suffer from oak regeneration failure and it has been noted as a problem in many stands in California. A lack of dispersers transferring acorns to safe (shaded) sites is the main recruitment limitation in dehesas, where shrub encroachment generally results in higher oak recruitment rates. In California, recruitment failure is due to a combination of factors. The effects of introduced Mediterranean annuals, heavy livestock grazing, fire suppression, and predation by native and non-native wildlife on acorns and seedlings are all implicated, depending on locale and time period. Afforestation has been the main instrument for addressing the regeneration problem, especially in Spain. Natural regeneration at local scales is favored by shrubs, but also supported by protecting seedlings and by modifying the environment so young oaks can grow to a safe height. Complete livestock exclusion is of limited value in California as it hampers seedling establishment due to increasing rodent density. In dehesas, however, natural regeneration can only occur in seasonally grazed or wholly ungrazed sites, though livestock-dependent landowners are generally reluctant to carry out these measures.

Keywords Oak afforestation · Oak restoration · Recruitment limitations · Regeneration failure · Tree regeneration

5.1 Ecological and Historical Background for Oak Regeneration

Little is known about the historical scale and impact of silvo-pastoral practices in Spanish oak forests and shrublands (Stevenson and Harrison 1992; Blondel et al. 2010) and the variety of landscape types they generated in the lowland areas of the

D. McCreary
UC Sierra Foothill Research and Extension, University of California , 8279 Scott Forbes Road, Browns Valley, California 95918, USA
e-mail: mcccreary@berkeley.edu

I. Cañellas
Forest Research Center, National Institute for Agriculture and Food Research and Technology, Ctra. de la Coruña km. 7,5 28040 Madrid, Spain
e-mail: canellas@inia.es

M. McClaran
School of Natural Resources and the Environment, University of Arizona, 325 Biosciences East Tucson, Tucson, AZ 85721, USA
e-mail: mcclaran@u.arizona.edu

T. Plieninger
Ecosystem Services Research Group, Berlin-Brandenburg Academy of Sciences and Humanities, Jägerstr 22/23 10117 Berlin, Germany
e-mail: plienint@geo.hu-berlin.de

Mediterranean Basin (López-Sáez et al. 2008; Pulido 2008). In the dehesa of southwestern Spain, historical records show that wooded pastures were developed as a complex management system not later than in early medieval times, at least near existing urban settlements (Linares and Zapata 2003; Ezquerra and Gil 2009; Chap. 2). Hereafter, the increase in the area covered by dehesas paralleled the growth of the human population, especially from the eighteenth century onwards, as a growing population required more and more arable and grazing lands (Linares and Zapata 2003). This process is considered to have been completed by the middle of the twentieth century, when almost all natural forest and shrublands in flat areas had been converted into open dehesas.

During the period 1940–1970 an intensification of agricultural practices and a number of other socioeconomic changes led to a crisis in the traditional dehesa system (Díaz et al. 1997). The dehesa suffered a sharp contraction due to tree cutting and lack of tree regeneration, a process that ceased during the 1980s as a result of new regulations and a rising environmental awareness. In Extremadura (the Spanish region comprising 30 % of all dehesa area; Chap. 1), around 5.7 million oak trees were lost and 9.6 % of the dehesa was converted to farmland due to a shift to intensive crop production between 1955 and 1985 (Elena-Roselló et al. 1987). The area covered by wooded dehesa has been stable during the last three decades, which probably reflects that tree mortality due to ageing in some areas is compensated for by encroachment and natural tree regeneration in abandoned farms (García del Barrio et al. 2004; Plieninger 2006; Figs. 5.1, 5.2).

Fig. 5.1 Stages in the conversion of forest to dehesa on a large property (Valero, northern Extremadura, southwestern Spain). Continuous green dark areas correspond to undisturbed forest-shrubland which are plowed to enhance tree growth and grassland production (savanna-like area in the centre), leaving selected trees to create the woodland. (Photograph by F. Pulido)

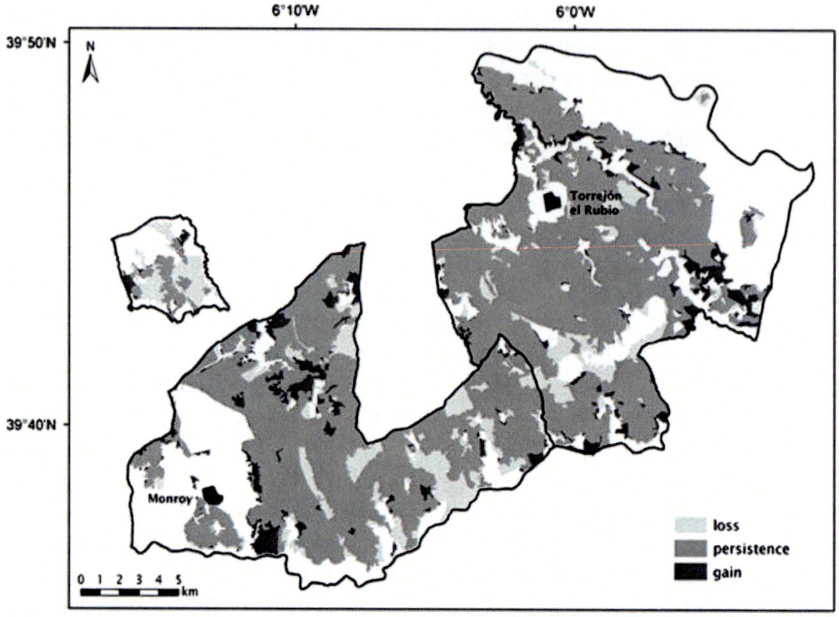

Fig. 5.2 Map illustrating land use changes in the period 1956–1998 in a dehesa area in Extremadura, southwestern Spain. Classification of dehesa cover according to temporal dynamics between 1956 and 1998. (Modified from Plieninger 2006)

Current dehesas are entirely human-made systems where continuous grazing is necessary to prevent shrub encroachment and maintain the savanna or woodland appearance. However, the trees are to a great extent the vestiges of natural forests and shrublands from which dehesas are derived. The transformation of dense forests into open dehesa did not necessarily involve the elimination of mature trees. Rather, adult trees were retained and young ones were thinned to accelerate the transition to adult size. Subsequently, individual trees have been pruned for decades to enhance the production of firewood and acorns (Montero et al. 2003) (Fig. 5.3).

In contrast, California oak woodlands have not been intensively managed. The earliest management activities were conducted by indigenous peoples who occupied California woodlands for at least 10,000 years prior to the arrival of Europeans (Chap. 2). Native Californians did extensively use fire (Blackburn and Anderson 1993), but there is little evidence that they, or the Europeans that followed them, intensely pruned trees as a source of firewood or to enhance acorn production. However, they did burn under trees to facilitate acorn harvesting, prevent shrub encroachment and oak crowding, and to remove acorn pests (Blackburn and Anderson 1993). During the pre-contact era oak trees were plentiful and served as a staple food source for most tribes. In the mid-1800s, with colonization and the suppression of indigenous management, substantial areas of

Fig. 5.3 After forest clearance, retained juvenile oak trees are thinned and pollarded to improve growth and acorn production in Spanish wooded dehesa. (Photograph by F. Pulido)

woodlands were converted to agriculture—especially in valleys where fertile soils supported valley oak (*Quercus lobata*) forests. Oaks were also harvested for charcoal, fuel for home heating, steamships, railroads and mining timbers in the nineteenth century. Throughout the twentieth century woodlands continued to be converted to cropland and, after the Second World War, there were massive efforts to remove oak trees from rangelands to enhance forage production. Between 1945 and 1975, approximately 800,000 ha were cleared, especially in the northern Sacramento Valley (Bolsinger 1988). Approximately 3,000 ha were converted annually for residential and commercial development in the 1970s and 1980s (Bolsinger 1988), and development continues at an even greater pace today. Recently there has also been an increase in the number of hectares of woodlands converted to vineyards, as demand for high quality wines has increased and the price of premium grapes has gone up. Such vines can be grown on hillsides that were previously left to livestock grazing. While it is difficult to accurately know the exact extent of losses, estimates suggest that roughly half of the oak woodlands in California have been lost compared to pre-European settlement levels (Burcham 1981), with species found on deep arable soils, like valley oak, reduced more than those commonly found on rougher terrain, such as blue oak (*Q. douglasii*).

5.2 Impacts of Oak Woodland Losses

The loss of oak woodlands in California and Spain has, no doubt, had many adverse impacts and ongoing losses continue to contribute to these problems. Oak woodlands are one of the most biodiversity-rich habitats and tree removal can adversely affect a wide range of wildlife species causing unprecedented losses in biodiversity (Chap. 8). Oak trees are necessary for specialized tree dwelling species in one or more phases of their life cycles (Aragón et al. 2010). Second, increasing tree density leads to higher density and richness of forest dwelling species (Díaz et al. 2003; Chap. 8). Third, as it has been shown for dehesas, the addition of forest specialists to the community as tree density increases does not imply the loss of species from open grasslands, resulting in a nested pattern of increased diversity (Díaz et al. 2003; Díaz 2009). In California more than half of the 600-plus species of terrestrial vertebrates utilize oak woodlands at some time during the year (CIWTG 2005; Chap. 8). Oaks are critical in protecting watersheds and ensuring the quality of water resources; they anchor the soil, preventing erosion and sedimentation (Chap. 4). Oak woodlands also provide the majority of forage that supports the livestock industry (Chaps. 10 and 13), as well as acorns, an important food source for wildlife (Chaps. 7 and 11). Finally, oak trees are also very desirable locales for recreation, including hunting and fishing, as well as increasingly popular public recreational activities (Chap. 12).

5.3 Poor Oak Recruitment and Regeneration

Calls for adequate tree replacement in dehesas date back to at least the middle of the twentieth century. Two more recent lines of evidence have been crucial to attracting the interest of land managers and policy makers. First, from the figures on dehesa land losses in 1955–1985, it was predicted that, if the rate of decline continued, the oak population of Extremadura would be completely lost in 80 years (Elena-Roselló et al. 1987). Second, recent research has demonstrated an almost complete lack of juvenile age classes in the demographic structure of most dehesa holm oak stands (Pulido et al. 2001; Plieninger et al. 2003; Ramírez and Díaz 2008; MARM 2008). More interestingly, a positive correlation between "dehesa age" (the time elapsed from forest clearance) and current mean age of stands has been observed, in such a way that older dehesas are formed by older trees and show a bell-shaped size structure (Fig. 5.4; Plieninger et al. 2003, 2004a). These results indicate that the lack of regeneration is an inherent problem, beginning from the time the dehesa is created from forest. Though this is the currently prevailing view of dehesa regeneration, some authors have pointed that old dehesas could have been self-regenerating by means of pulses of asexual propagation (Martín and Fernández-Alés 2006), though historic reports on successful long-term regeneration are virtually inexistent.

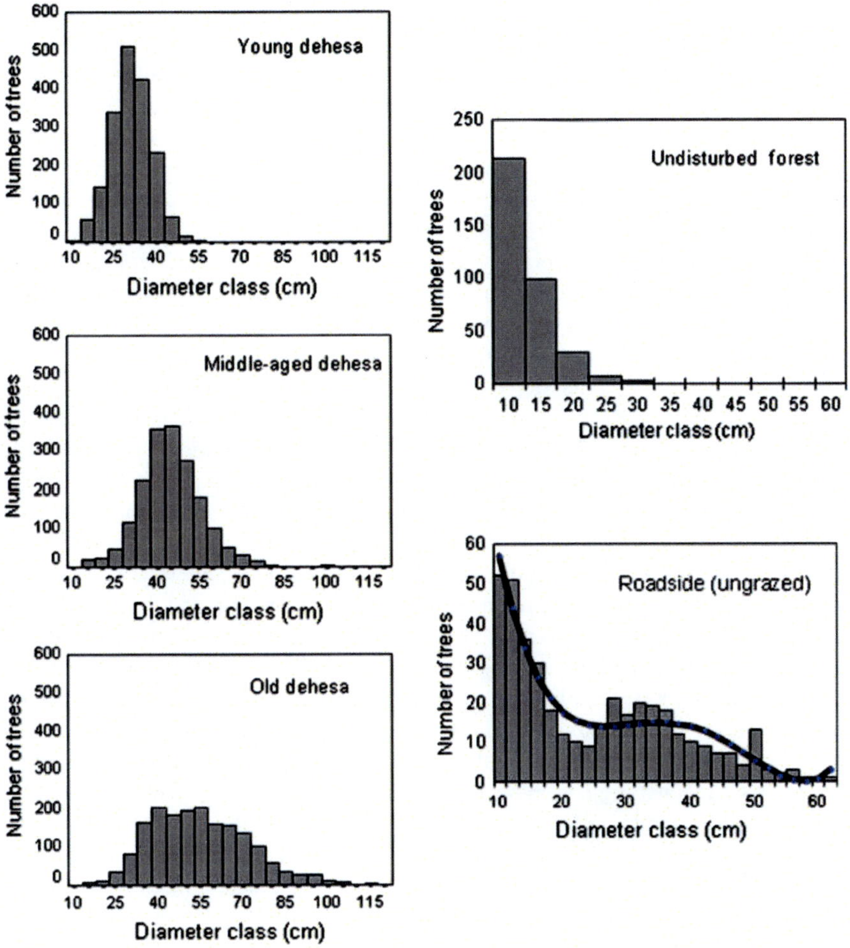

Fig. 5.4 Frequency distribution of tree diameter classes in stands with different management schemes in Spain: Young dehesas (60–100 years old), middle-aged dehesas (150–250), old dehesas (500–700), roadside stands (excluded since ca. 1970), and forest. (Modified from Plieninger et al. 2003)

Until the latter part of the twentieth century in California, there was little interest in conserving oaks, artificially regenerating oaks, or restoring degraded woodlands, in spite of the fact that there had been reports since the beginning of the century that several native oak species were not naturally regenerating adequately (Sudworth 1908). Since native oaks historically had little economic value to European settlers after the Gold Rush and were widely distributed throughout the state, they were often considered "weeds" and there was almost no research on how to propagate them or establish them in the field.

Three of California's native oak species have been reported to have insufficient natural regeneration to replace mortality. These include blue oak, valley oak, and Engelmann oak (*Q. engelmannii*) (Muick and Bartolome 1987; Bolsinger 1988), all deciduous white oaks. The determination that these species have insufficient natural regeneration to replace mortality has relied on inventories of the size-class distribution of oaks. This approach has limitations because the age of oaks is notoriously difficult to determine from size alone since trees of the same age can have vastly different sizes. McClaran (1986) went so far as to state that "tree size was found to be of no use in predicting tree age…" for blue oak (McClaran 1986). And Lawson (1993) pointed out that "highly variable growth rates even within the same site make size a poor correlate of age". In addition, even determining tree age by counting annual rings can be problematic because the core of large, mature trees is often decayed, making an accurate assessment of tree age impossible (Lathrop and Aret 1987).

Finally, fire or browsing can kill the aboveground portion of seedlings and saplings, making a true determination of their age after they sprout back impossible (Lawson 1993). The most commonly used practice to assess the success of regeneration for California oaks has been to identify trees in three general size classes: seedlings (<50 cm tall), saplings (>50 cm tall and <10 cm DBH—diameter at breast height), and mature trees (>50 cm tall and >10 cm DBH, Lawson 1993). While not perfect, this approach does provide a good general description of how a stand is progressing over time in that one can observe if seedlings are becoming saplings, and saplings mature trees. Since saplings are the trees that must be recruited into the mature-size class when the older trees die, insufficient sapling numbers suggest that current population densities will decline. Often, this "regeneration problem" is further exacerbated by land management practices that directly remove trees as well as by activities that make it more difficult for oak seedlings to become established and grow.

However, it is important to note that even for those oak species in California that have been shown to have poor regeneration, spatial patterns can be highly variable, and even over short geographical distances, the success of regeneration can be vastly different. Examples include better regeneration on north slopes vs. south slopes, and in swales (low-lying areas, dips) vs. ridges. Hence, generalizations about regeneration must be made cautiously and underscored by the point that in some locales there is not a "regeneration problem". These patterns, as well as the previously mentioned difficulties in determining the true age of oak trees, led Tyler et al. (2006), after an extensive review of the literature, to conclude that there was not enough information on mortality rates to support the conclusion of a generalized regeneration problem, though others have formed different conclusions (Muick and Bartolome 1987; Bolsinger 1988; Mensing 1991; Swiecki and Bernhardt 1998).

5.4 Dissecting Factors Responsible for Recruitment Failure

Plant regeneration is a dynamic process whereby new individuals are recruited into the adult population, compensating for losses due to natural mortality. For oak trees, this cycle encompasses several transitions between reproductive stages (flowers, seeds, seedlings and saplings) that depend on abiotic and biotic factors. This causes variable losses in reproductive potential (Table 5.1, Fig. 5.5). The probability of transition from any stage to the next in the reproductive cycle of oaks varies according to habitat and management (see review by Tyler et al. 2006).

In dehesa, practices related to tree production such as pruning and cork extraction can seriously affect flowering and fruiting success (Cañellas and Montero 2002; Alejano et al. 2008). At the understorey level, stocking density, shrub control or cereal cropping greatly determine rates of dispersal and the postdispersal fate of seeds, seedlings and saplings (Plieninger et al. 2004a; Pulido and Díaz 2005; Pulido et al. 2010). In a comprehensive analysis of among-habitat differences in holm oak recruitment, a 75-fold decrease in flower-to-sapling recruitment rates between holm oak forests and dehesas was found (Pulido and Díaz 2005). This whole-cycle disparity was the result of differences in the conversion rate from sound acorns to emerged seedlings. The inability to direct acorns to safe (shaded) sites by means of efficient dispersers has been shown to be the main recruitment limitation in dehesa (Cañellas et al. 2002; Pulido and Díaz 2005). Accordingly, various studies from the farm to the geographical scale showed that shrub encroachment generally results in higher oak recruitment rates (Ramírez and Díaz 2008; Smit et al. 2008; Plieninger et al. 2010; Pulido et al. 2010; Chap. 6). Nevertheless seedling recruitment may be hampered beneath certain shrub species. For example, in southern Portugal colonization of degraded cork oak stands by rockroses (*Cistus* spp.) resulted in arrested tree establishment (Acacio et al. 2007). Similarly, the highly water demanding *Cistus ladanifer* inhibits both adult and seedling performance as compared to several leguminous shrubs (Smit et al. 2008; Rolo and Moreno 2011).

Regeneration of oak woodlands in California has been extensively studied in the last two decades and, as a result, many different hypotheses have been proposed to explain recruitment failure. These can be grouped into three categories:

Table 5.1 Strength of different recruitment limitations in oak woodland ranches and dehesas

	California	Spain
Defoliation of flowering shoots	Unknown	Episodic
Fertilization failure	Unknown	Unknown
Acorn production	Probably not limiting	Probably not limiting
Germination rates	Not limiting	Not limiting
Seedling desiccation	Limiting	Severe
Seedling herbivory	Limiting	Not limiting
Sapling herbivory	Limiting	Severe

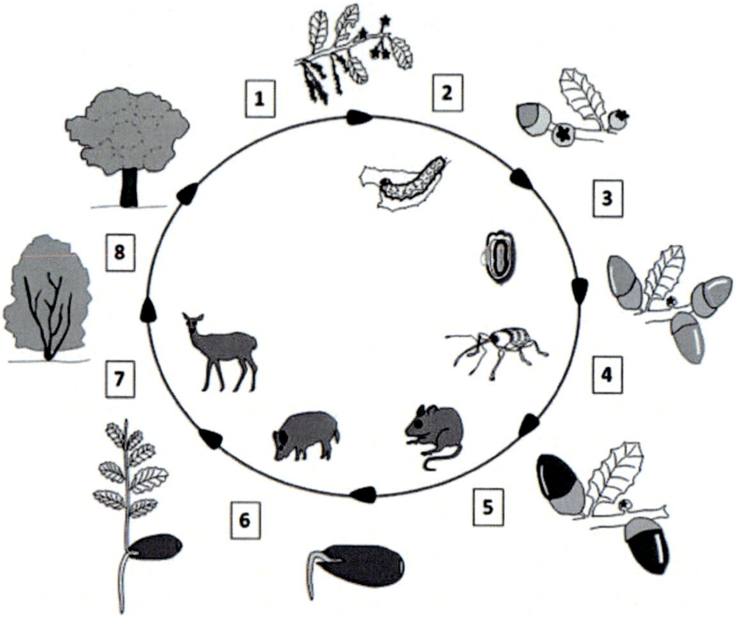

Fig. 5.5 Stages in the life cycle and biotic agents influencing recruitment in an idealized oak species in Spain. Drawings outside the circle represent oak stages, while organisms interacting with oaks are represented inside the circle. Numbers refer to the processes linking different stages: *1* flowering, *2* fertilization, *3* acorn growth, *4* acorn maturation, *5* acorn dispersal and germination, *6* seedling emergence *7* sapling establishment, and *8* growth from juvenile to adult tree. Organisms interacting in these processes are: *2* insect defoliators, *3* bacterial pathogens causing the drippy nut disease, *4* acorn borers, *5* acorn dispersers, *6* acorn consumers *7* browsing mammals (including livestock). (Drawings by F. Pulido; see also Pulido and Díaz 2002; Díaz et al. 2011)

negative direct or indirect effects of introduced Mediterranean annuals, livestock grazing, and microhabitat modification due to fuel buildup after fire suppression. These limitations can be exacerbated in marginal oak populations facing climate warming (Bayer et al. 1999). In addition, some authors suggest that gaps in age structure of oak populations are because recruitment occurs in infrequent pulses. At present, there is not much evidence to support this theory, since studies of blue oak stands demographics (White 1966; McClaran 1986; Mensing 1991; Kertis et al. 1993) tend to indicate that seedling recruitment has occurred over long intervals.

The change from predominantly perennial bunchgrasses to introduced Mediterranean annuals in California has created environmental conditions that hamper natural regeneration (Welker and Menke 1987; Gordon et al. 1989). Mediterranean annuals spread widely in California in the eighteenth and nineteenth centuries with the advent of widespread livestock grazing (Heady 1977). These plants often deplete soil moisture at a more rapid rate than perennials, especially in the early spring when acorns are sending down their roots (Welker and Menke 1990). Another consequence of the change in vegetation from predominantly perennials to annuals, has been an increase in the number and type of seeds produced from these annual plants. It is likely that this change in flora has been accompanied by an increase in certain rodent populations such as voles (*Microtus californicus*) that primarily feed on small seeds but that have been observed to also damage oak seedlings and even saplings up to 2 m tall (McCreary and George 2005).

Livestock grazing is also suspected as a primary cause of poor oak regeneration. Both cattle and sheep eat oak seedlings, as well as acorns and the foliage from tree branches. Browsed seedlings can remain stunted, repeatedly clipped back, for several decades before dying (White 1966; McClaran and Bartolome 1989). Heavy grazing in woodlands—especially over many years—can also indirectly affect oak recruitment by increasing plant density and soil compaction and reducing organic matter, all of which can make it more difficult for oak roots to penetrate downward and obtain moisture (Welker and Menke 1987). On the other hand, Swiecki and Bernhardt (1998) argue that blue oak recruitment is naturally dependent on advanced regeneration—a bank of persistent seedlings beneath the canopy—and will only occur when gaps are created in stands that allow sufficient light to reach the ground. They postulate that under current grazing management, even when gaps are created, there may simply not be enough good quality seedlings in many locations to respond to the new favorable conditions.

Another theory of poor regeneration has to do with fire. Certainly historical fire frequency rates are very different today than they were when Native Californians regularly burned the oak woodlands and there were no efforts to put out naturally-occurring fires (McClaran and Bartolome 1989). It has been suggested that since oaks clearly have evolved with, and are adapted to naturally occurring fire, the change in fire regimes may adversely affect the ability of oaks to successfully recruit. Since post-fire sprout growth can be rapid, fires in the past may have contributed to oak establishment (Plumb and McDonald 1981; McClaran and Bartolome 1989). Also, fuel buildup as a result of fire suppression may have created conditions unfavorable for recruitment (Mensing 1992) (Fig. 5.6).

5.5 Mitigating the Impacts of Poor Regeneration: Available Techniques

For Spanish holm oak and Californian Engelmann oak, a recent simulation of stand dynamics showed that, on average, mature trees could be maintained at current densities in old stands at levels of annual mortality around 0.4 % provided at least 4 new trees are recruited for each existing tree (Gibbons et al. 2008). Under these conditions, the number of mature trees will decline before they increase, even if restoration strategies are implemented immediately. It would be an improvement if such strategies also pursued a reduction of mortality rates and

Fig. 5.6 Naturally regenerating oak woodland in California. (Photograph by M. McClaran)

increased the number of recruited trees currently available (Gibbons et al. 2008). Unfortunately, restoration efforts conducted in dehesas and ranches have not generally relied on accurate estimates of mortality rates or frequency of artificial recruitment needed to reach a given age structure.

In Spain it has been shown that dehesa degradation due to lack of tree replacement can be reversed through livestock exclusion (Plieninger et al. 2003; Table 5.2; Fig. 5.5). In addition, a sigmoid function relating age structure with the time elapsed after grazing exclusion has been established (Ramírez and Díaz 2008). While more efforts should be devoted to investigate local effects on this relationship, it is generally accepted that, in order to ensure tree persistence, dehesa management should include periods of grazing exclusion assigned to different areas of the farm according to a rotational cycle (Montoya 1998; Plieninger et al. 2003; Ramírez and Díaz 2008; Moreno and Pulido 2009). As a low–

Table 5.2 List of restoration techniques and the degree of application in ranches and dehesas

	Ranches	Dehesas
Protection of naturally emerged saplings	Not used	Moderate use
Direct seeding	Rarely used	Rarely used
Planting with shelters and herb control	Moderate use	Widely used
Planting with large wire cages	Rarely used	Rarely used
Fencing of regenerating patches	Rarely used	Not used
Large-scale fencing	Not used	Not used
Directed seeding in safe sites	Rarely used	Rarely used

Fig. 5.7 Holm oak afforestation in marginal agricultural land in Extremadura. (Photograph by F. Pulido)

intervention alternative, direct seeding in safe sites (such as nurse shrub patches) could be prescribed provided the latter were available, which is unlikely due to very high stocking rates in most dehesas. Alternatively, seminatural safe sites can be easily created for direct seeding with piles of thin branches resulting from pruning operations, a novel method that is currently been assayed in some dehesas.

Despite the existence of a variety of methods enhancing natural regeneration, planting of nursery seedlings have been almost the only restoration technique used in the last 20 years (Campos et al. 2003). Through a highly EU-subsidized scheme intended to reduce marginal agricultural lands, afforestation of large areas with low initial tree density have been conducted after livestock exclusion. In these operations first- or second-year seedlings were planted at high densities (ca. 400 per ha) after discing to improve soil moisture and reduce competition. Subsequently, seedlings were protected with plastic shelters to reduce damages caused by small mammals and extreme temperatures. Finally, "densification", that is, the individual protection of existing saplings with wire cages, has also taken place in areas above a certain tree density. This method allows the maintenance of livestock grazing, thus compensating the high cost of large individual shelters (Fig. 5.7).

In California, tree loss—especially from construction activities—usually requires mitigation in accordance with local regulations. Tree removal commonly triggers requirements to plant acorns or seedlings at varying ratios to the number of trees removed depending on the enforcing agency. Practices to ensure that such plantings are successful have been slow to evolve, though it is now generally recognized that oaks can be successfully established if sufficient care and effort are exercised during planting and that the seedlings are maintained adequately (McCreary 2001; Table 5.2). In the past, however, many mitigation plantings were not successful, in large part because of inadequate maintenance, oversight and monitoring.

Determining what the most critical factors affecting the growth and survival of planted oaks are has relied heavily on observations of what limits natural or "volunteer" seedlings. As noted above, competition from dense annual grasses can prevent seedlings from becoming established. However, if this vegetation is removed and the areas are maintained weed-free for 2–3 years, there is often sufficient soil moisture to allow establishment (McCreary 2001; Chap. 6). Vegetation can be controlled using herbicides, physical weed removal, or by applying natural or synthetic mulches. Of course during drought years, or in the driest portions of oak woodlands, even this level of weed control may not be sufficient and supplemental irrigation may be necessary.

Evaluations of poor natural regeneration have shown that there are also a plethora of animals that damage oak seedlings aside from domestic livestock, including wild herbivores such as deer and rabbits, voles (*Microtus californicus*), pocket gophers (*Thomomys* spp.), and even insects such as grasshoppers (*Melanaplus* spp.), that can defoliate entire seedlings. It is essential to protect planted seedlings from these impacts if one expects artificial regeneration efforts to succeed. Several types of cages have been used to protect seedlings. Double-walled plastic tubes commonly called treeshelters have been particularly effective in protecting rangeland oaks from damage from a wide range of animals (McCreary 1997). These devices have the added benefit of stimulating rapid above-ground seedling growth by creating greenhouse-like conditions that promote stem elongation (Potter 1988).

Removing livestock has been suggested as means of enhancing natural oak regeneration in dehesas. However, this approach may not be as effective in California because of its positive impacts on voles, a common oak pest. In two studies in the Sierra Nevada foothills, the sudden exclusion of livestock caused a great increase in ground vegetation, resulting in a build-up of dead plant material known as *thatch* that promoted an increase in vole populations. This caused increased vole damage to seedlings, resulting in even poorer seedling performance than in the grazed pastures (Tecklin and McCreary 1993; Tecklin et al. 1997). A combination of livestock removal and weed control, however, could overcome this obstacle. Recent research suggests that grazing animals should be excluded until saplings are approximately 2 m tall (McCreary and George 2005).

Another promising approach for enhancing oak recruitment is to utilize naturally occurring seedlings and modify the environment where they grow, by way of weed control and treeshelters, such that they more rapidly grow to a height where they are less vulnerable to browsing and weed competition (McCreary et al. 2011). If such an approach is successful, it could greatly reduce the costs of recruiting oaks into the sapling stage, where it is generally believed that they have an excellent chance of becoming mature trees (Fig. 5.8).

Fig. 5.8 Repeated browsing by cattle, deer and small mammals suppressed the growth of this *Quercus douglasii* seedling. The dead stems are evidence of former attempts to grow and the blunt tips on those stems is evidence of browsing. This plant may be 10 years old and stunted by repeated defoliation. (Photograph by M. McClaran taken on University of California Sierra Foothill Research Center)

5.6 Oak Regeneration Efforts to Date

Attitudes concerning the management, use, and conservation of holm oaks in dehesas have changed considerably over time. Until the 1960s most landowners greatly appreciated holm oaks as a basic component of the traditional agroforestry system, supplying valuable products. However, active regeneration measures were unnecessary and rarely applied due to the relative youth of most dehesas.

When traditional agriculture entered into an era of crisis from the 1960s to the 1980s in Spain (Díaz et al. 1997), oak stands were often neglected and even uprooted in order to intensify agriculture (Elena-Roselló et al. 1987). A survey carried out among 59 dehesa farm managers in 2001 showed some common traits (Plieninger et al. 2004b). Most respondents valued the holm oak stands on their property for a number of reasons. Frequently mentioned arguments were to maintain land value, to prevent soil erosion, to create wildlife habitats, and to preserve scenic beauty. The more the managers depended on the dehesa as a source of income, the less appreciation they had for the oaks. Quite surprisingly, managers receiving grants for agri-environmental or afforestation schemes did not show a higher appreciation for oaks than other managers. Obviously, farmers participate in such schemes more for the financial benefit than due to an inner belief in resource conservation. But farm managers using and marketing oak products (e.g. acorns, firewood, charcoal) showed a higher appreciation for oaks. Thus, market incentives seem to be a more effective motive for oak appreciation than public schemes (Plieninger et al. 2004b).

In dehesas, planting of nursery seedlings in treeless areas has been promoted with subsidies since 1994 under EU regulations within the framework of the Common Agricultural Policy (CAP). According to official figures, in the period 1996–2002, when most afforestation took place, over 186,000 ha have been planted, mostly with holm oak and cork oak (MAPA 2004; Ovando et al. 2007). In these plantations landowners were initially committed to excluding livestock for

20 years after planting, with the resulting loss of income compensated for by EU subsidies. The goal of oak planting was the reduction of cropped area to reduce a cereal surplus, with no planned effect on dehesa regeneration and little control over the long-term consequences (Campos et al. 2003). The program did increase oak woodland area and decrease cereal production in the 1990s, but participation has dropped off in the last decade because of changes in how the subsidies are determined and provided.

Unfortunately, data on the long-term success of afforestation plans are very scarce, as no systematic monitoring effort has been conducted by regional agencies. Within the three main regions included in the dehesa in the period 1994–2006, 78 % of the CAP targeted land (125,669 ha) was planted in Andalucía, 67 % in Castilla-La Mancha (75,021 ha), and 70 % in Extremadura (53,855 ha). Within these programmes cork oak was planted mostly in pure stands, but also mixed with several pine species. Holm oak was planted in mono-specific stands or in coexistence with other broadleaved species.

The total acreage planted to oaks in California in the last 20 years has been relatively low—probably less than 5,000 ha. One reason for this is that there are few financial incentives for landowners, unlike Spain, to expand hardwood forests through planting. Much of the planting that has occurred has been the result of requirements for mitigation by local jurisdictions to plant trees to make up for those that were removed, especially accompanying development. These plantings rarely try to replace habitat and often there is inadequate monitoring to ensure that the seedlings survive longer than 3 years. As a result, they were rarely successful in restoring the many ecological values lost when the original trees were removed.

Local governments such as cities and counties are usually responsible for implementing rules or programs addressing oak woodland conservation and these approaches vary widely, depending on local threats to the resource (i.e. firewood harvesting, agricultural conversions, development pressures) and the political climate. In some locales there are ordinances, but these often focus on individual tree removal and rarely address oak woodland habitat conservation. In other places, counties have adopted language in their General Plans that promotes retention of woodlands and the values associated with them. Still other jurisdictions have voluntary oak conservation guidelines that are endorsed by County Boards of Supervisors and promoted by local oak conservation committees. These voluntary approaches seem to work best in rural areas where there are not large financial incentives to convert oak woodlands to other uses (i.e. vineyards or housing).

Probably the largest oak plantings in California have been undertaken by The Nature Conservancy, an international conservation organization that has considerable holdings in the State. There have been restoration plantings of several hundred hectares at reserves along the Sacramento and Cosumnes Rivers in Northern California. These have focused on restoring riparian forests with valley oaks and other hardwoods in areas where oaks historically grew, but were eliminated as a result of agricultural conversions, flood control, and/or fuel-wood harvest. Many of these plantings have been aided greatly by volunteers who are enthusiastic about helping to restore oaks when given the opportunity (Ballard et al. 2001). Some of The Nature Conservancy's plantings are now 20 years old and have produced young riparian forests.

5.7 What We Still Don't Know

Two review papers published on oak regeneration by authors working in savannas (Tyler et al. 2006) and dehesas (Pulido 2002) show that the "regeneration problem" has been dealt with in remarkably similar ways in both countries. The apparent lack of recruitment has been noticed for a long time but specific research and conservation measures have only been implemented in the last 20 years. While researchers and practitioners in California and Spain have come a long way in understanding the factors contributing to poor natural oak regeneration and developing successful approaches for artificial regeneration, there are still gaps in our knowledge that hamper our ability to predict future stand structure under different interventions. These gaps, and the associated management implications, are summarized below (Fig. 5.9).

Fig. 5.9 Rapid growth of *Quercus douglasii* saplings after fencing prevented browsing by cattle and deer. This plant is 2 m tall and is increasing in height each year, and is probably less than 20 years old. (Photograph by M. McClaran taken on University of California Sierra Foothill Research Center)

Assessment of insufficient regeneration. While several indicators of the fecundity component of the life cycle of oaks (such as survival rates of acorns and seedlings) have been intensively studied, the adult mortality component has been rarely quantified in the same stands. This implies that our assessment of regeneration failure is commonly done on a qualitative basis, usually by assuming that the observed birth rate in the population is "insufficient" to offset the current death rate. Qualitative assessment of regeneration is based on the observation that sapling recruitment to the juvenile or adult stage has not taken place over a long time period. While this rough procedure could be valid for most sites, it can lead to inaccurate conclusions if recruitment is highly episodic, as could be the case of oak stands where recruitment occurs in years with higher-than-average precipitation.

Significance of different recruitment limitations. Most studies conducted to date make a priori assumptions on the factors limiting oak recruitment. Thus, research efforts have concentrated on analyzing post-dispersal acorn and seedling survival, two conspicuous bottlenecks in the reproductive cycle. Nevertheless, recruitment could be equally hampered by insufficient seed production, inability to disperse seeds to safe sites, or low sapling survival due to chronic herbivory. These neglected sources of recruitment limitation are discussed below.

A possible role for seed limitation. Though acorn production has been the subject of many studies, the population response of oaks to variable fecundity levels is virtually unknown. While manipulating acorn production at the stand level may be logistically infeasible, critical demographic information could be gathered by measuring seedling recruitment following acorn crops in long-term studies. A positive correlation between acorn production and seedling recruitment would indicate that the limiting role of acorn production outweighs that of other factors.

Acorn dispersal as a critical step. In low density stands, where acorn crops are generally depleted beneath mother trees, most of the recruited seedlings originate from biotically dispersed acorns. Vertebrate dispersal may be a prerequisite for acorn to reach certain safe sites, such as shrub patches providing shade and protection against browsing mammals. Yet, previous studies on acorn dispersal have not addressed the demographic consequences of different spatial patterns of dissemination, though this information could be the basis for novel tools of assisted regeneration.

The importance of microhabitat for seed and seedling survival. It is still a bit of a mystery as to why saplings become successfully established (from planted acorns or seedlings) in some locales but not others. Our inability to predict the success of sapling recruitment under different scenarios of predation risk and abiotic parameters forces us to design homogeneous plantations with most of the acorns or seedlings perishing in inhospitable microhabitats.

Sapling survival and large-scale planting. Despite millions of seedlings having been planted in the last two decades, there is insufficient information on the factors driving the transition from sapling to juvenile tree. In the context of natural regeneration, we know that sapling growth is reduced by chronic herbivore damage and that plants may persist in the stunted form for several decades, but little is known on the response after herbivore exclusion. In artificially regenerated stands interventions to ensure adequate growth, such as weed removal or seedling protection, have been performed in a conservative, labor-intensive way. Better understanding of sapling requirements would also help to solve the dilemma of "many and widespread" versus. "few and aggregated" in large-scale plantations in heterogeneous environments.

Climate change and oak decline. Precipitation has a critical effect on acorn development, so that increasing drought in Mediterranean regions could compromise oak fecundity and regeneration (Pérez-Ramos et al. 2010). In their dendroecological study of Spanish holm oak stands, Gea et al. (2011) concluded that trees at warmer sites showed symptoms of growth decline, most likely explained by the increase in water stress in the last decades. Stands at colder locations did not show any negative growth trend and they may benefit from the current increase in winter temperatures. These results suggest that stands at warmer sites may be more threatened by climate change, as also suggested for oak populations in California (Bayer et al. 1999).

Though gaps in ecological and management issues should be addressed to reduce the uncertainty in the results of technical interventions, we are learning more all the time and we now have a range of tools to help us in our efforts to make sure that oak woodlands in California and Spain are conserved into the future.

Acknowledgments The authors want to thank two anonymous referees for constructive criticisms on an earlier draft.

References

Acacio V, Holmgren M, Jansen PA, Schrotter O (2007) Multiple recruitment limitation causes arrested succession in Mediterranean cork oak systems. Ecosystems 10:1220–1230

Alejano R, Tapias R, Fernández M, Torres E, Alaejos J, Domingo J (2008) The influence of pruning and climatic conditions on acorn production in holm oak (*Quercus ilex* L.) dehesas in SW Spain. Ann For Sci 65:209–215

Aragón G, López R, Martínez I (2010) Effects of Mediterranean dehesa management on epiphytic lichens. Sci Total Environ 409:116–122

Ballard H, Kraetch R, Huntsinger L (2001) How collaboration can improve a monitoring program. In: Standiford RB, McCreary D, Purcell KL (eds) Proceedings of the fifth symposium on oak woodlands: oaks in California's changing landscape. San Diego, pp 617–624

Bayer R, Schrom D, Schwan J (1999) Global climate change and California oaks. In: McCReary DD (ed) Proceedings of the second conference of the international oak society. San Marino, pp 154–165

Blackburn TC, Anderson K (1993) Introduction: managing the domesticated environment. In: Blackburn TC, Anderson K (eds) Before the wilderness: environmental management by native Californians. Ballena Press, Menlo Park, pp 15–24

Blondel J, Aronson J, Bodiou J, Boeuf G (2010) The Mediterranean region: biological diversity in space and time. Oxford University Press, UK

Bolsinger CL (1988) The hardwoods of California's timberlands, woodlands and savannas. USDA forest service Pacific Northwest research station resource bull. PNW–148, p 148

Burcham LT (1981) California range land: a historic-ecological study of the range resources of California. University of California, Davis

Campos P, Martín D, Montero G (2003) Economías de la reforestación del alcornoque y de la regeneración natural del alcornocal. In: Pulido F, Campos P, Montero G (eds) La gestión forestal de las dehesas. ICMC, Junta de Extremadura, pp 107–151

Cañellas I, Montero G (2002) The influence of pruning on the yield of cork oak dehesa woodland in Extremadura (Spain). Ann For Sci 59:753–760

Cañellas I, Pardos M, Montero G (2002) El efecto de la sombra en la regeneración natural del alcornoque (*Quercus suber* L.). Cuad Soc Esp Cienc For 15:107–112

CIWTG (California interagency wildlife task group) (2005) California wildlife habitat relationships (CWHR) system version 8.1, personal computer program. California department of fish and game, Sacramento. http://www.dfg.ca.gov/biogeodata/cwhr/. Accessed 5 Aug 2012

Díaz M (2009) Biodiversity in the dehesa. In: Mosquera MR, Rigueiro A (eds) Agroforestry systems as a technique for sustainable land management programme Azahar. AECID, Madrid, pp 209–225

Díaz M, Campos P, Pulido F (1997) The spanish dehesas: a diversity in land-use and wildlife. In: Pain DJ, Pienkowski MW (eds) Farming and birds in Europe. Academic Press, London, pp 178–209

Díaz M, Pulido F, Marañón T (2003) Diversidad biológica y sostenibilidad ecológica y económica de los sistemas adehesados. Ecosistemas 3, URL: www.aeet.org/ecosistemas/033/investigacion.htm

Díaz M, Alonso CL, Beamonte E, Fernández M, Smit C (2011) Desarrollo de un protocolo de seguimiento a largo plazo de los organismos clave para el funcionamiento de los bosques mediterráneos. In: Ramírez L, Asensio B (eds) Proyectos de investigación en parques nacionales: 2007–2010. Organismo Autónomo Parques Nacionales, Madrid, pp 47–75

Elena-Roselló M, López JA, Casas M, Sánchez del Corral A (1987) El carbón de encina y la dehesa. Madrid: Ministerio de Agricultura, Pesca y Alimentación. Instituto Nacional de Investigaciones Agrarias. Colección Monografías INIA, no 60, Madrid, p 113

Ezquerra FJ, Gil L (2009) La transformación histórica del paisaje forestal en Extremadura. Tercer Inventario Forestal Nacional, Ministerio de Medio Ambiente, Madrid

García del Barrio JM, Bolaños F, Ortega M, Elena-Roselló R (2004) Dynamics of land use and land cover change in dehesa Landscapes of the "Redpares" network between 1956 and 1998. Adv Geoecol 37:47–54

Gea G, Cherubino P, Cañellas I (2011) Tree-rings reflect the impact of climate change on *Quercus ilex* L. along a temperature gradient in Spain over the last 100 years. For Ecol Manage 262:1807–1816

Gibbons P, Lindenmayer D, Fischer J, Manning AD, Weinberg A, Seddon J, Ryan P, Barrett G (2008) The future of scattered trees in agricultural landscapes. Conserv Biol 22:1309–1319

Gordon DR, Welker JM, Menke JM, Rice KJ (1989) Competition for soil water between annual plants and blue oak (*Quercus douglasii*) seedlings. Oecologia 79:533–541

Heady HF (1977) Valley grassland. In: Barbourand MG, Major J (eds) Terrestrial vegetation of California. John Wiley and Sons, New York, pp 383–415

Kertis JA, Gross R, Peterson DL, Standiford RB, McCreary DD (1993) Growth trends of blue oak (*Quercus douglasii*) in California. Can J For Res 23:1720–1724

Lathrop EW, Arct MJ (1987) Age structure of engelmann oak populations on the Santa Rosa plateau. In: Proceedings of the Symposium on Multiple-use Management of California's Hardwood Resources. Berkeley, USA

Lawson DM (1993) The effects of fire on stand structure of mixed Quercus agrifolia and Q. engelmannii woodlands. MS Thesis, San Diego State University, p 122

Linares AM, Zapata S (2003) Una visión panorámica de ocho siglos. In: Pulido F, Campos P, Montero G (eds) La Gestión Forestal de las Dehesas. Instituto del Corcho, la Madera y el Carbón, Mérida, pp 13–25

López-Sáez JA, López P, López L, Cerrillo E, González A, Prada A (2008) Origen prehistórico de la dehesa en Extremadura: una perspectiva paleoambiental. Rev Estud Extr 63:493–503
MAPA (2004) Anuario de Estadística Agroalimentaria. MAPA, Madrid
MARM (2008) Diagnóstico de las Dehesa Ibéricas Mediterráneas. Tomo I. Available online. http://www.magrama.gob.es/es/biodiversidad/temas/montes-y-politica-forestal/anexo_3_4_coruche_2010_tcm7-23749.pdf
Martín A, Fernández-Alés R (2006) Long term persistence of dehesas. Evidences from history. Agrofor Syst 67:19–28
McClaran MP (1986) Age structure of Quercus douglasii in relation to livestock grazing and fire. Ph.D. Dissertation, University of California, Berkeley, p 119
McClaran MP, Bartolome JW (1989) Fire-related recruitment in stagnant *Quercus douglasii* populations. Can J For Res 19:580–585
McCreary DD (1997) Treeshelters: an alternative for oak regeneration. Fremontia 25:26–30
McCreary DD (2001) Regenerating rangeland oaks in California, University of California agriculture and natural resources publication 21601, p 62
McCreary D, George M (2005) Managed grazing and seedling shelters enhance grazing on grazed rangelands. Calif Agric 59:217–222
McCreary DD, Tietje WD, Davy J, Larsen R, Doran M, Flavell D, Garcia S (2011) Tree shelters and weed control enhance growth and survival of natural blue oak seedlings. Calif Agric 65:192–196
Mensing S (1991) The effect of land use changes on blue oak regeneration and recruitment. In: Standiford R (ed) Proceedings, symposium on oak woodlands and hardwood rangeland management. USDA forest service pacific southwest forest and range experiment station Gen. Tech. Rep. PSW–126, pp 230–232
Mensing S (1992) The impact of European settlement on blue oak (*Quercus douglasii*) regeneration and recruitment in the Tehachapi Mountains, California. Madroño 39:36–46
Montero G, Martín D, Cañellas I, Campos P (2003) Selvicultura y producción del alcornocal. In: Pulido F, Campos P, Montero G (eds) La gestión forestal de las dehesas, Instituto del Corcho, la Madera y el Carbón, Mérida, pp 63–106
Montoya JM (1998) Método de ordenación silvopastoral. In: Hernandez CG (ed) La dehesa. Aprovechamiento sostenible de los recursos naturales, Edit. Agricola Española S.A. Madrid
Moreno G, Pulido F (2009) Dehesa functioning, management, and persistence. In: Rigueiro A, Mosquera MR, McAdam J (eds) Agroforestry in Europe. Springer, Berlin, pp 127–160
Muick PC, Bartolome JW (1987) Factors associated with oak regeneration in California. In: Plumb TR, Pillsbury NH (eds) Proceedings of the symposium on multiple-use management of California's hardwood resources. USDA forest service pacific southwest forest and range experiment stat Gen Tech Rep PSW–100, pp 86–91
Ovando P, Campos P, Montero G (2007) Forestaciones con encina y alcornoque en el área de la dehesa en el marco del Reglamento (CE) 2080/92 (1993–2000). Revista Española de Estudios Agrosociales y Pesqueros 214:173–186
Pérez-Ramos IM, Ourcival JM, Limousin JM, Rambal S (2010) Mast seeding under increasing drought: results from a long-term data set and from a rainfall exclusion experiment. Ecology 91:3057–3068
Plieninger T (2006) Habitat loss, fragmentation, and alteration—quantifying regional landscape transformation in Spanish holm oak savannas (dehesas) by use of aerial photography and GIS. Landscape Ecol 21:91–105
Plieninger T, Pulido F, Konold W (2003) Effects of land use history on size structure of holm oak stands in Spanish dehesas: implications for conservation and restoration. Environ Conserv 30:61–70
Plieninger T, Pulido F, Schaich H (2004a) Effects of land-use and landscape structure on holm oak recruitment and regeneration at farm level in *Quercus ilex* L. dehesas. J Arid Environ 57:345–364

Plieninger T, Modolell J, Konold W (2004b) Land manager attitudes toward management, regeneration, and conservation of Spanish holm oak savannas (dehesas). Lands Urb Plan 66:185–198

Plieninger T, Rolo V, Moreno G (2010) Large-scale patterns of *Quercus ilex, Quercus suber, and Quercus pyrenaica* regeneration in central-western Spain. Ecosystems 13:644–660

Plumb TR, McDonald PM (1981) Oak management in California. General technical report. PSW–54. Pacific Sourchwest forest and Rangelan research station, p 12

Potter MJ (1988) Treeshelters improve survival and increase early growth rates. J For 86:39–41

Pulido F (2002) Biología reproductiva y conservación: el caso de la regeneración de los bosques templados y subtropicales de robles (*Quercus* spp.). Rev Chil Hist Nat 75:5–15

Pulido F (2008) Les processus de création des dehesas: brève synthèse historique. In: Systèmes agroforestiers comme technique pour la gestion durable du territoire. Programa Azahar, AECI, Lugo, pp 129–137

Pulido F, Díaz M (2002) Dinámica de la regeneración natural del arbolado de encina y alcornoque. Pulido F, Campos P, Montero G (eds) La gestión forestal de las dehesas. IPROCOR, Mérida, pp 39–62

Pulido F, Díaz M (2005) Recruitment of a Mediterranean oak: a whole-cycle approach. Ecoscience 12:99–112

Pulido F, Díaz M, Hidalgo S (2001) Size structure and regeneration of Spanish holm oak *Quercus ilex f*orests and dehesas: effects of agroforestry use on their long-term sustainability. For Ecol Manage 146:1–13

Pulido F, García E, Obrador JJ, Moreno G (2010) Multiple pathways for regeneration in anthropogenic savannas: incorporating abiotic and biotic drivers into management schemes. J Appl Ecol 47:1272–1281

Ramírez JA, Díaz M (2008) The role of temporal shrub encroachment for the maintenance of Spanish holm oak *Quercus ilex* dehesas. For Ecol Manage 255:1976–1983

Rolo V, Moreno G (2011) Shrub species affect distinctively the functioning of scattered *Quercus ilex* trees in Mediterranean open woodlands. For Ecol Manage 261:1750–1759

Smit C, den Ouden J, Díaz M (2008) Facilitation of holm oak recruitment by shrubs in Mediterranean open woodlands. J Veg Sci 19:193–200

Stevenson AC, Harrison RJ (1992) Ancient forest in Spain: a model for land-use and dry forest management in south-west Spain from 4000 bc to 1900 ad. Proc Prehist Soc 58:227–247

Sudworth GB (1908) Forest trees of the pacific slope. USDA forest service, p 441

Swiecki TJ, Bernhardt E (1998) Understanding blue oak regeneration. Fremontia 26:19–26

Tecklin J, McCreary DD (1993) Dense vegetation may encourage vole damage in young oak plantings. Restor Manage Notes 11:153

Tecklin J, Connor JM, McCreary DD (1997) Rehabilitation of a blue oak restoration project. In: Pillsbury NH, Verner J, Tietje WJ (eds) Proceedings of a symposium on oak woodlands: ecology, management and urban interface issues. USDA forest service pacific southwest research stat Gen Tech Rep PSW–160, pp 267–273

Tyler C, Kuhn B, Davis FW (2006) Demography and recruitment limitations of three oak species in California. Q Rev Biol 81:127–152

Welker JM, Menke JM (1987) *Quercus douglasii* seedling water relations in mesic and grazing-induced xeric environments. In: Proceedings, international conference on measurements of soil and plant water status, vol 2—Plants. 6–10 July 1987. Logan, pp 229–234

Welker JM, Menke JW (1990) The influence of simulated browsing on tissue water relations, growth and survival of *Quercus douglasii* (Hook and Arn.) seedling under slow and rapid rates of soil drought. Funct Ecol 4:807–817

White KL (1966) Structure and composition of foothill woodland in central-coastal California. Ecology 47:229–237

Chapter 6
Overstory–Understory Relationships

Gerardo Moreno, James W. Bartolome, Guillermo Gea-Izquierdo and Isabel Cañellas

Frontispiece Chapter 6. Holm oak dehesa managed for a grassland understory using periodic cultivation. Some shrubs have begun to invade. (Photograph by G. Moreno)

G. Moreno (✉)
Grupo de Investigación Forestal, Universidad de Extremadura, Plasencia 10600, Spain
e-mail: gmoreno@unex.es

J. W. Bartolome
Department of Environmental Science, Policy, and Management, University of California, Berkeley, CA 94720, USA
e-mail: jwbart@berkeley.edu

P. Campos et al. (eds.), *Mediterranean Oak Woodland Working Landscapes*, Landscape Series 16, DOI: 10.1007/978-94-007-6707-2_6,
© Springer Science+Business Media Dordrecht 2013

Abstract A key issue for sustainable management of oak woodlands is understanding the complex overstory-understory relationships that influence ecosystem productivity and stability. Oak removal is traditionally practiced in Californian ranches and Spanish oak dehesas to increase forage for grazing, but the response of the understory, and subsequently of the trees, is not fully understood. Existing knowledge of the effects of trees on understory forage production and the effects of the understory on tree production and recruitment is reviewed to synthesize from what is known and to identify knowledge gaps. Emphasizing the few published manipulative experiments to clarify the importance of facilitation and competition, plant to plant interactions are analyzed to examine three aspects of the relationship between trees and the understory: understory production, tree growth and production, and tree regeneration. First, we find that understory production is related to canopy-caused gradients of aboveground and belowground resources such as light, nutrients, and water. Second, the consequences of tree density and understory structure on oaks are analyzed, including competitive use of belowground resources. Third, the importance of the understory for oak seedling survival is discussed for its effect on the stability and sustainability of Spanish and Californian oak woodlands. While dehesa shrub encroachment is certainly favorable for oak seedling regeneration, it does not maintain longer-term stand functions and profitability from livestock, wildlife, and cork production. We conclude by proposing a future research agenda for the study of plant-to-plant relationships.

Keywords Canopy-caused gradients · Competition · Facilitation · Tree production · Seedling recruitment · Shrub encroachment

6.1 Introduction

Californian oak woodlands and Spanish dehesas are formed of evergreen and deciduous oaks within a grassland matrix dominated by annual grasses and forbs, where livestock production is integrated with oak and, on occasion, grain crop production (Huntsinger and Bartolome 1992; Campos et al. 2007; Marañón et al. 2009). In both Spain and California, oak woodland soils used for grazing tend to be shallow and infertile, unsuitable for intensive crop production. A mix of differing understory species and tree canopy densities provides a high degree of landscape structural diversity.

On the Iberian Peninsula, this diversity has been fostered by centuries, even millennia, of human influence and a combination of agricultural, pastoral, and forestry

G. Gea-Izquierdo · I. Cañellas
Forest Research Centre, National Institute for Agriculture and Food Research and Technology, Ctra A Coruña km 7.5 28040 Madrid, Spain
e-mail: gea.guillermo@inia.es

I. Cañellas
e-mail: canellas@inia.es

uses, where different vegetation structures depend on land use (Joffre et al. 1999; Marañón et al. 2009). Californian oak woodlands reflect a history of several thousand years of human influence, mostly through the use of fire as an element of management by indigenous peoples (Bartolome 1989), but with the addition since 1769 of European land use practices, including grazing, and wide-scale cultivation (Chap. 2). In addition, the California woodland has been hugely influenced and continues to be shaped by the introduction of non-native species that began with European–American colonization. In both systems, human impacts likely resulted in more open tree canopies and more of an herb-dominated understory (Marañón et al. 2009).

Currently, the Californian oak savanna type is considered a stable community that, in the absence of human intervention, changes slowly or not at all (Huntsinger et al. 1991) while Spanish dehesa, sometimes considered a natural part of the landscape in Southwestern Iberian Peninsula, has an unstable understory assemblage carefully maintained by land managers (Marañon 1988). Indeed, without direct human intervention, dehesas are rapidly invaded by aggressive shrubs (Campos et al. 2007) while a shrub understory is not common in Californian oak savanna (Table 6.1).

Table 6.1 Common vegetation in Californian oak woodland and Spanish dehesa

	California oak woodland	Spanish dehesa
Common oaks		
Evergreen	Coast live oak (*Quercus agrifolia*); interior live oak (*Q. wislizenii*)	Holm oak (*Q. ilex*) and cork oak (*Q. suber*) are the most common oaks in dehesa
Deciduous	Blue oak (*Q. douglasii*) and valley oak (*Q. lobata*). Blue oak is most widespread	Pyrenean oak (*Q. pyrenaica*) at higher elevations
Semi-deciduous	Englemann oak (*Q. engelmannii*)	Lusitanian oak (*Q. faginea*) and Algerian oak (*Q. canariensis*)
Soils	Varied volcanic, metamorphic, and sedimentary origins	Developed over acid slates and granites. Low contents of organic matter, mineral N and available P
Understory	Native perennial bunchgrasses invaded (emigrated from other Mediterranean regions) by non-natives such as annual grasses *Avena*, *Bromus*, and *Festuca* spp., and other herbaceous species	Annual grasses and other herbaceous species, subject to rapid invasion by rockrose (*Cistus* spp.) and leguminous brooms (*Retama*, *Genista* and *Cytisus* sp.), and less commonly by gorse (*Ulex* spp.) and heather (*Erica* spp.)
Tree management	Relictual, heterogenous spacing, extensive	Deliberate spacing, pruning, intensive

Trees intercept solar radiation and rainfall and usually compete more efficiently for belowground water and nutrients than understory plants. As a result, in most agroforestry systems the net effect of trees on herbaceous plant productivity is negative (Jose et al. 2004). In the mid-twentieth century, oak removal was promoted to increase forage production on oak woodlands in California (George 1987), and in the latter half of the twentieth century to facilitate mechanization in intercropped Spanish dehesas. Millions of oaks were lost (Elena et al. 1987; Fernández-Alés et al. 1992).

However, tree cover has been maintained because of the multiple positive benefits from oaks. For instance, in Spain and Portugal, acorns are of high value for feeding pigs, oak leaves provide a forage reserve during the dry season and during droughts, and trees protect livestock from extreme weather conditions. In California, acorns and leaves support wildlife, the trees shelter livestock, wood can be sold for firewood or chips, and most landowners prefer the look of a woodland to an undifferentiated open plain. Moreover, in both Spain and California the net effect of tree overstory on pasture understory varies depending on the site (Moreno 2008; Marañón et al. 2009) and is often neutral or positive, resulting in interest in replanting of oaks in formerly cleared Californian woodlands (Alagona 2008; UC-OWCW 2012). The relationship between oaks and forage production varies with abiotic factors and with the size and age of trees, and changes as the trees grow. For example, shrubs may enhance oak seedling recruitment, but later negatively affect tree growth and productivity. These temporal changes should be taken into account when defining structural goals for management practices for each specific site.

Reviewed here is how tree, shrub, and herb interactions affect ecosystem productivity and stability in Californian oak woodland and Spanish dehesa, in order to pull out integrative and comparative conclusions that can contribute to future management decisions. There is a lack of literature in some areas, especially for Californian oak woodlands, and suggestions for future research are included in the conclusions. An extensive recent review of Mediterranean-type savanna systems by Marañón et al. (2009) provides an excellent summary of species-environment interactions in the understory, with numerous examples from both Spain and California, so the precise details of understory species composition is not addressed comprehensively here (Fig. 6.1).

6.2 Tree Effects on Understory Production

Deciduous and evergreen oaks affect the production, species composition, chemical quality and phenology of the understory in Iberian dehesas (González Bernáldez et al. 1969; Alonso et al. 1979; Puerto et al. 1987; Calabuig and Gómez 1992; Moreno 2008; Gea-Izquierdo et al. 2009; Fernández-Moya et al. 2011; Rivest et al. 2011a) and Californian oak woodlands (Parker and Muller 1982; McClaran and Bartolome 1989; Marañón and Bartolome 1993; Callaway and Davis 1998).

Fig. 6.1 Matched photos taken in an un-grazed coast live oak woodland in 1982 and 1992 at Mt. Diablo State Park near San Francisco. Only the graduate student presence has changed. (Photographs by L. Huntsinger and J. Bartolome)

Fig. 6.2 The multiple colors of the grassy understory of a *blue* oak woodland near Hopland, California, reflect the interaction of soil, water, light, and species composition. (Photograph by L. Huntsinger)

This is a common feature of oak woodland and savanna communities worldwide (Rice and Nagy 2000; Marañón et al. 2009) and effects on understory can be explained by the spatial heterogeneity of resources created by the presence of scattered trees in these systems. Here, the canopy-caused resource gradients and the consequences for understory productivity and quality are analyzed (Fig. 6.2).

6.2.1 Canopy-Caused Resource Gradients

Evidence for the effects of trees on the spatial heterogeneity of light availability, microclimate affects, soil moisture, and nutrient distribution comes from comparing areas beneath and outside of the tree canopy (Dahlgren et al. 1997; Young 1997). Isolated oaks strongly reduce light availability for the plants beneath them. Montero et al. (2008) reported a 75 % reduction in light close to the trunks of evergreen holm oaks in Spanish dehesa. Light availability increased rapidly with distance from the trunk, with 70 % of the full sunshine reaching plants at the edge of the canopy, and 100 % out beyond about four times the canopy radius (Fig. 6.3a). As a consequence of tree shade and interception of long-wave radiation at night, daily and seasonal variations of temperature are buffered under the canopy (Moreno et al. 2007a). Researchers in California found similar results, with reductions in radiant energy under the evergreen coast live oak and deciduous blue oak ranging from 25 to 90 % (Jackson et al. 1990; Callaway et al. 1991). Marañón and Bartolome (1994) reported that light levels under coast live oaks were only 2 % of that in the open in mid-summer (Fig. 6.3b).

Oaks are long-lived trees, frequently more than 100 years old, and often over 300 years of age (McClaran and Bartolome 1989; Plieninger et al. 2003). Over an

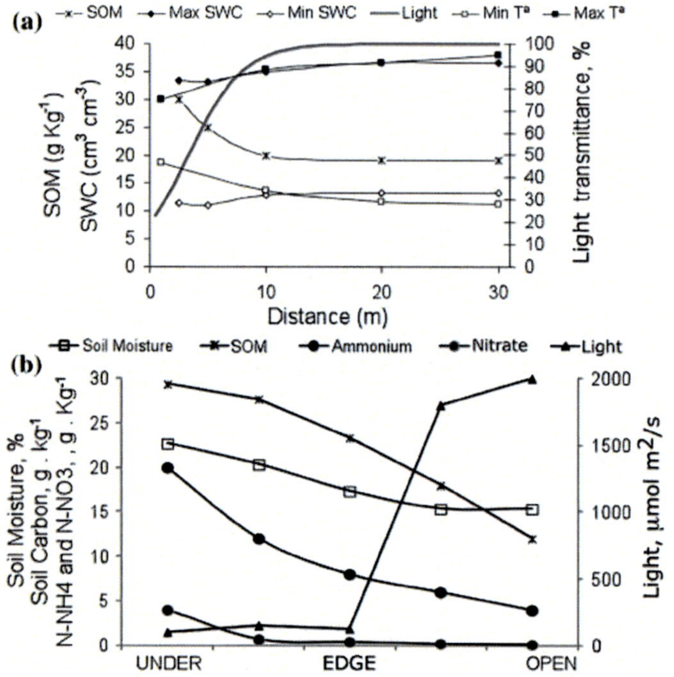

Fig. 6.3 Oak effects on resources. **a** Distribution of resources under and around isolated holm oaks (distances refer to from tree trunks), adapted from Moreno et al. (2007a). Soil organic matter (SOM; 0–30 cm depth); Maximum and minimum soil water content (SWC; measured over 3 years at 0–100 cm depth); Light (Percentage of light transmitted measured by fish eye photograph method); Min Tª and Max Tª (Mean values of daily minimum temperature measured in coldest month and mean values of daily maximum temperatures measured in hottest month, July). **b** Distribution of resources under and around coast live oaks, adapted from Marañón and Bartolome (1994) and Dahlgren et al. (2003). Light was measured at noon July 30; soil moisture was average from autumn to spring; organic carbon (0–15 cm depth); soil nitrogen (0–30 cm depth) was measured in March as ammonia and nitrate

extended period, trees significantly affect the fertility of the soil, mostly by recycling leaf litter and by the turnover of nutrients that are pumped through the root systems from deep in the soil and out beyond the canopy. In addition, trees are effective at retaining atmospheric solutes due to their high surface area and aerodynamic resistance (Moreno and Gallardo 2003), and throughfall and stemflow may contribute to soil nutrient inputs (McPherson 1997; Dahlgren et al. 1997, 2003). Moreover, trees reduce possible losses of nutrients by erosion and leaching (Young 1997). As a result, nutrients show higher values beneath oaks than in adjacent open areas (Dahlgren et al. 1997). Soil nutrient content generally decreases rapidly with distance and the influence of the trees disappears only a few meters beyond the canopy projection. In addition, part of the nutrient accumulation in the sub-canopy soil could occur at the expense of the adjacent area (McPherson 1997) given that animals tend to concentrate below the tree canopies and the wide lateral root system of trees in dehesas (Moreno et al. 2005) can bring nutrients from the interstitial area.

The positive effect of trees on soil fertility has been quantified for many dehesas (e.g., Vacher 1984; Puerto and Rico 1988; Escudero 1992; Gallardo et al. 2000; Gallardo 2003; Moreno et al. 2007b; Gea-Izquierdo et al. 2010) and Californian oak woodlands (Parker and Muller 1982; Marañón and Bartolome 1994; Dahlgren et al. 2003). The nutrient content in these savannoid soils depends largely on the build-up of soil organic matter (SOM) near the trees (Chap. 4). Although soil organic matter values are highly variable among and within sites, available data suggest that in general soil organic matter is higher in Californian oak woodlands, where carbon contents above 20 g/kg in the open and above 40 g/kg beneath oaks are frequent (Dahlgren et al. 2003). In Spanish dehesa values below 10 g/kg in the open and 20 g/kg beneath the canopy are frequent (Moreno et al. 2007b; Fernández-Moya et al. 2011). A common pattern is for soil organic matter to measure up to two times higher beneath the canopy (Fig. 6.3a). Nutrients determined by biological mechanisms, such as available nitrogen, reflect spatial distribution of soil organic matter. The same is true for other nutrients; but phosphorus, which is mostly determined by geochemical mechanisms, shows a highly variable spatial pattern more closely linked to physical variations in soils and parent material (Gallardo 2003).

Oaks significantly modify soil physical properties beneath the canopy in Spanish dehesas and Californian oak woodlands, increasing soil water-holding capacity, macroporosity and infiltration rates compared to open areas (Joffre and Rambal 1988; Puerto and Rico 1989; Frost and Edinger 1991). These changes are mostly explained by the increase in soil organic matter and the decreased bulk density near the trees (Cubera and Moreno 2007a). Changes in physical properties explain much of the observed increases in soil water content (SWC) under tree cover found by Puerto and Rico (1989) and Joffre and Rambal (1993) in subhumid (about 700 mm of annual rainfall) holm oak dehesa.

For California, Parker and Muller (1982), Marañón and Bartolome (1994), and Moody and Jones (2000) all found that in open coast live oak woodlands soil water content was lower and decreased more rapidly (Fig. 6.3b) outside the canopy, although the situation was reversed during extended droughts. Baldochi et al. (2004)

reported a positive effect of deciduous blue oaks on soil moisture, although other authors have reported no significant effect of this oak species on soil moisture in the rooting zone of annual understory plants during the time of year when these plants were phenologically most active (Jackson et al. 1990; Callaway et al. 1991).

In contrast, Cubera and Moreno (2007a), Gea-Izquierdo et al. (2009), and Moreno and Rolo (2011) found decreased soil water content near dehesa evergreen oaks, especially on the driest sites and/or during the driest years, similar what has been found for many other agroforestry systems (Young 1997; Jose et al. 2000). This phenomenon is attributed to decreased water input because of interception, and an increase in water loss through transpiration under the canopy, which could outweigh the positive effects of trees on water-holding capacity (Cubera and Moreno 2007a). Evergreen oaks intercept rainfall, in one holm oak example 30 % of rainfall (Mateos and Schnabel 2002), and absorb water from the soil continuously throughout the year with moderately high transpiration rates in winter and summer (Infante et al. 2003; David et al. 2004). The reasons for differences among sites, especially in Californian savanna and Spanish dehesa, are not yet clear. Jackson et al. (1990) reported much higher values of pasture root biomass outside the canopy in a California pasture than for an understory pasture from October to April, hampering the soil recharge in open areas because more water is transpired in that period. Only in May was root biomass higher beneath the canopy. A higher abundance of graminoid species in Californian savannas could partially explain the differences. Annuals tend to concentrate root growth and soil–water utilization in the upper soil profile, while the native perennial bunchgrasses of Californian savannas allocate a high proportion of their biomass to the development of a deep root system, allowing them to continue soil–water utilization well into the dry season and to contribute to the formation of a very dry soil profile (Holmes and Rice 1996). While soil water recharge is limited beneath trees in dehesas, in Californian oak savannas perennial grasses may limit this recharge.

6.2.2 Understory Composition, Nutrient Quality and Phenology

Savannas worldwide have proved similar in the way the tree canopy affects understory species composition, nutrient quality and phenology (Fig. 6.4). In dehesa, grasses are dominant beneath the canopy, while legumes and forbs become more abundant in the less fertile interspaces (Marañón 1986; Puerto 1992). This difference may be explained by the increased content of soil nitrogen and the nitrogen mineralization rate beneath oak canopy (Gallardo et al. 2000), which favors grasses as they need more soil nitrogen to thrive, while legumes and forbs are less dependent on soil nitrogen (Joffre 1990). The higher resistance of grasses to shading compared to legumes might explain this pattern (Nunes et al. 2005). Marañón and Bartolome (1993) demonstrated the importance of shade to the spatial location of species in a Californian example. They switched around intact

Fig. 6.4 Species composition, duration of *green* growth, and production may all differ under the oak canopy as compared to outside the canopy, as in these examples from California's central Sierra foothills in early summer (**a**) and from North Extremadura in Spain in mid-winter (**b**). (Photographs by L. Huntsinger and D.S. Howlett, respectively)

blocks of soil from under the canopy, at the edge, and in the open in a *Quercus agrifolia* savanna and found that shading caused high mortality of the herbaceous species that came from open areas (Fig. 6.5). Many other studies document changes in species composition for herbaceous communities depending on whether or not they are under the oak canopy in dehesa and woodland (Marañon et al. 2009). In regions with usually high plant α diversity or species diversity in habitats, the presence of a high number of species influenced by a tree canopy gradient of light, nutrients, and soil structure results in very high habitat diversity (β diversity or diversity of habitats) and total species diversity in a landscape (γ diversity or total species diversity; Chap. 8).

The herbaceous understory has a higher content of some nutrients (mainly N and K) in plants beneath than outside the canopy (González-Bernáldez et al. 1969; Puerto 1992; Moreno et al. 2007a, b). Herbaceous plants uptake nutrients located in the uppermost soil layer more easily than oaks, as Rivest et al. (2011b) demonstrated through experimental fertilizations in dehesa on different soil types. This helps explain why the chemical qualities of the understory reflect the heterogenous patterns of soil fertility around trees (Moreno et al. 2007b). However, the

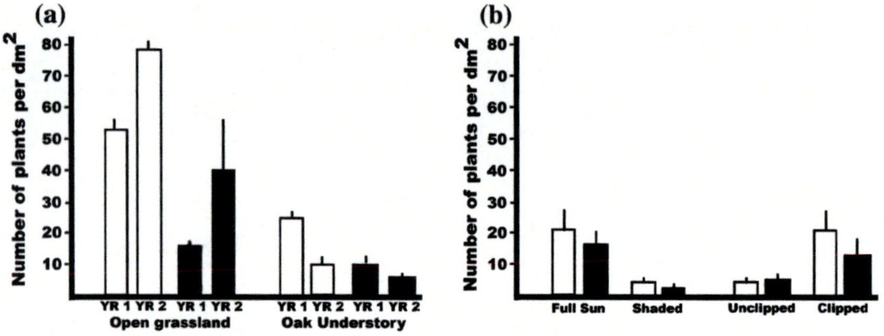

Fig. 6.5 Differences in density of understory plants in open grassland and under coast live oak canopy over 2 years. **a** Blocks of soil were transplanted from understory to open grassland (*shaded bars*) and vice versa (*unshaded bars*) and **b** subjected to *clipping* and *shading*. Adapted from Marañón and Bartolome (1993)

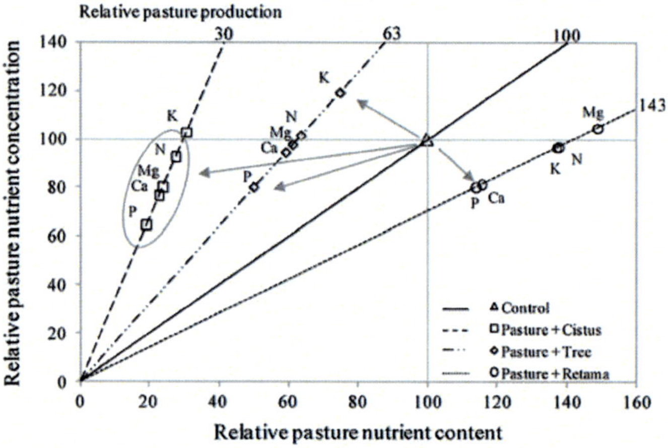

Fig. 6.6 This vector analysis of relative response in pasture production shows nutrient concentration and content (*N*, *P*, *K*, *Mg* and *Ca*) of vegetation under holm oak or shrubs (broom—*Retama*—or rockrose—*Cistus*–). Data for pasture growing in open areas was used as a reference (100 value for pasture production, nutrient content and concentration). Diagonal lines indicate pasture production (g m^2). *Arrows* and *circles* depict significant vector shifts. From Rolo et al. (2012)

understory responds to increased nutrient availability mostly though increased growth and changes in botanical composition and not less so to increases in plant nutrient concentrations (Gea-Izquierdo et al. 2010; Rolo et al. 2012; Fig. 6.6).

A longer growing season beneath the tree canopy, with an earlier start in winter and later drying in summer, is reported (Alonso et al. 1979; Puerto et al. 1987, 1990; Calabuig y Gómez 1992) (Fig. 6.4 and 6.7). Warmer temperatures beneath canopy would allow continued understory growth in winter compared to in open pasture

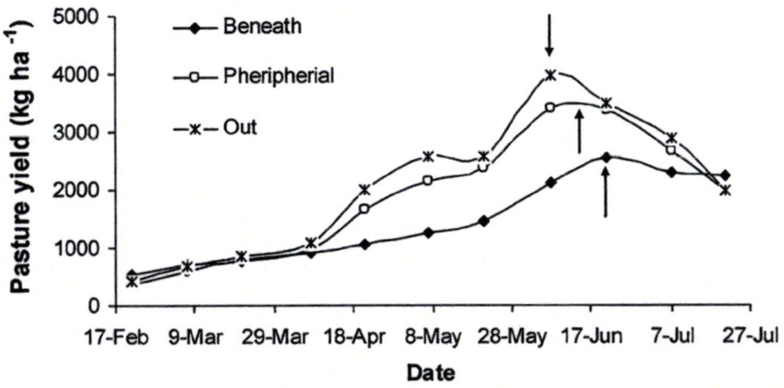

Fig. 6.7 Temporal evolution of forage yield at three distances from holm oak trees. Note the decrease of pasture yield beneath the canopy, and the temporal difference for the maximum yield. Adapted from Puerto (1992)

(Moreno et al. 2007a). Dominant grasses beneath a dehesa canopy dry out later in summer than forbs and legumes that are dominant outside of the canopy because grasses are capable of using water from deeper soil layers (Joffre et al. 1987).

6.2.3 Understory Production

The net effect of trees on understory production depends on the balance of positive, or facilitative effects and negative, or competitive effects (Marañón et al. 2009). Studies reveal that the effect of trees on the understory in open oak woodlands is highly variable, ranging from decreased to increased production (see examples in Callaway et al. 1991; Puerto 1992; Allen-Diaz et al. 1999). The direction and magnitude of these effects depends on environmental factors like precipitation, soil type and fertility as well as biological factors like the species in the understory and the kind of oaks, amount of canopy cover, tree age and the root architecture of the interacting plants in the community (Quilchano et al. 2007; Tyler et al. 2007). For instance, Frost and McDougald (1989) reported that herbaceous production was up to 115–200 % greater under scattered blue oak than on open grassland in California (Battles et al. 2008). Like this deciduous oak, evergreen oaks can increase pasture yield beneath their canopies, as reported for holm oak in Spain (Puerto 1992).

The positive response of understory production to moderate tree cover is generally attributed to, as reported above, more favorable physical and chemical soil properties and soil and air temperatures under the tree (Moreno et al. 2007a). In a manipulative experiment conducted in three dehesas, Moreno (2008) found that pasture yield was higher beneath the canopy. But in fertilized and watered plots pasture yield was significantly higher under artificial shade (50 % full-sunlight) than under the canopy (Fig. 6.8), showing that shade, despite the negative influence of reducing light for photosynthesis, probably played a greater positive

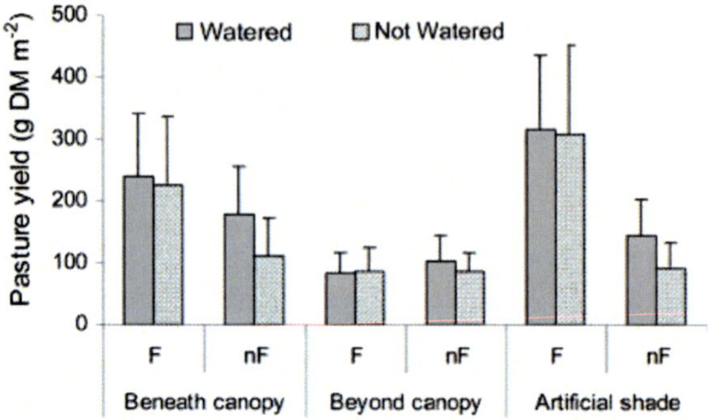

Fig. 6.8 Effects of artificial and natural shade, fertilization and watering on pasture yield (g m^{-2} ± S.D.) in Iberian dehesas. The treatments were: artificial shade (50 % of full sunlight), fertilization (200 g m^{-2} of NPK 15/15/15 in February), and watering (90 l m^{-2}, applied as 10 l m^{-2} every week from 1st April–31st May). Adapted from (Moreno 2008)

role by reducing damage to photosynthetic apparatus from too much light (photoinhibition, Valladares and Pugnaire 1999). In California, Frost and McDougald (1989) concluded that increased forage production under the canopy during drought conditions was, in large part, due to shading, which reduced moisture loss via evapotranspiration. Indeed, it has been pointed out that in a Mediterranean climate, maximum production of dehesa understory is obtained with around 30 % of overstory cover (Etienne 2005), and Allen-Diaz et al. (1999) reported that in Californian savannas evergreen oaks only inhibited production when canopies exceeded 25 %, whereas deciduous oaks did not consistently inhibit understory production until cover exceeded 60 %.

Although a sparse canopy can produce more understory growth, trees do intercept a certain proportion of solar radiation that could be used for photosynthesis (PAR; photosynthetically active radiation) and take up water and nutrients, making them unavailable for understory plants. As a consequence, many cases of significant reduction of pasture yield beneath oak canopy compared to open pasture have been reported, especially with evergreen oaks (Marañón and Bartolome 1993) for coast live oak; Puerto 1992; Nunes et al. 2005; Rivest et al. 2011a for holm oak), and under deciduous oaks with canopy cover of about 40–50 % (McClaran and Bartolome 1989; Battles et al. 2008).

These studies confirm that trees compete for resources with the understory. In the three dehesa experiments conducted by Moreno (2008), when the main nutrient (N, P, K) limitations were removed through fertilization, artificial shade produced a higher understory yield than tree shade, suggesting that negative effects, such as competition for soil water, limited production under the canopy (Fig. 6.8 and 6.9). By contrast, soil moisture does not seem to have a major role in net balance of effects of Californian oaks on pasture understory production (Callaway et al. 1991). The higher proportion of dehesa studies reporting a negative effect of oaks on understory production is understandable because soil moisture frequently is

Fig. 6.9 Livestock redistribute nutrients on rangelands. For example, they like to spend time in the shade, as shown by these Merino sheep and Spanish horses under cork and holm oaks in dehesas of Extremadura, Spain. They enrich the soils under trees with their manure. (Photograph by A. Hummer (**a**) and G. Moreno (**b**))

higher beneath oak canopy there than in the open in California, while it is often higher in open pasture than under the canopy in dehesa (Fig. 6.3).

Standiford and Howitt (1993) noted the contrasting effects of tree canopy on understory production in areas of higher and lower rainfall in California. McClaran and Bartolome (1989) found a rainfall-dependent tree/understory relationship, showing that canopy depresses forage yield more in higher rainfall areas (Fig. 6.10) and that tree facilitation, or benefits to understory production, increased with aridity and plant water stress. This fits the stress-gradient hypothesis, which posits that interactions among plants are context dependent, shifting from competition to facilitation as environmental stress decreases (Bertnes and Callaway 1994). Forage production was higher under the trees than in open areas where annual rainfall was below 500 L m^{-2} while the reverse is found in areas receiving more than 500 L m^{-2}.

By contrast, the stress gradient hypothesis has not been confirmed for dehesa. In fact, Moreno's (2008) experiment indicated the opposite. Understory yield beneath the canopy was higher than in the adjacent open grassland, but differences decreased with the aridity of the sites, with increases of 16.8, 34.0, and 33.4 % beneath canopy compared to open pasture in dehesas with annual rainfall around 450, 550 and 650 L m^{-2}, respectively. Similarly, Gea-Izquierdo et al. (2009) reported a positive effect of oak canopy on dehesa pasture yield in average climatic years, but the interaction changed with increasing abiotic water stress. In a dry year, the higher fertility beneath the canopy could not be used for plant growth because of the lack of moisture and the effect of the oak canopy was neutral. The decreased positive effect of trees with aridity in Spanish dehesas indicates that competition for soil water is an outstanding factor in the balance of positive and negative effects of trees on pasture. This kind of exception to the soil gradient hypothesis is common in Mediterranean ecosystems (Maestre et al. 2006, 2009), especially when the abiotic stress gradient is driven by a resource such as soil water in arid and semiarid ecosystems. The reasons for the differential behavior of

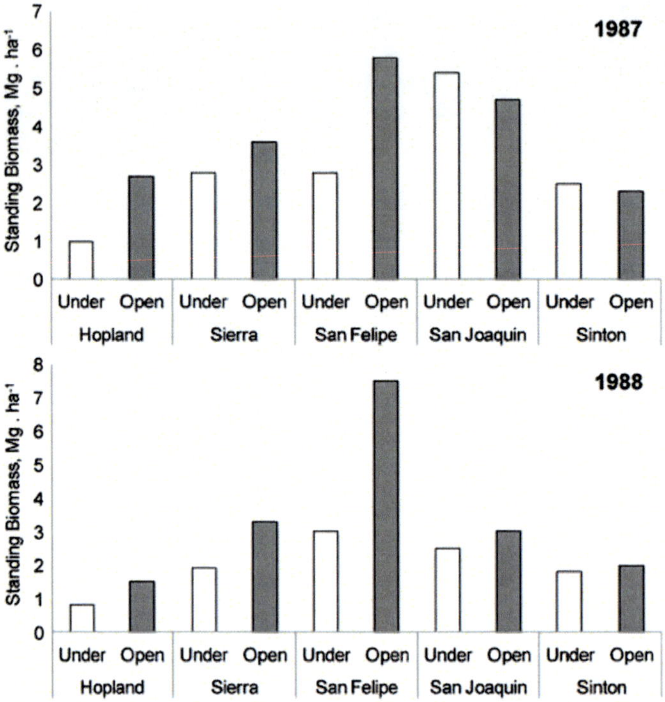

Fig. 6.10 Mean understory and open grassland peak aboveground biomass in 1986 and 1987 at five locations in Californian blue oak savanna. Annual rainfall in 1986 and 1987 was above and below average, respectively. The x-axis sites represent an annual rainfall gradient from 100 cm/year (Hopland) to 25 cm/year (Sinton). Adapted from McClaran and Bartolome (1989)

Californian oak woodlands and Spanish dehesas with respect to the stress gradient hypothesis need to be explored with consideration of the differences in pasture species life-forms and soil fertility and depth between the two systems.

6.3 Interactions Affecting Trees

6.3.1 Tree-Tree Competition: The Importance of Widely-Spaced Trees

Precipitation in Mediterranean systems is highly variable from year to year, and there is a summer drought of varying severity each year. The low tree density of oak woodlands and dehesas allows trees to survive and continue to produce even in severe drought conditions. Wider spacing between trees implies greater water availability for each tree, resulting in a reduction of the duration and intensity of tree water stress compared to trees growing in more closed forests of the same regions. Numerous authors report higher water potential and photosynthetic and transpiration rates at leaf and tree scales during the summer for holm and cork oaks in the dehesa, as compared to closed stands (Joffre and Rambal 1993; Infante et al. 2003; David et al. 2004; Moreno and Cubera 2008).

When trees are spaced further apart their roots can exploit a larger soil area and obtain more water and nutrients, explaining the improved physiological status of dehesa oaks. Oak tree roots expand outwards up to 7 times the projection of the canopy and as much as 25 times the canopy volume into the soil, allowing trees to meet their water needs during the dry Mediterranean summers (Moreno et al. 2005). In general, larger lateral root spread has been found in plants and trees growing at low densities in dry environments (Eastham et al. 1990; Schenk and Jackson 2002). Cubera and Moreno (2007a) showed during the summer, when herbaceous dehesa plants are dry and senescent, and unable to use water, soil water content continues decreasing as far as 20 m beyond the tree trunk and 200–300 cm in depth, indicating that holm oak trees were consuming the water accumulated there from winter rains. Similarly, for Californian savannas Baldochi and Xu (2007) conclude that Mediterranean oaks must meet their limited water supply by, among other mechanisms, constraining the leaf area index of the landscape by establishing a canopy with widely spaced trees. A strong relationship among tree density, water availability, and tree productivity is a common feature of semiarid savannas (e.g., Smith 1986). However, Battles et al. (2008) found that in a Californian oak woodland [blue oak (*Quercus douglasii*), interior live oak (*Quercus wislizenii*) and foothill pine (*Pinus sabiniana*)], with overall mean annual rainfall of 775 mm, tree productivity increased linearly with oak cover, while the total productivity (trees and understory) increased linearly with increasing canopy cover until it leveled off at approximately 55 % cover.

The spacing of trees is more critical in the driest open woodlands. Moreno and Cubera (2008) reported that in dry dehesas (annual rainfall <500 L m^{-2}), both predawn and midday water potentials, CO_2 accumulation, and sap flow density proportional to transpiration rates were significantly higher in trees growing in low tree density areas ($\sim$20 trees ha^{-1}) compared to those in high tree density areas ($\sim$100 trees ha^{-1}). By contrast, in humid dehesas (annual rainfall >700 L m^{-2}), differences in both water potentials and CO_2 accumulation among tree densities were very small and emerged only at the end of the dry season (Fig. 6.11). Indeed, Joffre et al. (1999) reported for Spanish dehesas that mean oak density increases with rainfall at a large geographical scale. This pattern seems a common feature for stable savannas as revealed by Sankaran et al. (2005) for African savannas.

Joffre and colleagues pointed out in 1999 that the dehesa structure follows an *ecohydrological equilibrium*, explained in the work of Eagleson and Segarra (1985) who hypothesized that water availability limits natural vegetation systems, resulting in a canopy density that produces both minimum water stress and maximum biomass. Natural savannas were defined as a biotic response to alternating wet and dry seasons, because the density of trees and grasses is controlled by the amount of soil water available during the vegetative season (Eagleson and Segarra 1985).

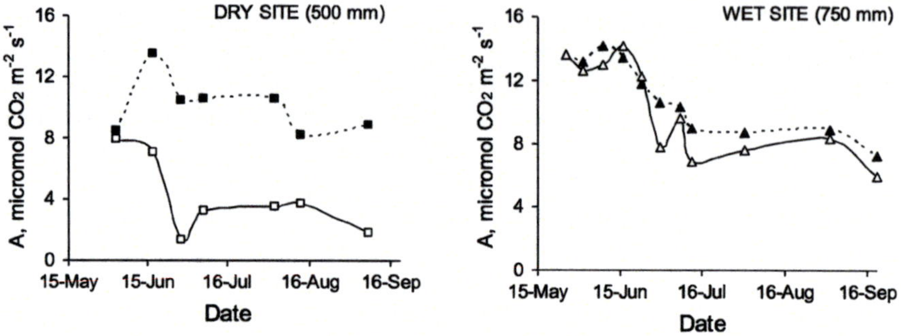

Fig. 6.11 Mean values for CO_2 accumulation rates in mature holm oak growing in dehesa with a canopy cover below 20 % (*black square* or *triangle*) and dense coppice with canopy cover above 90 % (*open square* or *triangle*). Adapted from Moreno and Cubera (2008)

Maximum woody cover in savannas with mean annual rainfall of less than ~650 mm is constrained by, and increases linearly with, mean annual rainfall. These arid and semi-arid savannas may be considered "stable" systems in which water constrains woody cover and permits grasses to coexist. Above an annual rainfall average of ~650 mm, savannas are "unstable" systems in which rainfall is sufficient for woody canopy closure, and disturbance such as fire and herbivory is required for the coexistence of trees and grass (Sankaran et al. 2005). While an ecohydrological equilibrium might play a role in the vegetative stability in most Californian open oak woodland (Marañón et al. 2009), it does not stabilize oak woodland dehesas, where if human intervention and grazing pressure is excluded, an immediate woody encroachment starts (Marañón et al. 2009; Rolo and Moreno 2011). The ecohydrological equilibrium noted in dehesa by Joffre et al. (1999) is an ecoystem "mimic", and reflects a management target of maximizing forage as well as tree production in the unnatural dehesa. More research is needed in this area to understand the diverse interactions of trees, shrubs, and water availability in California and Spain (Figs. 6.12, 6.13).

Apart from the direct positive effect of low tree density on tree water status, Úbeda et al. (2004) reported a clear benefit of forest clearance on the leaf nutrient content in cork oak. As a result of the improved hydric and nutritional status of trees in dehesas the production of acorns was 10 times higher in a managed holm oak dehesa compared to a dense holm oak forest (Pulido and Díaz 2005; Chap. 7).

6.3.2 *Trees and Shrubs: Competitive Use of Soil Resources in the Dehesa*

Very few studies concerning the effects of the shrub layer on trees are available in the dehesa, and there are even fewer for the California oak woodland, although shrubby infilling occur in both systems (Allen-Diaz et al. 2007; Plieninger et al. 2010).

Fig. 6.12 In California, the distribution of trees and shrubs in this blue and coast live oak woodland near San Francisco is largely a function of soil type, water, and past management and fire regimes. The flat area in the foreground was probably cultivated sometime in the twentieth century. The crisp lines between shrubs and grasslands reveal the underlying geology rather than a property or management border, and the trees are denser in the swales where soils are moister, as well as on the northern aspects of slopes (looking north). California oak woodlands occur on highly heterogenous soils. (Photograph by L. Huntsinger)

This section refers only to data compiled from studies conducted in Spanish and Portuguese dehesas.

It is to be expected that many shrub species compete with oaks for belowground resources, given that shrubs and oaks have similar root system structures (Canadell et al. 1996; Schenk and Jackson 2002; Rolo and Moreno 2012). The scanty information on the root systems of shrubs growing in dehesas has shown both shallow and deep root profiles, with roots reaching several meters in depth for heather, gorse, and broom (*Retama*), and less than 1 m for rockrose (Silva et al. 2002; Silva and Rego 2003; Rolo and Moreno 2012).

Regardless of the specific root profile, a shrub understory can decrease soil moisture, water potential, CO_2 assimilation rate, tree growth and the acorn production of evergreen oaks (Moreno et al. 2007a; Cubera and Moreno 2007b; Moreno and Rolo 2011; Rolo and Moreno 2011; Rivest et al. 2011a). Shrub encroachment influences the nutritional status of the trees with lower N, Ca and Mg leaf contents in trees growing with shrubs than in trees growing with native grasses (Moreno and Obrador, 2007; Rolo et al. 2012). By contrast, trees showed significantly higher values of foliar P in trees with shrubs than in trees with grasses. The increase of foliar P in trees with dehesa encroachment could indicate some level of positive interaction (facilitation) between trees and shrubs. Although beneficial interactions between woody plants are widely recognized (Scholes and

Fig. 6.13 In Spain, the distribution of trees and shrubs in this holm and cork oak woodland in Monfragüe National Park is largely a function of topography, land use history and fire. The flat area in the middle was cultivated for decades in the twentieth century, resulting in a very low tree density. Further up, a typical dehesa landscape resulted from deliberate tree clearing and continuous grazing. Below, in steeper areas, trees have almost been eliminated by periodic fires, and shrubs now dominate. (Photograph by G. Moreno)

Archer 1997; Barnes and Archer 1999; Marañón et al. 2009), possible facilitation in dehesas remains unexplained and unconfirmed, and requires further research given that P is commonly seen as one of the most limiting factors for many Mediterranean ecosystems (Vallejo et al. 2006).

Overall, studies indicate that shrubs have more negative (competition for water) than neutral (complementary use of water), or positive (facilitation for some nutrients) effects on tree growth and acorn production. Annual shoot elongation and acorn production per tree were significantly higher in dehesas with a pasture understory than in those with a shrub understory (Moreno et al. 2007a; Rivest et al. 2011a). Such belowground competition is known in other tree-shrub ecosystems (subtropical savanna parklands; Barnes and Archer 1999).

Nevertheless, leaf water potential of holm oak remains relatively high (> −1 MPa at dawn) even in shrub encroached dehesa plots during the summer (Cubera and Moreno, 2007b; Rolo and Moreno 2011) (Fig. 6.14), while holm oak leaf water potential in dense forests usually reaches under −4 MPa in semiarid regions (Moreno and Cubera 2008). In sum, the relative competitive effects are the following: dense forests >shrub-invaded plots >savannas. The maintenance of a relatively favorable water status for oaks in shrub encroached stands is essential for satisfactory acorn development in summer (Carevic et al. 2010). The limited negative effect of shrub understory is explained by the plasticity of the holm oak

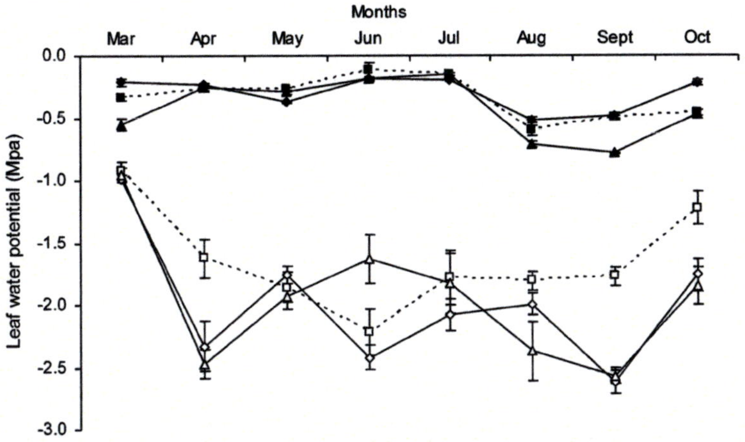

Fig. 6.14 Daily maximum (*filled symbol*) and minimum (*open symbol*) values of leaf water potential of mature holm oak measured through the summer in 2003. The understory consisted of cereal crop (*circles*), native grasses (*squares*) and shrubs (*triangles*). Adapted from Cubera and Moreno (2007b)

root system, allowing it to escape competition for soil water—holm oak has a deeper rooting system when growing with a shrub understory than when growing with a pasture understory (Rolo and Moreno 2012).

6.3.3 Tree and Pasture: Tree Dependence on Deep Soil Water

Spatial separation between herbaceous plants and tree root systems has been described in Californian blue oak savannas (Callaway et al. 1991) and in dehesas (Joffre et al. 1987; Gómez-Gutiérrez et al. 1989; Moreno et al. 2005). The latter authors found that roots of native grasses were located mostly in the upper 30 cm, and root length density (RLD) decreased exponentially with depth up to 70 cm (Fig. 6.15). In the same plots, holm oak had a lower root density in the first 10 cm of the soil, and oak root density remained almost uniform with depth at a given distance from the tree.

The limited vertical overlap of herb and oak root profiles suggests that competitive effects of understory herbs are unimportant for tree water uptake in dehesa. Several studies focusing on water dynamics in different agroforestry systems have shown different spatial partitioning of water resources between trees and the understory. Cubera and Moreno (2007a) reported spatial separation between herbaceous plants and trees in relation to soil water uptake. Soil dried uniformly beneath and outside the canopy only for the uppermost 50 cm of the soil, while at deeper layers soil water content increased with the distance from the tree trunk,

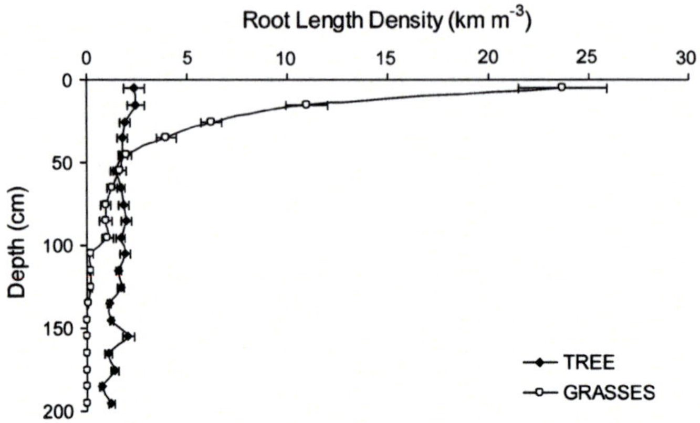

Fig. 6.15 Rooting profiles of trees and native grasses in holm oak dehesa. Adapted from Moreno et al. (2005)

indicating that herbaceous plants did not use water below 50 cm depth, as is consistent with their root system. Joffre et al. (1987) reported similar values, with annual and perennial grasses absorbing water from the uppermost 40 and 60 cm of the soil, respectively.

By contrast, during summer drought holm oak trees show a high dependence on water below 3 m depth (Cubera and Moreno 2007a). The low dependence of trees on water in the uppermost soil layer was shown in an experimental irrigation trial, where holm oak did not respond to irrigation in terms of fecundity, acorn production or shoot elongation (García-López 2005). Thus, while water limitation is an important feature in most dehesas, water consumed by grasses (and cereal crops) probably does not cause significant water stress to mature dehesa trees if tree roots can reach deep soil layers (Cubera and Moreno 2007a). Baldocchi et al. (2004) have demonstrated how oaks in California depend on the absorption of deep water, below the maximum rooting depth for understory grasses (Fig. 6.16).

6.4 Interactions Affecting Tree Seedling Performance and Recruitment

Seedling establishment and juvenile growth often limit recruitment of oaks, and morphological and physiological attributes during these periods are regarded as key factors for the recruitment and survival of tree populations. Adaptability to water stress at the seedling stage can be important because of high mortality and the need to survive in competition with understory grasses, shrubs and mature overstory trees for belowground resources (Mediavilla and Escudero 2004). Under the competitive conditions that the seedlings grow in, a non-conservative use of resources favors the quick root growth needed for establishment (Mediavilla and

Escudero 2003). The set of adaptive traits of oak seedlings determines differential recruitment success across the characteristic habitat and microhabitat heterogeneity of Iberian dehesas and Californian oak woodlands (Huntsinger et al. 2004).

6.4.1 Pasture and Oak Seedling Relationships

The intensity of competition for water and nutrients between understory plants and oak seedlings determines the survival of seedlings and thus oak recruitment. The negative effect of understory herbaceous plants on oak recruitment is well documented in Californian oak woodlands (Gordon and Rice 1993; McCreary 2001), but studies for dehesas are scant. Navarro-Cerrillo et al. (2005) found that holm oak survival increased 2.5 fold in the first summer after the transplanting of 1-year-old seedlings when the herbaceous plants were mechanically or chemically surpressed (Fig. 6.16 and 6.17). Enhancement of oak seedling survival and growth by the removal of competing understory grasses and forbs has proved successful with both natural and artificial regeneration in Californian oak woodlands and savannas (McCreary et al. 2011).

Pulido and Díaz (2005) have reported that most of the seedlings in dehesas (86 %) die during the first summer of life from desiccation due to lack of water. This pattern of seedling mortality has been repeatedly observed in Mediterranean

Fig. 6.16 Removing herbaceous vegetation from around an oak seedling can increase survival during the first year, as in the California example. This effect has been found in California and Spain

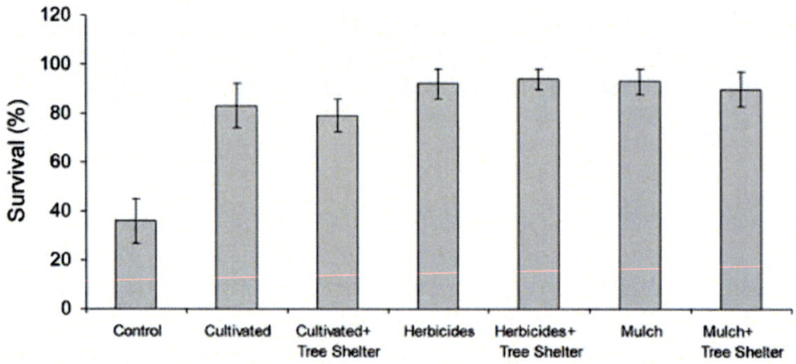

Fig. 6.17 Holm oak seedling survival with different methods of weed control. In the control treatment, where seedlings grew with competing weeds, there was significantly higher seedling mortality. Adapted from Navarro-Cerrillo et al. (2005)

climate regions, including California (Allen-Diaz and Bartolome 1992). In contrast, Pulido and Díaz (2005) found mortality in the second summer was negligible, indicating that presumably seedlings were using water too deep to be consumed by herbaceous plants. Hence, it seems that a noticeable proportion of the seedlings in dehesa can survive competition with native pasture plants, although perhaps not browsing by livestock and wildlife (Pulido et al. 2010). Survival may be lower when the herb layer is improved through seed sowing and/or fertilization (Olea and San Miguel 2006), affording the pasture understory a competitive advantage (Cubera et al. 2012).

The suppression of oak seedling growth by herbaceous species competition seems more acute in Californian oak woodlands. Collet et al. (2006) grew downy oak seedlings (*Q. pubescens*) for 4 years in bare soil or with grass competition and found that root system size was considerably reduced by grass competition. Gordon et al. (1989), examining the competitive effects of two annual species on Californian blue oak seedling growth in 1 m deep boxes, found that the type and density of grasses significantly effected seedling emergence and growth rate.

A high density of annual plants suppressed oak root growth and shoot emergence. Fibrous grass roots had a greater competitive effect than did tap-rooted forbs. Only 20 % of the acorns planted in high density smooth bromegrass (*Bromus hordeaceous)* neighborhoods showed aboveground shoot growth, 56 % of those planted in low density smooth bromegrass or the annual forb broad leaf filaree (*Erodium botrys*) emerged, while 19 % emerged in the control box with no forbs or grasses. The results suggested that competition for soil water with introduced annual species contributes to the high rate of blue oak seedling mortality observed in Californian woodland systems. Although the genetically determined strategy of oak seedlings to rapidly develop a deep taproot helps them escape the strong competition for soil water from herbaceous plants (Mediavilla and Escudero 2003), when seedlings grow in a dense herbaceous understory, the effectiveness of this strategy is limited, and the vertical root growth of the seedlings can be suppressed by neighboring pasture species (Gordon and Rice 1993).

6.4.2 Shrub and Oak Seedling Relationships

A shrubby understory is found in both dehesas and Californian oak woodlands (Marañón et al. 2009; Allen Diaz et al. 2007), but is much more common in Iberia. The extensive work on shrub-oak seedling relationships in dehesa contrasts with a meager published record in California. The higher interest shown by dehesa researchers is probably due to the worrying lack of oak regeneration that affects most managed and grazed dehesas, and the fact that without intensive management, most Spanish dehesa converts to a shrubland (Plieninger et al. 2010). Shrubs can use more resources than grasses, and what is more important, use both nutrients from the uppermost soil layer and water from deeper soil layers because of the dual (horizontal and vertical) rooting systems of many Mediterranean shrub species (Rolo and Moreno 2012). Indeed, as commented earlier shrub-encroached dehesas in general show lower soil resources than savanoid dehesas.

Although shrubs seem to use soil resources more exhaustively than herbaceous plants, shrub encroachment predictably results in a significant increase in oak seedling recruitment in dehesas (Pulido and Díaz 2005; Smit et al. 2008; Plieninger et al. 2010; Pulido et al. 2010) (Fig. 6.18). This happens for several reasons, including the attraction of acorn dispersers such as rodents and jays to shrubby areas, protection against browsing herbivores, improvement of soil fertility, and softening of the harmful effects of the Mediterranean summer drought (Retana et al. 1999; Gómez 2003; Gómez-Aparicio et al. 2004, 2005a, b; Puerta-Piñero et al. 2006; Plieninger et al. 2010), although acorn predation under some shrub species can limit regeneration (Callaway and Pugnaire 2007). Indeed, dispersers as rodents are acorn consumers that act as accidental dispersers (Muñoz and Bonal 2011). Pulido et al. (2010) conducted a controlled experiment where acorns were seeded into an enclosure to prevent acorn depredation, showing a direct positive effect of shrubs on acorn germination and holm oak seedling survival in Central-Spain dehesas (Fig. 6.19). This facilitative phenomenon, called the "nurse plant effect", is widespread among Mediterranean oak species seedlings (Castro et al. 2004, 2006; Gómez-Aparicio et al. 2005a; Marañón et al. 2009), although it is a species specific-dependent process (Rolo et al. 2013).

One question that arises is whether the facilitative effect of nurse shrubs on early recruitment of trees is caused by a "canopy effect" that creates a more favorable microclimate and protects the seedling from herbivores, or a "soil effect" because shrubs modify the soil properties, or both. Physical protection against wildlife and livestock has been described as the major mechanism that facilitates oak regeneration when highly competitive shrub species are present in managed dehesas (Puerta-Piñero et al. 2006; Smit et al. 2008; Plieninger et al. 2010). In addition, the provision of natural or artificial shade invariably results in a dramatic increase in seedling survival in dehesas and Californian woodlands (see Callaway 1992; Marañón et al. 2009). The effects of light limitation are negligible when compared with the increased survival and reduced photo-inhibition resulting from moderate shade (Gómez-Aparicio et al. 2004; Castro et al. 2006). After all,

Fig. 6.18 **a, b** Shrubs will encroach on dehesa within a few years if the understory is not intensively managed. Some species of shrubs have been shown to act as "nurse plants," fostering the recruitment of oak seedlings. (Photograph by G. Moreno)

seedling (1–3 years old) growth of Mediterranean oaks has been shown to be only moderately reduced even in 20 % sunlight (evergreen oaks; Cardillo and Bernal 2006) and 13 % sunlight (deciduous oaks; Gómez-Aparicio et al. 2006), which is comparable to the natural shade found in many managed Mediterranean climate forests characterized by an open structure (Gómez-Aparicio et al. 2006). Quero et al. (2006) demonstrated that shade ameliorated, or at least did not aggravate, drought impact on seedlings of four oak species (holm, cork, Pyrenean and Algerian oaks). Under drought conditions, deep-shaded seedlings were able to achieve higher photosynthetic rates, stomatal conductance, and N concentration than seedlings under full light. This apparent alleviation of drought impacts for seedlings growing in shade could explain the pattern of higher survival under shade of shrubs and trees commonly observed in Mediterranean systems.

An improvement in soil conditions under some shrub canopy is reported as positive for oak seedlings (Moro et al. 1997; Puerta-Piñero et al. 2006; Rolo et al. 2012, 2013). A reduction of soil compaction has a positive effect on shrubs (Verdú

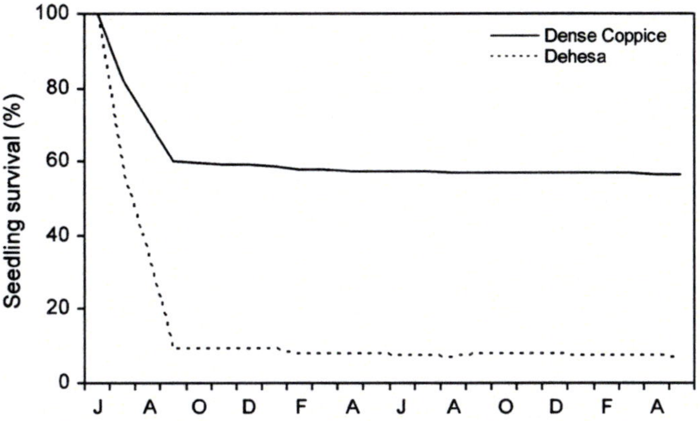

Fig. 6.19 A two year curve of seedling survival comparing a dehesa with a nearby shrub-encroached holm oak forest. Adapted from Pulido and Díaz (2005)

and García-Fayos, 1996). Gómez-Aparicio (2005a) found that shrub species significantly increased the depth at which maximum soil compaction was reached by 10 cm. With respect to soil moisture, results are less consistent, with one study reporting a significant increase in superficial soil moisture (Gómez-Aparicio et al. 2004; Moreno and Rolo 2011), but others failing to find this positive effect (Cubera and Moreno 2007b; Gómez-Aparicio et al. 2005a, b; Moreno and Rolo 2011). Nuñez et al. (2003) and Rolo et al. (2013), examined whether gum rockrose (*Cistus ladanifer*) has a nurse effect on holm oak seedlings, and found an overall neutral or negative effect on oak recruitment explained by the lower soil water content under gum rockroses.

In ecosystems characterized by a severe summer drought, pioneer shrubs represent a major safe site for early tree recruitment, improving seedling survival during summer by the modification of both the above and below-ground environment. Nevertheless, it is necessary to keep in mind that shrub-seedling interactions are species-specific and vary with the physical and biological environment (Marañón et al. 2009). In natural communities, only a subgroup of co-occurring species provides benefits, while the effects of the remaining species vary from competitive to neutral (Callaway and D'Antonio 1991; Puerta-Piñero et al. 2006; Callaway and Pugnaire 2007). Gómez-Aparicio et al. (2004), explored variation in the magnitude and direction of interactions along spatial gradients defined by altitude and aspect using 18,000 seedlings of 11 woody species planted under 16 different nurse shrubs. They found a consistent facilitative effect in all environmental situations explored, but with differences in the magnitude of the interaction, depending on the seedling species planted as well as the nurse shrub species involved. Additionally, shrub species can have an indirect effect on seedling recruitment through an increase or a reduction in pasture growth, with under-explored consequences for oak seedling survival.

6.5 Lessons Learned

A thorough understanding of the interactions among plants, soils, disturbance and other factors in dehesas and Californian oak woodlands is necessary to assess their sustainability. They share a Mediterranean climate, oaks, and many understory species. These shared characteristics are tempered by some significant differences in the amount of information available and some real differences in functional properties, especially those related to the roles of soil nutrients and water (Fig. 6.20).

Tree clearance as practiced in dehesa and Californian oak woodlands can positively affect the development of the understory and the remnant trees, which take advantage of the lower tree density. Tree roots can access a large volume of soil resources, especially water, unused by the understory layer. This allows the trees to maintain good water status over the summer and to grow more acorns, which is of importance for livestock rearing in dehesa and for wildlife in both places.

The limited overlap in the roots of the trees and the pasture together with the slow growth of many oak species and their capacity to thrive in poor soils make long-term management for livestock production, wildlife, and wood products in both Iberia and California possible. In the Iberian Peninsula oaks are actively managed as dehesa, without significant negative effects on pasture growth, and there is a growing interest in maintaining the trees. In California, the widespread thinning or clearing of oaks has been greatly reduced in the last few decades with the recognition of their limited impact on forage production at lower densities, and diverse oak conservation initiatives have been launched.

Fig. 6.20 **a, b** Shrub encroachment occurs distinctively in Californian and Spanish oak woodlands. Above, a shrub understory is comparatively uncommon in California but does occur, here on volcanic soils in the northern Sierra foothills. Blue and valley oaks intermix with various shrub species including buckbrush (*Ceanothus cuneatus*) (photo unattributed). Below, by contrast, shrub encroachment occurs quickly in Spanish dehesa unless shrubs are controlled by periodic cultivation and grazing. (Photographs by G. Moreno)

We have described how oak trees can facilitate understory forage production through the positive effect of shade and improved soil fertility. In Californian oak woodland pasture productivity increases up to as much as 60 % tree cover, especially for drier areas where rainfall is below 500 L m^{-2}. By contrast, in dehesa the effect of trees on pasture understory growth becomes more negative with aridity. This phenomenon in California oak woodlands needs to be confirmed taking into account different pasture taxa and soil types. To our knowledge, a conclusive study comparing the effect of evergreen and deciduous oaks on pasture understory yield is still missing, but from partial studies conducted in Californian oak woodlands, it can be said that positive effects are less evident with evergreen than with deciduous oaks.

The net effect of trees on understory yield varies considerably, with some studies reporting understory yield increased in the vicinity of the trees and others finding the opposite. More consistent is the positive effect of trees on understory nutrient content, landscape and habitat diversity, and phenology through the lengthening of the growing season, although evidence for this latter phenological effect is not as strong for Californian oak woodlands.

The two-layered dehesa system has been called unsustainable because of the absence of natural tree regeneration. Although oak seedlings have physiological adaptations for overcoming understory competition during summer drought, efforts made by managers to favor understory yield and reduce shrub understory could have a negative effect on oak seedling establishment. In addition, the current grazing pressure seems to be a definitive cause of a lack of oak regeneration in Spain, as presented in this volume (Chap. 5). Shrub encroachment in Californian oak woodlands has not been adequately researched and is poorly understood. In contrast, dehesa shrub encroachment has been extensively researched and shown as a way to potentially increase and sustain oak recruitment (Ramírez and Díaz 2008). Different Mediterranean shrubs seem to play multiple positive effects on microclimate and soil that favor tree establishment.

Although dehesa shrubs compete with trees for soil resources more strongly than herbaceous plants, the nutritional and hydric status of mature trees is not substantially affected by shrub unerstory. Hence, the rotation of grazing zones to generate a mosaic with some temporal ungrazed shrubby patches where oak can regenerate can be recommended as a mechanism to favor dehesa persistence without compromising the short-term productivity of trees. Although some practical methods for ensuring adequate oak recruitment have been developed, researched, and implemented in Californian oak woodlands, providing for natural oak regeneration is still a challenge in California because of the many influences on it, the numerous oak species involved, and inconsistent linkages to management activities including grazing.

6.6 Future Research

Explicit long-term strategies should be designed to promote management practices that ensure dehesa and Californian oak conservation. However, in order to convince landowners, administrators and policy-makers, better knowledge is needed.

For instance, although tree effects on soil and understory have been widely studied, studies focusing on the conditions and specific sites that result in positive or negative interactions are still needed. In this sense, the optimal tree density of dehesas under different uses and ecological constraints has not been adequately studied. The optimum tree density must be determined as a function of the effect of trees on understory and as a function of the tree–tree competition for soil resources, primarily for water. It would be useful to link these results to the specific characteristics of ecological sites such that they could guide management goals in both dehesas and Californian oak woodlands.

The nurse effect of shrubs is a species-specific phenomenon. For example, studies on the effects of gum rockrose on holm oak seedlings, the most common tree-shrub combination in the dehesa, have failed to show this effect. Further studies focusing on specific combinations of tree and shrub species on specific ecological sites will be needed. These studies should include developing a better knowledge of the root systems of shrubs. Similarly, the effects of changes in understory composition and production on seedling survival must be studied. And while dehesa shrub encroachment is certainly favorable for oak seedling regeneration, it does not maintain longer term stand functions and profitability from livestock, wildlife, and cork production. How to manage this temporal change in the relationships of shrubs and trees is an important topic (Ramírez and Díaz 2008). Finally, basic shrub studies in California are needed, with an emphasis on the effects of fire and grazing intensity on vegetation change (Fig. 6.21).

Finally, the stability of dehesa and Californian oak woodland systems in the face of long-term climatic change will need further study. The expected increase in the

Fig. 6.21 Once shrubs like gum rockrose have taken over a dehesa for the long term, it is no longer able to function for the production of multiple agricultural products, and is no longer really a dehesa woodland. As trees and shrubs get larger, the inter-relationship between them changes too, towards stronger competition. (Photograph by G. Moreno)

probability of extreme events could have dramatic consequences in dehesa (Joffre et al. 1999) and oak woodlands (Allen-Diaz et al. 2007). Indeed, apparently higher oak tree mortality is presently occurring in dehesas (Sanchez et al. 2002) and in California (Rizzo and Garbelotto 2003), including mortality caused by invasive pathogens. The effect of global warming on this increased mortality is uncertain and it deserves more research, as do the consequences of global warning for many dehesas and Californian oak woodlands undergoing woody encroachment.

References

Alagona PS (2008) Homes on the range: cooperative extension and environmental change on California's privately owned hardwood rangelands. Envir Hist 13:325–349

Allen-Diaz BH, Bartolome JW (1992) Survival of *Quercus douglasii* (Fagaceae) seedlings under the influence of fire and grazing. Madroño 39:47–53

Allen-Diaz BH, Bartolome JW, McClaran MP (1999) California oak savanna. In: Anderson RC, Fralish JS, Baskin JM (eds) Savannas, barrens, and rock outcrop plant communities of North America. Cambridge University Press, Cambridge

Allen-Diaz BH, Standiford R, Jackson RD (2007) Oak woodlands and forests. In: Barbour M, Keeler-Wolf T, Schoenherr A (eds) Terrestrial vegetation of California. University of California Press, Berkeley

Alonso H, Puerto A, Cuadrado S (1979) Efecto del arbolado sobre el suelo en diversas comunidades de pastizal. Anuario del CEBA de Salamanca 5:263–277

Baldocchi DD, Xu L (2007) What limits evaporation from Mediterranean oak woodlands—The supply of moisture in the soil, physiological controlby plants or the demand by the atmosphere? Adv Water Resour 30:2113–2122

Baldocchi DD, Xu L, Kiang N (2004) How plant functional-type, weather, seasonal drought, and soil physical properties alter water and energy fluxes of an oak-grass savanna and an annual grassland. Agr Forest Meteorol 123:13–39

Barnes PW, Archer S (1999) Tree-shrub interactions in a subtropical savanna parkland: competition or facilitation? J Veg Sci 10:525–536

Bartolome JW (1989) Ecological history of the California Mediterranean-type landscape. In: Clawson WJ (ed) Landscape ecology: study of Mediterranean grazed ecosystems, Proceedings of MAB symposium and XVI International grasslands congress, Nice, France

Battles JJ, Jackson RD, Shilsky A, Allen-Diaz BH, Bartolome JW (2008) Net primary production and biomass distribution in the blue oak savanna. In: Merenlender A, McCrearyD, Purcell K (eds) Proceedings of the sixth California oak symposium: today's challenges, tomorrow's opportunities, USDA Forest Service, General Technical Report PSW–GTR–217

Bertness M, Callaway RM (1994) Positive interactions in communities. Trends Ecol Evol 9:191–193

Calabuig E, Gómez JM (1992) Calidad del pasto. In: Gómez-Gutiérrez JM (ed) El libro de las dehesas salmantinas. Junta de Castilla-León, Salamanaca

Callaway RM (1992) Effect of shrubs on recruitment of *Quercus douglasii* and *Quercus lobata* in California. Ecology 73:2118–2128

Callaway RM, D'Antonio CM (1991) Shrub facilitation of coast live oak establishment in central California. Madroño 38:158–169

Callaway RM, Davis FW (1998) Recruitment of *Quercus agrifolia* in central California: the importance of shrub-dominated patches. J Veg Sci 9:647–656

Callaway RM, Pugnaire FI (2007) Facilitation in plant communities. In: Pugnaire FI, Valladares F (eds) Handbook of functional plant ecology, 2nd edn. CRC Press, Boca Raton

Callaway RM, Nadkarni NM, Mahall BE (1991) Facilitation and interference of *Quercus douglasii* on understory productivity in central California. Ecology 72:1484–1499

Campos P, Caparrós A, Cerdá E, Huntsinger L, Standiford RB (2007) Modeling multifunctional agroforestry systems with environmental values: dehesa in Spain and woodland ranches in California. In: Weintraub A, Romero C, Bjorndal T, Epstein R (eds) Handbook of operations research in natural resources, International series in operations research and management, vol 99. Springer, Berlin

Canadell J, Jackson RB, Ehleringer JR, Mooney HA, Sala OE, Schulze E-D (1996) Maximum rooting depth of vegetation types at the global scale. Oecologia 108:583–595

Cardillo E, Bernal CJ (2006) Morphological response and growth of cork oak (*Quercus suber* L.) seedlings at different shade levels. Forest Ecol Manag 222:296–301

Carevic FS, Fernández M, Alejano R, Vázquez-Piqué J, Tapias R, Corral E, Domingo J (2010) Plant water relations and edaphoclimatic conditions affecting acorn production in a holm oak (*Quercus ilex* L. ssp. *ballota*) open woodland. Agrofor Syst 78:299–308

Castro J, Zamora R, Hódar JA, Gómez JM (2004) Benefits of using shrubs as nurse plants for reforestation of Mediterranean mountains: a 4-year study. Restor Ecol 12:352–358

Castro J, Zamora R, Hódar JA (2006) Restoring the forest of *Quercus pyrenaica*. Appl Veg Sci 9:137–142

Collet C, Löf M, Pagès L (2006) Root system development of oak seedlings analysed using an architectural model. Effects of competition with grass. Plant Soil 279:367–383

Cubera E, Moreno G (2007a) Effect of single *Quercus ilex* trees upon spatial and seasonal changes in soil water content in Dehesas of central western Spain. Ann Forest Sci 64:355–364

Cubera E, Moreno G (2007b) Effect of land use on soil water dynamic in Dehesas of Central-Western Spain. Catena 71:298–308

Cubera E, Moreno G, Solla A, Madeira M (2012) Root system of *Quercus suber* L. seedlings in response to herbaceous competition and different watering and fertilisation regimes. Agroforest Syst. doi:10.1007/s10457–012–9492–x

Dahlgren RA, Singer MJ, Huang X (1997) Oak tree and grazing impacts on soil properties and nutrients in California oak woodlands. Biogeochemistry 39:45–64

Dahlgren RA, Horwath WR, Tate KW, Camping TJ (2003) Blue oak enhances soil quality in California oak woodlands. Calif Agric 57:42–47

David TS, Ferreira MI, Cohen S, Pereira JS, David JS (2004) Constraints on transpiration from an evergreen oak tree in southern Portugal. Agr Forest Meteorol 122:193–205

Eagleson PS, Segarra RI (1985) Water-limited equilibrium of savanna vegetation systems. Water Resour Res 21:1483–1493

Eastham J, Rose CW, Cameron DM, Rance SJ, Talsma T, Charles-Edwards DA (1990) Tree/understory interactions at a range of tree densities in an agroforestry experiment. II. Water uptake in relation to rooting patterns. Aust J Agr Res 41:697–707

Elena M, López JA, Casas M, Sánchez del Corral A (1987) El carbón de encina y la dehesa. Colección Monografías INIA no. 60, Ministerio de Agricultura, Pesca y Alimentación, Madrid

Escudero A (1992) Intervención del arbolado en los ciclos de los nutrientes. In: Gómez-Gutiérrez JM (ed) El libro de las dehesas salmantinas. Junta de Castilla-León, Salamanaca

Etienne M (2005) Silvopastoral management in temperate and Mediterranean areas. Stakes, practices and socio-economic constraints. In: Mosquera-Losada MR, Riguero-Rodriguez A, McAdam J (eds) Silvopastoralism and sustainable land management. CAB International, Oxfordshire

Fernández Alés R, Martín A, Ortega F, Alés EE (1992) Recent changes in landscape structure and function in a mediterranean region of SW Spain (1950–1984). Landscape Ecol 7:3–18

Fernández-Moya J, San Miguel-Ayanz A, Cañellas I, Gea-Izquierdo G (2011) Variability in Mediterranean annual grassland diversity driven by small-scale changes in fertility and radiation. Plant Ecol 212:865–877

Frost WE, Edinger SB (1991) Effects of tree canopies on soil characteristics of annual rangelands. J Range Manage 44:286–288

Frost WE, McDougald NK (1989) Tree canopy effect on herbaceous production of annual rangeland during drought. J Range Manage 42:281–283
Gallardo A (2003) Effect of tree canopy on the spatial distribution of soil nutrients in a Mediterranean dehesa. Pedobiologia 47:117–125
Gallardo A, Rodríguez-Saucedo JJ, Covelo F, Fernández-Alés R (2000) Soil nitrogen heterogeneity in a dehesa ecosystem. Plant Soil 222:71–82
García-López E (2005) Efectos del manejo sobre la producción y regeneración del arbolado en dehesas de encina (*Quercus ilex*). Dissertation, U of Salamanca, Spain
Gea-Izquierdo G, Montero G, Cañellas I (2009) Changes in limiting resources determine spatio-temporal variability in tree–grass interactions. Agrofor Syst 76:375–387
Gea-Izquierdo G, Allen-Díaz B, San Miguel A, Cañellas I (2010) How do trees affect spatio-temporal heterogeneity of nutrient cycling in Mediterranean annual grasslands? Ann Forest Sci 67:112–122
George MR (1987) Management of hardwood range: a historical review. Range Science Report No. 12, UC Davis, University of California
Gómez JM (2003) Spatial patterns in long-distance dispersal of *Quercus ilex* acorns by jays in a heterogeneous landscape. Ecography 26:573–584
Gómez Gutiérrez JM, Barrera Mellado I, Fernández Santos B (1989) Fitomasa subterránea en pastizales semiáridos de dehesa. Estudio comparativo de cuatro transecciones. Pastos 18–19:95–107
Gómez-Aparicio L, Zamora R, Gómez JM, Hódar JA, Castro J, Baraza E (2004) Applying plant facilitation to forest restoration: a meta-analysis of the use of shrubs as nurse plants. Ecol Appl 14:1128–1138
Gómez-Aparicio L, Valladares F, Zamora R, Quero JL (2005a) Response of tree seedlings to the abiotic heterogeneity generated by nurse shrubs: an experimental approach at different scales. Ecography 28:757–768
Gómez-Aparicio L, Gómez JM, Zamora R, Boettinger JL (2005b) Canopy vs. soil effects of shrubs facilitating tree seedlings in Mediterranean montane ecosystems. J Veg Sci 16:191–198
Gómez-Aparicio L, Valladares F, Zamora R (2006) Differential light responses of Mediterranean tree saplings: linking ecophysiology with regeneration niche in four co-occurring species. Tree Physiol 26:947–958
González-Bernáldez F, Morey M, Velasco F (1969) Influences of *Quercus ilex* rotundifolia on the herb layer at El Pardo woodland. Boletín de la Sociedad Española de Historia Natural 67:265–284
Gordon DR, Rice KJ (1993) Competitive effects of grassland annuals on soil water and blue oak (*Quercus douglasii*) seedlings. Ecology 74:68–82
Gordon DR, Menke JM, Rice KJ (1989) Competition for soil water between annual plants and blue oak (*Quercus douglasii*) seedlings. Oecologia 79:533–541
Holmes TH, Rice KJ (1996) Patterns of growth and soil-water utilization in some exotic annuals and native perennial bunchgrasses of California. Ann Bot Lond 78:233–243
Huntsinger L, Bartolome JW (1992) A state–transition model of the ecological dynamics of *Quercus* dominated woodlands in California and southern Spain. Vegetatio 99–100:299–305
Huntsinger L, Bartolome JW, Starrs PF (1991) A comparison of management strategies in the oak woodlands of Spain and California. USDA Forest Service Gen. Techical Report PSW-126
Huntsinger L, Sulak A, Gwin L, Plieninger T (2004) Oak woodland ranchers in California and Spain: conservation and diversification. In: Schnabel S, Ferreira A (eds) Sustainability of agrosilvopastoral systems—Dehesas, Montados, advances in geoecology 37. Catena Verlag, Reiskirchen
Infante JM, Domingo F, Fernández-Aléz R, Joffre R, Rambal S (2003) *Quercus ilex* transpiration as affected by a prolonged drought period. Biol Plant 46:49–55
Jackson LE, Strauss RB, Firestone MK, Bartolome JW (1990) Influence of tree canopies on grassland productivity and nitrogen dynamics in a deciduous oak savanna. Agr Ecosyst Environ 32:89–105

Joffre R (1990) Plant and soil nitrogen dynamics in Mediterranean grasslands: a comparison of annual and perennial grasses. Oecologia 85:142–149
Joffre R, Rambal S (1988) Soil water improvement by trees in the rangelands of southern Spain. Acta Oecol: Oec Plant 9:405–422
Joffre R, Rambal S (1993) How tree cover influences the water balance of Mediterranean rangelands. Ecology 74:570–582
Joffre R, Leiva Morales MJ, Rambal S, Fernández Alés R (1987) Dynamique racinaire et extraction de l'eau du sol par des graminées pérennes et annuelles méditerranéennes. Acta Oecol: Oec Plant 8:181–194
Joffre R, Rambal S, Ratte JP (1999) The dehesa system of southern Spain and Portugal as a natural ecosystem mimic. Agrofor Syst 45:57–79
Jose S, Gillespie AR, Seifert JR, Biehle DJ (2000) Defining competition vectors in temperate alley cropping system in the midwestern USA: 2. Competition for water. Agrofor Syst 48:41–59
Jose S, Gillespie AR, Pallardy SG (2004) Interspecific interactions in temperate agroforestry. Agrofor Syst 61:237–255
Maestre FT, Valladares F, Reynolds JF (2006) The stress-gradient hypothesis does not fit all relationships between plant–plant interactions and abiotic stress: further insights from arid environments. J Ecol 94:17–22
Maestre FT, Callaway RM, Valladares F, Lortie CJ (2009) Refining the stress-gradient hypothesis for competition and facilitation in plant communities. J Ecol 97:199–205
Marañón T (1986) Plant species richness and canopy effect in the savannalike "dehesa" of SW–Spain. Ecologia Mediterranea 12:131–141
Marañón T (1988) Agro-sylvo-pastoral systems in the Iberian Peninsula: Dehesas and Montados. Rangelands 10:255–258
Marañón T, Bartolome JW (1993) Reciprocal transplants of herbaceous communities between *Quercus agrifolia* woodland and adjacent grassland. J Ecol 81:673–682
Marañón T, Bartolome JW (1994) Coast live oak (*Quercus agrifolia*) effects on grassland biomass and diversity. Madroño 41:39–52
Marañón T, Pugnaire FI, Callaway RM (2009) Mediterranean-climate oak savannas: the interplay between abiotic environment and species interactions. Web Ecol 9:30–43
Mateos B, Schnabel S (2002) Rainfall interception by holm oaks in Mediterranean open woodland. In: Garcia-Ruiz JM, Jones JAA, Arnaez J (eds) Environmental change and water sustainability, Spanish National Research Council and the University of La Rioja Press, Logroño
McClaran MP, Bartolome JW (1989) Effects of Quercus *douglusii* (Fagaceae) on herbaceous understory along a rainfall gradient. Madroño 36:141–153
McCreary DD (2001) Regenerating rangeland oaks in California. Univ Calif Div Agric Nat Res Pub. p 21601
McCreary DD, Tietje WD, Davy JS, Larsen R, Doran MP, Flavell DK, Garcia S (2011) Tree shelters and weed control enhance growth and survival of natural blue oak seedlings. Calif Agric 65:192–196
McPherson GR (1997) Ecology and management of North American savannas. University of Arizona Press, Tucson
Mediavilla S, Escudero A (2003) Mature trees *versus* seedlings: differences in leaf traits and gas exchange patterns in three co-occurring Mediterranean oaks. Ann Forest Sci 60:455–460
Mediavilla S, Escudero A (2004) Stomatal responses to drought of mature trees and seedlings of two co-occurring Mediterranean oaks. Forest Ecol Manag 187:281–294
Montero MJ, Moreno G, Bertomeu M (2008) Light distribution in scattered-trees open woodlands in Western Spain. Agrofor Syst 73:233–244
Moody A, Jones JA (2000) Soil response to canopy position and feral pig disturbance beneath *Quercus agrifolia* on Santa Cruz Island, California. Appl Soil Ecol 14:269–281
Moreno G (2008) Response of understory forage to multiple tree effects in Iberian dehesas. Agr Ecosyst Environ 123:239–244

Moreno G, Cubera E (2008) Response of *Quercus ilex* to tree density in terms of tree water status, CO_2 assimilation and water transpiration. Forest Ecol Manag 254:74–84

Moreno G, Gallardo JF (2003) Atmospheric deposition in oligotrophic *Quercus pyrenaica* forest: implications for forest nutrition. Forest Ecol Manag 171:17–29

Moreno G, Obrador JJ (2007) Effects of trees and understory management on soil fertility and nutritional status of holm oaks in Spanish dehesas. Nutr Cycl Agroecos 78:253–264

Moreno G, Rolo V (2011) Dinámica del suso del agua edáfica entre estratos vegetales en dehesas matorralizadas del Suroeste de la Península Ibérica. Estudios en la Zona no Saturada del Suelo 10:53–58

Moreno G, Obrador JJ, Cubera E, Dupraz C (2005) Fine root distribution in Dehesas of Central-Western Spain. Plant Soil 277:153–162

Moreno G, Obrador J, García A (2007a) Impact of evergreen oaks on soil fertility and crop production in intercropped dehesas. Agr Ecosyst Environ 119:270–280

Moreno G, Obrador JJ, García E, Cubera E, Montero MJ, Pulido FJ, Dupraz C (2007b) Driving competitive and facilitative interactions in oak dehesas with management practices. Agrofor Syst 70:25–40

Moro MJ, Pugmaire FI, Haase P, Puigdefábregas J (1997) Effect of the canopy of *Retama sphaerocarpa* on its understory in a semiarid environment. Funct Ecol 11:425–431

Muñoz A, Bonal R (2011) Linking seed dispersal to cache protection strategies. J Ecol 99:1016–1025

Navarro Cerrillo RM, Fragueiro B, Ceaceros C, Campo A, Prado R (2005) Establishment of *Quercus ilex* L. subsp. ballota [Desf.] Samp. using different weed control strategies in southern Spain. Ecol Eng 25:332–342

Nunes J, Madeira M, Gazarini L (2005) Some ecological impacts of *Quercus rotundifolia* trees on the understory environment in the "montado" agrosilvopastoral system, Southern Portugal. In: Mosquera-Losada MR, Riguero-Rodriguez A, McAdam J (eds) Silvopastoralism and sustainable land management. CAB International, Oxfordshire

Núñez JJ, Pulido FJ, Moreno G (2003) The role of shrubs in holm oak regeneration: dissecting competitive vs. facilitative effects in dry Mediterranean environment. International Symposium on Dehesas and other silvopastoral Systems, Cáceres

Olea L, San Miguel-Ayanz A (2006) The Spanish dehesa. A traditional Mediterranean silvopastoral system linking production and nature conservation. 21st General meeting of the European Grassland Federation, Badajoz

Parker VT, Muller CH (1982) Vegetational and environmental changes beneath isolated live oak trees (*Quercus agrifolia*) in a California annual grassland. Am Midl Nat 107:69–81

Plieninger T, Pulido FJ, Konold W (2003) Effects of land-use history on size structure of holm oak stands in Spanish dehesas: implications for conservation and restoration. Environ Conserv 30:61–70

Plieninger T, Rolo V, Moreno G (2010) Large-Scale Patterns of *Quercus ilex*, *Quercus suber*, and *Quercus pyrenaica* Regeneration in Central-Western Spain. Ecosystems 13:644–660

Puerta-Piñero C, Gómez JM, Zamora R (2006) Species-specific effects on topsoil development affect *Quercus ilex* seedling performance. Acta Oecol 29:65–71

Puerto A (1992) Síntesis ecológica de los productores primaries. In: Gómez-Gutiérrez JM (ed) El libro de las dehesas salmantinas. Junta de Castilla-León, Salamanaca

Puerto A, Rico M (1988) Influence of tree canopy (*Quercus rotundifolia* Lam. and *Quercus pyrenaica* Willd.) on field succession in marginal areas of Central-Western Spain. Acta Oecol: Oec Plant 9:337–358

Puerto A, Rico M (1989) Influence of tree canopy (*Quercus rotundifolia* Lam.) on content in surface soil water in Mediterranean grasslands. Ecology (CSSR) 8:225–238

Puerto A, García JA, García A (1987) El sistema de ladera como elemento esclarecedor de algunos efectos del arbolado sobre el pasto. Anuario del CEBA de Salamanca 12:297–312

Puerto A, Rico M, Matías MD, García JD (1990) Variation in structure and diversity in Mediterranean grasslands related to trophic status and grazing intensity. J Veg Sci 1:445–452

Pulido FJ, Díaz M (2005) Recruitment of a Mediterranean oak: a whole-cycle approach. Ecoscience 12:99–112

Pulido FJ, García E, Obrador JJ, Moreno G (2010) Multiple pathways for tree regeneration in anthropogenic savannas: incorporating biotic and abiotic drivers into management schemes. J Appl Ecol 47:1272–1281

Quero JL, Villar R, Marañón T, Zamora R (2006) Interactions of drought and shade effects on seedlings of four *Quercus* species: physiological and structural leaf responses. New Phytol 170:819–834

Quilchano C, Marañón T, Pérez-Ramos IM, Neojovich L, Valladares F, Zavala M (2007) Patterns and ecological consequences of abiotic heterogeneity in managed cork oak forests of Southern Spain. Ecol Res 23:127–139

Ramírez JA, Díaz M (2008) The role of temporal shrub encroachment for the maintenance of Spanish holm oak Quercus ilex dehesas. Forest Ecol Manag 255:1976–1983

Retana J, Espelta JM, Gracia M, Riba M (1999) Seedling recruitment. In: Rodà F, Retana J, Gracia CA, Bellot J (eds) Ecology of mediterranean evergreen Oak forests. Springer, Berlin

Rice KJ, Nagy ES (2000) Oak canopy effects on the distribution patterns of two annual grasses: the role of competition and soil nutrients. Am J Bot 87:1699–1706

Rivest D, Rolo V, Lopez-Díaz ML, Moreno G (2011a) Shrub encroachment in Mediterranean silvopastoral systems: *Retama sphaerocarpa* and *Cistus ladanifer* induce contrasting effects on pasture and *Quercus ilex* production. Agr Ecosyst Environ 141:447–454

Rivest D, Rolo V, Lopez-Díaz ML, Moreno G (2011b) Belowground competition for nutrients in shrub-encroached Mediterranean dehesas. Nutr Cycl Agroecos 90:347–354

Rizzo DM, Garbelotto M (2003) Sudden oak death: endangering California and Oregon forest ecosystems. Front Ecol Envir 1:197–204

Rolo V, Moreno G (2011) Shrub species affect distinctively the functioning of scattered *Quercus ilex* trees in Mediterranean open woodlands. Forest Ecol Manag 261:1750–1759

Rolo V, Moreno G (2012) Interspecific competition induces asymmetrical rooting profile adjustments in shrub encroached open oak woodlands. Trees-Struct Funct 26:997–1006

Rolo V, López–Díaz ML, Moreno G (2012) Shrubs affect soil nutrients availability with contrasting consequences for pasture understory and tree overstory production and nutrient status in Mediterranean grazed open woodlands. Nutr Cycl Agroecos 93:89–102

Rolo V, Plieninger T, Moreno G (2013) Facilitation of holm oak recruitment through two contrasted shrubs species in Mediterranean grazed woodlands: Patterns and processes. J Veg Sci 24:344–355

Sanchez ME, Caetano P, Ferraz J, Trapero A (2002) *Phytophthora* disease of *Quercus ilex* in south-western Spain. Forest Pathol 32:5–18

Sankaran M, Hanan NP, Schole RJ et al (2005) Determinants of woody cover in African savannas. Nature 438:846–849

Schenk HJ, Jackson RB (2002) The global biogeography of roots. Ecol Monog 72:311–328

Scholes RJ, Archer SR (1997) Tree-grass interactions in savannas. Annu Rev Ecol Syst 28:517–544

Silva SS, Rego FC (2003) Root distribution of a Mediterranean shrubland in Portugal. Plant Soil 255:529–540

Silva SS, Rego FC, Martins-Louçao MA (2002) Belowground traits of mediterranean woody plants in a Portuguese shrubland. Ecologia Mediterranea 28:5–13

Smit C, den Ouden J, Díaz M (2008) Facilitation of *Quercus ilex* recruitment by shrubs in Mediterranean open woodlands. J Veg Sci 19:193–200

Smith TM, Goodman PS (1986) The effect of competition on the structure and dynamics of Acacia savannas in Southern Africa. J Ecol 74:1031–1044

Standiford RB, Howitt RE (1993) Multiple use management of California's hardwood rangelands. J Range Manage 46:176–181

Tyler CM, Odin DC, Callawway R (2007) Dynamic of woody species in the Californian grassland. In: Stromberg MR, Corbin JD, D'Antonio CM (eds) California grasslands. Ecology and management. University of California Press, Berkeley

Úbeda X, Ferreira A, Sala M (2004) The nutritive status of *Quercus suber* L. in the province of Girona, Spain: a foliar analysis. In: Schnabel S, Ferreira A (eds) Sustainability of agrosilvopastoral system—Dehesas, Montados, Advances in GeoEcology 37. Catena, Reiskirchen

UC-OWCW (2012) University of California Oak Woodland Conservation Workgroup.http://ucanr.edu/sites/oak_range/ (accessed on October 2012)

Vacher J (1984) Analyse phyto et agro-ecologique des dehesas pastorales de la Sierra Norte (Andalousie. Espagne). Ecoteque Mediterraneene du CEPE/CNRS. Montpellier. France

Valladares F, Pugnaire FI (1999) Tradeoffs between irradiance capture and avoidance in semiarid environments simulated with a crown architecture model. Ann Bot Lond 83:459–470

Vallejo VR, Aronson J, Pausas J, Cortina J (2006) Restoration of mediterranean woodlands. In: Van Andel J, Aronson (eds) Restoration ecology. The New Frontier, Blackwell Publications, Oxford

Verdú M, García-Fayos P (1996) Nucleation processes in a Mediterranean bird-dispersed plant. Funct Ecol 10:275–280

Young A (1997) Agroforestry for soil management. CAB International, Wallingford

Chapter 7
Acorn Production Patterns

Walter D. Koenig, Mario Díaz, Fernando Pulido, Reyes Alejano, Elena Beamonte and Johannes M. H. Knops

Frontispiece Chapter 7. Forest–dehesa transition in central Spain (Photograph by M. Díaz)

W. D. Koenig (✉)
Lab of Ornithology and Department of Neurobiology and Behavior, Cornell University,
159 Sapsucker Woods Road, Ithaca, NY 14850, USA
e-mail: wdk4@cornell.edu

P. Campos et al. (eds.), *Mediterranean Oak Woodland Working Landscapes*,
Landscape Series 16, DOI: 10.1007/978-94-007-6707-2_7,
© Springer Science+Business Media Dordrecht 2013

Abstract Acorns—the fruits of oaks—are a key resource for wildlife in temperate forests throughout the Northern Hemisphere. Acorns are also economically important for extensive livestock rearing, and as a staple food have supported indigenous human populations. Consequently, differences in how individual trees and populations of oaks invest in acorn production, both in terms of the size of the acorn crop and of the size of individual acorns, are of interest both ecologically and economically. Acorn production by oaks in both California and Spain tends to be highly variable and spatially synchronous. We summarize studies conducted in the two regions that investigate the factors influencing acorn production. One hypothesis explored is that, as a consequence of management, acorn production tends to be affected by different environmental factors in the two regions; another hypothesis is that acorn production in oaks in Spanish dehesas produce larger and more predictable acorn crops than trees in less managed Spanish forests or in California woodlands. Other factors potentially influencing acorn production are summarized, including biotic factors, trade-offs with growth, trade-offs with acorn size, and pollen limitation. We conclude with a discussion of spatial synchrony and acorn production at the community level. There remain many questions concerning the mating systems of oaks, trade-offs between different oak life-history characters, and the patterns and drivers of spatial synchrony. Environmental conditions in the two regions are similar, but understanding how their subtle differences influence acorn production is likely to yield important insights about the proximate and ultimate factors affecting acorn production and masting behavior.

Keywords Acorns · Acorn production · Acorn size · Dehesa · Masting · Oak savanna · Spatial synchrony

---

M. Díaz · E. Beamonte
Department of Biogeography and Global Change, Museo Nacional de Ciencias Naturales (BGC-MNCN), Spanish National Research Council (CSIC), Serrano 115bis E-28006 Madrid, Spain
e-mail: Mario.Diaz@ccma.csic.es

E. Beamonte
e-mail: ele.beamonte@gmail.com

F. Pulido
Grupo de Investigación Forestal, Universidad de Extremadura, E-10600 Plasencia, Spain
e-mail: nando@unex.es

R. Alejano
Dpto. CC. Agroforestales, Universidad de Huelva, Campus de La Rábida 21819 Palos de la Frontera, Huelva, Spain
e-mail: ralejan@dcaf.uhu.es

J. M. H. Knops
School of Biological Sciences, University of Nebraska, 348 Manter Hall, Lincoln, NE 68588, USA
e-mail: jknops2@unl.edu

7.1 Introduction

Acorns—primarily the fruits of oaks (genus *Quercus*)—are a key resource for wildlife in temperate forests throughout the Northern Hemisphere. Acorns are also economically important for extensive livestock rearing, as well as for a staple food for some indigenous human populations, at least historically. Consequently, differences in how individual trees and populations of oaks invest in acorn production, both in terms of numbers—the size of the acorn crop—and of the size of individual acorns themselves, are of interest both ecologically and economically.

What makes oaks particularly exciting scientifically is the propensity of many, if not all, populations to engage in the phenomenon of "masting" or "mast-fruiting"; that is, they produce acorn crops that vary markedly from year to year and do so more or less synchronously over what, in at least some cases, can be tens or hundreds of millions of individuals across large geographic areas. How and why they accomplish this feat, at both the proximate and ultimate levels, are questions of considerable evolutionary interest (Kelly and Sork 2002).

Spain and California are comparable in size (Spain: 505,000 km^2; California: 411,000 km^2) and both have oak-dominated, foothill landscapes with scattered trees over a grassland matrix (savannas in California and dehesas in Spain) that cover nearly 10 % of their land area (Chapter opening photograph and Figs. 7.1, and 7.2). The fact that a Mediterranean climate, with cool wet winters and hot, dry summers (Hobbs et al. 1995) characterizes oak habitats in both regions renders comparisons of oaks and acorn production in the two regions particularly appealing and scientifically valuable (Huntsinger and Bartolome 1992). Making a comparison even more intriguing is the fact that the scattered spatial configuration of oak tree populations is man–made in Spanish dehesas but apparently natural in Californian savannas. Given that the spatial distribution of trees is likely to affect reproductive effort because spacing limits competition (Chap. 6), intercontinental comparisons linked to comparisons between dehesas and nearby oak forests in Spain could help determine management practices and environmental factors that have the capacity to change patterns of acorn production by oak trees, as well as the likely mechanism causing these changes.

In this chapter, we summarize what is known and not known about acorn production—including both acorn crop size and acorn size—in Spain and California. Our ultimate goals are to use similarities and differences between the two regions to help understand the evolution of this poorly understood phenomenon and to improve our understanding of the ecological effects of variability in this important natural resource.

Fig. 7.1 Coastal oak woodland intermixed with savanna and chaparral (*the dark patches*) adjacent to Hastings Reservation, California. Although considerable clearing took place between the late 1800s and early 1900s, the scattered distribution of the dominant oak species is mostly natural, whereas in the Spanish dehesas this distribution is created and maintained from continuous tree populations in forests by human management (see chapter opening frontispiece). (Photograph by W. D. Koenig)

7.2 Acorn Crop Size

There are three major classes of factors potentially affecting acorn crop size that are particularly relevant for a comparison of California and Spain. First are environmental factors, including rainfall and temperature. Second are biotic factors, including birds and mammals that eat or collect acorns and herbivores that live in them prior to acorn fall. Third involves differences in habitat and management such as whether trees are in forests or open habitats and the effects of pruning, soil treatments, and other landscape management practices that are virtually universal in dehesa. After briefly summarizing work on these three sets of factors, we consider the evidence for there being differences in one or more of the components of acorn production between California and Spain. Next we discuss several issues related to acorn production currently being investigated in both regions, including trade-offs between acorn production and acorn size, trade-offs between acorn production and growth, and pollen limitation. We end with a discussion of spatial synchrony and acorn production at the community level, questions currently being investigated in both regions.

Fig. 7.2 Forest-dehesa transition in the National Park of Cabañeros in Spain, where long-term studies of acorn production in paired holm oak populations in forest and dehesa are being conducted. (Photograph by M. Díaz)

7.2.1 Environmental Factors

A summary of some of the environmental factors that have been found to correlate with acorn production in Californian and Spanish oaks (Table 7.1) suggests some intriguing differences. In California, conditions during the spring appear to be particularly important for valley oak (*Q. lobata*) and blue oak (*Q. douglasii*), two deciduous species that mature acorns in a single year, and, when lagged appropriately, for California black oak (*Q. kelloggii*), a deciduous species that requires two years to mature acorns. Rainfall in a prior year is important to two of the evergreen species, coast live oak (*Q. agrifolia*) and canyon live oak (*Q. chrysolepis*), as well as for California black oak. In Spain, three species have been studied in this regard including holm oak (*Q. ilex*), cork oak (*Q. suber*) and downy oak (*Q. humilis*). A reoccurring factor affecting both the size of the acorn crop and, in a few cases, other variables including acorn mass and synchrony, is water stress during the summer and early fall as acorns mature, as indicated by xylem water potential, measures of summer drought, and even canopy foliage (NDVI or the "normalized difference vegetation index"; Camarero et al. 2010). Although some evidence for a similar effect of summer drought on acorn production in Missouri oaks, including red oak (*Q. rubra)* and black oak (*Q. velutina*) has been reported (Sork et al. 1993), summer conditions do not appear to play an important role in acorn crop size of any of the species of California oaks for which there are currently data, a result we confirmed for the same five populations studied by

Table 7.1 A summary of the environmental variables correlating with acorn production in California and Spanish oak populations.

Environmental variables	Effect	Species	Reference
Spanish oaks			
Summer water stress (drought)	–	*Q. ilex*	Pérez-Ramos et al. (2010)
Torrential rain in spring	+		
Min temp, rel. humidity, rainfall (January)	+	*Q. ilex*	García-Mozo et al. (2001)
Rainfall (March)	+		
Relative humidity (April)	+		
Mean temp (June)	+		
Rainfall (September)	+		
Spring rainfall	+	*Q. ilex*	Alejano et al. (2008)
Autumn rainfall	+		
Xylem water potential (mid-summer)	+		
Xylem water potential (mid-summer)	+	*Q. ilex*	Carevic et al. (2010)
Maximum canopy foliage	+	*Q. ilex*	Camarero et al. (2010)
Spring temp	+	*Q. suber*	Pons and Pausas (2012)
Summer water stress (drought)	–		
Spring frost	–	*Q. suber*	García-Mozo et al. (2001)
Mean temp (September)	– (acorn mass)	*Q. ilex*	Alejano et al. (2011)
Summer water stress (drought)	+ (synchrony)	*Q. ilex and Q. humilis*	Espelta et al. (2008)
California oaks			
Mean temp (April)	+	*Q. lobata*	Koenig et al. (1996)
Mean fall temp (year –1)	–		
Mean temp (April)	+	*Q. douglasii*	Koenig et al. (1996)
Rainfall (year –1)	+	*Q. agrifolia*	Koenig et al. (1996)
Rainfall (year –2)	+	*Q. chrysolepis*	Koenig et al. (1996)
Mean temp (winter, year –1)	–		
Rainfall (year –1)	+		
Rainfall (spring, year –1)	–	*Q. kelloggii*	Garrison et al. (2008)
Mean temp (spring, year –1)	+		

Correlations are with the size of the annual acorn crop except where noted. Data for *Q. suber* is for trees maturing acorns in one year

Koenig et al. (1996) using the summer drought index of Espelta et al. (2008) and 32 years of data through 2011 (correlation between the drought index and subsequent acorn production ranged from −0.18 to 0.13, all $P > 0.3$).

At this stage, the cause of this apparent difference remains speculative. However, one possibility is that it is related to climatological differences between the two regions. Although both are unambiguously Mediterranean in that winters are relatively cool and wet while summers are warm and dry, there is a notable difference in terms of the length and relative dryness of the summers, which are apparently shorter in Spain (Jackson 1985), where summer precipitation occurs as

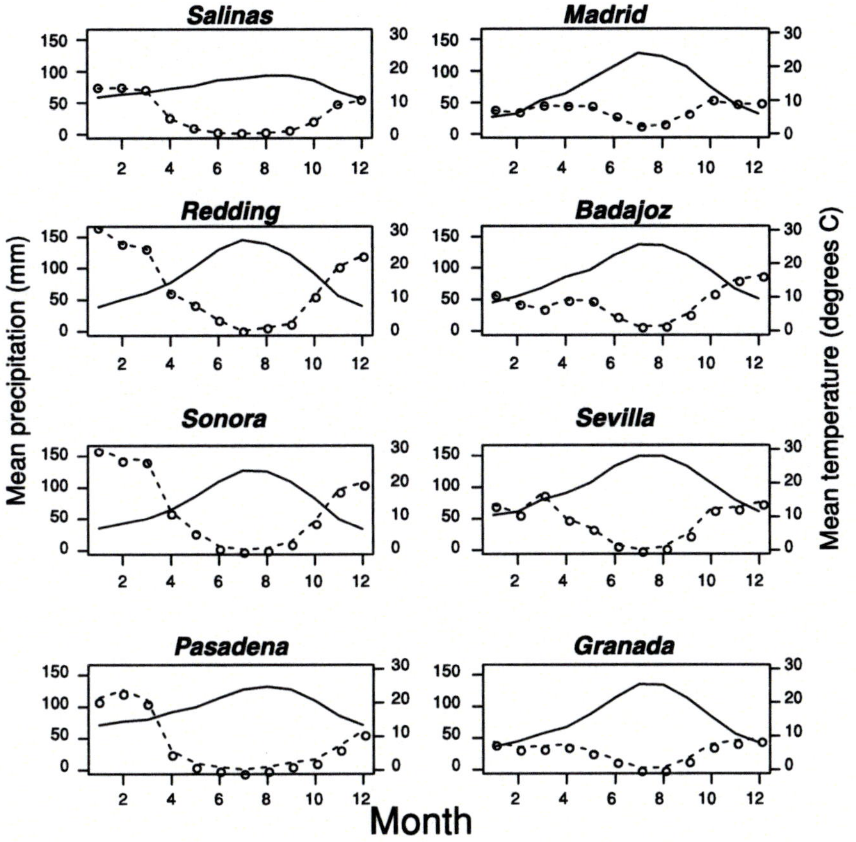

Fig. 7.3 Representative climate graphs for Mediterranean regions in California (*left*) and Spain (*right*). *Broken lines and circles* are mean monthly precipitation; *solid lines* are mean monthly temperatures. The summer dry season is more compressed in Spain

summer storms that are unpredictable from year to year (Fig. 7.3). One way to quantify this difference is to compare the percent of total annual precipitation falling in the four months from June through September, the main months during the summer when acorns are maturing. For the four arbitrary Spanish sites depicted in Fig. 7.3, this value is 11.9 ± 5.0 %, whereas in the California sites, only 3.7 ± 0.8 % of rain occurs during this period.

This suggests that the summer dry season is longer and drier in California than in Spain. To the extent that this is true, one might predict that summer water stress would be even more important in California than Spain, but this does not appear to be the case for oaks. The more cogent difference, however, may be that there is very little variation in the environmental conditions during the period of acorn maturation in California compared to Spain. For example, from daily weather records going back to 1939 data at Hastings Reservation in central coastal

California, where W. Koenig and J. Knops have studied acorn production since 1980, the mean (±SD) precipitation falling between 1 June and 30 September was only 1.2 ± 1.4 cm with 62 of 72 years (86 %) having <2 cm of rain during this 4-month period. Given this lack of variability, it is not surprising that summer conditions appear to have little effect on the acorn crop in California. It would be of interest to make additional such comparisons in order to better understand the relationship between environmental variability and the ecological factors affecting acorn crop size in specific populations.

Despite this difference, it is notable that some of the most common environmental factors correlating with acorn production in both California and Spain take place while trees are flowering in the spring or during acorn development (summer drought). This indicates that factors other than resources available to trees at the start of the season are important to acorn production, including pollen limitation, fertilization success, and resources that become available during acorn development itself (Espelta et al. 2008; Pérez-Ramos et al. 2010).

7.2.2 Biotic Factors

Although not studied as intensively as abiotic factors, biotic (herbivory-related) factors can have important effects on acorn production. Working with holm oak, Pulido and Díaz (2005) found biotic factors caused 29 % of predispersal losses to acorns in forests and 10 % in dehesas. In a comparative analysis of holm oak recruitment in grazed, cropped, and encroached dehesas, Pulido et al. (2010) showed that resource-mediated effects overrode the effects of insect predation and pathogens on tree fecundity in all habitats, primarily by causing acorn abortion. These results suggest that in holm oak, production of sound acorns is environmentally rather than biotically determined, in the absence of population peaks of its natural enemies.

In some cases, however, herbivores and pathogens can clearly affect fecundity in oaks. In both California and Spain, elongating shoots bearing male catkins and pistillate flowers are potentially defoliated by insects, mainly moth caterpillars in the families Noctuidae, Tortricidae, and Lymantridae (Fig. 7.4). By feeding upon leaf tissues, caterpillars not only reduce carbon assimilation in the growing shoots, they also interfere in shoot elongation and development of the pistillate flowers in the distal portion of the shoots.

Thus far, few studies have tested for the effects of shoot defoliation on oak fecundity, and those that have been performed report differing results. Insecticide spraying suppressing herbivory increased fecundity in pedunculate or English oak (*Q. robur*) in England (Crawley 1985) but not in holm oak in Spain (Díaz et al. 2004). The latter study showed that holm oak overcompensated for the tissue lost at the shoot level, thus stressing the importance of carbohydrate stores and the timing of resource allocation for growing acorns to predict the impact of folivores.

Fig. 7.4 Images of the three main biotic agents causing predispersal losses in acorn production. **a** The leaf rolling tortricid moth *Tortrix viridana* feeding on new shoots of holm oak (photograph by F. Pulido). **b** Sugar-rich exudation dropping from a holm oak acorn infested by bacterial pathogens (photograph by M. Díaz). **c** Cross section of a holm oak acorn showing consumption of cotyledons by larvae of *Curculio* weevils (photograph by F. Pulido)

A more realistic way to look at the effects of shoot defoliation is to compare acorn crops among control sites and sites where large-scale spraying for pest control has been carried out. A preliminary study comparing 12 paired dehesa sites with and without spraying showed a non-significant 1.2-fold increase in the acorn crop index in treated sites (F. Pulido, unpublished data). Since acorn production partly depends on resources stored in previous years, however, spraying in a given year might still be expected to result in increased acorn production the year following treatment.

After fertilization, growing acorns can be infested by bacterial pathogens (mostly in the genus *Brenneria* [=*Erwinia*]) causing the so-called "drippy nut" disease (Fig. 7.4; Hildebrand and Schroth 1967; Biosca et al. 2003). Bacteria enter acorns through holes or crevices, so that borer insects, especially acorn weevils (*Curculio* spp.), are potential vectors of this poorly known disease. As a result of bacterial activity inside the acorn, a sugar-rich exudation is produced that leads to cessation of acorn growth. In holm oak dehesa local losses of developing acorns due to this disease range from 16 to 24 % in one study (Pulido and Díaz 2005) and from 4 to 16 % in another site (Pulido et al. 2010). In a large-scale survey including 89 sites in 14 counties in southwestern Spain, the mean occurrence of the disease ranged from 0 to 60 % of infested trees (Vázquez et al. 2000). Although it is believed that the prevalence of bacterial infection is triggered by summer storms, further studies are needed to clarify the origin and economic impact of this important disease.

The third cause of predispersal acorn damage in savannas and dehesas is infestation by borer insects. This is a conspicuous phenomenon resulting in potentially important economic losses in Spain due to rejection of infested acorns by livestock (Rodríguez-Estévez et al. 2009). Briefly, acorns can be infested by moth larvae (mostly *Cydia* spp.) that reach the cotyledons after boring by themselves through the acorn cap or, alternatively, they can be occupied by weevil larvae that emerge from eggs previously deposited by the adult female by perforating the pericarp (Bonal et al. 2010; Díaz et al. 2011). Infestation rates of acorns are variable but they are reasonably well predicted by the size of acorn crops, both across individual trees (Bonal et al. 2007) and between years (Díaz et al. 2011). In holm oak dehesas infestation rates remain below 20 % in good acorn years, while more than 60 % of the acorns can be attacked in poor acorn years (Leiva and Fernández-Alés 2005; Pulido and Díaz 2005; Pulido et al. 2010).

In California, Koenig et al. (2002) estimated that a mean of 39–100 % of acorns from individual valley oak were removed by arboreal predators—primarily birds and squirrels—prior to acorn fall, with the proportion removed being inversely correlated to the overall mean acorn crop. Similarly, the proportion of remaining acorns damaged by insects decreased with focal tree productivity in two of three species (valley oak and blue oak, but not coast live oak), with the mean annual proportion of acorns infested with insects varying from 0 to 63 %. In neither case were neighborhood effects detected; that is, trees outproducing local conspecifics did not appear to attract a disproportionate number of arboreal seed removers (predators but also potential seed dispersers) or insect predators.

Reviewing studies of the same three Californian oaks, Tyler et al. (2006) found mean infestation rates of canopy-collected acorns ranged from 0 to 31 %. Acorns parasitized by weevils tend to occur with higher frequency than moth-infested acorns, especially when there were late summer rains, which favored the emergence of adult weevils from the ground underneath oak trees. The fraction of cotyledon tissue eaten by these larvae before exiting acorns determines the chance for germination and seedling establishment. As a result, the effect of such parasitism on seedling recruitment from large acorns produced in dehesas is less pronounced than the effect on recruitment from small acorns produced in dense stands (Siscart et al. 1999; Leiva and Fernández-Alés 2005).

In California, Dunning et al. (2002) found that the majority of ground-collected acorns had some insect damage in *Q. agrifolia* and *Q. engelmannii* (Engelmann oak). The level of insect damage was less than 20 % of the entire acorn, and the portions of the acorn most likely to be damaged were the cotyledons rather than the embryo, again suggesting that infested acorns should be taken into account when analyzing oak recruitment prospects.

7.2.3 Management and Habitat

A third class of factors potentially influencing acorn crop size is management, habitat, and site differences, including whether trees are growing in forests where competition may be considerable or in more open habitats, and whether trees are pruned or otherwise managed. Such factors have been examined in some detail in Spain, where dehesas are intensively managed both for acorn production as a food source for livestock (Parsons 1962) and, in the case of cork oak, for their unique bark (i.e., cork production). Acorn production measured in Spanish sites are highly variable, with productivity ranging from 0.5 to 147.0 kg acorns/tree for holm oak forests and dehesas, and 0.5–135.0 kg/tree for cork forests, scaling up to an estimated 79.3–469.6 kg acorns/ha for holm and 256.9–448.5 kg/ha for cork oak (Carbonero 2008; Díaz and Pulido 2009). To what extent is such variability due to differences in management, habitat, or sites?

In a study of regeneration of holm oak, Pulido et al. (2010) found that trees in "cropped" habitats—that is, plots that are fenced and used for cereal production—produced more female flowers and larger acorn crops in each of two years compared to trees in grazed and shrub-encroached plots, indicating an important role for management. Habitat and/or sites can also be important, as indicated by studies of the differences in acorn crop size of holm oak in forest and nearby dehesa sites in Cabañeros National Park where acorn production, but not acorn size, were significantly greater in the dehesas (87.0 ± 49.1 vs 34.7 ± 27.9 acorns m^{-2}; $P = 0.03$; means for acorn crops between 2003 and 2009; Beamonte 2009). Such differences are most likely due to differences in resources available in the two different habitat types (Díaz et al. 2011).

Carevic et al. (2010) failed to find significant effects of two soil treatments (ploughing; ploughing and sowing of European yellow lupine (*Lupinus luteus*)) on acorn production patterns in holm oak dehesa. Xylem water potential in ploughed soils was higher than in control areas, but the unusually wet summer in both years of the study may have reduced the importance of water for acorn development (Alejano et al. 2008).

Pruning—a widespread procedure conducted mainly to produce firewood and increase browse production that varies from modest thinning of small branches to more drastic opening up of the canopy (Huntsinger et al. 1991)—has also been shown to affect acorn production, although the effect appears to be variable. Cañellas et al. (2007) found no effect of moderate pruning (removing 30 % of crown biomass) in a mixed holm and cork oak dehesa when acorn production was poor, but pruning at this level apparently decreased acorn production when it was good.

Studies by Alejano et al. (2008) investigated the effects of pruning on holm oak in more detail, comparing oaks that had been subjected to light, moderate, and heavy traditional pruning along with a non-traditional method of "crown-regeneration pruning" in which the outermost branches of the tree crown were removed, thereby shortening water transport distances and resulting in a more compact crown that was hypothesized to improve water balance. Results over five years failed to indicate any significant overall effect of the traditional pruning method on acorn production. Similar results were obtained by Carbonero (2011) studying the influence of moderate pruning on acorn production in holm oak dehesas in Cordoba, Spain. There was, however, evidence that the non-traditional pruning method tested by Alejano et al. (2008) significantly enhanced acorn production, indicating that although traditional methods of pruning have questionable effects, new methods conducted taking into consideration the architecture of the trees may increase productivity. Parallel work by Alejano et al. (2011) investigating the factors influencing acorn mass in holm oak has found significant effects of location and year but not pruning, tree size, topography, or crowding (interspecific competition).

7.2.4 Components of Acorn Production

Most species of oaks that have been studied thus far, including all those for which there are data in either California or Spain, exhibit both considerable variation in seed production from year to year and a great deal of individual variation within and often between populations. One approach to understanding the causes of this variation is to quantify variability in a way that can be compared across populations.

Herrera (1998) was possibly the first to use a series of metrics to quantify the components of masting behavior in a comparative way, including variables measuring annual and individual variability, between-individual synchrony, and the endogenous cycles of temporal autocorrelation—that is, the degree to which acorn production by individuals and populations is correlated with production in a prior year. In general, oaks conform to the pattern predicted by "normal masting" (Kelly 1994; Koenig and Knops 2002) in which there is significant, but not complete, bimodality in seed production across years and for which there is evidence for resource switching. The latter is important because it demonstrates that reproductive effort is not simply being driven by variation in annual resource abundance (the "resource tracking" hypothesis), but rather is an evolutionary strategy that involves diverting resources from acorn production to other functions in some years and overinvesting in reproduction in others (Sork et al. 1993; Koenig et al. 1994b). Are similar patterns exhibited by Spanish and California oaks, and if not, what is driving the differences?

Although the data available to make such a comparison are limited, we are able to summarize data from 49 populations of eight species of California oaks studied at various sites around the state for up to 32 years (a total of 1,065 individuals) by W. Koenig and J. Knops and 42 populations of three species of Spanish oaks (primarily the *ballota* subspecies of holm oak (*Q. ilex* subsp. *ballota*) but also two populations of cork oak, one of downy oak and one of the *ilex* subspecies of holm oak (*Q. ilex* subsp. *ilex*), studied over 4–12 years (2,112 individual trees). For each study, masting metrics were calculated for each subpopulation and then averaged for all populations of the same species surveyed in the same study. This yielded data for a total of 16 studies, including nine for Spanish oaks (7 for holm oak and 1 each for cork and downy oak), and eight for California oaks (1 each for valley, blue, canyon live, coast live, California black, interior live (*Q. wislizenl*), Engelmann (*Q. engelmannii*), and Oregon (*Q. garryana*) oaks). Methods for quantifying the acorn crop (see Box 1) involved visual surveys in California and for three of the Spanish studies (Koenig et al. 1994a) and crown or branch sampling for the other Spanish studies (Carbonero 2008; Espelta et al. 2008; Díaz et al. 2011). Analyses were conducted using untransformed data and are summarized in Table 7.2.

Five metrics were compared, including mean population coefficient of variation (CV_p), which provides an index of the mean annual variability of acorn production in the population, and the mean individual coefficient of variation ($\overline{CV_i}$), which

Table 7.2 Mean (±SD) population and individual characteristics of acorn production by Californian and Spanish oaks

Variable	(1) California	(2) Spain	(2a) Spain (dehesas)	(2b) Spain (forests)	*P* value (1 vs. 2)	*P* value (1 vs. 2a)	*P* value (1 vs. 2b)	*P* value (2a vs. 2b)
Mean population CV (CV_p)	96.4 ± 14.2	84.4 ± 58.0	52.4 ± 27.3	124.4 ± 64.3	0.58	0.01	0.25	0.06
Mean individual CV ($\overline{CV_i}$)	158.1 ± 24.2	117.9 ± 59.6	79.4 ± 15.3	165.9 ± 60.3	0.10	0.00	0.75	0.02
Mean pairwise synchrony ($\bar{r}_p$)	0.45 ± 0.09	0.45 ± 0.18	0.36 ± 0.16	0.54 ± 0.19	0.94	0.21	0.31	0.20
Mean population temporal autocorrelation (1-yr lag) ($ACF1_p$)	−0.28 ± 0.20	−0.14 ± 0.18	−0.12 ± 0.22	−0.19 ± 0.14	0.20	0.23	0.54	0.73
Mean individual temporal autocorrelation (1-yr lag) ($\overline{ACF1_i}$)	−0.18 ± 0.10	−0.17 ± 0.09	−0.21 ± 0.05	−0.09 ± 0.10	0.87	0.56	0.29	0.10

Number of populations included was nine for Spain (five in dehesas and four in forests) and eight for California, each population representing the averaged results of 1–18 subpopulations of the same species studied in different parts of the region. California data from W. Koenig and J. Knops (unpublished). Spanish dehesa data from R. Alejano (unpublished data), Carbonero et al. (2008) and Díaz et al. (2011); Spanish forest data from Espelta et al. (2008), M. Díaz and F. Pulido (unpublished data) and Díaz et al. (2011). Data from pairwise synchrony was not available for the dehesa study by Carbonero et al. (2008) and temporal autocorrelation data was not available from the Espelta et al. (2008) forest study measures the mean individual annual variation and provides an upper limit to CV_p. The third measure, mean pairwise synchrony between all individuals in the population ($\bar{r}_p$), is an index of how synchronous acorn production is among the trees sampled in the population. Also calculated when possible are $\mathrm{ACF1_p}$ and $\overline{\mathrm{ACF1_i}}$, measures of temporal autocorrelation, or the extent to which acorn production of the population ($\mathrm{ACF1_p}$) and of individual trees ($\overline{\mathrm{ACF1_i}}$) correlates with acorn production the prior (or the next) year. These provide an index of the degree to which acorn production is driven by endogenous factors such as stored resources, since they are indicative of the extent to which trees "switch" resources to or away from reproduction from one year to the next (Sork et al. 1993; Koenig et al. 1994b).

Results indicate no significant differences between measures of masting in the two regions as a whole (Table 7.2). However, standard deviations were quite large for the Spanish data due to an apparent difference between values for populations from the dehesas compared to those from a higher-density forest. Dividing the Spanish data into these two groups revealed significantly lower CV_p and $\overline{CV_i}$ for trees in Spanish dehesas compared to either California or Spanish forests. There were no significant differences, however, in either pairwise synchronies ($\bar{r}_p$) or temporal autocorrelations, although these comparisons were based on smaller sample sizes.

These results, although preliminary, at least suggest that management practices may significantly influence acorn production patterns in Spain. Specifically, oaks in managed dehesas appear to exhibit reduced masting behavior, yielding acorn crops that are more predictable at the stand level and subject to greater external (environmental) influence (Koenig et al. 2003).

More data are clearly needed, however. For example, analyses of four nearby forest and dehesa stands in the National Park of Cabañeros (Díaz et al. 2011) suggests that dehesas exhibit CV_p values at least as large as those from forest stands (99.3 ± 27.5 % vs. 79.6 ± 89.7 %, respectively); trees in the dehesa sites also exhibited higher synchrony than those in the forest sites. Such findings suggest that differences between dehesas and forests may be due less to differences in management and more to areas managed as dehesas being located in higher-quality sites than remnant forest stands (see Sect. 7.2.3).

It has also been suggested that larger and more predictable acorn crops by trees in dehesas are due to an active selection of individual trees, either by retaining only the best trees during dehesa formation or by planting acorns of better-producing trees in open land (Montero et al. 2000). Whether such artificial selection has taken place or not is unknown, although it seems unlikely given what is known about the history of dehesas and the normal practices of land managers (Díaz et al. 1997; Moreno and Pulido 2009)

Regardless of whether or not the acorn production patterns of trees in dehesas are altered by management, endogenous influences are apparently still important, as seen in the strong negative individual temporal autocorrelations ($\overline{\mathrm{ACF1_i}}$ values) found in all populations including those in Spanish dehesas and forests and in

California (Table 7.2). Ongoing studies are designed to clarify the ways management has or has not altered the inherent acorn production patterns of oaks in dehesas.

7.2.5 Trade-Offs with Acorn Size

Although not studied as intensively as acorn crop size, a second key aspect of acorn production is size of the acorns themselves. Three studies are relevant, including two on holm oak in Spain and a third on valley oak in California. The first was particularly detailed, examining the effects of tree size, topographic position, crowding and interspecific competition, climatic factors, pruning, and size of the acorn crop over six years in trees growing in a dehesa (Alejano et al. 2011). As with crop size, there was considerable variation among trees. Drought during September, the key month for acorn growth, was particularly important, whereas no factor related to tree size or position was significant. They also found that the size of the acorn crop correlated negatively with acorn size and concluded that there appeared to be a trade-off between acorn size and number, as expected from life-history theory (Smith and Fretwell 1974; Wilbur 1977).

An eight-year study carried out in forests and dehesas of Cabañeros (Beamonte 2009; Beamonte and Díaz, unpublished data) showed quite different results. As in the above study, there were significant between-habitat differences in crop size but not in seed size, with crops being larger in dehesa than forest stands. However, no correlation was detected between seed size and crop size in the dehesa, while there was a positive, rather than a negative, correlation between these variables in the forest ($r = 0.30$, $P = 0.02$). Apparently forest trees are able to invest simultaneously in large seeds and large seed crops, an unexpected finding given that positive covariations between life-history characters are expected to be found when resources are not limiting (Venable 1992), whereas environmental conditions are relatively poor in forest stands due to competition for light and nutrients (Díaz et al. 2011).

This study also estimated the repeatability over a four-year period of seed and seed crop size, a measure of consistency that provides an upper limit to its heritability (Falconer and Mackay 1996). Repeatability (R) of both mean seed size and crop size was significant but moderate, especially for the dehesa (seed size: $R = 0.339 \pm 0.002$; crop size $R = 0.382 \pm 0.002$), but also for forest trees (seed size: $R = 0.227 \pm 0.007$; crop size $R = 0.010 \pm 0.005$), indicating moderate heritability of these traits. The hierarchical partitioning of seed size variation between habitats, among-trees within habitats, among branches within trees, and within branches (seed traps) indicates that majority of variance in seed size occurs within trees—particularly within branches—and among trees within habitats. Variation between habitats was small and not significant. Moderate repeatability of seed size between years and low variance related to environmental (among-habitat) factors suggest that neither seed size nor crop size are controlled by environmental factors, and that processes affecting variability in seed size operate primarily among plants by promoting variable rather than optimal seed sizes (Herrera 2009).

Lack of consistent selective pressure for optimal seed size undermines the theoretical basis for size-number trade-offs (Smith and Fretwell 1974). This trade-off was the focus of a study by Koenig et al. (2009a), who examined acorn mass in valley oak over a four-year period. They found that trees produced larger acorns when they had larger acorn crops, again failing to confirm a trade-off between seed size and number.

7.2.6 *Trade-Offs with Growth*

A second commonly studied trade-off is that between growth and reproduction. Although the intensity of a growth-reproduction trade-off is again expected to be more apparent in habitats with low nutrient availability or other environmental stresses (Reznick 1985), these costs can be difficult to detect among long-lived organisms such as oaks in poor environments because reproductive failure is likely to be relatively frequent.

Analysis of 70 holm oak trees over a period of nine years in Cabañeros, Spain, revealed no correlation between radial growth and acorn production during the same year (Díaz et al. 2011; Beamonte and Díaz, unpublished data). Growth was negatively correlated with reproduction the prior year and positively correlated with reproduction the following year, while reproduction was negatively correlated with reproduction the following year (a negative lag-1 autocorrelation). These results suggest the existence of stronger trade-offs in life-history characters acting across years rather than within years, as also found in California oaks (Knops et al. 2007).

Much more work needs to be done before we achieve a full understanding of how long-lived organisms partition their resources between the classic trade-offs of seed size and number, growth and reproduction, and male and female effort. Ongoing long-term studies of oaks in both California and Spain are making considerable headway on these evolutionarily important issues, yielding results that continue to challenge traditional life-history theory.

7.2.7 *Pollen Limitation*

All oaks are wind-pollinated, but determining how this key feature of their reproductive biology affects patterns of acorn production has proved difficult. One of the main problems has been to determine how far pollen travels. It has sometimes been assumed that pollen in such species was abundant and capable of traveling long distances, thus resulting in extensive gene flow (Koenig and Ashley 2003; Davis et al. 2004; Friedman and Barrett 2009), but a growing body of empirical and theoretical work has indicated that pollen limitation may play a key role in masting (Kelly et al. 2001; Satake and Iwasa 2002). Recent studies employing modern molecular methods capable of determining paternity of acorns have begun to address this issue, which is important due to the potential for pollen abundance to be limiting acorn production both within and among years.

In Spain, Garcia-Mozo et al. (2007) addressed this issue by measuring pollen emissions and environmental correlates of acorn production. They found rates of pollen emission were the most important factor determining mature acorn yields, indicating that pollen limitation is a key factor influencing acorn production in this species. Although pollen emissions have yet to be quantified in California, studies on blue and valley oaks have indicated that pollen dispersal may be far more restricted than previously thought in a way that could have an important influence on acorn production (Knapp et al. 2001; Sork et al. (2002). The latter study, based on results of molecular analyses of *Q. lobata* acorns in combination with a statistical model of paternity and genetic structure, is particularly notable as it found that the effective number of pollen donors per tree was strikingly small ($N_{ep} = 3.68$) and the average pollen dispersal distance was extremely short (64.8 m). Based on these results, these authors concluded that ongoing demographic attrition could reduce neighborhood size in this species to the extent that there could be a risk of reproductive failure and genetic isolation.

An alternative approach, taken by Abraham et al. (2011) on a different population of valley oak in California, is to directly determine paternity of acorns. Based on their analyses, N_{ep} was determined to be 219 and only 30 % of acorns were apparently fertilized by pollen coming from trees within 200 m, indicating significantly farther gene flow than estimated by the Sork et al. (2002) study. It would clearly be of interest to obtain comparable data from dehesas where trees are regularly spaced and intensively managed.

Regardless of how this controversy plays out, it would appear that pollen limitation plays a key role in acorn production. For example, recent work by Koenig et al. (2012) examining the relationship between phenology and acorn production in valley oak has found evidence that trees flowering in the middle of the season, when the majority of other trees are flowering and producing pollen, produce more acorns than trees flowering early or late in the season. The potential for differences in phenology playing a role in driving annual differences in the acorn crop has yet to be investigated, however.

7.2.8 Spatial Synchrony

Masting is a population-level phenomenon: a single tree may produce a variable acorn crop, but masting occurs by virtue of the fact that trees throughout the population do so more or less synchronously. Only recently, however, have researchers begun to investigate exactly how large that population is through the study of what is known as spatial synchrony (Liebhold et al. 2004).

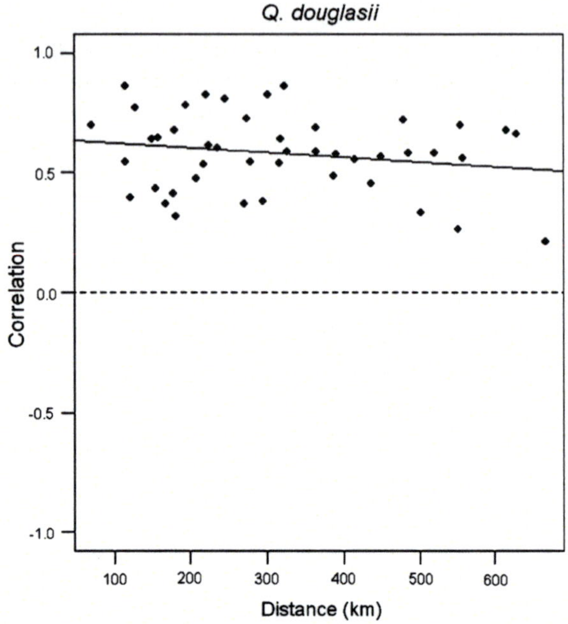

Fig. 7.5 Spatial synchrony in acorn production of *Q. douglasii* based on data from 10 sites in California studied over 17 years (Koenig and Knops, unpublished data). Plotted are the pairwise correlation coefficients versus the distance (in km) between sites. Note that all pairwise correlations are positive

Spatial synchrony is currently being investigated in California oaks by means of a statewide survey conducted since 1994 by W. Koenig and J. Knops. Preliminary results indicate relatively high spatial synchrony in at least some cases extending throughout the state. As an example, results for blue oak measured at 10 sites (Fig. 7.5) demonstrate (1) a decline in synchrony with distance, a pattern expected under most circumstances, and (2) significant spatial synchrony between sites across the entire geographic range of the species. These results indicate that acorn production in blue oak, perhaps the most abundant oak in California dominant across an area of over 50,000 km^2, is highly synchronous, providing wildlife with vast quantities of food in a mast year and leaving large areas with few acorns in a poor year. Comparable results have been found for holm oak by R. Alejano (unpublished) based on data acquired over six years at 18 sites up to nearly 500 km distant in Spain.

What drives such geographically widespread synchrony? One possibility is the "Moran effect," the hypothesis that environmental factors drive spatial synchrony (Ranta et al. 1997; Koenig 2002). In the case of oaks, ongoing analyses suggest that spatial synchrony in the variables correlating with acorn production within populations—in the case of blue oak, mean April temperature (Table 7.2)—may drive spatial synchrony among populations as well (Koenig and Knops, unpublished data).

The primary alternative to the Moran effect is the hypothesis that trees are synchronized by their mutual dependence on pollen produced by surrounding trees for fertilizing their flowers, a phenomenon known as "pollen coupling" (Satake and Iwasa 2000). Thus far, the evidence for pollen coupling as a driver of spatial synchrony in oaks is mostly indirect, but theoretical considerations have shown that even if pollen does not usually travel large distances, pollen coupling is capable of synchronizing reproduction over relatively large areas (Satake and Iwasa 2002). Resolving this issue will require not only more data on acorn production gathered over large geographic areas—the acquisition of which may in the future be facilitated by remote sensing (Yao et al. 2008)—but also by a greater understanding of the pattern and process of pollen dispersal itself.

Evidence thus far suggests that variability in flowering effort in oaks is relatively small compared to the high annual variation in the acorn crop (Pérez-Ramos et al. 2010). To the extent this is true, this further emphasizes the importance of pollen flow and successful fertilization—factors likely to be influenced by environmental factors during flowering and seed development—in determining the size of the acorn crop.

7.2.9 Acorn Production at the Community Level

Although the above analyses suggest the possibility of key differences in patterns of acorn production within populations, many communities of predators—particularly of vertebrates—tend to be generalists eager to depredate acorns of any species. Consequently, for some questions the relevant variable is overall acorn production by all species of oaks in the community rather than production by any individual species.

We currently know little about patterns of overall community acorn production either in California or Spain. In California, different species of oaks generally do not produce acorns synchronously, and thus annual variability in acorn abundance decreases with oak species diversity, a phenomenon that facilitates persistence by at least two acorn-dependent species, the acorn woodpecker (*Melanerpes formicivorus*) and western scrub-jay (*Aphelocoma californica*) (Koenig and Haydock 1999; Koenig et al. 2009b). Whether similar dependences exist among Spanish species and oak diversity has not been explored, although acorn-eating species such as the Eurasian jay (*Garrulus glandarius*) and European magpie (*Pica pica*) would be likely candidates.

There are, however, reasons to suspect that there might be intriguing differences between the two regions. One of the major factors facilitating asynchrony in acorn production by different species of oaks is the length of time needed for acorns to mature. In species in the white oak subgenus *Quercus* ("1-year" species), flowers produced in the spring are generally fertilized and mature into acorns the following fall, 5–7 months later. In contrast, species in the intermediate and black oak subgenera *Protobalanus* and *Erythrobalanus* generally, although not always, require an additional year to mature acorns ("2-year" species); that is, flowers produced in the spring of year x do not mature and produce acorns until the fall of year $x + 1$. As we have already seen, acorn production by many populations is influenced by environmental conditions during the period that flowers are produced and/or fertilized. As a result, acorn production between 1-year species of oaks is often at least somewhat synchronous, where there tends to be little or no synchrony between 1-year and 2-year species. For example, based on the five species Koenig and Knops have studied in central coastal California since 1980, the mean (±SD) population synchrony between the four combinations of species that require the same number of years to mature acorns is 0.57 ± 0.22, whereas mean synchrony for the six combinations of species that require a different number of years to mature acorns is only -0.23 ± 0.11, a significant difference (Wilcoxon rank sum test, $W = 0.24$, $P = 0.01$). Similarly, Espelta et al. (2008) reported high synchrony in acorn production between holm and downy oaks (both 1-year species) in Northeastern Spain.

This is potentially significant because California oaks are fairly evenly divided between 1-year and 2-year species, whereas Spanish oaks are not. Of the seven widespread species of California tree oaks (blue, Oregon white, valley, canyon live, coast live, California black, and interior live oaks), four are 1-year and three are 2-year species. In addition, there are at least 10 shrub species, of which seven are 1-year and three are 2-year species. In contrast, of the four widespread Mediterranean species of Spanish tree oaks (Pyrenean (*Q. pyrenaica*), Portuguese (*Q. faginea*), holm, and cork oaks), three are 1–year species while one, cork, is primarily a 1-year species but sometimes matures acorns in two years, with the frequency of the two types varying geographically (Díaz-Fernández et al. 2004). In addition, there is but a single shrub oak, the Kermes oak (*Q. coccifera*), which is the only consistent 2-year species in the region. This greater diversity in both oak species and time for acorns to mature is likely to reduce variability in annual acorn production at the community level in California compared to Spain, with considerable potential consequences on wildlife populations that have yet to be investigated.

7.3 Conclusions

There is clearly much more to be learned from comparisons of acorn production in California and Spain. The intensive management of oaks in dehesas provides an outstanding opportunity to learn more about the role of endogenous compared to abiotic factors such as temperature and rainfall in influencing acorn production at both the individual and population level. There also remain many questions concerning the mating systems of oaks, trade-offs between different oak life-history characters, and the patterns and drivers of spatial synchrony. Environmental conditions in the two regions are similar, but understanding how their subtle differences influence acorn production is likely to yield important insights about the proximate and ultimate factors affecting acorn production and masting behavior.

Box 1. Methods for Estimating Acorn Production

Despite decades of attention from wildlife managers and forest researchers, there is still no consensus as to the best way to quantify acorn production. As result, researchers use many different techniques, not all of which yield data that are readily comparable. Here we provide a brief review of these methods, dividing them into "direct" and "indirect" methods.

Direct Methods

Direct methods involve sampling in the crown or harvesting from the ground. They are more accurate for calculating real (or absolute) acorn production than indirect methods, although they suffer from the disadvantage of potentially ignoring acorns removed by birds or other wildlife prior to maturing. These methods include the following.

1. *Knocking down the acorns and collecting them under the crown*—This traditional method is also used for harvesting olives and some other fruits. Its primary disadvantages are that it is labor intensive, time consuming, and, in the case of large trees or of dense tree stands where individual canopies grow entangled, logistically difficult. It also potentially underestimates the crop by missing immature acorns that are not yet ready to fall. This is generally not a viable option if assessing many trees is desired, which is often the case due to large within-population variation and among population differences.
2. *Containers or traps method*—This method consists of placing containers or traps under the crown of the trees where acorns are removed on a regular basis (Fig. 7.B1). Many different kinds of containers have been used, varying in shape and construction. Containers may be on or attached to the ground, or hung from branches with ropes or wire to avoid consumption of acorns by large herbivores (wild ungulates or livestock). Typically, several containers are placed either regularly at different orientations or under the crown in a randomized design. Total acorn production per tree is obtained by adding, at the end of the dissemination period, the fruits periodically counted or weighed and then multiplying by the estimated fraction of the crown cover sampled by the traps.

Livestock and wild ungulates can be a problem for using containers since cattle and deer can easily knock over most traps. When livestock are present it is therefore a good idea to plan on protecting traps with fencing or use a design such as hanging containers in the tree that will minimize their impact.

The container method is also labor-intensive requiring considerable setup and repeated maintenance. Only a small proportion of the canopy is sampled, and only acorns that fall into the containers are counted or weighed, so arboreal acorn removal by animals is not considered—something that can be a serious problem in

Fig. 7.B1 Containers for estimating acorn production under a flowering holm oak (*Q. ilex*) in Huelva, Spain. Note the dendrometers on the oaks for measuring radial growth. (Photograph by R. Alejano)

certain years (Koenig et al. 1994a). If the goal, however, is to determine the acorn crop available for livestock, ground predators such as deer, or ground dispersers such as mice, this method should be seriously considered.

Acorn production measured with containers was quite consistent with the total acorn yield (measured by knocking down all acorns in the tree) divided by the crown surface ($R^2 = 0.82$, $F_{1,\ 39} = 184$, $P < 0.001$; Alejano et al. 2008).

3. *Visual surveys*—This method, which may involve a timed or complete survey of acorns on individual trees, is a nondestructive method allowing the subsequent harvesting of fruits. Other advantages include:
   (a) Counts are made just once during the dissemination period, so it is quicker and far less labor intensive than other direct methods.
   (b) Depending on the species and area, it can be performed one to two months before acorns mature, and thus to some extent allows crop prediction. It is important not to delay counting until after acorns start falling, since the method will then underestimate the crop unless caps remain on the tree and can be included in the survey.
   (c) Assuming the timing is right, counts will include most acorns that might later be removed from the crown by seed predators prior to acorn fall, and thus it potentially provides a more accurate measure of overall productivity than methods that quantify acorns that fall, such as the container method.

Visual surveys have been found to be consistent with the acorns harvested by using the container method (Koenig et al. 1994a), and has been widely used both in California and in Spain.

There are, however, disadvantages: counts are likely to be affected by factors influencing the ease with which acorns are seen such as light conditions, canopy cover, leaf density, and acorn coloration. The main disadvantage of timed visual surveys that do not completely sample the acorn crop, however, is that it only provides a measure of the relative, rather than the absolute, crop size. Counts are typically performed in an unknown area of the crown, so transforming this number into total number of acorns per tree or even total weight of acorns per tree is not an easy task.

Despite this caveat, however, tests of this method have generally been favorable. Perry and Thill (1999) tested five visual surveys methods and found the Koenig et al. (1994a) method to be the most efficient. Carevic et al. (2009; Fig. 7.B2) compared visual surveys and containers and obtained a regression that would be the starting point for estimating acorn weight from acorns counted for a particular species and geographical area. Residuals tended to deviate from expectations when many acorns were counted, and, to a lesser extent, when few acorns were seen. Counting for a longer period when acorns are rare or hard to see might improve the relationship between visual surveys and the "real" acorn crop when acorns are sparse; it is less clear how to distinguish between acorn crops at the upper end of the spectrum. To the extent that the acorn crop is good and such separation is desirable, an alternative method is probably needed.

The visual survey method proposed by Espárrago et al. (1992) and later modified by Vázquez (1998) has been used in Spanish dehesas as well. For its application, acorns within a 20 cm^2 wooden frame placed in front of different areas of the crown are counted. The average of at least 50 such counts per tree are done and used as an index of tree production. Several models have been proposed to translate the resulting acorn number into the total acorn crop assuming the crown to be a cylinder. Fernandez et al. (2008) checked the consistency of the method obtaining good results. A training period was desirable, however, since the experience of observers was found to influence the results.

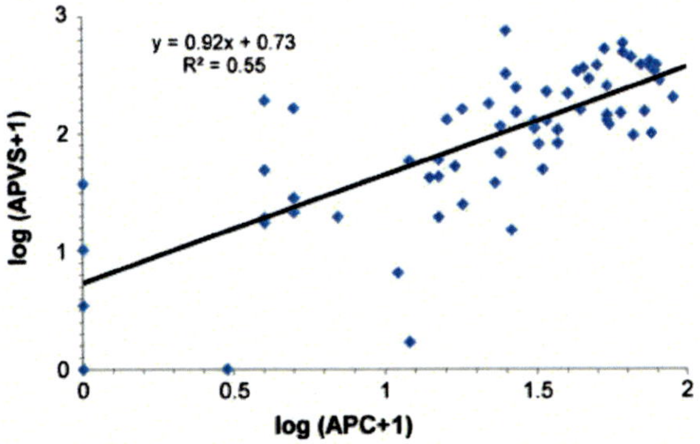

Fig. 7.B2 Regression of acorn production estimated from visual surveys (APVS) on acorn production estimated from container traps (APC, measured in g m^{-2} of crown area) for dehesas of holm oak in Huelva, Spain (from Carevic et al. 2009)

4. *Ranking methods*—Several methods have been used to evaluate acorn crops in dense oak forests in the USA. Sharp and Chisman (1961), studying white oak (*Q. alba*), proposed a qualitative method consisting of classifying a tree as a poor, good or extraordinary producer. Acorns in the end of the branches in the upper third of the crown were counted and averaged to yield acorn production per tree or per stand. A second method was proposed by Whitehead (1969) involving three qualitative parameters: the percentage of the crown containing seeds (0–3), the percentage of shoots within the crown producing seeds (also 0–3), and the average number of acorns per shoot (0–4). The Whitehead index is then obtained by adding the three values (thus 0–10), and was found by Perry and Thill (1999) to be highly correlated with the total number of acorns m^{-2} of crown area.

In Spain, Pulido and Díaz have developed a ranking method for long-term monitoring of acorn and pollen production of holm oak populations (see www.globimed.net/investigacion/Veceria01.htm, Díaz et al. 2011). Production is ranked into five categories: 0: no acorns or catkins; 1: <10% of the canopy covered by acorns/catkins; 2: 10–50%; 3: 50–90% and 4: >90%. Catkins are estimated in spring, when most trees are in full bloom, and acorns are estimated in early fall, after aborted seeds and those infested by insects have fallen. Several tests have demonstrated strong among-observers consistency in rank estimates after a short training period. Data taken in 2007–2010 from 145 trees provided with seed traps in Cabañeros National Park showed a strong correlation between this index and measures of the production of acorns in terms of the number of sound seeds m^{-2} ($r = 0.55$, $P = < 0.001$, $N = 374$; Díaz et al. 2011). This method enables rapid estimates of the among-years and among-individuals variation in the production of acorns and catkins, and also of the production of new shoots and leaves in spring and of the proportion of the canopy with leaves dry or lost for large number of trees, either isolated or growing in dense stands.

Indirect Methods

Several indirect methods have been described or mentioned for estimating acorn crops. We mention them here for completeness.

1. *Pollen*—A positive correlation has been reported between the amount of airborne Mediterranean oak pollen released to the atmosphere and the size of the acorn harvest (García-Mozo et al. 2007). This finding supports the hypothesis that pollen may be limiting, at least under some conditions, and have an important effect on subsequent acorn production in these wind-pollinated species, similar to its effects in many anemophilous species (Galán et al. 2004). To the extent this is true, integration of aerobiological, phenological and meteorological data could represent an important step forward in forest fruit production research (García-Mozo et al. 2007).

2. *Remote sensing*—Several recent studies have employed remote sensing techniques, including hyperspectral imaging, to estimate acorn yields (Yao et al. 2008; Panda et al. 2010; Yao and Sakai 2010). Such methods can at least in theory allow the mapping of acorn production over large geographic areas so as to yield within-stand abundance and spatial synchrony of acorn production. Remote sensing methods have yet to be applied to studies in either California or Spain, although they may eventually offer a powerful and less labor-intensive tool for assessing acorn production in our Mediterranean oak forests.
3. *Dendrochronology*—Based on the assumption of a tradeoff between growth and reproduction, Speer (2001) proposed a technique for mast reconstruction using dendrochronology for non-Mediterranean oaks. Although his results provided some optimism for this approach, it has not been used or tested by later authors. One problem is that in some cases it is likely that a negative correlation between growth and acorn production may be due to correlated effects of environmental variables rather than a trade-off per se (Knops et al. 2007). Nonetheless, the strong negative correlation between growth and reproduction observed in many species (Drobyshev et al. 2010) means that growth can potentially provide information useful for predicting subsequent acorn production in some species, regardless of the mechanism involved.
4. *Fattening of pigs* (for Spanish dehesas)—A traditional way to estimate acorn crops in Spanish dehesas is based on the degree to which pigs fatten during the dissemination period when they feed almost exclusively on acorns. Historical records with yearly controls would be required for this method to be practical.

Acknowledgments This paper is a contribution to the Spanish projects PAC–02–008 (Junta de Castilla–La Mancha), REN2003–07048/GLO and CGL20098-08430 (MCYT), 09/2002 (MMA) and 003/2007 (MMA), MONTES (Consolider–Ingenio CSD 2008–00040), P07 RNM02688 (Junta de Andalucía– FEDER, UE), and S UM 2006–00026–00–00 (MEC). WDK's work on California oaks has been supported by NSF grant DEB–0816691 and the University of California's Integrated Hardwoods Range Management Program.

References

Abraham ST, Zaya DN, Koenig WD, Ashley MV (2011) Interspecific and intraspecific pollination patterns of valley oak, *Quercus lobata*, in a mixed stand in coastal central California. Int J Plant Sci 172:691–699

Alejano R, Tapias R, Fernández M, Torres E, Alaejos J, Domingo J (2008) Influence of pruning and the climatic conditions on acorn production in holm oak (*Quercus ilex* L.) dehesas in SW Spain. Ann Forest Sci 65:209. doi:10.1051/forest:2007092

Alejano R, Vázquez-Piqué J, Carevic F, Fernández M (2011) Do ecological and silvicultural factors influence acorn mass in holm oak (southwestern Spain)? Agrofor Syst 83:25–39

Beamonte E (2009) Variabilidad espacial y temporal en el tamaño de semilla de la encina *Quercus ilex* subsp. *ballota*: repetibilidad y compromisos. MsC Thesis, Complutense University, Madrid, Spain

Biosca EG, González R, López-López MJ, Soria S, Montón C, Pérez-Laorga E, López MM (2003) Isolation and characterization of *Brenneria quercina*, causal agent for bark canker and drippy nut of *Quercus* spp. in Spain. Phytopathology 93:485–492

Bonal R, Muñoz A, Díaz M (2007) Satiation of predispersal seed predators: the importance of considering both plant and seed levels. Evol Ecol 21:367–380

Bonal R, Muñoz A, Espelta JM, Pulido F (2010) Los coleópteros perforadores de los frutos de encinas, robles, castaños y avellanos: biología, daños y tratamientos. Hojas Divulgadoras, MARM. Madrid, pp 46

Camarero JJ, Albuixech J, López-Lozano R, Auxiliadora Casterad M, Montserrat-Martí G (2010) An increase in canopy cover leads to masting in *Quercus ilex*. Trees 24:909–918

Cañellas I, Roig S, Poblaciones M, Gea-Izquierdo G, Olea L (2007) An approach to acorn production in Iberian dehesas. Agrofor Syst 70:3–9

Carbonero MD, Fernández Ranchal A, Fernández Rebollo P (2008) La producción de bellota en la dehesa. In: Fernández P, Carbonero MD, Blázquez A (eds) La dehesa del norte de Córdoba. Univ Córdoba, Córdoba, Spain, pp 185–204

Carbonero MD (2008) Los métodos de aforo de la producción de bellota en encina. Un análisis comparativo. In: Reunión científica de la Sociedad Española para el estudio de los pastos, 47. 575–581

Carbonero MD (2011) Evaluación de la producción y composición de la bellota de encina en dehesas. Ph.D. thesis, Univ Córdoba, Córdoba Spain

Carevic F, Alejano R, Fernández-Martínez M, Martín D (2009) Validación del método de conteos visuales para la cuantificación de la producción de bellota en dehesas de encina (*Quercus ilex* ssp. *ballota*). V Congreso Forestal Español, Avila, Spain

Carevic FS, Fernández M, Alejano R, Vázquez-Piqué J, Tapias R, Corral E, Domingo J (2010) Plant water relations and edaphoclimatic conditions affecting acorn production in a holm oak (*Quercus ilex* L. ssp. *ballota*) open woodland. Agrofor Syst 78:299–308

Crawley MJ (1985) Reduction of oak fecundity by low-density herbivore populations. Nature 314:163–164

Davis HG, Taylor CM, Lambrinos JG, Strong DR (2004) Pollen limitation causes an Allee effect in a wind-pollinated invasive grass (*Spartina alterniflora*). Proc Nat Acad Sci (USA) 101:13804–13807

Díaz M, Pulido FJ (2009) Vecería en la encina: primeros resultados. Third workshop of the ENCINA project. Parque Nacional de Cabañeros

Díaz, M, Alonso CL, Beamonte E, Fernández M, Smit C (2011) Desarrollo de un protocolo de seguimiento a largo plazo de los organismos clave para el funcionamiento de los bosques mediterráneos. In: Ramírez L, Asensio B (eds) Proyectos de investigación en Parques Nacionales: 2007–2010. Organismo Autónomo Parques Nacionales, Madrid, pp 47–75. Available at http://www.marm.es/es/ministerio/organizacion/organismos-publicos/03_INVESTIGACION_OK_tcm7-180265.pdf

Díaz M, Campos P, Pulido FJ (1997) The Spanish dehesas: a diversity in land-use and wildlife. In: Pain DJ, Pienkowski MW (eds) Farming and birds in Europe: the common agricultural policy and its implications for bird conservation. Academic Press, San Diego, pp 178–209

Díaz M, Pulido FJ, Møller AP (2004) Herbivore effects on developmental instability and fecundity of holm oaks Oecologia 139:224–234

Díaz-Fernández PM, Climent J, Gil L (2004) Biennial acorn maturation and its relationship with flowering phenology in Iberian populations of *Quercus suber*. Trees 18:615–621

Drobyshev I, Övergaard R, Saygin I, Niklasson M, Hickler T, Karlsson M, Sykes MT (2010) Masting behaviour and dendrochronology of European beech (*Fagus sylvatica* L.) in southern Sweden. Forest Ecol Manage 259:2160–2171

Dunning CE, Paine TD, Redak RA (2002) Insect-oak interactions with coast live oak (*Quercus agrifolia*) and Engelmann oak (*Q. engelmannii*) at the acorn and seedling stage. In: Standiford RB, McCreary D, Purcell KL (tech coords) Proceedings of 5th symposium on oak woodlands: oaks in California's changing landscape. Pacific SW Forest & Range Exp Station General Technical Report PSW–GTR–184

Espárrago F, Vázquez FM, Pérez MC (1992) Métodos de aforo de la montanera de *Quercus rotundifolia* Lam. In: II Coloquio sobre el cerdo Mediterráneo: 55

Espelta JM, Cortés P, Molowny-Horas R, Sánchez-Humanes B, Retana J (2008) Masting mediated by summer drought reduces acorn predation in Mediterranean oak forests. Ecology 89:805–817

Falconer DS, Mackay TFC (1996) Introduction to quantitative genetics, 4th edn. Longman, New York

Fernández P, Carbonero MD, Blázquez A (2008) La dehesa en el Norte de Córdoba. Perspectivas futuras para su conservación. Córdoba University, Córdoba, Spain

Friedman J, Barrett SCH (2009) Wind of change: new insights on the ecology and evolution of pollination and mating in wind-pollinated plants. Ann Botany 103:1515–1527

Galán C, Vázquez L, García-Mozo H, Domínguez E (2004) Forecasting olive (*Olea europaea*) crop yield based on pollen emission. Field Crops Res 86:43–51

García-Mozo H, Hidalgo PJ, Galán C, Gómez-Casero MT, Domínguez E (2001) Catkin frost damage in Mediterranean cork-oak (*Quercus suber* L.). Israel J Plant Sci 49:41–47

García-Mozo H, Gómez-Casero MT, Domínguez E, Galán C (2007) Influence of pollen emission and weather-related factors on variations in holm-oak (*Quercus ilex* subsp. *ballota*) acorn production. Environ Exper Bot 61:35–40

Garrison BA, Koenig WD, Knops JMH (2008) Spatial synchrony and temporal patterns in acorn production of California black oaks. In: Merenlender A, McCreary D, Purcell KL (tech coords) Proceedings of 6th symposium on oak woodlands: today's challenges, tomorrow's opportunities. Pacific SW Forest & Range Exp Station General Technical Report PSW–GTR–217

Herrera CM (1998) Population-level estimates of interannual variability in seed production: what do they actually tell us? Oikos 82:612–616

Herrera CM (2009) Multiplicity in unity: plant subindividual variation and interactions with animals. University of Chicago Press, Chicago, IL, USA

Hildebrand DC, Schroth MN (1967) A new species of *Erwinia* causing the drippy nut disease of live oaks. Phytopathology 57:250–253

Hobbs RJ, Richardson DM, Davis GW (1995) Mediterranean-type ecosystems: opportunities and constraints for studying the function of biodiversity. In: Davis GW, Richardson DM (eds) Mediterranean–type ecosystems: the function of biodiversity. Springer, New York, pp 1–42

Huntsinger L, Bartolome JW (1992) Ecological dynamics of *Quercus* dominated woodlands in California and southern Spain: a state-transition model. Vegetatio 99–100:299–305

Huntsinger L, Bartolome JW, Starrs PF (1991) A comparison of management strategies in the oak woodlands of Spain and California. In: Standiford RB (tech coord) Proceedings of symposium on oak woodlands and hardwood rangeland management. Pacific SW Forest & Range Exp Station Gen Tech Rep PSW126, 300–306

Jackson LE (1985) Origins of California's Mediterranean grasses. J Biogeogr 12:349–361

Kelly D (1994) The evolutionary ecology of mast seeding. Trends Ecol Evol 9:465–470

Kelly D, Sork VL (2002) Mast seeding in perennial plants: Why, how, where? Annu Rev Ecol Syst 33:427–447

Kelly D, Hart DE, Allen RB (2001) Evaluating the wind-pollination benefits of mast seeding. Ecology 82:117–126

Knapp EE, Goedde MA, Rice KJ (2001) Pollen-limited reproduction in blue oak: implications for wind pollination in fragmented populations. Oecologia 128:48–55

Knops JMH, Koenig WD, Carmen WJ (2007) A negative correlation does not imply a treadeoff between growth and reproduction in California oaks. Proc Nat Acad Sci (USA) 104:16982–16985

Koenig WD (2002) Global patterns of environmental synchrony and the Moran effect. Ecography 25:283–288

Koenig WD, Ashley MV (2003) Is pollen limited? The answer is blowin' in the wind. Trends Ecol Evol 18:157–159

Koenig WD, Haydock J (1999) Oaks, acorns, and the geographical ecology of the acorn woodpecker. J Biogeogr 26:159–165
Koenig WD, Knops JMH (2002) The behavioral ecology of oaks. In: McShea WJ, Healy WM (eds) Oak forest ecosystems. The Johns Hopkins University Press, Baltimore, pp 129–148
Koenig WD, Knops JMH, Carmen WJ, Stanback MT, Mumme RL (1994a) Estimating acorn crops using visual surveys. Can J For Res 24:2105–2112
Koenig WD, Mumme RL, Carmen WJ, Stanback MT (1994b) Acorn production by oaks in central coastal California: variation in and among years. Ecology 75:99–109
Koenig WD, Knops JMH, Carmen WJ, Stanback MT, Mumme RL (1996) Acorn production by oaks in central coastal California: influence of weather at three levels. Can J For Res 26:1677–1683
Koenig WD, Knops JMH, Carmen WJ (2002) Arboreal seed removal and insect damage in three California oaks. In: Standiford RB, McCreary D, Purcell KL (tech coords) Proceedings of 5th symposium on oak woodlands: oaks in California's changing landscape. Pacific SW Forest & Range Exp Station Gen Tech Rep PSW–GTR–184, pp 193–204
Koenig WD, Kelly D, Sork VL, Duncan RP, Elkinton JS, Peltonen MS, Westfall RD (2003) Dissecting components of population-level variation in seed production and the evolution of masting behavior. Oikos 102:581–591
Koenig WD, Knops JMH, Carmen WJ, Sage RD (2009a) No trade-off between seed size and number in the valley oak *Quercus lobata*. Am Nat 173:682–688
Koenig WD, Krakauer AH, Monahan WB, Haydock J, Knops JMH, Carmen WJ (2009b) Mast-producing trees and the geographical ecology of western scrub-jays. Ecography 32:561–570
Koenig WD, Funk KA, Kraft TS, Carmen WJ, Barringer BC, Knops JMH (2012) Stabilizing selection for within-season flowering phenology confirms pollen limitation in a wind-pollinated tree. J Ecol 100:758–763
Leiva MJ, Fernández-Alés R (2005) Holm-oak (*Quercus ilex* subsp. *ballota*) acorns infestation by insects in Mediterranean dehesas and shrublands: Its effect on acorn germination and seedling emergence. Forest Ecol Manage 212: 221–229
Liebhold A, Koenig WD, Bjørnstad ON (2004) Spatial synchrony in population dynamics. Annu Rev Ecol Evol Syst 35:467–490
Montero G, San-Miguel A, Cañellas I (2000) Systems of Mediterranean silviculture: the dehesa. Mundiprensa, Madrid
Moreno G, Pulido FJ (2009) The functioning, management, and persistence of dehesas. In: Rigueiro A, McAdam J, Mosquera R (eds) Agroforestry in Europe. Springer, Amsterdam, pp 127–160
Panda SS, Hoogenboom G, Paz JO (2010) Remote sensing and geospatial technological applications for site-specific management of fruit and nut crops: a review. Remote Sens 2:1973–1997
Parsons JJ (1962) The acorn-hog economy of the oak woodlands of Southwestern Spain. Geogr Rev 52:211–235
Pérez-Ramos IM, Ourcival JM, Limousin JM, Rambal S (2010) Mast seeding under increasing drought: results from a long-term data set and from a rainfall exclusion experiment. Ecology 91:3057–3068
Perry RW, Thill RE (1999) Estimating mast production: an evaluation of visual surveys and comparison with seed traps using white oaks. South J Appl For 23:164–169
Pons J, Pausas JG (2012) The coexistence of acorns with different maturation patterns explains acorn production variability in cork oak. Oecologia 169:723–731
Pulido FJ, Díaz M (2005) Regeneration of a Mediterranean oak: a whole-cycle approach. Écoscience 12:92–102
Pulido F, García E, Obrador JJ, Moreno G (2010) Multiple pathways for tree regeneration in anthropogenic savannas: incorporating biotic and abiotic drivers into management schemes. J Appl Ecol 47:1272–1281
Ranta E, Kaitala V, Lindström J, Helle E (1997) The Moran effect and synchrony in population dynamics. Oikos 78:136–142

Reznick D (1985) Costs of reproduction: an evaluation of the empirical evidence. Oikos 44:257–267

Rodríguez-Estévez V, García A, Gómez AG (2009) Characteristics of the acorns selected by free range Iberian pigs during the montanera season. Livestock Sci 122:169–176

Satake A, Iwasa Y (2000) Pollen coupling of forest trees: forming synchronized and periodic reproduction out of chaos. J Theor Biol 203:63–84

Satake A, Iwasa Y (2002) Spatially limited pollen exchange and a long-range synchronization of trees. Ecology 83:993–1005

Sharp WM, Chisman HH (1961) Flowering and fruiting in the white oaks. I. Staminate flowering through pollen dispersal. Ecology 42:365–372

Siscart D, Diego V, Lloret F (1999) Acorn ecology. In: Rodà R, Retana J, Gracia CA, Bellot J (eds) Ecology of Mediterranean evergreen oak forests. Springer, Berlin, pp 75–86

Smith CC, Fretwell SD (1974) The optimal balance between size and number of offspring. Am Nat 108:499–506

Sork VL, Bramble J, Sexton O (1993) Ecology of mast-fruiting in three species of North American deciduous oaks. Ecology 74:528–541

Sork VL, Davis FW, Smouse PE, Apsit VJ, Dyer RJ, Fernandez-M JF, Kuhn B (2002) Pollen movement in declining populations of California valley oak, *Quercus lobata*: Where have all the fathers gone? Mol Ecol 11:1657–1668

Speer JH (2001) Oak mast history from dendrochronology: a new technique demonstrated in the Southern Appalachian region. Ph.D. thesis, Univ Tennessee, Knoxville, TN, USA

Tyler CM, Kuhn B, Davis FW (2006) Demography and recruitment limitations of three oak species in California. Q Rev Biol 81:127–152

Vázquez FM (1998) Producción de bellotas en *Quercus*. I. Métodos de estimación. Solo Cerdo Ibérico 1:59–66

Vázquez FM, Balbuena E, Doncel E, Ramos S (2000) Distribución del melazo en la provincia de Badajoz para la cosecha de bellotas de *Quercus rotundifolia* Lam. durante 1999. Boletín de Sanidad Vegetal Plagas 26:287–296

Venable DL (1992) Size-number trade-offs and the variation of seed size with plant resource status. Am Nat 140:287–304

Whitehead CJ (1969) Oak mast yields on wildlife management areas in Tennessee. Tennessee Wildlife Resources Agency, Nashville

Wilbur HM (1977) Propagule size, number, and dispersion pattern in *Ambystoma* and *Asclepias*. Am Nat 111:43–68

Yao Z, Sakai K (2010) Mapping spatial variation in acorn production from airborne hyperspectral imagery. For Stud China 12:49–54

Yao Z, Sakai K, Ye X, Akita T, Iwabuchi Y, Hoshino Y (2008) Airborne hyperspectral imaging for estimating acorn yield based on PLS B-matrix calibration technique. Ecol Inform 3:237–244

Part III
Management, Uses, and Ecosystem Response

Chapter 8
Effects of Management on Biological Diversity and Endangered Species

Mario Díaz, William D. Tietje and Reginald H. Barrett

Frontispiece Chapter 8. Large flock of griffon (*Gyps fulvus*) and black (*Aegypius monachus*) vultures in the forest-dehesa transition of the National Park of Cabañeros (central Spain). (Photograph by M. Díaz)

M. Díaz (✉)
Department of Biogeography and Global Change, Museo Nacional de Ciencias Naturales (BGC-MNCN), Spanish National Research Council (CSIC),
Serrano 115bis, E-28006Madrid, Spain
e-mail: Mario.Diaz@ccma.csic.es

P. Campos et al. (eds.), *Mediterranean Oak Woodland Working Landscapes*, Landscape Series 16, DOI: 10.1007/978-94-007-6707-2_8,
© Springer Science+Business Media Dordrecht 2013

Abstract High biodiversity in Spanish and California woodlands is due to the intermixing of habitat types and habitat elements. Dehesa management in Spain creates a mosaic of vegetation that includes trees, shrubs, and grasslands. Maintaining this diversity requires control of invasive shrubs, but sustaining the woodlands calls for periodic management to permit an encroachment of shrubs that foster oak regeneration. Californian oak woodlands are also high in biodiversity, but have been managed far less intensively, largely for acorns and game in the pre-contact period and for livestock grazing and game in current times. Shrub invasion is slower and less common than in Spain. The impacts of livestock on oak regeneration seems to vary across California's very heterogeneous climatic and soil conditions. Just as biodiversity supports the multifunctional dehesa economy, the possibilities of income generation from biodiversity may be crucial to the sustenance of California oak woodland ranches, reducing conversion to intensive agriculture and urbanization.

Keywords Biodiversity · Habitat mixing · Oak woodland · Spatial and temporal scales · Species richness

8.1 Introduction

Spanish and California oak woodlands are unique landscapes with high levels of biological diversity. Like other Mediterranean climate regions, these woodlands qualify as biodiversity hotspots (Myers et al. 2000; López-López et al. 2011). Many species of economic interest and conservation concern co-exist with common land uses and management practices.

Diversity in natural ecosystems is believed to support the maintenance of several key ecological functions, because redundancy within functional groups increases the probable survival of at least one species per group. However, the specific role of biodiversity in the support of many ecological functions, such as pollination or seed dispersal, is not well understood (García and Martínez 2012), nor is its role fully known in sustaining managed semi-natural ecosystems that

---

W. D. Tietje
Department of Environmental Science, Policy, and Management,
University of California, Berkeley, 2156 Sierra Way, Suite C,
San Luis Obispo, CA 93401, USA
e-mail: tietje@berkeley.edu

R. H. Barrett
Department of Environmental Science, Policy, and Management,
University of California, Berkeley, Berkeley, CA 94720-3114, USA
e-mail: rbarrett@berkeley.edu

have been created or maintained by human activity, like ranch and dehesa woodlands. Conservation of species-rich systems or endangered species is of increasing relevance in the design of land-use policies aimed at maintaining agricultural uses in developed countries (Mattison and Norris 2005). The high levels of biodiversity in dehesa and California oak woodland support multifunctional land uses that can increase the stability of income for landowners by providing opportunities for income from wildlife-oriented products, including hunting, fishing, and birdwatching, many types of recreation, and various forms of conservation markets and payments, in addition to agricultural income.

This chapter reviews biological diversity patterns found in oak woodlands in Spain and California and analyzes how biodiversity levels relate to management at distinct spatial scales. Reported levels of species richness of groups such as vascular plants, butterflies, or terrestrial vertebrates (especially birds and mammals) are first compared between oak woodlands and nearby habitats. Second, we analyze how richness varies within oak woodlands, in relation to effects of management on habitat structure. Finally, we address how other land uses affect levels of biodiversity. The aim is to address how and why habitat mixtures, biological diversity, and land uses are related in Spanish and Californian oak woodlands, and how such relationships contribute to the long-term sustainability of these working landscapes.

8.2 Vertebrate Wildlife in California Oak Woodlands

The great diversity of plants and animals in the 3 million ha of California oak woodland evolved in a physically complex, heterogeneous landscape under the favorable influence of the relatively mild and predictable Mediterranean climate. Since the Pleistocene, Native Americans and much later-arriving settlers of European heritage, further shaped the mix of habitats (Chap. 2).

Among the six major woody vegetation types in California identified by the California Wildlife Habitat Relationships System (CWHR; Airola 1988), oak woodlands harbor the greatest diversity of terrestrial vertebrate species (Fig. 8.1). Out of the 1,030 terrestrial species statewide, the database shows that 391 use oak woodlands. All of the six major California oak woodland types are predicted to have similar numbers of terrestrial vertebrates. However, due to different geological histories and adaptations to regional abiotic and biotic conditions, each woodland type is unique and therefore wildlife community composition varies among types.

Of the 391 terrestrial vertebrate species that occur in California oak woodland, the CWHR system indicates that 226 (58 %) have no special legal status, 47 (12 %) are game species, and 122 (31 %) are listed in some category of special concern (Table 8.1). The majority of species (95 %) are native. Species of concern include the Tehachapi slender salamander (*Batrachoceps stebbinsi*), yellow-blotched salamander (*Ensatina eschscholtzii croceater*), San Joaquin pocket mouse (*Perognathus inornatus*), golden eagle (*Aquila chrysaetos*), and Cooper's hawk (*Accipiter cooperii*) (Fig 8.2).

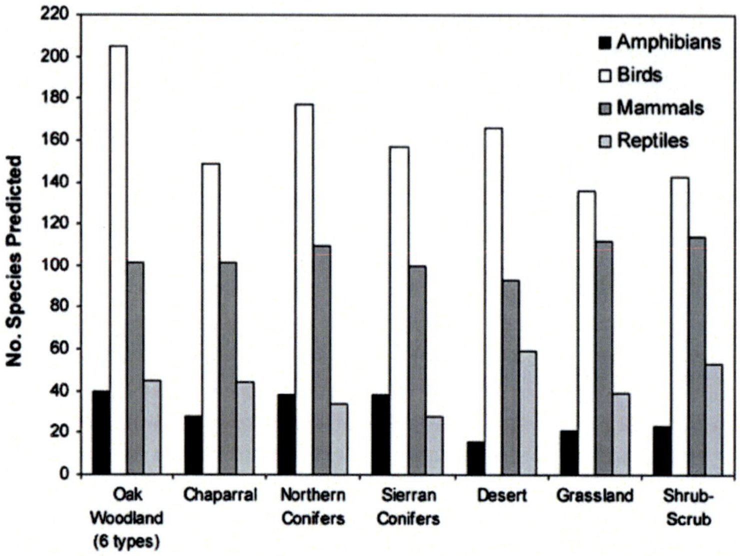

Fig. 8.1 Terrestrial wildlife species richness predicted by the California Wildlife Habitat Relationships System (CWHR; Airola 1988) for oak woodlands and other major habitat types in the state. The CWHR is an information system for California's wildlife that contains life history, geographic range, habitat relationships, and management information on the species of amphibians, reptiles, birds, and mammals known to occur in the state. On the basis of its habitat requirements, each species is predicted to occupy one or several habitat types, including oak woodlands

8.2.1 Wildlife Diversity at Landscape and Patch Scales

8.2.1.1 Landscape-Level Heterogeneity and Vertebrate Diversity

California oak woodlands are a mosaic of grassland, shrubland, and woodland. Landscape ecologists refer to distinct areas of grassland, shrubland, and woodland as "patches." The landscape surrounding a woodland patch is termed "matrix." Contemporary landscape ecology principles predict that the occurrence and persistence of wildlife in the landscape depends on the size, habitat quality, and arrangement of patches within the matrix (Forman 1995). For those species with habitat needs that can be met by a variety of habitats (generalist species), the matrix can be an alternative or secondary habitat that can sustain them, at least temporarily. Kie et al. (2002) documented a higher abundance of mule deer (*Odocoileus hemionus*) in a landscape where the matrix was of sufficient quality to permit safe movement among woodland patches when local food was depleted. Klinger et al. (1989) reported that for several years following a prescribed fire, mule deer moved from wooded patches to feed in burned chaparral during the wet

Table 8.1 Legal status by class of the 391 terrestrial vertebrate species predicted by the California Wildlife Habitat Relationships System (CWHR: Airola 1988; Fig. 8.1)

Status[1]	Amphibians	Reptiles	Birds	Mammals	Total species
Federal endangered	3	2	4	10	19
Federal threatened	2	3	4	0	9
California endangered	1	2	8	4	15
California threatened	4	3	3	2	12
California fully protected	2	2	6	2	12
California species of special concern	11	13	33	25	82
Federal candidate	0	0	1	0	1
BLM sensitive	4	5	6	14	29
USFS sensitive	12	8	7	6	33
CDF sensitive	0	0	9	0	9
All special concern (above)	18 (45 %)	20 (44 %)	47 (23 %)	37 (37 %)	122 (31 %)
Harvest (game) species	1 (3 %)	0	18 (9 %)	28 (28 %)	47 (12 %)
No special status	21 (53 %)	25 (56 %)	141 (69 %)	39 (39 %)	226 (58 %)
Introduced	2 (5 %)	0	5 (2 %)	13 (13 %)	20 (5 %)
Total species in class[2]	40	45	205	101	391

Percentages given are relative to the total number of species within each class

[1] An entire species is counted within a status category if one or more of its subspecies have been so designated

[2] Totals may be less than the sum of the above rows due to some species being classified in more than one category

seasons; deer preferentially used shrubby patches in close proximity (<200 m) to patches of woodland and grassland.

The intrinsic properties of patches influence patch occupancy in a species-specific way. Consequently, each patch tends to have a unique suite of species, resulting in high overall biodiversity at the landscape level (Sisk et al. 1997). Key conclusions from research conducted during the last two decades in woodland landscapes (Radford and Bennett 2007; Prevedello and Vieira 2010; Thornton et al. 2011) are (a) habitat quality of the patch and of the matrix, together, influence wildlife diversity in the landscape; (b) the extent of tree cover across a landscape disproportionately and positively influences animal diversity, because trees increase the functional connectivity across the landscape either as treed linkages or as individual trees that provide stepping stones through inhospitable areas (Beier 1993; Manning et al. 2006); and (c) the effects of landscape pattern on species occurrences are species-specific; i.e., species respond differently to the same landscape pattern.

Based on these conclusions, management at the patch scale (small-scale management) and at the landscape scale (large-scale management) are complementary in their influence on species diversity, arguing for the management of oak woodland at large scales. Bolger et al. noted in 1997: "Our results also underscore the need for landscape-scale variation to be explicitly incorporated in wildlife-habitat relationship models, particularly if they are to be applied to heterogeneous

Fig. 8.2 The California golden eagle *Aquila chrysaetos*, which is federally protected in the U.S. as a "Bird of Conservation Concern," nests in large oak trees, here in a valley oak (*Quercus lobata*). (Photograph by B. Lyon)

landscapes." A general guideline for management: maintain the mix of habitats, especially tree cover, in the working landscape.

8.2.1.2 Patch-Level Heterogeneity and Vertebrate Diversity

Most studies of oak woodland diversity have been carried out at the patch scale, and in patches dominated by trees. Tree canopy, through its effect on light penetration, creates microsite conditions in which various floral assemblages develop. Floristic diversity, together with the site's physical attributes, provides the varied and unique habitat needs of wildlife species (habitat elements or habitat structures). The CWHR system lists 124 distinct habitat elements statewide (Airola 1988), for example standing dead trees or "snags". When key habitat elements are removed, the number of predicted vertebrate species declines unless a functionally equivalent element is present (Table 8.2). The relative importance of each element varies widely among taxa, however, as does the effect of management practices on the occurrence and abundance of each element.

Table 8.2 The number of vertebrate oak woodland species that depend on each of 5 selected habitat elements as predicted by the California Wildlife Habitat Relationships System (CWHR; Airola 1988), and therefore would not be present unless the element (or a functional equivalent) was present

	No. of species dependent on the habitat element				
Habitat element	Birds	Mammals	Amphibians	Reptiles	Overall
Shrubs	54 (26 %)	25 (25 %)	0	2 (4 %)	81 (21 %)
Riparian	42 (20 %)	5 (5 %)	7 (17 %)	1 (2 %)	55 (14 %)
Cavities	32 (16 %)	13 (13 %)	0	0	45 (11 %)
Snags (standing dead trees)	25 (12 %)	8 (8 %)	2 (5 %)	0	35 (9 %)
Coarse woody debris	2 (<1 %)	8 (8 %)	7 (17 %)	1 (2 %)	18 (5 %)

Numbers are based primarily on expert opinion incorporated in the CWHR system's species occurrence models

Shrubs in the understory of treed patches increase the diversity of vertebrate wildlife (Table 8.2; Tietje and Vreeland 1997). Among seven California oak woodland types, average shrub cover varies between 9 % for blue oak (*Quercus douglasii*) and 38 % for coast live oak (*Quercus agrifolia*), partly explaining the differences in vertebrate communities in different woodland types. In California's central coast, the presence of a well-developed, structurally complex understory layer is positively associated with winter densities of numerous oak woodland birds (W. Tietje unpublished data).

Second only to a shrub layer, riparian areas and other wetlands contribute most to the maintenance of biological diversity in oak woodlands (Table 8.2). Bird density and species richness in riparian areas are typically much higher than in the surrounding oak woodland (e.g., Laymon 1984). Most of the rare species in California oak woodland have some habitat needs that must be met by wetland habitat. California tiger salamander (*Ambystoma californiense*) requires vernal pools or stockponds for breeding. As many vernal pools have succumbed to development, stockponds installed by ranchers have become more important to the species (USFWS 2004). Least Bell's vireo (*Vireo bellii pusillus*) is an obligate of southern California riparian habitat dominated by willow (*Salix* spp.). Giusti et al. (2003) detected 52 resident and 29 migrant species during 10 years of bird monitoring of a stream in north-central California; 12 of them are listed as protected, sensitive, or of conservation concern. There is relatively little study of the management, wildlife values, and ecosystem services provided by intermittent streams, a notable gap in oak woodland research.

Compared to other forest types, snags are rare in oak woodland because mature oak trees gradually disintegrate, losing dead branches little by little, while retaining living branches. Living oaks, in which most nesting cavities occur, may actually be more suitable than snags for cavity-nesting birds. Cavities in living oaks have harder, thicker, wood that provides more insulation and nest protection, and living oaks persist for many years. Purcell and Drynan reported in 2008 that nearly 71 % of cavity nesting birds using hardwoods nested in live trees.

Results of a cavity-blocking experiment in the western Sierra Nevada foothills during the mid-1980s suggested that nesting cavities were not limiting to cavity-nesting birds (Waters et al. 1990), but since then numbers of European starlings (*Sturnus vulgaris*) have increased (Purcell et al. 2002). Starlings usurp nest cavities and predate nests (Olson et al. 2008), but the long-term effects of starling invasion are unclear. Koenig (2003), using long-term datasets to compare densities of 27 native cavity-nesting species before and after starling invasion, found no compelling evidence that starlings were responsible for a decline in any species. He cautioned, however, that it may be too early to detect strong effects on some species. Indeed, the best cavity nest sites providing protection from both bad weather and nest predators might now be a limited resource for some bird species (K. Purcell, personal communication). More subtly, competition for nest cavities might ultimately cause some native species to delay nesting until later in the season, with unknown long-term consequences (Ingold 1994).

Coarse woody debris such as downed trees, large limbs, and large broken pieces of wood on the woodland floor, is used by all vertebrate groups, but is especially important for small mammals and amphibians (Table 8.2). Woodland management may remove such debris to increase forage, reduce fire risk, or simply to clean things up. Coarse woody debris is most abundant on the California central coast and least abundant in northeastern California, but is largely lacking over half of California oak woodland (Tietje et al. 2002). On lightly managed intermixed blue oak and coast live oak woodland in coastal central California, coarse woody debris volume ranged from 1–21 m^3 per ha in patches with 40–90 % canopy cover (W. Tietje unpublished data). In mixed California black oak (*Quercus kelloggii*)-ponderosa pine (*Pinus ponderosa*) stands in the central Sierra Nevada, woody debris was rare; however, dead limbs were common on the large living black oak trees, probably helping to mitigate for the deficiency of it on the ground (Garrison et al. 2002).

8.2.1.3 Heterogeneity at Intermediate Scales: Isolated Trees and Mixed Oak Stands

Large trees, especially large deciduous oak trees, are important for the maintenance of wildlife diversity. Acorn woodpecker (*Melanerpes formicivorus*) typically selects larger trees for use as "granary trees" in which they cache large numbers of acorns (Wilson et al. 1991). The authors attributed much of the breeding bird diversity in woodland of lower tree density to the characteristically larger trees in those stands. Large (>104 cm dbh) valley oak (*Q. lobata*) and blue oak trees are used by Western purple martins (*Progne subis arboricola*) at what is likely the last location in California where the species regularly nests in oak woodland (Williams 2002). The large valley oak is also the tree species most used by mammals (Barrett 1980). With the alteration of oak woodland by intensive agricultural development and urban sprawl, large, isolated trees become an increasingly important feature of the fragmented landscape. Manning et al. (2006)

Fig. 8.3 The acorn woodpecker (*Melanerpes formicivorus*), a common inhabitant of California oak woodland, occurs only in woodlands with at least two species of oak trees. (Photograph by B. Lyon)

acknowledged the ecological value of these trees, identifying them as "keystone" structures in human-modified landscapes. By adding structural diversity to the landscape, isolated trees enhance bird diversity, particularly for species that would not occur in crop fields without them (DeMars et al. 2010) (Fig 8.3).

Because each oak species offers unique and temporally variable resources, mixed-oak stands can be a precondition for the occurrence of some wildlife. The acorn woodpecker is the classic example, its distribution restricted to oak woodland where at least two species of oak trees occur, thereby increasing the chance that at least one tree species will produce a crop of acorns each year (Koenig and Haydock 1999). Further underscoring the importance of a mix of tree species, three oak woodland parids, the oak titmouse (*Baeolophus inornatus*), chestnut-backed chickadee (*Poecile rufescens*), and bushtit (*Psaltriparus minimus*), co-occur in mixed stands of evergreen and deciduous oaks where the three species are able to segregate foraging sites (Hertz et al. 1976). Still further, a bark forager, Nuttall's woodpecker (*Picoides nuttallii*), forages on blue oak during the breeding season, but then turns to gray pine (*Pinus sabiniana*) and an evergreen oak during the winter (Block 1991). Clearly, the maintenance of mixed stands is a prudent management guideline.

8.2.2 Land Use and Wildlife

Urban et al. (1987) discussed important differences between natural, Native American, and historic human disturbances versus today's anthropogenic activities. These differences may present challenges to plant and animal species. Contemporary land disturbances, such as intense wildfire following long periods of suppression, can be "large-frequent" compared to most historic disturbances. Historic disturbances were more likely to be either "large-infrequent" or "small-frequent", for example frequent lightning-caused wildfires and fires promoted by Native Americans, herders, and farmers. Further, human activities today tend to remove important habitat elements over broad areas (e.g., large trees and shrubs in woodland landscapes), and therefore reduce structural complexity. Finally, the most intensive anthropogenic disturbances in California woodlands (vineyards and urban development) are relatively recent, and it is not yet clear whether wildlife assemblages will be able to adapt and cope with these habitat modifications.

8.2.2.1 Extensive Land Use (Grazing)

The effects of livestock grazing on the oak woodland community vary widely according to timing, intensity, and livestock type. Not surprising, responses to grazing are highly variable among plant and animal species. Vegetative elements important to wildlife (e.g., grass, woodland shrub cover, and litter biomass) may be reduced, changed structurally, or otherwise altered by livestock grazing (Jones 2000). Grazing can reduce small mammal abundance (Johnson and Horn 2008) and can alter the foraging behavior of wild herbivores such as deer (Kie 1996). Despite clear changes in understory vegetation, Verner et al. (1997) found few differences in the number of bird territories on grazed and ungrazed areas of California oak-gray pine woodland. Several studies conducted in the Southwestern US and in California oak woodland suggest that well-managed grazing systems can benefit native plant and wildlife species. For example, native perennial grasses were found to benefit from being grazed by cattle and sheep (Edwards 1992), and Marty (2005) reported higher richness of aquatic invertebrates and native plants in continuously-grazed vernal pool grasslands than at ungrazed sites in Central California oak woodland. Several endangered species are known to benefit from grazing, including San Joaquin kit foxes (*Vulpes macrotis mutica*) (USFWS 2010), and Stephen's kangaroo rats (*Dipodomys stephensi*) (USFWS 1997). Weiss (1999) found that grazing was necessary to maintain habitat suitability for the endangered Bay checkerspot butterflies (*Euphydryas editha bayensis*). It has been argued that if livestock are properly managed, desired goals of animal production, economic sustainability, and wildlife conservation can usually be achieved (George 1991).

8.2.2.2 Intensive Agriculture (Vineyards)

The installation of new vineyards, oftentimes in undeveloped oak woodland, increased dramatically during the 1990s in coastal California. With the exception of several habitat specialists, avian community composition can be similar in remnant woodland patches adjacent to vineyards and in undeveloped woodlands

(Tietje et al. 2008). Reynolds et al. (2008) reported comparable nest densities and nesting success of birds in wooded patches surrounding vineyards and in nearby undeveloped areas. The authors, however, cautioned that it may simply have been too early to detect effects. Combining models of mesocarnivore occurrence with projected vineyard development indicated that in a cropland-oak woodland mosaic, mesocarnivores were less likely to occupy large vineyard blocks than small, isolated, vineyards (Hilty et al. 2006).

Growers often erect artificial roosting boxes or nesting boxes in vineyards to increase wildlife diversity, help with pest control, and to enhance aesthetics (Heaton et al. 2008). Songbird nest boxes have been successful in attracting some species, especially Western bluebirds (*Sialia mexicana*), which forage on insects potentially damaging to grape production (Jedlicka et al. 2011) and breed successfully in the vineyards (Fiehler et al. 2006). Nonetheless, vineyards should not be viewed as a replacement of lost native habitat. The entry of bluebird fledglings into a modified, perhaps treeless, environment raises the question of whether vineyards can be ecological traps (Battin 2004) with low juvenile survival. Because bluebirds depend on mistletoe (*Phoradendron villosum*) during the winter, there may not be enough food to support fledglings wintering in vineyards largely devoid of the large trees on which mistletoe occurs (Dickinson and McGowan 2005).

8.2.2.3 Urban Development

At the turn of the century, thousands of hectares of oak woodland were converted annually to commercial and residential development, making urban development the leading cause of oak woodland fragmentation and conversion (USDA Forest Service 2012). Most studies on urbanization have targeted birds. Avian response to urbanization is guild-specific and species-specific. At one end of the development spectrum are urban-adapted species such as Northern mockingbirds (*Mimus polyglottos*) and Western scrub-jays (*Aphelocoma californica*) that are favored by development (Blair 1996; Merenlender et al. 2009). Cassin's vireos (*Vireo cassini*) and lark sparrows (*Chondestes grammacus*) disassociate with urbanization, but still other species, for example Bewick's wrens (*Thryomanes bewickii*), occur as long as preferred habitat is present locally (Bolger et al. 1997; Merenlender et al. 2009). Total bird diversity, but not diversity of native birds, is greater at moderate levels of development (i.e., in areas that include golf courses, parks, large lots) apparently due to the occurrence of those birds (mostly non-native) that exploit exotic vegetation, plus those that exploit native vegetation (Blair 1996; Merenlender et al. 2009). Development effects can extend at least 1,000 m beyond developed areas (Bolger et al. 1997), supporting recent findings that the occurrence of a species in the broader landscape is influenced by the characteristics of the local area. Yard trees, treed remnant patches, greenways, and parks may function as stepping stones, facilitating movement within the developed area and across the larger landscape (Watson et al. 2005). Finally, suburban development and, notably, the activities associated with human development, can impoverish amphibian diversity (Riley et al. 2005). Negative impacts to amphibian populations from changes in habitat structure, stream water quality, and stream permanence seem to begin when 10–20 % of the watershed is developed (Paul and Meyer 2001).

8.2.2.4 Fire

Although direct mortality from exposure to heat or smoke is quite rare, fire indirectly affects animals by its effects on habitat. As Wirtz et al. (1988) documented for small mammals, longer-term response to fire varies among species and is proportional to the alteration of the habitat. In the post-burn years, wildlife composition moves toward species adapted to the level of complexity of the habitat created by the fire. Careful use of prescribed fire in oak woodland to mimic historical fire intensity and fire regimes can promote biodiversity at the landscape level (Martin and Sapsis 1991). Possible similarity between the effects of fire and of livestock grazing on habitat was proposed recently as a research hypothesis by Purcell and Stephens (2005). Much will be gained by a more thorough understanding of how grazing can be managed to influence woodland habitat structure in a manner similar to that of prescribed burning. Air quality regulations, risks to property, and cost are making prescribed burning more difficult to implement. In summary, prescribed fires of low intensity can help to maintain woodland habitat mixes and their associated diversity, either alone or in combination with low-intensity grazing. More research on this topic is clearly needed, however.

8.2.2.5 Firewood

Recent studies on the effects of firewood harvesting on birds demonstrate that cutting at only moderate levels and taking care to maintain habitat elements such as large trees and shrubs results in few changes to avian community composition (Aigner et al. 1998; Garrison et al. 2005). Open-canopy species such as Western wood-pewees (*Contopus sordidulus*), evening grosbeaks (*Coccothraustes vespertinus*), phainopeplas (*Phainopepla nitens*), and Bullock's orioles (*Icterus bullockii*) were favored by the cutting, whereas two dense-woodland specialists, Pacific-slope flycatchers (*Empidonax difficilis*) and Hutton's vireos (*Vireo huttoni*), declined significantly after the cutting.

8.3 Biological Diversity of Spanish Dehesas

The Mediterranean Basin was populated by humans soon after humanity's movement beyond Africa (>500,000 BP; Finlayson and Carrión 2007). Direct human influences on wildlife and habitat are widespread (Blondel et al. 2010). The dehesa land-use system maintains levels of biological diversity which are similar or even higher than of the original Mediterranean forests, to the point that dehesas are one of the few anthropogenic habitats that qualify for protection under the European Habitats Directive (Díaz and Pulido 2009). Díaz et al. (1997) and Díaz (2009) hypothesize that high levels of biological diversity in dehesas are caused by the coexistence of animal and plant species of two kinds of contrasting habitat types, forest and grassland, which rarely mix naturally at close spatial scales (Jeltsch et al. 1996; Manning et al. 2006)—a kind of extended ecotone. To the extent that human management has created and maintained such a mix, management has led to high levels of biological diversity in dehesas (Díaz et al. 1997, 2001, 2003; Díaz 2009).

The maintenance of biodiversity of dehesas is linked to a handful of keystone species of animals and plants (Díaz et al. 2003, 2011). Biological diversity in these keystone-driven systems would not be related to ecological stability unless the keystone species are associated with high diversity (Díaz 2002, 2009; Naeem et al. 2009). However, biological diversity underlies the economic sustainability of low-intensity, multifunctional systems because specialization in one or a few products does not provide enough income to sustain a dehesa. The exploitation of products from wildlife such as game or wild plants is essential to landowner income in many cases (Díaz et al. 1997; Campos et al. 2005). Biodiversity conservation, either focused on species-rich systems or on endangered species, is of increasing economic relevance for the design of land-use policies aimed at maintaining agricultural uses in developed countries (Mattison and Norris 2005), and systems and practices that contribute to biodiversity conservation and their associated values usually qualify for receiving governmental subsidies (Tscharntke et al. 2005). Subsidies or other payments to landowners for ecosystem services could be essential for the economic sustainability of low-intensity, species rich land-use systems such as dehesas (Díaz et al. 1997, 2003; Campos et al. 2005; Bugalho et al. 2011; Chaps. 12, 13).

Reviewing biological diversity in dehesas at spatial scales from the patch to the landscape, we focus on the "habitat-mix" hypothesis as it explains observed patterns. We then examine how a high level of dehesa diversity is maintained, analyzing spatial scales at which habitat mixes support key dehesa ecological processes, and see how these sustain the long-term profitability, or economic sustainability, of this anthropogenic system.

8.3.1 *Species Richness in Dehesas and Nearby Habitats and its Conservation Value*

Species densities, or species richness per unit area, tend to be higher in dehesas than in other nearby habitats, whether minimally disturbed or anthropogenic (Fig. 8.4). The trend is clear-cut for species-rich groups such as passerine birds and

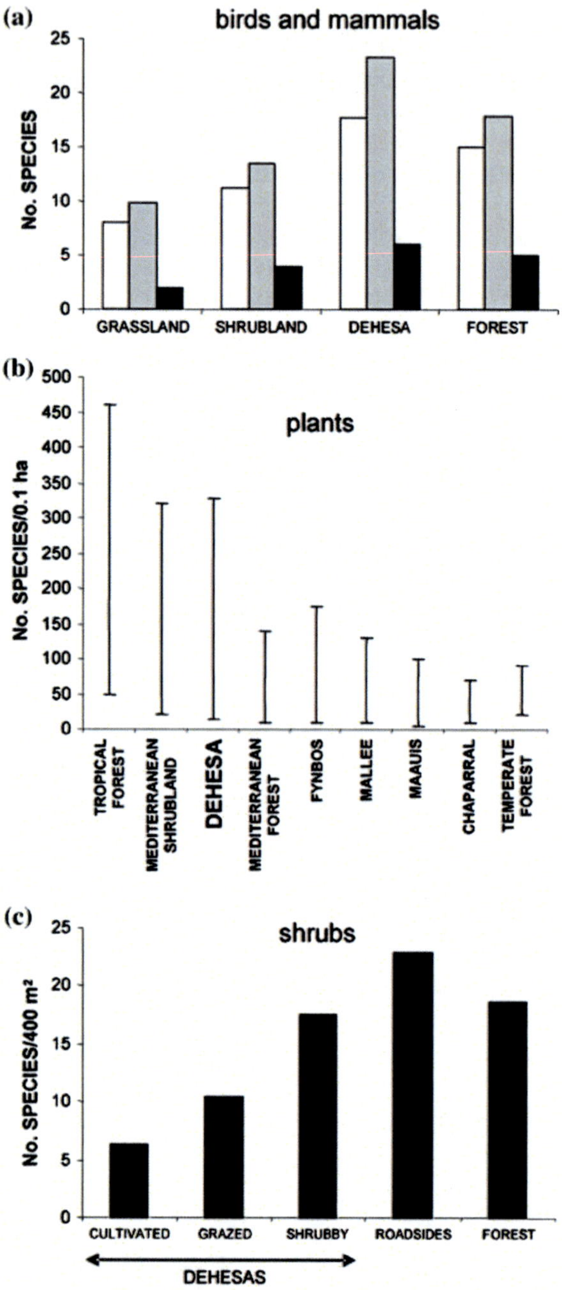

Fig. 8.4 Species densities found in dehesas as compared with other habitat types. (**a**) Passerine birds (open bars: breeding season; grey bars: wintering) and medium-sized and large mammals (closed bar) in an area of 800 km^2 in central Spain. Species densities are for 10 ha for passerines and 150 ha for mammals (from data in Tellería et al. 1992). (**b**) Higher plants in forests and shrublands worldwide (after Marañón 1986, including data from dehesas given by Naveh and Whittaker 1979 and Ojeda et al. 2000; whiskers indicate the range of values reported). (**c**) Shrubs in cork oak (*Quercus suber*) forests and dehesas of the southwestern quarter of Spain (Domínguez et al. 2007)

herbaceous plants (Marañón 1986; Tellería 2001), but less marked for medium-sized and large mammals such as rabbits, hares, and ungulates, and may even be the opposite for species-poor groups like shrubs that may be eliminated by dehesa management (Díaz et al. 2003; Díaz 2009). The trend of having higher species densities in dehesas than in nearby habitats varies locally for groups of intermediate diversity such as diurnal butterflies, whose species densities are higher in the open oak woodlands of central Spain than in nearby closed forests or grasslands (Viejo et al. 1989), but lower than in holm oak (*Quercus ilex*) forests in south-western Spain (Jiménez-Valverde et al. 2004). The higher species densities of species-rich groups makes up for the lower densities of species-poor groups, so overall species densities are usually higher in open oak woodlands than in other habitats nearby (Fernández and Pérez 2004 offer a plant-based case study).

Species found in dehesas are distributed more or less evenly with high levels of diversity. Species densities (α-diversity, the number of different species within a patch; Magurran 1988) are usually close to total species richness (γ-diversity, the overall diversity or the sum of all species in all patches) in dehesas due to low species turnover between patches (β-diversity, the number of species that are unique among patches; Pineda and Montalvo 1995; Tellería 2001; Jiménez-Valverde et al. 2004). This trend is demonstrated in particular for small-sized species that belong to species-rich groups. Large-sized species tend to depend on habitat elements or structural characteristics found both within dehesa and among habitat types in dehesa landscapes, and should therefore be sampled at large spatial scales. Such studies are not yet available (but see González et al. 1990, that provide a landscape-scale model for Spanish imperial eagles *Aquila adalberti*) but such results can be expected as relevant habitat elements and structures are still common in dehesa areas.

Up to 140 species of conservation concern that inhabit dehesas are listed in the Annexes of the European Birds and Habitats Directives, although very few of them are highly dependent on these working landscapes (Díaz et al. 2006; Table 8.3). These figures represent 9–34 % of the species of terrestrial vertebrates present in Spain, 14 % of the plants and 69 % of the mammals of the Spanish species listed in the Directives, and 6–42 % of the European species listed in Directives. For nine species (6 %), more than 75 % of their Spanish populations are found in dehesas (Table 8.3). Most species use mosaic landscapes where dehesas coexist with both undisturbed forest and open habitats (Díaz et al. 2003, 2006). Overall, then, dehesas alone do not maintain a large proportion of critically endangered species; instead, the coexistence of dehesas with other habitat types at the landscape scale contributes to the maintenance of a large proportion of the species of European conservation concern. Several game species inhabit dehesas, both small game (European rabbits (*Oryctolagus cuniculus*), hares (*Lepus granatensis*), red-legged partridges (*Alectoris rufa*), pigeons (*Columba palumbus*), (*C. oenas*), big game (red deer (*Cervus elaphus*), and wild boar (*Sus scrofa*); González and San Miguel 2005). The influence of dehesa management on game, sometimes benign, can at other times have strong negative effects on diversity by changing game

Table 8.3 Status and dependency on dehesas of the 140 species inhabiting dehesas protected by the European Birds and Habitats Directives (after Díaz et al. 2006)

	Mammals	Birds	Reptiles	Amphibians	Freshwater fish	Invertebrates	Plants	Total
Status								
Critically endangered	0	2	0	0	0	0	2	4
Endangered	6	5	0	0	2	1	7	21
Vulnerable	14	7	1	0	4	4	7	37
Near or not threatened	11	30	5	10	2	5	15	78
Total	31	44	6	10	8	10	31	140
Dependence on dehesa (% population)								
100 %	0	0	0	0	0	0	2	2
75–99 %	0	4	0	0	1	0	2	7
50–74 %	1	3	0	0	1	1	10	16
<50 %)	30	37	6	10	6	9	17	115
No. species protected by Directives (Europe)[1]	73	219	85	50	61	112	483	1083
No. species protected by Directives (Spain)[1]	45	152	37	18	19	40	217	528
No. species in Spain[2]	90	361	69	29	50	–	–	–

Status was obtained from the corresponding Red Data Books (Díaz et al. 2006). The dependency of each species on dehesas (A: almost 100 % of populations; B: 75–100 %; C: 50–75 %; D < 50 % of populations living in dehesas) was determined using Atlas and Red Book data (Díaz and Pulido 2009)

[1] Díaz et al. (2006)

[2] M. Díaz and G.G. Nicola, unpublished

Fig. 8.5 The Iberian lynx (*Lynx pardinus*) is one of the most endangered mammals of the world. It hunts for rabbit (*Oryctolagus cuniculus*), its staple prey, in the open dehesa close to its shrubby breeding and resting places. (Photograph by H. Garrido/EDB–CSIC)

populations due to overhunting, the vegetation profile due to livestock grazing, or animal communities due to predator control (Chap. 11; Díaz et al. 2009) (Fig. 8.5).

8.3.2 Patterns of Species Richness and Species Distribution Within Open Oak Woodlands

8.3.2.1 Effects of Tree Cover

In open woodlands high values of diversity at small spatial scales (α-diversity) are due to the mix, at even a few square meters, of contrasting habitat types: forests, represented by the canopy of scattered trees, and open sites, represented by the grasslands over which trees are distributed (Díaz et al. 1997, 2003, 2009; Manning et al. 2006). This allows the close coexistence of forest and open-country species, as demonstrated for small-sized groups such as herbs, ants, and breeding passerines (Díaz 2009).

Species densities of herbaceous plants are lower under tree canopies than in open grassland in dehesa, but the species composition differs between these two locations. This partitioning of available space allows for higher species densities of herbaceous vegetation within a few square meters in dehesa as compared to both open grasslands and forests (Marañón 1986; Fernández and Pérez 2004). The opposite pattern of relative richness has been found for ants (Reyes-López et al.

2003), with most species nesting in or under canopies and just a few open-habitat specialists occupying grassland between trees, but the partitioning of space here too supports a diverse assemblage of species in the dehesa overall.

Species densities of breeding passerines increased with tree cover at the scale of 0.25–ha plots in dehesas with low shrub covers (Pulido and Díaz 1992; Díaz et al. 2001). This occurs in a nested pattern, so the species assemblages that occupy plots with low tree cover are subsets of the larger assemblages of high-cover plots (Fig. 8.6). Assemblages in low-cover plots are dominated by open-country birds such as larks. Forest generalist species such as finches and sparrows and forest specialists such as tits and *Sylvia* warblers are added to the bird community as tree cover increases (Fig. 8.6). Tellería (2001) compared the bird assemblages of dehesas with those of closed forest. He found that total bird richness (γ-diversity) in dehesas was higher than in nearby forests, and that this increase was due to the addition of forest generalist and open-country birds. These results support the "mixed habitat" hypothesis at intermediate scales of a few hectares. In fact, the relationship between tree cover and diversity of breeding birds at small scales

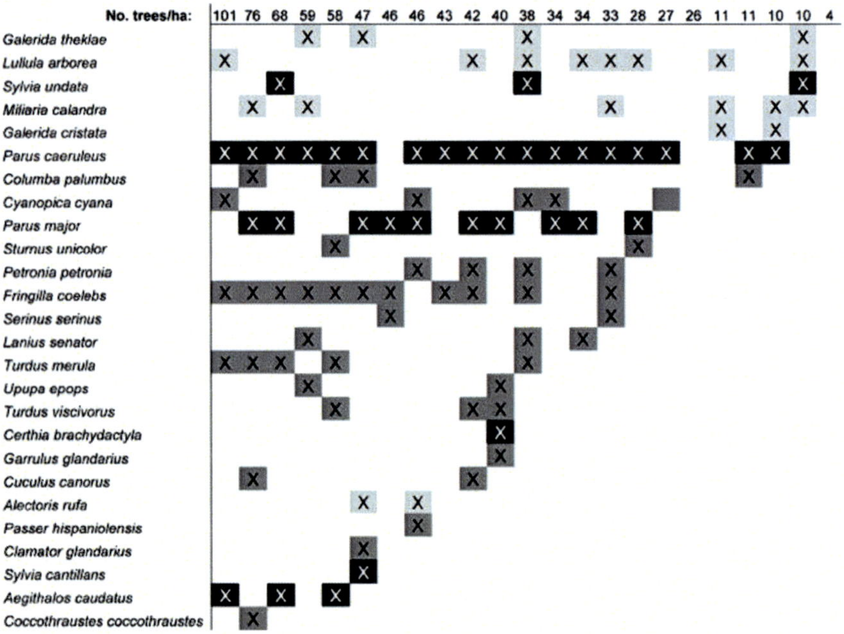

Fig. 8.6 Nestedness of breeding bird communities according to tree density (upper line) in 23 plots of 50 × 500 m in grazed dehesas with less than 10 % cover of shrubs located around the National Park of Monfragüe (Cáceres province, central Spain). Light grey open-habitat species (ground nesting and foraging); dark grey forest generalists (nesting in trees and shrubs and ground foraging); black forest specialists (tree and shrub nesting and foraging; Pulido and Díaz 1992 and unpublished data). The nestedness value of the matrix was 8.22 ($P = 0.08$; Lomolino 1996)

(smaller than 1 ha) only holds for dehesas with low shrub cover, disappearing for encroached dehesas with shrub covers above 25 % (Díaz et al. 2001), the higher shrub cover apparently rendering these dehesas unsuitable for open-country birds.

8.3.2.2 Effects of Understory Vegetation

Species richness of a variety of groups varies according to differences in understory vegetation in dehesa. Traditional management included small areas cultivated with cereals (typically 10 % of the farm on average) grown in long-term rotations (two years or more) and small patches invaded by shrubs in areas of low livestock grazing pressure (Díaz et al. 1997). This management scheme produced a mix of patches with different understory vegetation at the spatial scale of 100–500 ha, which is the typical size of a dehesa (Díaz et al. 1997). The responses of species to understory vegetation types (grassland, cereal crops, or shrub) are in general independent of, or additive to, the positive responses to tree cover described above (Pulido and Díaz 1992; Fernández and Pérez 2004). Peak species densities of breeding birds and earthworms are usually found in the dominant configuration of open grassland with scattered trees (Fig. 8.7), as is true for herbaceous plants (Pérez 2006), and the cause is the same as for the higher diversity of these groups in open woodlands as compared to closed forest (i.e., the addition of open-habitat species). Shrubs and small mammals tend to reach peak species densities (and abundances; Díaz et al. 1993;

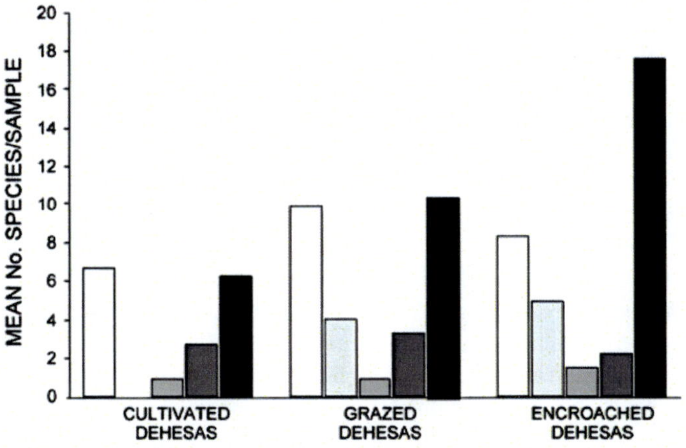

Fig. 8.7 Species densities found in dehesas according to type of understory vegetation (cereal crops, open grassland or shrubs). Open bars are breeding passerines in 2.5 ha plots (Pulido and Díaz 1992). Light grey bars are medium-sized and large mammals (rabbits and small carnivores, and ungulates, respectively) in a sampling area of 150 ha (Tellería et al. 1992). Medium grey bars are small mammals caught with a sampling effort of 240 traps/night per sample (Díaz et al. 1995). Dark grey bars are earthworms in 0.5 m^2 plots (Díaz and González unpublished data). Black bars are shrubs in 400 m^2 plots (Domínguez et al. 2007). Differences were statistically significant for birds and shrubs, but not for mammals and earthworms (Díaz et al. 2001; Domínguez et al. 2007)

Pérez 2006; Domínguez et al. 2007) in shrub-invaded (encroached) dehesas, because small mammals depend for foraging and reproduction on cover that hides them from predators (Díaz et al. 1993; Muñoz et al. 2009). Cultivated dehesas showed the lowest species densities (Fig. 8.7), probably because plowing has negative short-term effects on most open-habitat species.

Although peak species densities in dehesas occur in grassland patches (the most common type of patches), shrubby and cultivated patches maintain species that are not found in grassland (Pulido and Díaz 1992). Several bird species depend on shrub-encroached patches for nesting or foraging (*Sylvia* warblers and European blackbirds (*Turdus merula*)). Corn buntings (*Miliaria calandra*) and crested larks (*Galerida cristata*) depend on cereal crops (Pulido and Díaz 1992). Shrubby patches close to open grasslands are preferred by some lizards (*Psammodromus hispanicus* and *Lacerta lepida*), apparently to balance antipredatory and thermo-regulatory requirements (Martín and López 2002). Dehesas that include cultivated, open and encroached patches have higher numbers of species than properties dominated by only one of the varied types of understory vegetation. For this reason, land uses that decrease vegetation diversity and patchiness within a property, such as abandonment of grazing or cultivation, specialization on big-game hunting, or intensification of grazing or cultivation have negative effects on biological diversity (Díaz et al. 1997; González and San Miguel 2005; Moreno and Pulido 2009).

8.3.2.3 Landscape-Scale Effects

Landscape features such as pools, streams, small rocky outcrops, or human constructions such as paths, roads, small buildings, and brush piles, are as essential for the maintenance of several species or species groups as scattered trees within grasslands are for forest species, and offer small game such as rabbits and birds suitable permanent refuges (Tellería et al. 1992). Pools and streams are occupied by fish, amphibians, birds and plants (Díaz et al. 2006) and provide livestock and wildlife drinking water during the summer drought (González and San Miguel 2005). Buildings are occupied by species closely linked to humans such as house sparrows (*Passer domesticus*), swallows (*Hirundo rustica*), (*H. daurica*), house martins (*Delichon urbica*) and house mice (*Mus domesticus*) (Díaz et al. 2006) (Fig. 8.8).

Dehesas coexist with grasslands, croplands, and patches of undisturbed forests at regional scales of thousands of hectares. Large-sized and highly mobile species use dehesas as foraging grounds, utilizing nearby habitat as nesting or roosting sites. The endangered common crane (*Grus grus*) and the woodpigeon (*Columba palumbus*), a game bird, require roosting places in reservoirs or woodlots during winter (Díaz et al. 1996). Both species select cultivated, grazed, or encroached dehesas for foraging, balancing food and safety requirements (Tellería et al. 1994; Díaz et al. 1995; Díaz and Martín 1998). Big game (red deer and wild boar) require medium to large shrub patches for daytime or seasonal refuge, but graze in dehesas at nighttime or during the summer drought (Carranza et al. 1991). Finally, large

Fig. 8.8 Pools, either natural or constructed to provide water to livestock and wildlife, are landscape elements occupied by fish, amphibians, birds and plants that are not found elsewhere in oak woodlands. (Photograph by M. Díaz)

and endangered predators or scavengers such as the Spanish imperial eagle, the black stork (*Ciconia nigra*), the Iberian lynx (*Lynx pardina*), the Iberian wolf (*Canis lupus*) and the black vulture (*Aegypius monachus*) need large undisturbed patches of Mediterranean forest or shrubland for nesting and resting, and yet rely on dehesa areas for foraging (Díaz et al. 1997, 2006; González and San Miguel 2005). A recent study of individual black vulture nests located in undisturbed Mediterranean forests demonstrates that birds do not forage in forests close to colonies; instead, they move over hundreds of hectares daily, using dehesas as foraging grounds (Carrete and Donázar 2005). Preferential use of dehesas increases with distance from the colony, suggesting that dehesas are optimal foraging grounds and as important for the conservation of black vultures as the undisturbed patches the birds use for nesting (Carrete and Donázar 2005).

8.3.3 Management, Diversity and Long-Term Sustainability of Spanish Dehesas at Multiple Spatial Scales

High levels of biological diversity in the dehesa are not maintained solely by natural processes, but are a consequence of all the factors we have reviewed above. Many are a result of land use decisions, including the mix of forest, woodland edge, and grassland species; the patchiness of the distribution of shrubs, cereal cultivation, pools and rocky outcrops within dehesa properties; and the landscape-scale coexistence of dehesas with landscape elements such as reservoirs, streams and larger areas of undisturbed forest and shrubland (Díaz et al. 2001, 2003; Díaz and Pulido 2009). Unfortunately the long-term sustainability of this quasi-artificial system is in question.

The regeneration of tree populations is almost lacking in the dehesas that typically host peak diversity levels (Pulido et al. 2001, 2010; Pulido and Díaz 2005; Chap. 5). Recruitment of dehesa trees depends on the activity of scatter-hoarding animal dispersers such as wood mice (*Apodemus sylvaticus*) and Algerian mice (*Mus spretus*), birds such as European jays (*Garrulus glandarius*) (Gómez 2003; Pulido and Díaz 2005; Díaz et al. 2007, 2011), and on nurse shrubs that protect seedlings from desiccation during summer drought (Pulido and Díaz 2005; Smit et al. 2008; Pulido et al. 2010; Chaps. 5, 6). The scarcity of keystone dispersers and shrubs in grazed dehesas (review in Díaz et al. 2011) explains tree regeneration failure and diminished ecological sustainability, whereas encroachment by shrubs and subsequent colonization by keystone dispersers would explain dehesa regeneration after abandonment (Ramírez and Díaz 2008; Pulido et al. 2010). Shrub encroachment may reduce dehesa productivity, but encourages nurse shrubs and the activity of animal dispersers that depend on shrub and tree cover for nesting and foraging (Díaz et al. 1993; Alonso 2006; Muñoz et al. 2009).

Management is essential to maintain the coexistence of forest and open habitats in Spain. Biological diversity can contribute to the long-term maintenance of dehesas because of the diversity of production that can be created, and through society's support for land-use systems that maintain high levels of biodiversity or endangered species (Díaz et al. 1997, 2003; Campos et al. 2005). Paradoxically, however, the enhancement of the economic sustainability of these systems through measures aimed at increasing biological diversity or populations of endangered species decreases the long-term ecological sustainability of the system by reducing tree recruitment. Critical life cycle stages of a key dehesa element, the trees, depend on the presence of low-diversity, encroached dehesa patches that are habitat for species that facilitate oak regrowth. The resolution of this paradox (Fig. 8.9) depends on the development of landscape-scale measures subsidizing the rotation of areas managed as dehesa with high biodiversity but low tree recruitment (the most typical savanna-like landscape) and shrub encroached areas with low diversity and high recruitment (Ramírez and Díaz 2008; Moreno and Pulido 2009).

Landscape-scale thinking is essential for biodiversity conservation in anthropogenic systems (Concepción et al. 2008, 2012). Preliminary data suggests around 20 years of abandonment is needed to promote tree regeneration (Ramírez and Díaz 2008). Encroachment of shrubs on some 10 % of a large dehesa seems to offer significant benefits to biodiversity (Moreno and Pulido 2009). But more accurate estimates of these key parameters and their spatial variation is essential to drive the political and economic changes necessary to ensure the long-term economic and ecological sustainability of dehesa systems.

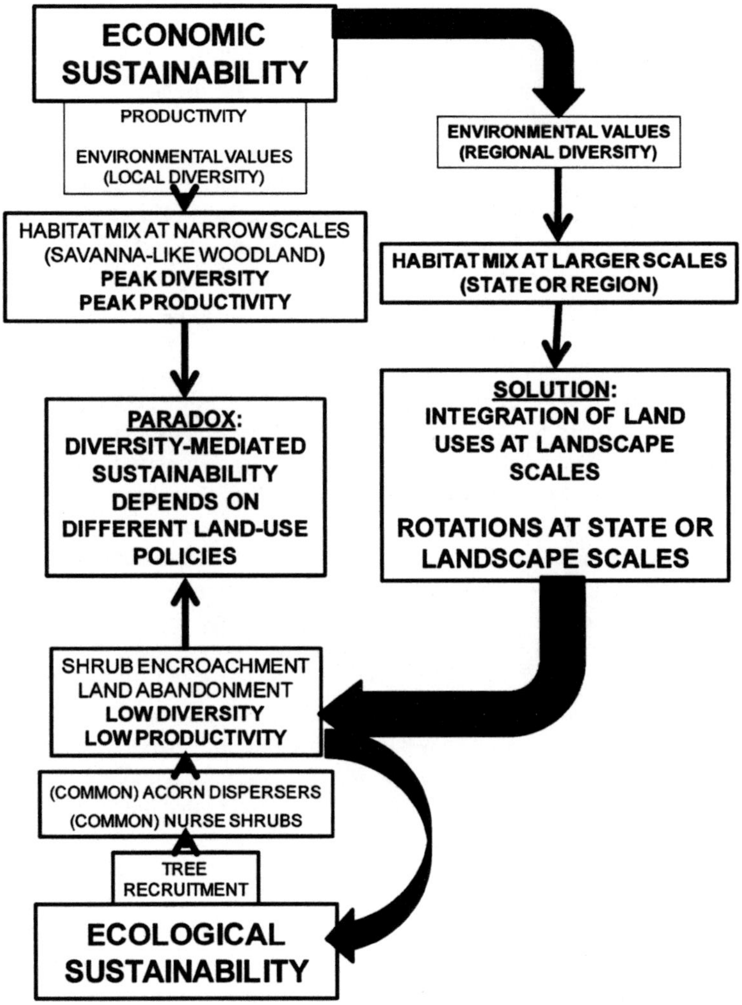

Fig. 8.9 The paradox of the value of biological diversity for maintaining the ecological and economic sustainability of dehesa systems. Local diversity and productivity, and hence economic sustainability, depends on local habitat management that promotes a close mix of habitat types (forest and grassland) at local scales. Long term ecological sustainability requires oak regeneration, and depends on land uses that decrease local diversity and productivity through shrub encroachment. The solution to the paradox is to develop land-use policies that maintain habitat mixes at larger scales (farm or region) through land-use rotations, including abandonment. The cessation of dehesa management allows shrub encroachment, seed dispersal, and tree recruitment, even as it reduces the local diversity of habitat types

8.4 Conclusions

Species richness of a wide range of taxa tends to be higher in California and Spanish oak woodlands than in other nearby systems. These working landscapes are in fact intimate mixtures of habitat types at several spatial scales. Such mixtures allow the close coexistence of large numbers of species with different habitat requirements. The habitat mixtures that characterize working oak woodland landscapes in California are largely a result of the physical and biotic characteristics of the land, including topography, soil type, rainfall pattern, the natural and anthropogenic fire regime, and perhaps the allopathic properties of some plants. Intense management, including tree thinning and pruning, livestock grazing, shrub removal, and cropping, has had a profound effect in shaping the mix of habitat types of the Spanish dehesas and a more minor influence in California. California oak woodland matrices are not sustained by intensive human management in the same way that Spanish dehesas are. If all anthropogenic activity was excluded today, California's natural flammability and the relatively slow encroachment of shrubs and trees would ensure that a patch matrix of woodland, grassland, and shrubs would be present 200, 300, and perhaps 500 years from now. Recent human influence in the California woodlands has two conflicting and interacting types of impacts: fire suppression that encourages woody encroachment, and grazing, firewood harvest, catastrophic fire risk, and tree and shrub removal that has had the opposite effect.

Because of the temporal and spatial scales at which they are occurring, the main threats to biodiversity in California oak woodlands currently come from the increase in high-disturbance land uses such as the transformation of woodland by intensive agriculture, urbanization, large wildfires, and to the extent that the problem exists in various localities and situations (Tyler et al. 2006), poor regeneration of several species of oak trees (Chap. 5). Long-term effects of these threats to biodiversity are still mostly unknown. Research indicates that specialist species such as golden eagles, purple martins, or salamanders can be extirpated from impacted areas. Generalist species such as mule deer or mockingbirds may become more common and exotic species may fill niches vacated by the original residents as the woodlands are transformed by such changes. Wildlife community composition shifts accordingly. These processes enfold slowly and will be revealed only by longer-term study. Management will need to adapt as more is learned.

Similar threats (except fire, that spreads with difficulty in dehesas due to the open canopy layer), together with deforestation and development of irrigation technologies, have threatened the maintenance of biodiversity in the Spanish dehesa. Today, however, the main threat to dehesa is the consumption of oak seedlings and acorns by grazing and foraging animals, exacerbated by unrestricted livestock grazing, and the abandonment of productive agricultural land (Díaz et al. 1997; Díaz and Pulido 2009; Moreno and Pulido 2009). High levels of anthropogenic disturbance are essential in dehesas to avoid encroachment by shrubs and trees that recolonize following land abandonment. Ironically, however, shrub encroachment is essential for successful tree regeneration (Chap. 5). Therefore in Spain, preventing land-use and grazing intensification alone will not suffice to preserve dehesas in the long term. In this system, management practices that increase biological diversity decrease the shrub encroachment necessary for tree recruitment, and therefore the ecological sustainability of the system. To put it plainly, tree recruitment depends on a low-diversity condition, but tree recruitment is needed to create a high diversity condition. This paradox can be solved at the landscape scale (Fig. 8.9). The promotion of rotation of areas with high diversity and productivity (but low recruitment), with encroached areas that manifest low diversity and high recruitment would enhance ecological sustainability. And such management would encourage diversity and associated economic values in time (area use by different species as rotations progress) and in space (the preservation of large-sized species dependent on landscape-scale land-use mixtures).

As throughout the long history of Spanish and California oak woodlands, human activities can support oak woodland biodiversity at multiple temporal and spatial scales.

Acknowledgments The authors thank M Bugalho, KL Purcell, WD Koenig, and GA Giusti for reviewing an earlier version of the manuscript. Constructive discussion during the last ten years with F Pulido, J Carranza and P Campos, among others, was essential for the development of the ideas presented here on the future of dehesas. MD Hardy, WD Koenig, KL Purcell and and TJ Swiecki provided valuable discussion on the topics presented in the California oak woodland section. MA Hardy prepared the final tables and figures, and helped with the editing of a late draft of the manuscript. RE Larsen provided literature for the livestock grazing paragraph. This paper is a contribution to the projects QLK5–CT–2002–01495 and FP7–KBBE–2008–227161 (European Union V and VII Framework Programs), 096/2002 and 003/2007 (Spanish Organismo Autónomo Parques Nacionales) and REN2003–07048/GLO, CGL2009-08430, CGL2010-22180-03-03 and CSD2008–00040 (Spanish Ministerio de Ciencia y Tecnología), as well as to the thematic network GlobiMed. Appreciation is expressed to the University of California, Berkeley, Department of Environmental Science, Policy, and Management, for allowing the time to complete the manuscript, and to the University of California Cooperative Extension Office, San Luis Obispo, for supporting the writing of the manuscript.

References

Aigner PA, Block WM, Morrison ML (1998) Effect of firewood harvesting on birds in a California oak-pine woodland. J Wildl Manage 62:485–496

Airola DA (1988) Guide for the California wildlife habitat relationships system. California Dep Fish and Game, Sacramento, p 74

Alonso CL (2006) Arrendajo—*Garrulus glandarius*. In: Carrascal LM, Salvador A (eds) Enciclopedia virtual de los vertebrados españoles (http://www.dfg.ca.gov/biogeodata/cwhr/). Museo Nacional de Ciencias Naturales, Madrid. http://www.vertebradosibericos.org/

Barrett RH (1980) Mammals of California oak habitats—management implications. In: Plumb TR (tech coord) Proceedings Symposium ecology, management, and utilization of California oaks, US Department Agriculture, Berkeley, pp 275–291

Battin J (2004) When good animals love bad habitats: ecological traps and the conservation of animal populations. Conserv Biol 18:1482–1491

Beier P (1993) Determining minimum habitat areas and habitat corridors for cougars. Conserv Biol 7:94–108

Blair RB (1996) Land use and avian species diversity along an urban gradient. Ecol Appl 6:506–519

Block WM (1991) Foraging ecology of Nuttall's woodpecker. Auk 108:303–317

Blondel J, Aronson J, Bodiou J, Boeu G (2010) The mediterranean region: biological diversity in space and time. Oxford University Press, Oxford

Bolger DT, Scott TA, Rotenberry JT (1997) Breeding bird abundance in an urbanizing landscape in coastal southern California. Conserv Biol 11:406–421

Bugalho MN, Caldeira MC, Pereira JS, Aronson J, Pausas JG (2011) Mediterranean cork oak savannas require human use to sustain biodiversity and ecosystem services. Front Ecol Environ 9:278–286

Campos P, Caparrós A, Sanjurjo E (2005) Spain. In: Merlo M, Croitoru L (eds) Valuing mediterranean forests: towards total economic value. CAB International, Wallingford, pp 319–330

Carranza J, Hidalgo SJ, Medina R, Valencia J, Delgado J (1991) Space use by red deer in a Mediterranean ecosystem. Appl Anim Behav Sci 30:363–371

Carrete M, Donázar JA (2005) Application of central-place foraging theory shows the importance of mediterranean dehesas for the conservation of the cinereous vulture, *Aegypius monachus*. Biol Conserv 126:582–590

Concepción ED, Díaz M, Baquero RA (2008) Effects of landscape complexity on the ecological effectiveness of agri-environment schemes. Lands Ecol 23:135–148

Concepción ED, Díaz M, Kleijn D, Báldi A, Batáry P, Clough Y, Gabriel D, Herzog F, Holzschuh A, Knop E, Marshall JP, Tscharntke T, Verhulst J (2012) Interactive effects of landscape context constrains the effectiveness of local agri-environmental management. J Appl Ecol 49:695–705

DeMars CA, Rosenberg DK, Fontaine JB (2010) Multi-scale factors affecting bird use of isolated remnant oak trees in agro-ecosystems. Biol Conserv 143:1485–1492

Díaz M (2002) Elementos y procesos clave para el funcionamiento de los sistemas naturales: las medidas con significado funcional como alternativa a los indicadores clásicos. In: Ramírez L (coord) Indicadores ambientales. Situación actual y perspectivas. Organismo Autónomo Parques Nacionales, Madrid, pp 229–264

Díaz M (2009) Biodiversity in the dehesa. In: Rigueiro A, Mosquera MR (eds) Agroforestry systems as a technique for sustainable land management. Programme Azahar. AECID, Madrid, pp 209–225

Díaz M, Alonso CL, Arroyo L, Bonal R, Muñoz A, Smit C (2007) Desarrollo de un protocolo de seguimiento a largo plazo de los organismos clave para el funcionamiento de los bosques mediterráneos. In: Ramírez L, Asensio B (eds) Proyectos de investigación en parques nacionales: 2003–2006. Organismo Autónomo Parques Nacionales, Madrid, pp 29–51

Díaz M, Alonso CL, Beamonte E, Fernández M, Smit C (2011) Desarrollo de un protocolo de seguimiento a largo plazo de los organismos clave para el funcionamiento de los bosques mediterráneos. In: Ramírez L, Asensio B (eds) Proyectos de investigación en parques nacionales: 2007–2010. Organismo Autónomo Parques Nacionales, Madrid, pp 47–75

Díaz M, Baquero RA, Carricondo A, Fernández F, García J, Yela JL (2006) Bases ecológicas para la definición de las prácticas agrarias compatibles con las Directivas de Aves y de Hábitats. Ministerio de Medio Ambiente-Universidad de Castilla-La Mancha, Toledo. http://www.mma.es/portal/secciones/biodiversidad/desarrollo_rural_paisaje/naturaleza_rural/estudios_bases_ecologicas.htm

Díaz M, Campos P, Pulido FJ (1997) The Spanish dehesas: a diversity of land use and wildlife. In: Pain D, Pienkowski M (eds) Farming and birds in Europe: The common agricultural policy and its implications for bird conservation. Academic Press, London, pp 178–209

Díaz M, Campos P, Pulido FJ (2009) Importancia de la caza en el desarrollo sustentable y en la conservación de la biodiversidad. In: Sáez de Buruaga M, Carranza J (coords) Gestión

Cinegética en Ecosistemas Mediterráneos. Consejería de Medio Ambiente de la Junta de Andalucía, Sevilla, pp 21–33

Díaz M, González E, Muñoz-Pulido R, Naveso MA (1993) Abundance, seed predation rates, and body condition of rodents wintering in Spanish Holm–oak *Quercus ilex* L. dehesas and cereal croplands: effects of food abundance and habitat structure. Z Saugetierkd 58:302–311

Díaz M, González E, Muñoz-Pulido R, Naveso MA (1995) Habitat selection patterns of common cranes *Grus grus* wintering in Holm-oak *Quercus ilex* dehesas of central Spain: effects of human management. Biol Conserv 75:119–124

Díaz M, Martín P (1998) Habitat selectivity by wintering woodpigeons (*Columba palumbus*) in Holm-oak Quercus ilex dehesas of central Spain. Gibier Faune Sauvage 15:167–181

Díaz M, Pulido FJ (2009) Dehesas perennifolias de *Quercus* spp. In: Bases ecológicas preliminares para la conservación de los tipos de hábitat de interés comunitario presentes en España. Dirección General de Medio Natural y Política Forestal, Ministerio de Medio Ambiente, y Medio Rural y Marino, Madrid

Díaz M, Pulido FJ, Marañón T (2001) Diversidad biológica en los bosques mediterráneos ibéricos: relaciones con el uso humano e importancia para la sostenibilidad de los sistemas adehesados. In: Campos P, Montero G (eds) Beneficios comerciales y ambientales de la repoblación y la regeneración del arbolado del monte mediterráneo. CIFOR-INIA, Madrid, pp 269–296

Díaz M, Pulido FJ, Marañón T (2003) Diversidad biológica y sostenibilidad ecológica y económica de los sistemas adehesados. Ecosistemas 2003/3. http://www.aeet.org/ecosistemas/033/investigacion4.htm

Díaz M. Asensio B, Tellería JL (1996) Aves Ibéricas. I. No paseriformes. JM Reyero, Madrid

Dickinson JL, McGowan A (2005) Resource wealth drives family group living in western bluebirds. Proc R Soc London B 272:2423–2428

Domínguez F, Sáinz H, Sánchez R (2007) Tipificación, biodiversidad y conservación de alcornocales ibéricos. ADENA-WWF, Madrid, unpublished report

Edwards SW (1992) Observations on the prehistory and ecology of grazing in California. Fremontia 20:3–11

Fernández F, Pérez R (2004) El bosque mediterráneo: Flora y vegetación. In: Canseco V, Asensio B (coords) La red española de Parques Nacionales. Canseco Editores, Talavera de la Reina, pp 251–271

Fiehler CM, Tietje WD, Fields WR (2006) Nesting success of Western Bluebirds (*Sialia mexicana*) using nest boxes in vineyard and oak savannah habitats of California. Wilson J Orn 118:552–557

Finlayson C, Carrión JC (2007) Rapid ecological turnover and its impact on Neanderthal and other human populations. Trends Ecol Evol 22:213–222

Forman RTT (1995) Land mosaics: the ecology of landscapes and regions. Cambridge University Press, Cambridge, p 632

García D, Martínez D (2012) Species richness matters for the quality of ecosystem services: a test using seed dispersal by frugivorous birds. Proc R Soc B 279:3106–3113

Garrison BA, Triggs ML, Wachs RL (2005) Short-term effects of group-selection timber harvest on landbirds in montane hardwood-conifer habitat in the central Sierra Nevada. J Field Orn 76:72–82

Garrison BA, Wachs RL, Giles TA, Triggs ML (2002) Dead branches and other wildlife resources on California black oak (*Quercus kelloggii*). In: Laudenslayer WF Jr, Shea PJ, Valentine BE, Weatherspoon CP, Lisle TE (tech coord) Proceedings Symposium Ecology and Management of Dead Wood in Western Forests. USDA Forest Service, Reno, pp 593–604

George MR (1991) Grazing and land management strategies for hardwood rangelands. In: Standiford RB (ed) Proceedings Symposium Oak Woodlands and Hardwood Rangeland Management. US Dep Agriculture, Davis, pp 315–319

Giusti GA, Keiffer RJ, Vaughn CE (2003) The bird community of an oak woodland stream. Cal Fish and Game 89:72–80

Gómez JM (2003) Spatial patterns in long-distance dispersal of *Quercus ilex* acorns by jays in a heterogeneous landscape. Ecography 26:573–584

González LM, Bustamante J, Hiraldo F (1990) Factors influencing the present distribution of the Spanish imperial eagle *Aquila adalberti*. Biol Conserv 51:311–319

González LM, San Miguel (coord) A (2005) Manual de buenas prácticas de gestión de fincas de monte mediterráneo de la Red Natura 2000. Ministerio de Medio Ambiente, Madrid

Heaton E, Long R, Ingels C, Hoffman T (W.D. Tietje, Technical Coordinator and Editor) (2008) Songbird, bat, and owl boxes—vineyard management with an eye toward wildlife. UC ANR Publication 21636, p 51

Hertz PE, Remsen JV Jr, Zones SI (1976) Ecological complementarity of three sympatric parids in a California oak woodland. Condor 78:307–316

Hilty JA, Brooks C, Heaton E, Merenlender AM (2006) Forecasting the effect of land-use change on native and non-native mammalian predator distributions. Biodiv Conserv 15:2853–2871

Ingold DJ (1994) Influence of nest-site competition between European starlings and woodpeckers. Wilson Bull 106:227–241

Jedlicka JA, Greenberg R, Letourneau DK (2011) Avian conservation practices strengthen ecosystem services in California vineyards. PLoS ONE 6(11):e27347

Jeltsch F, Milton SJ, Dean WRJ, Van Rooyen N (1996) Tree spacing and coexistence in semiarid savannas. J Ecol 84:583-595

Jiménez–Valverde A, Martín J, Munguira ML (2004) Patrones de diversidad de la fauna de mariposas del Parque Nacional de Cabañeros y su entorno (Ciudad Real, España central) (Lepidoptera, Papilionoidea, Hesperioidea). Anim Biodiv Conserv 27:15–24

Johnson MD, Horn CM (2008) Effects of rotational grazing on rodents and raptors on a coastal grassland. West N Amer Natur 68:444–452

Jones A (2000) Effects of cattle grazing on North American arid ecosystems: a quantitative review. West N Amer Natur 60:155–164

Kie JG (1996) The effects of cattle grazing on optimal foraging in mule deer (*Odocoileus hemionus*). Forest Ecol Manage 88:131–138

Kie JG, Bowyer RT, Nicholson MC, Boroski BB, Loft ER (2002) Landscape heterogeneity at differing scales: effects on spatial distribution of mule deer. Ecology 83:530–544

Klinger RC, Kutilek MJ, Shellhammer HS (1989) Population responses of black-tailed deer to prescribed burning. J Wildl Manage 53:863–871

Koenig WD (2003) European starlings and their effect on native cavity-nesting birds. Conserv Biol 17:1134–1140

Koenig WD, Haydock J (1999) Oaks, acorns, and the geographical ecology of acorn woodpeckers. J Biogeogr 26:159–165

Laymon SA (1984) Riparian bird community structure and dynamics: Dog Island, Red Bluff, California. In: Warner RE, Hendrix KM (eds) California riparian systems: ecology, conservation, and productive management. University California Press, Berkeley, pp 587–597

Lomolino MV (1996) Investigating causality of nestedness of insular communities: selective immigrations or extinction? J Biogeogr 23:699–703

López-López P, Luigi M, Alessandra F, Emilio B, Luigi B (2011) Hotspots of species richness, threat, and endemism for terrestrial vertebrates in SW Europe. Acta Oecol 37:399–412

Magurran AE (1988) Ecological diversity and its measurement. Princeton University Press, Princeton

Manning AD, Fischer J, Lindenmayer DB (2006) Scattered trees are keystone structures—implications for conservation. Biol Conserv 132:311–321

Marañón T (1986) Plant species richness and canopy effect in the savanna-like "dehesa" of S.W. Spain Ecol Medit 12:131–141

Martín J, López P (2002) The effect of mediterranean dehesa management on lizard distribution and conservation. Biol Conserv 108:213–219

Martin RE, Sapsis DB (1991) Fires as agents of biodiversity: pyrodiversity promotes biodiversity. In: Proceedings Symposium biodiversity of northwestern California. University California Wildl Res Center, Berkeley, pp 150–157

Marty JT (2005) Effects of cattle grazing on diversity in ephemeral wetlands. Conserv Biol 19:1626–1632
Mattison EHA, Norris K (2005) Bridging the gaps between agricultural policy, land-use and biodiversity. Trends Ecol Evol 11:610–616
Merenlender AM, Reed SE, Heise KL (2009) Exurban development influences woodland bird composition. Landscape Urban Plann 92:255–263
Moreno G, Pulido FJ (2009) The functioning, management and persistence of dehesas. In: Rigueiro A, Mosquera MR, McAdams J (eds) Agroforestry in Europe: current status and future prospects. Springer Science, Berlin, pp 127–160
Muñoz A, Bonal R, Díaz M (2009) Ungulates, rodents, shrubs: Interactions in a diverse mediterranean ecosystem. Basic Appl Ecol 10:151–160
Myers N, Mittermeier RA, Mittermeier CG, da Fonseca GAB, Kent J (2000) Biodiversity hotspots for conservation priorities. Nature 403:853–858
Naeem S, Bunker DE, Hector A, Loreau M, Perrings (eds) (2009) Biodiversity, ecosystem functioning adn human well-being. Oxford University Press, Oxford
Naveh Z, Whittaker RH (1979) Structural and floristic diversity of shrublands and woodlands in northern Israel and other mediterranean areas. Vegetation 41:179–190
Ojeda F, Marañón T, Arroyo J (2000) Plant diversity patterns in the Aljibe Mountains (S. Spain): a comprehensive account. Biodiv Conserv 9:1323–1343
Olson G, Purcell KL, Grubbs D (2008) Nest defense behaviors of native cavity–nesting birds to European starlings. In: Merenlender A, McCreary D, Purcell K (tech cords) Proceedings California oak symposium: today's challenges, tomorrow's opportunities. USDA Forest Service, Albany, pp 457–470
Paul MJ, Meyer JL (2001) Streams in the urban landscape. Annu Rev Ecol Syst 32:333–365
Pérez IM (2006) Factores que condicionan la regeneración natural de especies leñosas en un bosque mediterráneo del sur de la Península Ibérica. PhD Thesis, University of Sevilla, Sevilla
Pineda FD, Montalvo J (1995) Biological diversity in dehesa systems. In: Gilmour D (ed) Biological Diversity outside Protected Areas. Overview of traditional agroecosystems. IUCN, Forest Conservation Programme, Gland, pp 107–122
Prevedello JA, Vieira MV (2010) Does the type of matrix matter? A quantitative review of the evidence. Biodiv Conserv 19:1205–1223
Pulido FJ, Díaz M (1992) Relaciones entre la estructura de la vegetación y las comunidades de aves nidificantes en las dehesas: influencia del manejo humano. Ardeola 39:63–72
Pulido FJ, Díaz M (2005) Regeneration of a Mediterranean oak: a whole-cycle approach. EcoScience 12:92–102
Pulido FJ, Díaz M, Hidalgo SJ (2001) Size-structure and regeneration of holm oak (*Quercus ilex*) forests and dehesas: effects of agroforestry use on their long-term sustainability. Forest Ecol Manage 146:1–13
Pulido FJ, García E, Obrador JJ, Moreno G (2010) Multiple pathways for tree regeneration in anthropogenic savannas: incorporating biotic and abiotic drivers into management schemes. J Appl Ecol 47:1272–1281
Purcell KL, Drynan DA (2008) Use of hardwoods by birds nesting in ponderosa pine forests. In: Merenlender A, McCreary D, Purcell K (tech cords) Proceedings California oak symposium: today's challenges, tomorrow's opportunities. USDA Forest Service, Albany, pp 417–431
Purcell KL, Stephens SL (2005) Changing fire regimes and the avifauna of California oak woodlands. Studies Avian Biol 30:33–45
Purcell KL, Verner J, Mori SR (2002) Factors affecting the abundance and distribution of European starlings (*Sturnus vulgaris*) at the San Joaquin Experiment al Range. In: Standiford RB, McCreary D, Purcell K (tech cords) Proceedings Symposium oak woodlands: an oak in California's changing landscape. USDA Forest Service, Albany, pp 305–321
Radford JQ, Bennett AF (2007) The relative importance of landscape properties for woodland birds in agricultural environments. J Appl Ecol 44:737–747

Ramírez JA, Díaz M (2008) The role of temporal shrub encroachment for the maintenance of Spanish holm oak Quercus ilex dehesas. Forest Ecol Manage 255:1976–1983

Reyes-López J, Ruiz N, Fernández-Haeger J (2003) Community structure of ground-ants: the role of single trees in a Mediterranean pastureland. Acta Oecol 24:195–202

Reynolds M, Gardali T, Merrifield M, Hirsch-Jacobsen R, Armstrong A, Wood D, Smith J, Heaton E, LeBuhn G (2008) Reproductive success of oak woodland birds in Sonoma and Napa counties, California. In: Merenlender A, McCreary D, Purcell K (tech cords) Proceedings California oak symposium: today's challenges, tomorrow's opportunities. USDA Forest Service, Albany, pp 433–445

Riley SPD, Busteed GT, Kats LB, Vandergon TL, Lee LFS, Dagit RG, Kerby JL, Fisher RN, Sauvajot RM (2005) Effects of urbanization on the distribution and abundance of amphibians and invasive species in Southern California streams. Conserv Biol 19:1894–1907

Sisk TD, Haddad NM, Ehrlich PR (1997) Bird assemblages in patchy woodlands: modeling the effects of edge and matrix habitats. Ecol Appl 7:1170–1180

Smit C, den Ouden J, Díaz M (2008) Facilitation of holm oak recruitment by shrubs in Mediterranean open woodlands. J Veget Sci 19:193–200

Tellería JL (2001) Passerine bird communities of Iberian dehesas: a review. Anim Biodiv Conserv 24:67–78

Tellería JL, Alcántara M, Asensio B, Cantos FJ, Díaz JA, Díaz M, Sánchez A (1992) Evaluación del Impacto Ambiental del Embalse de Monteagudo (Avila-Toledo) sobre la Fauna de Vertebrados Terrestres. Ministerio de Obras Públicas y Urbanismo, Madrid, unpublished report

Tellería JL, Santos T, Díaz M (1994) Effects of agricultural practices on bird populations in the Mediterranean region: the case of Spain. In: Hagemeijer EJM, Verstrael TJ (eds) Bird Numbers 1992. Distribution, monitoring and ecological aspects. Statistics Netherlands and SOVON, Beek–Ubbergen, The Netherlands, pp 57–75

Thornton DH, Branch LC, Sunquist ME (2011) The influence of landscape, patch, and within-patch factors on species presence and abundance: a review of focal patch studies. Landscape Ecol 26:7–18

Tietje WD, Isaacs J, Bavrlic K, Rein S (2008) Breeding bird assemblages in wooded patches in vineyard and undeveloped oak woodland landscapes in coastal-central California. In: Merenlender A, McCreary D, Purcell K (tech cords) Proceedings California oak symposium: today's challenges, tomorrow's opportunities. USDA Forest Service, Albany, pp 447–456

Tietje WD, Vreeland JK (1997) Cover story: vertebrates diverse and abundant in well-structured oak woodland. California Agric 51:8–14

Tietje WD, Waddell K, Vreeland JK, Bolsinger C (2002) Coarse woody debris in oak woodland of California. Western J Appl Forestry 17:139–146

Tscharntke T, Klein AM, Kruess A, Steffan-Dewenter I, Thies C (2005) Landscape perspectives on agricultural intensification and biodiversity—ecosystem service management. Ecol Lett 8:857–874

Tyler CM, Kuhn B, Davis FW (2006) Demography and recruitment limitations of three oak species in California. Q Rev Biol 81:127–152

USDA Forest Service. (2012). Forest inventory and analysis national program. http://www.fia.fs.fed.us/tools-data/default.asp

USFWS (United States Department of Interior Fish and Wildlife Service) (1997) Draft recovery plan for the Stephen's kangaroo rat, Region 1. United States Department of Interior Fish and Wildlife Service, Portland, Oregon. http://ecos.fws.gov/docs/recovery_plan/970623.pdf. Accessed 30 Oct 2011

USFWS (United States Department of Interior Fish and Wildlife) (2004) Endangered and threatened wildlife and plants; determination of threatened status for the California tiger salamander; and special rule exemption for existing routine ranching activities; final rule. Federal Register 69, No. 149, 50 CFR Part 17, RIN 1018–A168, 69 FR 68568

USFWS [United States Department of the Interior Fish and Wildlife Service] (2010) San Joaquin kit Fox (*Vulpes macrotis mutica*) 5-Year review: summary and evaluation, pp 37–38. http://ecos.fws.gov/docs/five_year_review/doc3222.pdf. Accessed 3 July 2010

Urban DL, O'Neill RV, Shugart HH Jr (1987) Landscape ecology: a hierarchical perspective can help scientists understand spatial patterns. Bioscience 37:119–127

Verner J, Purcell KL, Turner JG (1997) Bird communities in grazed and ungrazed oak-pine woodlands at the San Joaquin Experimental Range. In: Pillsbury NH, Verner J, Tietje WD (tech. cords) Proceedings Symposium oak woodlands: ecology, management, and urban interface issues. US Department of Agriculture, San Luis Obispo, pp 381–390

Viejo JL, de Viedma MG, Martínez E (1989) The importance of woodlands in the conservation of butterflies (Lep: Papilionoidea and Hesperioidea) in the centre of the Iberian Peninsula. Biol Conserv 48:101–114

Waters JR, Noon BR, Verner J (1990) Lack of nest site limitation in a cavity-nesting bird community. J Wildl Manage 54:239–245

Watson JEM, Whittaker RJ, Freudenberger D (2005) Bird community responses to habitat fragmentation: how consistent are they across landscapes? J Biogeogr 32:1353–1370

Weiss SB (1999) Cars, cows, and checker spot butterflies: Nitrogen deposition and management of nutrient–poor grasslands for a threatened species. Conserv Biol 13(6):1476–1486

Williams BDC (2002) Purple martins in oak woodlands. In: Standiford RB, McCreary D, Purcell K (tech cords) Proceedings Symposium oak woodlands: oaks in California's changing landscape. USDA Forest Service, Albany, pp 323–334

Wilson RG, Manley P, Noon BR (1991) Covariance patterns among birds and vegetation in a California oak woodland. In: Proceedings Symposium oak woodlands and hardwood rangeland management. Standiford RB (tech coord). USDA Forest Service, Albany, pp 126–135

Wirtz WO II, Hockman D, Muhm JR, Souza SL (1988) Post fire rodent succession following prescribed fire in southern California chaparral. In: Management of amphibians, reptiles, and small mammals in North America. Szaro RC, Severson KE, Patton DR (tech coord) USDA Forest Service, Fort Collins, pp 333–339

Chapter 9
Models of Oak Woodland Silvopastoral Management

Richard B. Standiford, Paola Ovando, Pablo Campos and Gregorio Montero

Frontispiece Chapter 9. A thinned blue oak stand in the Northern Sacramento Valley of California, shows coppice regeneration and forage growth. (Photograph by R. Standiford)

R. B. Standiford (✉)
Department of Environmental Science, Policy and Management, University of California, Berkeley, 130 Mulford Hall MC 3110, Berkeley, CA 94720, USA
e-mail: standifo@berkeley.edu

P. Campos et al. (eds.), *Mediterranean Oak Woodland Working Landscapes*, Landscape Series 16, DOI: 10.1007/978-94-007-6707-2_9,
© Springer Science+Business Media Dordrecht 2013

Abstract Spanish dehesas and California ranchlands provide a diverse array of woodland-produced commodities, including forage, wood, acorns, habitat, game, and amenities. Several silvopastoral models exist for analyzing such production. An examination of management scenarios that include encouraging natural regeneration in dehesa is offered, and then compared with management where no extra inputs are provided and the tree overstory is gradually lost over time. A significant issue in Spain and California alike is sustaining production while making certain there is natural regeneration and recruitment of the oaks. A sensitivity analysis of public inputs, product prices, and discount factors is provided. Silvopastoral models for California woodlands illustrate the importance of incorporating actual landowner behavior in policy analysis to accurately represent the future trajectory of oak woodlands.

Keywords Silvopastoral systems · Multi-functionality · Oak natural regeneration · Market and non-market incomes · Positive mathematical programming · Bioeconomic models · Optimal control

9.1 Introduction

Silvopastoral management of oak woodlands in California and Spain commonly provides fuelwood from oak and shrub clearing or tree pruning (Fig. 9.1), fodder (acorns, grass and browses), cereal fodder in long rotations, wild game, honey, and other diverse private goods and services (Moreno et al. 2007). In addition to these traditional uses, California oak woodlands and the Spanish dehesa provide ecosystem services of growing interest to the public and policymakers, including recreational opportunities, carbon storage, and wildlife habitat (Chaps. 8, 11, 12). A continued supply of such goods from private oak woodlands in California and Spain depends on owners receiving monetary and non-monetary benefits greater than the opportunity costs of forgoing competing land uses.

P. Ovando · P. Campos
Institute of Public Goods and Policies (IPP), Spanish National Research Council (CSIC), Albasanz 26–28 28037 Madrid, Spain
e-mail: paola.ovando@cchs.csic.es

P. Campos
e-mail: pablo.campos@csic.es

G. Montero
Forest Research Centre, National Institute for Agriculture and Food Research and Technology, Ctra. de la Coruña km. 7,5 28040 Madrid, Spain
e-mail: montero@inia.es

Fig. 9.1 Firewood, even in fairly recent history (1960s), was a staple fuel in Spain, especially when converted to charcoal, and wood pruned from holm and cork oaks provided a ready source of fuel for energy. While that market decreased in the 1960s with propane and oil-based heating, thanks to a rising Spanish interest in barbecuing and wood-based cooking, firewood retains some value (Elena-Roselló et al. 1987) (Photograph by P. F. Starrs)

In this chapter two general approaches are described for assessing silvopastoral management systems. A silvopastoral model for western Spanish holm oak (*Quercus ilex* L.) dehesa allows evaluation—using an extended cost-benefit approach—of a managed, or facilitated, natural regeneration project over the entire productive cycle (rotation) of woodland in Monfragüe shire in Extremadura. This woodland is part of the buffer zone that surrounds Monfragüe National Park. For California oak woodlands, an optimal control multi-objective silvopastoral model is presented, showing the influence of the interrelationships of grazing, firewood harvest, and hunting on optimal economic outputs.

9.2 A Spanish Dehesa Silvopastoral Model

There are two important trends in the dehesa of southwestern Spain that are important to dehesa conservation. First is a marked recent decline in dehesa productivity and profitability due to poor oak regeneration, which is most commonly attributed to unrestricted grazing (Díaz et al. 1997; Pulido et al. 2001; Pulido and Zapata 2006; Plieninger 2006, 2007). Estimates based on the Spanish National Forest Inventory show that in dehesa areas of Andalucía, Extremadura, and Castilla-La Mancha, natural regeneration is insufficient or nonexistent in more than 60 % of holm oak and in 95 % of cork oak woodlands (MARM 2008, 2011). The second trend is a moderate long-term appreciation in real land price. From 1994 to 2010 the nominal and real average cumulative rate of Spanish dry natural grassland price change was 5.68 and 2.65 %, respectively (MARM 2011). With low commercial profitability from traditional silvopastoral management, land price appreciation is tied to increases in the amenities enjoyed by private owners (Campos and Riera 1996; Campos 1997; Campos et al. 2009; Chap. 13).

The lack of natural oak regeneration in the dehesa is unsustainable from the perspective of tree-related yield of goods and services such as acorns, firewood, browse and wildlife. While the current management regime with poor oak regeneration appears to generate competitive private profitability rates, especially when considering increasing prices for dehesa properties, private amenity values, rental fees for hunting of wild game, livestock grazing, and various government subsidies, there are long periods of negative cash flow that accompany afforestation efforts or facilitated natural regeneration of oaks (Campos et al. 2008a, b). This makes regeneration of the dehesa commercially unattractive to many private landowners, who base their decisions on past—and even historical—trends. That may be short-sighted with respect to opportunities for the production of future goods and services (Martín et al. 2001; Campos et al. 2008a).

Economic analysis of facilitated natural regeneration for holm oak woodlands requires development of a management model, incorporating growth and yield functions from the beginning of the regeneration treatments to the end of holm oak production cycle. For this purpose, we use the set of forestry operations described in a model developed by Montero et al. (2000). This offers information on diameter growth, acorn and firewood yields, implemented in a management scheme that is tracked through an entire rotation cycle (Fig. 9.2).

The holm oak production cycle is modeled here over a 250–year rotation for the Monfragüe holm oak dehesas. This shire has a surface area of 133,282 ha, covering the territories of seven municipalities. The useful agricultural land (UAL) is mainly dehesa (43 % of UAL), and dryland pasture and temporary grain fields (19 % of UAL), as tabulated by Campos et al. (2008a). Extensive livestock production is important in the Monfragüe shire, although increasing livestock grazing pressure hinders natural regeneration (Chap. 5; Campos et al. 2001; Pulido et al. 2001; Rodríguez et al. 2004; Pulido and Zapata 2006; Plieninger 2006, 2007).

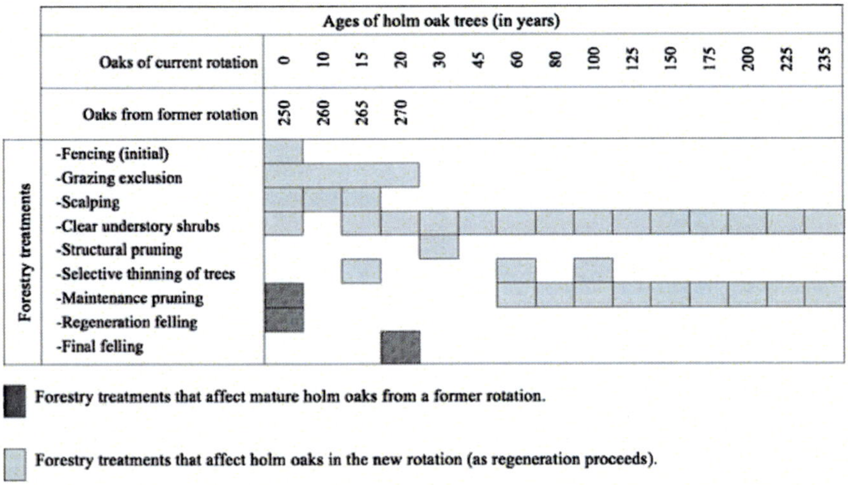

Fig. 9.2 A forestry operations schedule for the dehesa holm oak management model of facilitated natural regeneration

9.2.1 Dehesa Holm Oak Growth Functions

Diameter growth throughout the life cycle of an oak tree is a good indicator of stand development, and of its capacity to produce timber and firewood. Normal diameter growth determines the basal area growth in even-aged holm oak stands.

Holm oak diameter growth is estimated from annual growth data from 34 holm oak trees collected from oak rounds measured at breast height (dbh = 1.30 m). This data was used to fit a Richards-Chapman function, with a mean error of 4.0 cm and adjustment coefficient (R^2) of 0.86:

$$d_{sc}(cm) = 115.528 \cdot \left(1 - e^{-0.00644568 \cdot t}\right)^{\left(\frac{1}{0.987524}\right)} \tag{9.1}$$

where d_{sc} is the diameter (in cm) without bark at breast height (1.3 m), and t is the estimated oak age (normal age). It is assumed that the real oak age equals to t + 10, assuming that holm oaks take 10 years to reach 1.3 m in height, after being planted or recruited.

9.2.2 Holm Oak Silvopastoral Management Model

The prescribed treatments for facilitating natural regeneration include 20-years of grazing exclusion (using fences of a height sufficient to exclude deer), structural tree pruning, selective tree thinning, shrub clearing, and regeneration felling (Fig. 9.2). During the 20-year grazing exclusion period, both the quantity of forage

Table 9.1 Key yield and input indicators of facilitated natural regeneration investment and non investment scenarios for the entire cycle of holm oak

Class	Unit	Quantity (units hectare^{-1})	
		Total cycle	Annual
Investment scenario (250-year)			
Firewood	kg	428,453	1,714
Maintenance pruning	kg	16,206	65
Oak trees felling	kg	412,247	1,649
Forage estimated consumption	FU[a]	41,952	168
Acorns (total biological yield)	kg	140,110	560
Acorns (montanera)	kg	67,253	269
Acorns (big game)	kg	33,561	134
Working hours	hour	2,437	9.7
Machinery and equipment	hour	816	3.3
Non-investment scenario (70-year)			
Firewood	kg	80,267	1,147
Maintenance pruning	kg	1,427	149
Oak trees felling	kg	6,840	998
Forage estimated consumption	FU[a]	12,768	182
Acorns (total biological yield)	kg	10,454	149
Acorns (montanera)	kg	5,018	72
Acorns (big game)	kg	2,509	36
Working hours	hour	441	6.3
Machinery and equipment	hour	162	2.3

[a] *FU* forage unit

units (FU)[1] consumed by livestock in oak woodlands and acorn consumption in the *montanera* period (Chap. 10) must be reduced under the natural regeneration scenario (Table 9.1). This exclusion period is needed, and required under government subsidy policy, to reduce browsing damage to regenerating oaks.

Regeneration felling is the initial treatment to facilitate natural regeneration. This involves cutting a large percentage of aging holm oak trees to enhance on-site seeding under the tree canopy, without completely forgoing ongoing firewood yields from the remaining trees through the regeneration and recruitment period. The most productive trees are left standing until the final clear-cut of remaining mature trees takes place. The model posits three consecutive felling operations to assist regeneration. At first felling, left in place are 20 or more well-distributed high acorn yielding trees per ha. After a decade, a second regeneration felling is scheduled, leaving at least 15 older trees/ha. A third felling is scheduled twenty years after the first felling, and removes the remaining mature holm oaks. This is timed to match an end to the grazing exclusion period (Figs. 9.2). By then, the number of oaks per ha is considerable lower and more typical of dehesa (Figs. 9.3).

[1] A forage unit (FU) represents the energy contained in a kilogram of barley at 14.1 % humidity, or 2,723 kilocalories of metabolic energy (INRA 1978).

Fig. 9.3 Young holm oak trees from natural regeneration with recent structural pruning to transform them from shrubs to a tree-like form (Badajoz, Spain) (Photograph by A. Adamez)

This prescription results in a seedling density of some 3,000 stems per hectare, which will decrease to 2,000 oaks per hectare after 16 years following the first thinning and a shrub clearing. Thinning treatments reduce competition from weak trees, favoring the growth of the residual trees. It is assumed that diseased trees are removed simultaneously with periodic shrub clearing every 25 years to favor oak tree growth by reducing competing vegetation and disease spread.

Finally, the model has one formation pruning of the regenerating oaks when they reach the age of 31 years (real age), coinciding with the second thinning and a periodic clearing of understory shrubs (Fig. 9.4). In addition, the model includes cyclical maintenance pruning to encourage acorn yield (although efficacy of such treatments are now questioned, Chaps. 5, 7), to balance the coppice form (as semi-round), to obtain firewood, and to provide livestock browse. Pruning operations should not affect more than one-third of the coppice biomass.

9.2.3 Firewood and Acorn Yields

Firewood is a byproduct of forest management operations that comes from maintenance pruning, sanitary felling, and thinning (Figs. 9.4, 9.5, 9.6). The linear function estimates published by Montero et al. (2000) relate firewood yield to oak tree diameter, both for firewood resulting from pruning and from tree felling treatments. Thinned trees are assumed to have a diameter 35–40 % lower than the average diameter of the holm oak stand, with a 75 % firewood yield, based on lower intensity management than the empirical data used by Montero et al. (2000).

Fig. 9.4 Recently pruned holm oak dehesa (Salamanca province, Spain). The wood obtained as byproduct of maintenance pruning can be used for posts, firewood, or in charcoal preparation (Photograph by P.F. Starrs)

Fig. 9.5 Wood posts gleaned from the thinning of wild olive trees in the dehesa Montes de Propios of the Jerez de la Frontera Municipality. After curing, these are used for fencing (Photograph by P. Campos)

For felling of diseased trees (sanitary felling) it is further assumed that only 50 % of the resulting firewood will be commercialized. Firewood yield estimates for the entire holm oak natural regeneration scenario and its alternative non-regeneration scenario are displayed in Table 9.1.

Fig. 9.6 **a, b** The traditional method of charcoal preparation consists of covering a woodpile with soil and straw, allowing only a small amount of air to enter. The wood sticks burn very slowly in a "cold fire" and become charcoal. **c, d** Charcoal loaded in bags for transport and sale (Photographs by P. Ovando)

Acorn yield in Mediterranean areas is characterized by considerable annual variation (Chap. 8). There are periods with high acorn production, followed by a varying number of years with low to moderate yields (Chap. 7; Pulido and Díaz 2003). Montero et al. (2000), in an article based on existing literature and experimental data, provide a set of hypotheses for estimating acorn yield along the entire productive cycle of a holm oak. It is assumed that holm oak acorn yield starts having commercial value when oak trees are 21 years old at about the time the grazing exclusion period comes to an end. During the early stages of oak growth, from years 21–49, data used are from Rupérez (1957). From year 50 to 99, acorn yield based on the estimates provided by González and Allue (1982). From year 100 to the end of tree's productive cycle data is from Vázquez et al. (1999). Furthermore, 20 % of acorn biological yield is assumed not to be available for livestock and game animals due to insect and rodent depredation and other environmental effects (Díaz et al. 2011; Chap. 8). Figure 9.7 provides information on average acorn yield per oak tree and the number of productive oak trees per hectare throughout the holm oak rotation.

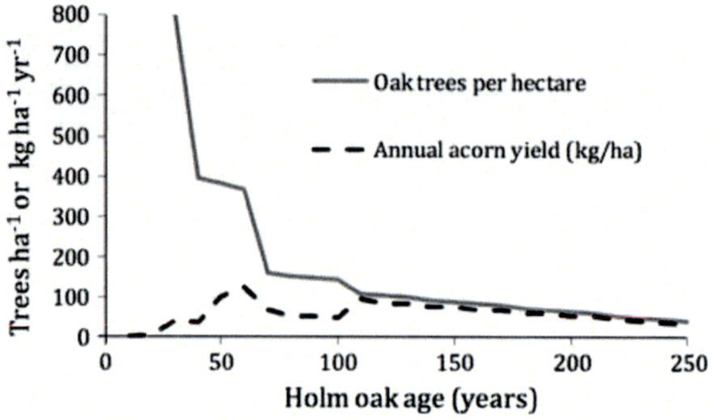

Fig. 9.7 Average oak tree density and acorn yield per hectare and year through the entire holm oak productive cycle

9.2.4 Cost-Benefit Analysis: Alternative Holm Oak Dehesa Management Scenarios

The facilitated natural regeneration economic cycle lasts 250 years, the time between two regeneration fellings. If no supplemental regeneration treatments are applied at the time of the regeneration felling, holm oak will gradually disappear. If no grazing exclusion is implemented, the recruitment is unsuccessful and oaks will slowly decline and the dehesa will be converted into a treeless pastureland (Fig. 9.8).

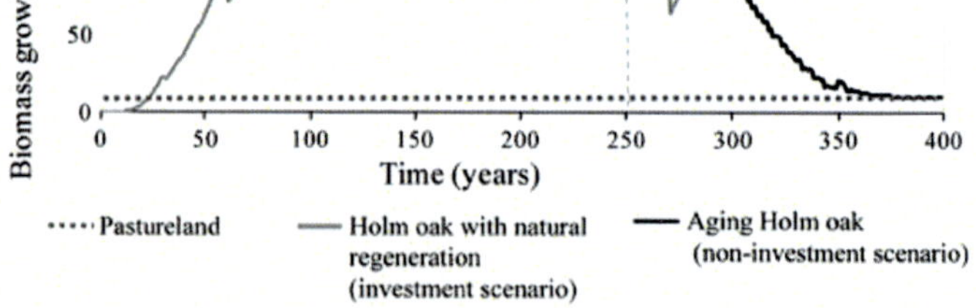

Fig. 9.8 Facilitated holm oak natural regeneration investment and non-investment scenarios

9.2.5 Private Market and Non-market Outputs from Alternative Management Scenarios

Our scenario is for a nonindustrial holm oak woodland private owner with mixed low-risk investor-consumer rationality. The private landowner is assumed to require commercial returns, while also deriving private amenities from the woodland environmental services (Ovando et al. 2010). In our models, the landowner is assumed to obtain private capital income from sales of firewood (F), acorns (A), rent from grazing resources other than acorns (GR) and hunting (HR), supplemented by government net subsidies for forest management, and the enjoyment of private amenities (PA) (non-market private consumption of environmental services internalized in woodland market prices). Constant prices are assumed to correspond with actual prices (sales, net subsidies, and costs) or with estimated prices for private amenities in 2002. We present a sensitivity analysis of the effect of discount rates, market prices for acorns, grazing resources, and government subsidies. To aggregate commercial and environmental benefits in a consistent manner, we use simulated exchange prices for private amenity values (Campos et al. 2009).

Except for private amenity value, woodland private benefits is based on direct market prices. Big game hunting income is determined by what landowners would be paid for leasing their land for hunting, net of costs and taxes. Grazing income (only for forage) reflects the market prices for leasing one hectare of holm oak woodland, or leasing open pastureland for livestock forage (Campos et al. 2001; Rodríguez et al. 2004). Acorns are valued based on the price a dehesa owner gets for each kilogram of Iberian pig weight gain during the *montanera* period and the acorn consumption needed to yield that gain (Table 9.2). Of the acorn yield, 32 % is assumed to be consumed by wildlife (especially red deer and wild boar), and 48 % feeds Iberian pigs. Acorn consumption by big game is valued at 60 % of the price of acorns in the montanera period (Chap. 10).

Private amenity value is a non-market ecosystem service that the dehesa landowner might consume, having the right to exclude other potential users or consumers. These environmental uses include private recreational services, the ability to house and entertain friends, enjoyment of a countryside quality of life, and a number of passive uses (legacy, option, and existence values). Future income streams from private amenities are capitalized into land market prices since owners and buyers have that in mind when they consider owning and maintaining a dehesa, and they are willing to pay for these private uses when they decide to maintain a property or to buy a piece of land. Indeed, private amenities have been recognized by the scientific literature as a factor in land prices (Campos and Riera 1996; Campos 1997; Torell et al. 2001; Lange 2004; Campos et al. 2009).

In this study, the private amenity value comes from a contingent valuation survey applied to a sample of 19 dehesa owners in the Monfragüe area (Campos and Mariscal 2003), updated to 2002 prices.[2] Private amenities reflect the maximum cash losses that Monfragüe owners would be willing to accept (WTA) compared with the private environmental uses provided by owning a dehesa. It is assumed that the private amenity value aggregates all dehesa owner environmental unpriced uses without differentiating any single use (Table 9.2).

9.2.6 Holm Oak Management Costs and Government Grants

Facilitated natural regeneration[3] and alternative non-investment scenarios consider only forest management costs (Table 9.2). The economic information used to estimate the work units (labor and machinery) and costs related to diverse forestry operations have been collected in the Monfragüe shire (Rodríguez et al. 2004). This study also considers government grants for forestry treatments in the Extremadura region (Table 9.2) net of taxes on production (DOE 2002).

Cost items that are taken into account are the same as those the conventional System of National Accounts (SNA) considers to estimate forest total and capital incomes (Eurostat 2000). Total cost is estimated as the aggregation of labor cost (LC) and the intermediate consumption of raw materials (RM) and services (SS). Consumption of fixed capital (CFC) is not accounted for since annual investment in fixed capital goods (only fences, in this case) match consumption.

9.2.7 Capital and Total Income Net Present Value, and Holm Oak Investment Versus Non-investment Scenarios

We apply cost-benefit analysis tools to estimate the net present values (NPV) of all future streams of private capital income resulting from the silvopastoral management scenarios analyzed. The streams of private benefits and costs are discounted using the estimated real profitability rates that landowners get from land uses prior to afforestation (Ovando et al. 2010). This represents the private opportunity cost of capital for land investment at the study sites. The estimated annual private profitability rate is 5.5 % in the Monfragüe shire, although a sensitivity analysis is provided showing discount rates ranging from 1 to 8 %.

---

[2] The contingent valuation survey was conducted in 2000 with results subsequently updated to 2002 prices. We assume that private amenity value has the same temporary variation as the market price of non-irrigated pastureland in the Extremadura region (MAPA 2003).

[3] Facilitated natural regeneration for oak trees is required in dehesa to build a tree layer. Regeneration of oaks based only on natural processes does not occur in open spaces. The dehesa is a fragile working landscape maintained by livestock, avoiding overcutting of biomass to meet human demand.

Table 9.2 Market and environmental benefits prices, government grants and forestry management costs in Monfragüe

Class	Unit	Price (2002 \$ unit^{-1})
Market benefits		
Firewood	kg	0.06
Grazing rent (additional to montanera)	ha	53.30
Acorn[a] (1.71 \$ kg hwg^{-1}/9 kg acorn kg hwg^{-1})	kg	0.19
Hunting rent	ha	6.46
Environmental benefits		
Private amenity	ha	95.19
Government grants[b]		0.00
Oaks structural pruning	tree	0.71
Oaks maintenance pruning	tree	2.57
Selective thinning (trees with a diameter < 18 cm)	tree	0.21
Shrub clearing	ha	114.00
Forestry management cost		
Fencing	ha	714.59
Shrubs clearing	ha	116.95
Structural pruning	tree	2.37
Selective thinning (trees with a diameter < 18 cm)	tree	0.27
Selective thinning (trees with a diameter > 18 cm) [c]	tree	15.68
Maintenance pruning[b]	tree	12.26
Regeneration felling	ha	3,035.92
Final felling	ha	1,281.74

[a] hwg: pig weight gain in montanera period
[b] DOE (2002)
[c] Average value

9.2.8 Capital and Total Annual Incomes

We estimate the present values of capital income gains or losses as the difference between discounted capital incomes obtained from the facilitated holm oak dehesa natural regeneration investment scenarios and the discounted capital incomes generated by an aged holm oak non-investment scenario (defined earlier).

Capital income (CI_{pp}) at producer prices is an annual income indicator that reflects the difference between total benefits (TO), derived from market output sales, and private amenity consumption and total cost (TC) associated with holm oak dehesa management. The dehesa landowner business objective is the capital income at basic prices (CI_{bp}). The difference between producer price[4] and basic price indicators is that the latter includes subsidies (S) net of taxes (T) on production:

[4] Prices before government intervention via subsidies and taxes on products.

$$\mathrm{CI}_{\mathrm{pp}} = \mathrm{TO} - \mathrm{TC} = \mathrm{F} + \mathrm{A} + \mathrm{GR} + \mathrm{HR} + \mathrm{PA} - \mathrm{LC} - \mathrm{RM} - \mathrm{SS} \quad (9.2)$$

$$\mathrm{CI}_{\mathrm{bp}} = \mathrm{CI}_{\mathrm{pp}} + \mathrm{S} - \mathrm{T}. \quad (9.3)$$

Total annual income at producer prices ($\mathrm{TI}_{\mathrm{pp}}$) and at basic prices ($\mathrm{TI}_{\mathrm{bp}}$), are estimated by adding labor cost to $\mathrm{CI}_{\mathrm{pp}}$ and $\mathrm{CI}_{\mathrm{bp}}$, respectively:

$$\mathrm{TI}_{\mathrm{pp}} = \mathrm{CI}_{\mathrm{pp}} + \mathrm{LC}. \quad (9.4)$$

$$\mathrm{TI}_{\mathrm{bp}} = \mathrm{CI}_{\mathrm{pp}} + \mathrm{LC} \quad (9.5)$$

9.2.9 Net Present Value Indicators

Net present value (NPV) of the expected stream of private capital incomes considers an infinite sequence of holm oak facilitated natural regeneration cycles. For non-investment scenarios, once oak trees disappear due to mortality and lack of regeneration, treeless pastureland is assumed to be the permanent land use. The capital value of a hectare of holm oak dehesa that is managed as scheduled by the facilitated natural regeneration model ($V_{n,\infty}$) is estimated considering the following equations:

$$V_n = \sum_{j=t}^{T_n} \delta^{t-1} y_n(t), \quad (9.6)$$

$$V_{n,\infty} = \left[\left(1 + \delta^{T_n} + \delta^{T_n \cdot 2} + \delta^{T_n \cdot 3} \cdots\right) V_n\right], \quad (9.7)$$

$$V_{n,\infty} = \left(\frac{1}{1 - \delta^{T_n}}\right) V_n \quad (9.8)$$

where y_n represents the value of any income variable (in one hectare) in any year of the economic cycle of facilitated natural regeneration where rotation length is defined by T_n ($\mathrm{T_n} = 250$ years); and δ represents the intertemporal discounting function: $\delta = 1/(1 + \mathrm{r})$, being r the annual discount rate.

For a non-investment scenario with no regeneration treatment, after 70 years, holm oaks are assumed to disappear. The net present value of a 250 year old holm oak stand with no regeneration treatment is estimated according the following equation:

$$V_{wr} = \sum_{j=t}^{T_n+70} \delta^{j-t} y_{wr}(t), \quad (9.9)$$

Without regeneration, an aging holm oak stand will be replaced by treeless pastureland used for livestock rearing with no grazing exclusion periods, which have annual incomes defined by y_j. Thus, the net present value of an aging holm oak result with no regeneration, converting to pastureland ($V_{n \to j,\infty}$), is estimated according to:

$$V_{wr \to j,\infty} = V_{wr} + \delta^{T_n+1}(1 + \delta^1 + \delta^2 + \ldots)y_j(t), \quad (9.10)$$

$$V_{wr \to j,\infty} = V_{wr} + \delta^{T_n+1}\left(\frac{1}{1 - \delta^{T_n}}\right)y_j(t). \quad (9.11)$$

9.2.10 Results of the Dehesa Silvopastoral Scenarios

The present values of the expected stream of capital and total private incomes from the facilitated holm oak natural regeneration scenario, and the alternative non-investment scenario of gradual depletion of holm oak woodland, shows the present value of an infinite series of facilitated natural regeneration cycles compared to the present value of aging holm oak woodland that is permanently replaced by bare pastureland once oak trees disappear due to mortality and lack of natural regeneration (Table 9.3).

Two NPV indicators are considered: (1) the net present value of the stream of expected market and non-market outputs minus the expected costs from the facilitated natural regeneration scenario; and (2) the net benefits that show the difference between the NPV of the investment scenario of facilitated natural regeneration and non-investment scenario. These indicators are useful in the analysis—given current market benefits and governmental grants to holm oak management—to determine if renewing an old holm oak stand with treatments to encourage natural regeneration is an attractive investment for dehesa landowners.

Table 9.3 Net present value (NPV) of capital and total private incomes for investment and non-investment scenarios for facilitated holm oak natural regeneration ($ per hectare, year 2002)[a]

Class	Investment scenario: Facilitated holm oak natural regeneration A	Non-investment scenario: Aging holm oak B	Net benefits NPV C = A − B
Net present values	$V_{n,\infty}$	$V_{wr \to j,\infty}$	
Capital income at producer's prices (CI_{pp})	920.17	2,401.13	−1,480.96
Capital income at basic prices (CI_{bp})	1,326.30	2,547.81	−1,221.51
Total income at producer's prices (TI_{pp})	5,011.06	3,732.36	1,278.70
Total income at basic prices (TI_{bp})	5,417.28	3,879.04	1,538.24

[a] Present discounted values for an infinite time horizon frame. Discount rate: 5.5 %

The present value of the expected stream of outputs from the scenario for facilitating holm oak natural regeneration exceeds the value of the expected costs of those treatments (Table 9.3). The value of $V_{n,\infty}$ is positive, even if a dehesa owner receives no governmental grants for holm oak management (CI_{pp}). Nonetheless, letting a holm oak stand gradually decline is more profitable to the dehesa owner ($V_{wr \to j, \infty}$), and requires no initial investments for grazing exclusion fences and regeneration felling treatments (Table 9.2). In the non-investment scenario, landowners do not have a 20-year grazing exclusion period, and benefit from the revenues of leasing the land for hunting or using the land for livestock production.

These results suggest that facilitating holm oak natural regeneration offers a positive present value, even in a situation without governmental support for forest management operations. Nevertheless, under current market prices and governmental support conditions, facilitated holm oak natural regeneration cannot compete with a scenario of gradual forest decline due to aging and no recruitment.

Since private amenities are assumed to have the same value for all the land uses, it follows that the benefit from investment projects are entirely due to the discounted value of private commercial capital incomes. The effect of gradual depletion of holm oak on private amenity values is unstudied for the Monfragüe region. Nonetheless, recent research suggests that private amenity value is positively related to the proportion of forest area in southwestern Spanish cork oak woodlands (Ovando et al. 2010). A decrease in the holm oak tree population in dehesa as result of insufficient investment in holm oak renewal is likely to affect private amenity consumption.

Over the last fifteen years, the European Union and the Spanish government have strongly encouraged holm oak afforestation in pastureland, shrubland, and cropland, under the European Regulations 2080/92 and 1257/99 (Ovando et al. 2007). Those government aids promulgated in 2002 cover plantation costs, including fencing, 5 years of plant maintenance payment, and 20 years of financial compensation for grazing exclusion from the regenerating area. The European Union's (EU) ongoing policy reform in rural development focuses on multifunctional agriculture in compliance with the EU's environmental goals, which include mitigating biodiversity losses and climate change. This new rural development scheme may add government support to natural woodland regeneration practices in European agroforestry systems. Facilitated natural regeneration in the dehesa could be an efficient option for maintaining and even increasing the dehesa's current carbon stock and biodiversity (Díaz et al. 1997; Campos et al. 2008a).

Our results indicate that the 20-year compensation for grazing exclusion, which in 2002 rose to \$165 ha^{-1} year^{-1} for dehesa and treeless pastureland, would generate enough incentives to pursue a facilitated natural regeneration project for holm oak (BOE 2001). Natural regeneration investment would still be competitive when compared to the non-investment scenario, even considering 20 years of compensation for grazing exclusion that is 25 % lower than that for afforestation projects in pastureland and dehesa under EU Regulation 1257/99 in Spain (Fig. 9.9).

Livestock and wildlife grazing pressure can seriously hinder holm oak regeneration capacity (Díaz et al. 1997; Pulido et al. 2001; Pulido and Zapata 2006;

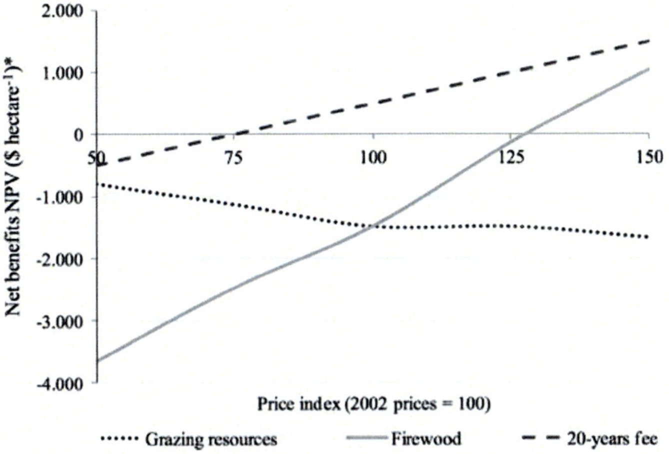

Fig. 9.9 Sensitivity of net benefits at basic prices, which includes subsidies net of taxes on products, of facilitated natural regeneration investment in holm oak to variation in prices of grazing resources, firewood and 20-year payments. *Note* *Average euro/dollar exchange ratio in 2002: 1 euro (€) = 0.95 US dollar ($) (BDE 2012)

Pulido and Díaz 2003; Plieninger 2006, 2007). Since the regeneration investment scenario implies 20 years of grazing exclusion, simulation of the effect of changes in the price for grazing resources (forage and acorns) on dehesa owner NPV capital incomes (CI_{bp}) may be crucial in driving this investment decision. Indeed, a decrease of 50 % in montanera (acorn) and grazing rent prices is not even enough to make facilitated natural regeneration scenario more competitive than letting the holm oak woodlands move instead toward open pasture land (Fig. 9.9).

However, the net benefits of the holm oak facilitated natural regeneration scenario are quite sensitive to firewood price variability. An increase of 27 % over current firewood prices would make the investment scenario a more attractive alternative than the non-investment one, since firewood is a byproduct of regeneration felling (Fig. 9.9). It is worth mentioning that under current dehesa rules, an increase in firewood prices may not be an incentive for harvesting dehesa oaks since those trees are highly protected and the dehesa owner requires special authorization to harvest trees for firewood.

The net benefits (in terms of CI_{bp}) of the holm oak facilitated natural regeneration scenario are slightly sensitive to variations on real discount rate. The facilitated holm oak natural regeneration scenario would be the preferred option for a landowner that demands an interest rate lower than 2 % from this investment scenario (Fig. 9.10). Discount rates lower than 2 % seem to be far from the rates that dehesa and other Mediterranean oak woodland owners use for discounting the stream of future expected private capital incomes from silvopastoral uses and private amenity consumption (Ovando et al. 2010).

Total income NPV indicators are less relevant for a dehesa landowner, but may be a key factor for designing forest conservation policies, since employment

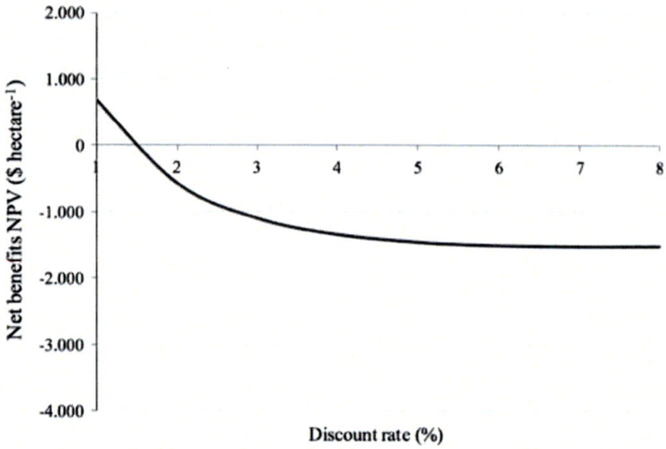

Fig. 9.10 Sensitivity of net benefits at basic prices, which includes subsidies net of taxes on products, of facilitated holm oak natural regeneration investment to discount rates

generation is one of the aims of European Common Agricultural Policy reforms. Facilitating natural regeneration of an aged holm oak stand delivers higher labor demand (Table 9.1) and labor incomes that offset the negative benefits of the present value of private capital incomes (Table 9.3).

9.3 California's Silvopastoral Management System

Silvopastoral management of oak woodlands in California relies on tree, forage, and livestock management to produce diverse economic and environmental values. Silvicultural, range production, and livestock models exist to assess silvopastoral management. The general approach develops an optimal control model to link biological, environmental, and economic components. The objective function is to maximize discounted net value by landowners over a planning horizon for livestock, firewood harvest, and fee hunting enterprises. Equation (9.12) below shows the general framework for this model, based on forage production models, oak growth models, and hunting revenue models (Standiford and Howitt 1992, 1993).

$$\max NPV = \int_{t=0}^{T} e^{-rt}\{WR_t(WDSEL_t) + HR_t(WD_t, HRD_t, exog.) + LR_t[HRD_t, CS_t, FOR_t(WD_t, exog.)]\} \tag{9.12}$$

such that:

$$\dot{WD} = f(WD_t, exog.) - WDSEL_t \qquad \text{(equation of motion for oaks)}$$

$$\dot{HRD} = G(HRD_t, exog.) - CS_t \qquad \text{(equation of motion for livestock)}$$

Initial Conditions: $WD_0 = INITWD$ and $HRD_0 = INITHRD$ where WD and HRD are the stock of wood volume and livestock numbers (cows); WR, HR and LR are the net revenues of firewood, hunting and livestock respectively; WDSEL is the volume of firewood sold and CS is a vector of the classes of livestock sold; FOR is the forage quantity available by season; r is the interest rate; and exog. are exogenous site factors (soil productivity, annual rainfall and temperature).

Solving this equation with existing prices and climatic data shows that in the last decade, the optimal solution was for landowners to completely clear the oak trees on their property because of the low growth rates of the trees, and the reduced forage production under tree canopies. This represents a "normative" approach to making recommendations to landowners on a maximum return from management of their oak woodlands.

However, these models did not reflect the actual behavior of oak woodland managers during this time period. Scenarios calculated for the early 1990s in California concluded that markets at that time would lead landowners to clear their oaks to increase forage yield for livestock production (Standiford and Howitt 1992). Although common in the 1940s–1970s, this behavior was actually rare in the nineties, contradicting the prediction of the model (Standiford et al. 1996).

These normative models have the drawback of omitting a landowner's amenity value from oak stands. To more realistically model landowner behavior in the current market, policy and climatic regimes, a positive mathematical programming (PMP) approach (Howitt 1995) was used to derive missing elements of the true costs and returns of oak harvest and retention for landowners. The dynamic optimization model is enhanced with a constraint for actual landowner behavior. The actual amount of firewood harvest and tree removal by landowners (Bolsinger 1988) was a constraint added to the model, and recalculated incorporating the actual behavior. The shadow prices derived from the behavior constraint represents the marginal benefit of retaining trees. That value was integrated to calculate a "hedonic" quadratic cost function for account for the apparently negative utility to landowners from overcutting oaks on their property.

Figure 9.11 compares firewood stumpage price to the "apparent" hedonic price. The difference between the two curves represents the "cost" of overcutting firewood, or the private amenity consumption value of retaining trees. Figure 9.12 shows the trajectory of optimum oak cover of the normative model, excluding the hedonic cost, and the positive model, which is calibrated to actual producer behavior.

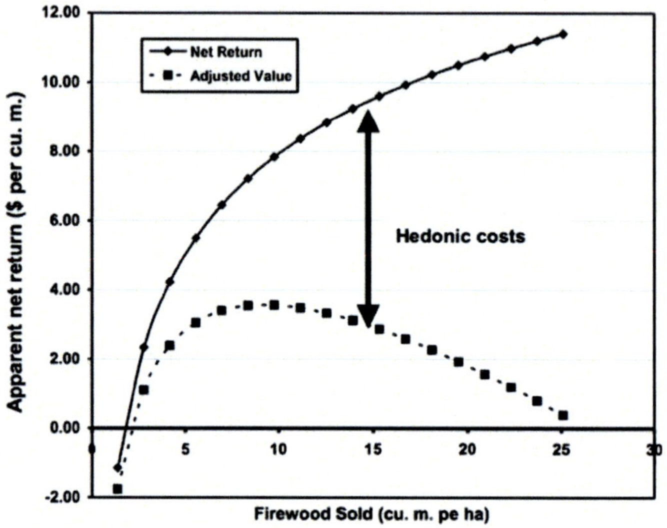

Fig. 9.11 Net firewood return per cubic meter as function of amount of wood harvested (reproduced from Standiford and Howitt 1992)

9.3.1 Commercial Production from Woodlands

A positive mathematical programming approach was used to model the trajectory of oak canopy cover, firewood harvest, and cattle stocking for different risk and land productivity conditions (Standiford and Howitt 1992). Figures 9.13 and 9.14 shows the contribution of the three major commercial enterprises to total net

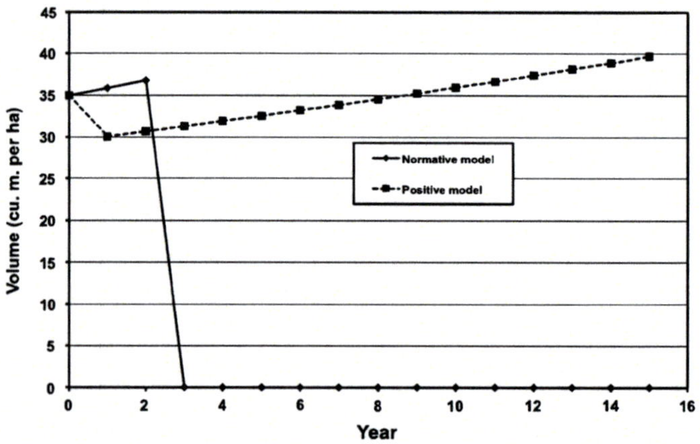

Fig. 9.12 Comparison of positive versus normative solutions to oak silvopastoral model

Fig. 9.13 Stacked firewood on a ranch in Tulare County, California, where there is a commercialized program marketing oak firewood (Photograph by M. McClaran)

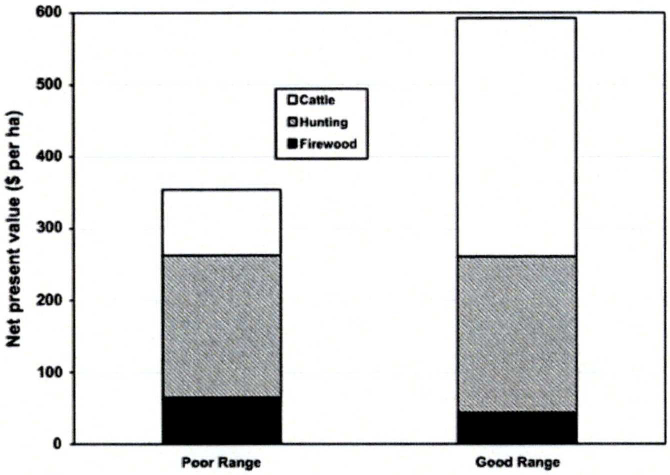

Fig. 9.14 Net present value (*NPV*) of California oak woodlands from various commercial enterprises on poor, or low productivity grazing land, and good, or high productivity grazing land (Standiford and Howitt 1993)

present value for an oak landowner with an initial condition of 30 percent oak canopy and a risk factor assuming that a loss can be tolerated in one year out of ten for a high and low productivity forage production area (Standiford and Howitt

1993). Fee hunting is an important enterprise, contributing 40–70 percent of total woodland value. Firewood, the only major wood product on California's oak woodlands, has low value compared to cattle or hunting enterprises. The marginal value of retaining oak tree cover for hunt club habitat often exceeds the marginal value of the extra forage or firewood harvest value resulting from tree harvest (Standiford and Howitt 1992). The model shows diversification of silvopastoral enterprises reduces tree harvesting and cattle grazing.

The model was evaluated for different risk scenarios for landowners, using a Cooper-Charnes chance-constrained approach (Charnes and Cooper 1959). In general, the higher the risk aversion, the more likely that firewood harvest would be relied on to reduce the probability of economic loss during low livestock price years, or poor forage production years. The capital value of the trees is a hedge against years with low livestock profitability. Inclusion of a risk term shows that firewood harvest and livestock grazing intensity both increase. Policies reducing landowner risk, such as a subsidized loan program during poor forage production or low livestock price years, might reduce the need to cut the trees for an infusion of capital (Standiford and Howitt 1993).

9.3.2 Tree Growth and Modeling

The approach used in the silvopastoral model described above was a whole stand oak growth model, based on 81 sample plots with 1,013 trees, located in seven different geographic regions throughout California oak woodlands (Standiford and Howitt 1988). This model derived a site index relationship for oaks on rangelands based on height and index diameter, rather than height-age relationships because of the difficulty in determining oak age. It also developed a simple basal area growth model for different initial stocking levels by site class. The model also provides correlations between basal area stocking and overstory canopy cover percent and tree height, as well as a site index relationship.

Another promising modeling approach for future silvopastoral modeling in California uses a distance independent, individual tree model (Standiford 1997). This approach provides an opportunity to model stand structure changes over time, with different thinning prescriptions. Stand structure changes can also incorporate coppice management, using the relationships derived in McCreary et al. (2008) and Standiford et al. (2010a, b). The probability of natural seedling regeneration can also be incorporated into these models (Standiford et al. 1997).

The USDA Forest Service supplemented its National Forest Inventory, focused on commercial forestlands, to include oak woodlands in the 1980s. The first report on the growth, harvest, regeneration status and mortality of the series of permanent plots on California oak woodlands was reported in Bolsinger (1988) and formed the basis for the positive programming approach described above (Standiford and Howitt 1992). The series of plots was expanded in the past decade to provide a more robust statistical survey of the state, with additional emphasis on wildlife habitat elements such as woody debris and snags, and exotic diseases such as Sudden Oak Death and pests such as the Golden Spotted Oak Borer (Waddell and Barrett 2005). The data from the U.S. Forest Service's Forest Inventory and Analysis unit can be used for additional calibration of actual landowner behavior.

Additional work on tree growth can be derived from controlled thinning experiments. Almost 20 years of stand development has been evaluated for three thinning levels for coast live oak (*Q. agrifolia*) throughout the Central Coast of California (Bonner et al. 2008). Over 15 years of stand structure change, sprouting, and acorn production have been measure for three thinning levels for blue oak (*Q. douglasii*) and interior live oak (*Q. wislizenii*) in the southern Sierra Nevada (Standiford et al. 2010b).

Ecologically based state and transition models have also been derived for various oak woodland cover types (George and Alonso 2008). These provide probabilities of different ecological pathways with different management and disturbance regimes. This approach can link tree cover, range productivity, and other vegetation species together, and provide input to economic and management models.

9.3.3 Incorporating Other Products of Silvopastoral Management

Cattle production, firewood harvest and fee hunting are the products currently dominant in California's silvopastoral management system. The models described above can be expanded to include other emerging markets as additional information on values and management costs are derived.

California has an emerging market for biomass energy, mainly using cogeneration facilities throughout the state. There have been opportunities for utilization of solid wood for cogeneration through various incentive programs (BIWG 2006). The overall wood volume from oak woodlands is substantially lower than on commercial conifer forestlands, which are only break-even at best at this time. Delivered wood prices are currently quite low, with high transportation costs.

Most California livestock production on oak woodlands is from cow-calf operations, with the sale of calves as the primary economic product (Standiford and Howitt 1993). These markets have been subject to extreme variability. There has been interest in evaluating value-added cattle products, with expanding demand for grass-fed beef and new meat packing facilities proposed to utilize grass-fed cattle. There has also been an interest in utilizing more stocker operations to manage the risk of annual forage fluctuations resulting from rainfall and temperature variability. Several studies point to the possible markets for these new livestock management and marketing strategies (Harper et al. 2005; Blank et al. 2006).

With the passage in 2006 of California's Global Warming Solutions Act (Assembly Bill 32) by the Legislature, the state set limits on greenhouse gas (GHG) emissions (ARB 2006). The law reduces GHG to 1990 levels by 2020–a reduction of

30 %–and another 80 % reduction by 2050. The new law establishes a cap-and-trade program to develop markets designed to encourage the sequestering of carbon. Preliminary analysis of the implications for oak woodlands showed that only $0.70 per hectare per year for central Sierra Nevada oak woodlands based on current markets is expected (Forero et al. 2010). However, as the implications of AB 32 for California's economy develop, the prices for sequestering carbon in oak woodlands may increase, and create new market opportunities for silvopastoral management.

9.4 Synthesis and Conclusions

These two silvopastoral modeling efforts in the Spanish dehesa and California oak woodlands reveal the important linkage of multiple outputs with realistic cost and return data. For the Spanish dehesa, silvopastoral modeling indicates that even if natural oak regeneration is not as profitable as grazing alone, given current social preferences and the shortcomings of the government's land use policy, investment in tree regeneration and development is needed in order to maintain future options for providing commodities and amenities for future generations. Long-term holm oak dehesa conservation may depend on implementing accurate compensation schemes, since private landowners are often unable to accept the short-term cash losses required to invest in dehesa regeneration. This work gives insights into the income losses that private owners may incur from natural oak regeneration treatments and grazing restrictions. We strongly suggest that future research is needed to improve scientific and policy knowledge regarding the minimum payments and the appropriate compensation schemes needed to induce dehesa owners to invest in the regeneration of aging oak woodlands (Ramírez and Díaz 2008), which would simultaneously help mitigate long-term biodiversity loss (Chap. 8) and potentially boost landowner amenity and financial benefits from dehesa improvement and afforestation.

For California oak woodlands, the modeling effort shows the importance of incorporating actual landowner behavior into findings derived from current cost and return data. Landowners do receive value from maintaining certain levels of oak stands, and any policy analysis needs to carefully take this into account. Enhancements in modeling efforts are possible as the interrelationships between the various products from silvopastoral systems become better understood. In addition, new markets are anticipated, especially for ecosystem services and carbon sequestration, which will create new opportunities for sustainable silvopastoral management outcomes.

Acknowledgments The investigation carried out in dehesas is a contribution to the project "Análisis prospectivo de las rentabilidades social y privada de las forestaciones de encinas y alcornoques del periodo 1994–2000 en el marco del Reglamento 2080/92 (AREA)", supported by funding from the BBVA Foundation. The authors thank an anonymous referee for reviewing an earlier version of the manuscript.

References

ARB (Air Resources Board) (2006) California Global Warming Solutions Act of 2006, California State Assembly Bill 32. http://www.arb.ca.gov/cc/docs/ab32text.pdf

BIWG (Bioenergy Interagency Working Group) (2006) Recommendations for a bioenergy plan for California. State of California, California energy commission CEC–600–2006–004–F. p 52

Blank SC, Boriss H, Forero LC, Nader GA (2006) Western cattle prices vary across video markets and value-adding programs. Calif Ag 60(3):160–165

BDE (Banco de España) (2012) Official exchange rates. Available online: www.bde.es/webbde/es/estadis/ccff/0305.pdf

BOE (Boletín Oficial del Estado) (2001) Real Decreto 6/2001, de 12 de enero, sobre fomento de la forestación de tierras agrícolas. BOE 12:1621–1630

Bolsinger CL (1988) The hardwoods of California's timberlands, woodlands, and savannas. USDA Forest Service Pacific Northwest Research Station Resource Bulletin PNW–RB–148, p 149

Bonner LE, Pillsbury NH, Thompson RP (2008) Long-term growth of coast live Oak in three California counties—17–year results. In: Merenlender A, McCreary D, Purcell, KL, (tech. eds): Proceedings of the sixth California oak symposium: today's challenges, tomorrow's opportunities. U.S. Department of Agriculture, Forest Service, Pacific Southwest Research Station, CA, Gen. Tech. Rep. PSW–GTR–217:69–78

Campos P (1997) Análisis de la rentabilidad económica de la dehesa. Situación. Serie de Estudios Regionales Extremadura, 111–121

Campos P, Mariscal P (2003) Preferencias de los propietarios e intervención pública: el caso de las dehesas de la comarca de Monfragüe. Investig Agrar: Sist Recursos Forest 12:407–422

Campos P, Riera P (1996) Rentabilidad social de los bosques. Análisis aplicado a las dehesas y los montados ibéricos. Información Comerc Españ 751:47–62

Campos P, Rodríguez Y, Caparrós A (2001) Towards the Dehesa total income accounting: theory and operative Monfragüe study cases. Investig Agrar: Sist y Recursos Forest. Spec Issue New Forestlands Econ Account: Theor Appl 1:45–69

Campos P, Daly H, Oviedo P, Ovando P, Chebil A (2008a) Accounting for single and aggregated forest incomes: Application to public cork oak forests of Jerez in Spain and Iteimia in Tunisia. Ecol Econ 65:76–86

Campos P, Ovando P, Montero G (2008b) Does private income support sustainable agroforestry in Spanish dehesa? Land Use Pol 25:510–522

Campos P, Oviedo JL, Caparrós A, Huntsinger L, Seita-Coelho I (2009) Contingent valuation of Woodland-Owner private amenities in Spain, Portugal, and California. Rangel Ecol Man 62:240–252

Charnes A, Cooper WW (1959) Chance-constrained programming. Man Sci 6:70–79

Díaz M, Campos P, Pulido FJ (1997) The Spanish dehesas: a diversity in land-use and wildlife. In: Pain DJ, Pienkowski W (eds) Farming and birds in Europe. Academic Press, London, pp 178–209

Díaz M, Alonso CL, Beamonte E, Fernández M, Smit C (2011) Desarrollo de un protocolo de seguimiento a largo plazo de los organismos clave para el funcionamiento de los bosques mediterráneos. In: Ramírez L, Asensio B (eds) Proyectos de investigación en parques nacionales: 2007–2010. Organismo Autónomo Parques Nacionales, Madrid, pp 47–75

DOE (Diario Oficial de Extremadura) (2002) Orden de 10 de abril de 2002, por la que se convocan ayudas para la gestión sostenible de los montes en el marco del desarrollo rural y se regula el procedimiento para su concesión. DOE 45:4908–4937

Elena-Roselló M, López JA, Casas M, Sánchez del Corral A (1987) El carbón de encina y la dehesa. Ministerio de Agricultura, Pesca y Alimentación. Instituto Nacional de Investigaciones Agrarias, Madrid. Colección Monografías INIA, n° 60. Madrid, pp 113

Eurostat (2000) Manual on economic accounts for agriculture and forestry—EAA/EAF 97 (Rev.1.1). European Communities, Luxemburg

Forero LC, Standiford RB, Stewart WC (2010) Carbon sequestration on the California Oak Woodland. Unpublished Cooperative Extension handout

George MR, Alonso MF (2008) Oak Woodland Vegetation Dynamics: a State and transition approach. In: Merenlender A, McCreary D, Purcell K (tech. eds): Proceedings of the sixth California oak symposium: today's challenges, tomorrow's opportunities. U.S. Department of Agriculture, Forest Service, Pacific Southwest Research Station, CA, Gen. Tech. Rep. PSW–GTR–217:93–104

González A, Allue JL (1982) Producción, persistencia y otros estudios en una dehesa extremeña. MAPA, INIA. Anal Inst Nac Invest Agrar, Serie: Forestal, 5, Separata 6

Harper JM, Klonski KM, Livingston P (2005) Sample costs for an organic cow-calf operation. University of California Cooperative Extension BF-NC-05-O. http://coststudies.ucdavis.edu/

Howitt R (1995) Positive mathematical programming. Amer J Agric Econ 77:329–342

Institut National De La Recherche Agronomique (INRA) (1978) Principes de la nutrition et de d'alimentation des ruminants. Besoins alimentaires del animaux. Valeur nutritive des aliments. INRA, Versailles

Lange GM (2004) Manual for environmental and economic accounts for forestry: a tool for cross-sectoral policy analysis. FAO Working Paper. FAO, Rome

Martín D, Campos P, Montero G, Cañellas I (2001) Extended cost benefit analysis of holm oak dehesa multiple use and cereal grass rotations. Spec Issue New Forestlands Econ Account: Theor Appl 1:109–124

McCreary DD, Tietje WD, Frost WE (2008) Stump sprouting of blue oaks 19 years after harvest. In: Proceedings of Sixth Symposium on Oak Woodlands: Today's challenges, tomorrow's opportunities. USDA Forest Service Gen Tech Rep PSW– GTR–217: 333–341

Ministerio de Agricultura, Pesca y Alimentación (MAPA) (2003) Encuesta de precios de la tierra 2002 (base 1997). Secretaría General Técnica, MAPA. Available online: http://www.magrama.es/es/estadistica/temas/encuesta-de-precios-de-la-tierra/

Ministerio de Medio Ambiente y Medio Rural y Marino (MARM) (2008) Diagnóstico de las Dehesa Ibéricas Mediterráneas. Tomo I. Available online: http://www.magrama.gob.es/es/biodiversidad/temas/montes-y-politica-forestal/anexo_3_4_coruche_2010_tcm7-23749.pdf

Ministerio de Medio Ambiente y Medio Rural y Marino (MARM) (2011) Encuesta precios de la tierra 2010. Available online: http://www.magrama.gob.es/es/estadistica/temas/encuesta-de-precios-de-la-tierra/

Montero G, Martín D, Campos P, Cañellas I (2000) Selvicultura y producción del encinar (Quercus ilex L.) en la comarca de Monfragüe. CSIC-INIA-UCM Internal working paper. CSIC-INIA-UCM, Madrid, p 47

Moreno G, Obrador JJ, García E, Cubera E, Montero MJ, Pulido F, Dupraz C (2007) Driving competitive and facilitative interactions in oak dehesas through management practices. Agrofor Sys 70:25–40

Ovando P, Campos P, Montero G (2007) Forestaciones con encina y alcornoque en el área de la dehesa en el marco del Reglamento (CE) 2080/92 (1993–2000). Rev Españ Estud Agrosoc Pesquer 214:173–186

Ovando P, Campos P, Oviedo JL, Montero G (2010) Private net benefits from afforesting marginal crop and shrublands with cork oaks in Spain. For Sci 56:567–577

Plieninger T (2006) Las dehesas de la penillanura cacereña. Universidad de Extremadura, Caceres, Origen y evolución de un paisaje cultural

Plieninger T (2007) Compatibility of livestock grazing with stand regeneration in Mediterranean holm oak parklands. J Nature Conserv 15:1–9

Pulido F, Díaz M (2003) Dinámica de la regeneración natural del arbolado de encina y alcornoque. In: Pulido F, Campos P, Montero G (eds) La gestión forestal de la dehesa. Junta de Extremadura/IPROCOR, Mérida, pp 39–62

Pulido F, Zapata S (2006) Prólogo. In: Plieninger T (ed) Las dehesas de la penillanura cacereña. Origen y evolución de un paisaje cultural. Cáceres, Universidad de Extremadura, pp 13–14

Pulido FJ, Díaz M, Hidalgo SJ (2001) Size structure and regeneration of Spanish holm oak Quercus ilex forests and dehesas: effects of agroforestry use on their long-term sustainability. For Ecol Man 146:1–13
Ramírez JA, Díaz M (2008) The role of temporal shrub encroachment for the maintenance of Spanish holm oak *Quercus ilex* dehesas. Forest Ecol Manage 255:1976–1983
Rodríguez Y, Campos P, Ovando P (2004) The commercial economics of a public dehesa in the Monfragüe shire. In: Schnabel S, Ferreira A (eds.) Sustainability of agro-silvo-pastoral systems. Dehesas, Montados. Catena Verlag, Reiskirchen. Adv Geoecol 37:85–96
Rupérez A (1957) La Encina y sus tratamientos. Madrid. Quoted by Montero et al (2000). Gráficas Manero. Madrid
Standiford RB (1997) Growth of blue oak on California's hardwood rangelands. In: Proceedings of a symposium on Oak Woodlands: ecology, management, and urban interface issues, March 19–22, 1996, USDA Forest Service Research Paper, CA, PSW–GTR–160:169–176
Standiford RB, Howitt RE (1988) Oak stand growth on California's hardwood rangelands. Calif Ag 42:23–24
Standiford RB, Howitt RE (1992) Solving empirical bioeconomic models: a rangeland management application. Am J Agric Econ 74:421–433
Standiford RB, Howitt RE (1993) Multiple use management of California's hardwood rangelands. J Range Man 46:176–181
Standiford RB, McCreary D, Gaertner S, Forero LC (1996) Impact of firewood harvesting on hardwood rangelands varies with region. Calif Ag 50:7–12
Standiford RB, McDougald NK, Frost WE, Phillips R (1997) Factors influencing the probability of oak regeneration on southern Sierra Nevada woodlands in California. Madroño 44:170–183
Standiford RB, McCreary DD, Barry S, Forero LC (2010a) Blue Oak stump sprouting on California's Northern Sacramento Valley Hardwood Rangeland. Calif Ag 65:148–154
Standiford RB, McDougald NK, McCreary DD (2010b) Thinning: a tool for restoration of California's oak woodlands. Soc Am For Natl Conven, Albuquerque, New Mexico, Oct 30, 2010
Torell LA, Rimbe NN, Tanaka JA, Bailey SA (2001) The lack of a profit motive for ranching: implications for policy analysis. In: Torell LA, Barlett T, Larrañaga R (eds) Current issues in rangeland resource economics. Ann Meet Soc Range Mgmt 47–58
Vázquez FM, Doncel E, Martín M, Ramos S (1999) Estimación de la producción de bellotas de los encinares de la provincia de Badajoz en 1999. Número 3:67–75
Waddell KL, Barrett TM (2005) Oak woodlands and other hardwood forests of California, 1990s. USDA Forest Service Resource Bulletin PNW–RB–245, p 94

Chapter 10
Raising Livestock in Oak Woodlands

Juan de Dios Vargas, Lynn Huntsinger and Paul F. Starrs

Frontispiece Chapter 10. Among dehesa livestock, the Iberian pig offers a signature product. Denomination of origin classifications certify the breed of a pig and its diet, with the most valuable *jamón* coming from Iberian pigs fattened on acorns. (Photograph by P. F. Starrs)

J. de. D. Vargas (✉)
Facultad de Veterinaria, Universidad de Estremadura, Avda Universidad S/N 10003 Cáceres, Spain
e-mail: jdvargas@unex.es

L. Huntsinger
Department of Environmental Science, Policy, and Management, University of California, Berkeley, 130 Mulford Hall, Berkeley, MC 3110 CA 94720, USA
e-mail: huntsinger@berkeley.edu

P. Campos et al. (eds.), *Mediterranean Oak Woodland Working Landscapes*, Landscape Series 16, DOI: 10.1007/978-94-007-6707-2_10,
© Springer Science+Business Media Dordrecht 2013

Abstract In Spain and California, oak woodlands are often used for livestock grazing and husbandry. Livestock raised on the range include sheep, goats, cattle, and horses. In addition, two distinctive Spanish dehesa products are only rarely seen in California, the free-range domestic pig and the fighting bull. The grazing and land-use systems are markedly distinct in each country, even though strong historical and economic connections exist, and climate and ecology are similar. Discussion of the use by livestock of mast and other forage emphasizes the agrosilvopastoral roots of livestock husbandry in Spain. A facet especially worthy of note is the distinctive rearing of the Iberian pig, and the pig's role in modern-day Spanish food culture and iconography. In California, land use change is a major threat to traditional oak woodland ranching, while in Spain the declining economic value of some dehesa products has caused losses in the diversity of management. In Spain, markets for unique local livestock products are well developed, with denominations of origin and certification of livestock breeds and production systems. In California, attempts to develop niche markets for unique rangeland livestock products are in early development, but fit a growing interest in local, sustainable, forms of agricultural production, and demand for grass-fed, organic, or "natural" meats (with no hormones or sub-therapeutic antibiotics). The conclusions highlight commonalities and differences between Spanish and Californian livestock production. It remains to be seen whether changing markets can make grazing in the woodlands an enterprise profitable enough to sustain itself in the decades ahead.

Keywords Jamón · Livestock production · Swineherding · Animal husbandry · Dehesa · Ranch

10.1 Livestock in the Oak Woodlands

In California, Spanish colonists found grazing land that resembled what they left behind in southern and western Spain. California owes the origin of its livestock production to Spanish-Mexican practice, brought to California in the late eighteenth century. But the complexities of grazing among oaks on the Iberian peninsula have not entirely transferred: animals grazed on the dehesa include cattle, sheep, goats, horses, fighting bulls, and the singular Iberian pig, and acorns, forage, and browse are valued components of livestock diet. Domesticated animals raised in California are less diverse, mainly cattle, although sheep and goats

---

P. F. Starrs
Department of Geography, University of Nevada, Reno, MS 0154,
Reno, NV 89557-0048, USA
e-mail: starrs@unr.edu

remain locally significant. The dominant understory grasses of today's California oak woodland arrived along with the Spanish livestock (Chap. 2).

There are superficial similarities and vast differences in the keeping of livestock in Spain and California. The climate and forage resources are similar, which influences livestock production. Yet acorns and mast, for example, play a markedly different role in Spain, where they support the Iberian pig production system, than in California where acorns can be a hazard to livestock and mostly are left to wildlife. This chapter explains and explores livestock production in the Spanish dehesa and the California oak woodland ranches.

10.2 Mediterranean Oak Woodlands in Spain and California as Livestock Range

In the woodlands of Spain and California understory grasses and broad-leaved plants, oak leaves, twigs, and acorns provide a rich source of nutrients for foraging animals. Understory grasses are mostly annuals from the Mediterranean region, cool season species that sprout in the fall in response to autumn rains, grow only slowly during the coolest temperatures in winter, and grow rapidly and for as long as soil moisture lasts in the spring (Jackson 1985). In Spain, an estimated 20 % of forage production occurs in fall, 10 % in winter, and 70 % in spring (Olea et al. 1991). With the onset of the dry summer season, annual grasses and forbs turn brown and die (Fig. 10.1). Highly variable yearly rainfall patterns, affected by

Fig. 10.1 During the yearly summer drought the annual grasses that predominate in the understory of Mediterranean oak woodlands are brown. The dry period tends to run longer into the fall in California, as shown here. (Photograph by P. F. Starrs)

amount, timing, and frequency, alter the production of understory grasses and forbs by orders of magnitude.

Oak woodland livestock producers in dehesas and ranches face unpredictable and varying forage production. The latest work on annual forage production in the dehesa quantifies a wide variation in productivity, ranging from 200 to 5,372 kg/ha in field-based studies (Gonzalez et al. 2012). In three years of study, forage production averaged 2,031 kg/ha, but with notable year-to-year variation: 996 kg/ha (2008), 2,595 kg/ha (2009), and 2,500 kg/ha (2010) (Gonzalez et al. 2012). Dry weight production of forage in California varies from 500 to 3,000 kg/ha (Huntsinger and Starrs 2006). Production depends also on geographical location and topography. Spain and California are drier and warmer in the south; northern or high elevation areas are colder and wetter. In California, the cold Pacific Ocean results in a moderated, but cooler and moister, climate near the coast. In some areas shrubs are common and are browsed by livestock, although many are not palatable.

In addition to high and unpredictable variation in amount of forage, there is seasonal variation in nutrient values that is subject to weather. New fall grass may be rich in nutrients on a dry weight basis, but it is high in water content and low in biomass. As grasses dry in late spring and summer, nutrient quality drops off, and drops further still when fall rains leach nutrients and cause decay (Fig. 10.2). The amount and nutrient content of forage is not optimum for supporting livestock year-round (Fig. 10.3), with protein and energy periodically in short supply. Additionally, annual grasses may at some times be deficient in selenium, copper, potassium, and zinc, at times requiring supplementation. At some locations high amounts of molybdenum aggravate copper deficiency (George et al. 2001).

Livestock production cycles follow these annual patterns, and must accommodate unpredictability and variation within and between years. In California,

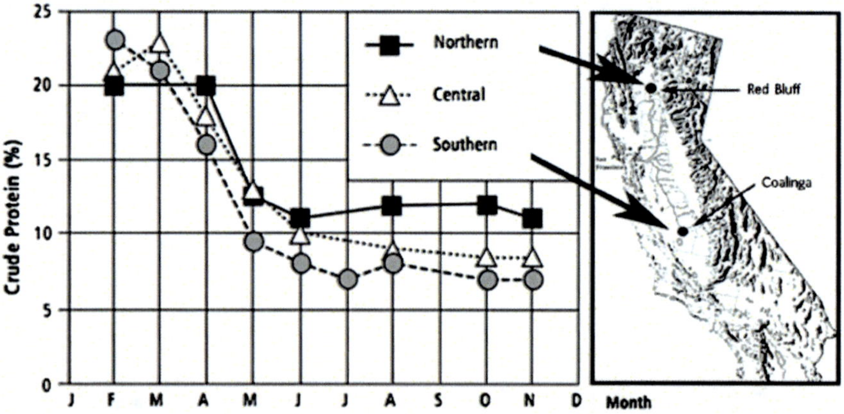

Fig. 10.2 Seasonal crude protein content of composite samples taken from 17 ranches along a north to south gradient from Red Bluff to Coalinga, California (Hart et al. 1932, adapted in George et al. 2001)

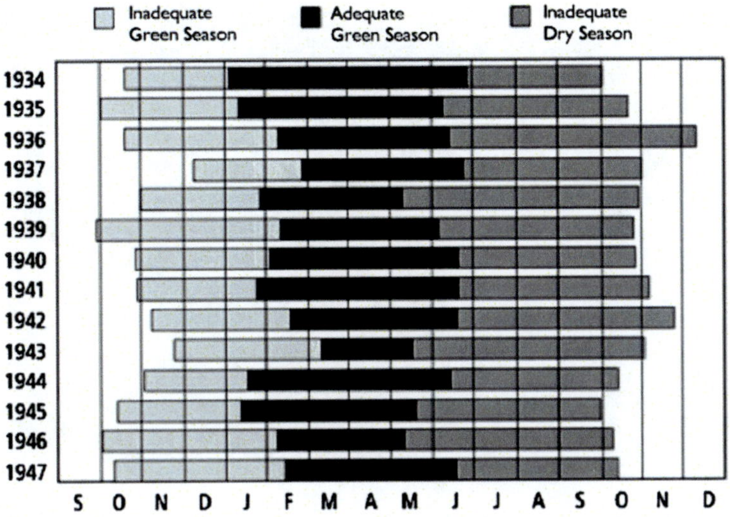

Fig. 10.3 Yearly variation in forage production in the foothill oak woodlands of the southern Sierra Nevada in California (Bentley and Talbot 1951, as adapted in George et al. 2001). "Inadequate green" indicates that forage is insufficient in quantity and/or of low quality. "Inadequate dry" is when annual grasses have died, and the "adequate green" season is when good quality forage is plentiful. During periods of inadequate forage, supplementation of protein and/or energy, or reduced stocking rates may be required to maintain herd condition. Similar seasonal patterns shape livestock production in Spain. In California, synchronous calving may be timed so that the herd's maximum forage demand is during the adequate green season

calving is timed to take advantage of spring growth. In Spain, pigs range the woodlands in fall when acorns have dropped. Supplemental feeds, selling of calves, transhumance, irrigated pasture, leasing of distant pastures, and movement of stock are used to cope with unpredictability and the summer dry period (Ferrer and Broca 2001; Huntsinger et al. 2010b). While a dehesa or oak woodland ranchlands of uniform topography may make management easier, diverse terrain can be used to advantage by the land manager, since it will peak in forage production at different times in different areas (González et al. 1998). Transhumance uses topographical differences on a still larger scale (Huntsinger and Starrs 2006).

The oak canopy in Mediterranean oak woodlands is another influence on forage production (Chap. 6), creating a moister microclimate that can increase production, influence species composition, and extend the period when forage is green. For very dense oak canopies, understory forage production can be suppressed, which may necessitate the pruning or thinning of trees to maintain adequate production levels. Oaks provide feed for livestock in the form of mast (Fig. 10.4). Acorns, twigs, and leaves are consumed from the trees or from the ground. In California, acorns are used incidentally by livestock, but mainly are valued as food for game and other wildlife. Dehesa acorns are the focus of a unique system that culminates in "acorn ham," with woodlands deliberately manipulated not only to

Fig. 10.4 Acorns are prime feed for free-ranging pigs in the dehesa, and the supply can be plentiful, as here near Córdoba. In California there is some tradition of domestic swine being turned loose in woodlands, but most acorns are consumed by wildlife including feral pigs. (Photograph by P. F. Starrs)

maximize production of grass, but also acorn consumption by pigs. In fact, the diversity of Spanish production enables the livestock industry to better capture the nutrient flow from oak woodlands. Acorns are rich in carbohydrates, protein, and oils, in the minerals calcium, phosphorus, and potassium, and the vitamin niacin (Rodríguez-Estévez et al. 2008). Oak leaves, particularly from live oak species, retain relatively high levels of protein through the summer drought when grasses are dead—though the tannins can at times pose a problem in California (Sullins and Maas 2011). The native pigs and goats of the Spanish dehesa are especially adept at making use of these resources.

In Spain, a dehesa of holm oaks (*Quercus rotundifolia* or *Q. ilex ballota* L.) at a density of 40–50 trees/ha (Hernández 1998) is considered adequate to allow for grass and acorn production (Fuentes 1991), although this depends on multiple factors including the desire of managers to emphasize pig or livestock production, and willingness to use tractors or arduous hand labor to clear shrubs on a regular basis and grow crops among trees. Cork oaks (*Q. suber*), with bark stripped every 9–12 years, are managed for a considerably higher tree density. Holm oak woodlands acorn drop is concentrated between October and December, but in amounts that vary widely (Montoya 1993; Cañellas et al. 2007). Cork oaks can produce acorns into early the next calendar year. Oak density and consequently acorn production has declined through the late twentieth century on many sites (Beaufoy 1995; Montero et al. 1998; Plieninger 2006), variously attributed to poor regeneration, high stocking rates, and historical tree cutting (Fig. 10.5).

Costly replanting efforts, requiring a long cessation of grazing and fencing of woodlands to prevent wildlife from eating young oaks, have recently been

Fig. 10.5 Although dehesa oak distribution is more uniform than in California, oaks are found at varying densities depending on management, soil and climatic conditions, and the extent of historical tree removal for firewood, increased fodder, or understory cultivation. (Photograph by P. F. Starrs)

undertaken with EU support. Acorn production in California likewise varies year to year, and comes in the fall (Chaps. 5 and 7). In the 1940s California agricultural advisory services urged a complete removal of oaks to increase forage production, but now only cautious thinning of oaks for firewood sales and forage increase is recommended. The trees are promoted for providing wildlife habitat, shade, watershed protection, and scenery that augments land prices, among other things. In very general terms, at about 50 % canopy cover or more, understory forage production is reduced. Trees are today distributed in response to soils, water, and happenstance, and as relicts of past management. Areas considered grazeable in the blue oak woodlands generally have 175–325 trees/ha (Huntsinger and Starrs 2006), but distribution is less regular in California than in Spain, with a cover of under 10 % in some areas (Chap. 1), while other sites feature a completely closed canopy.

Dehesa acorn production is affected by tree density, by trees bearing heavily only one year out of two or three, by dehesa management, tree health, grazing practices, climatic conditions, and the idiosyncrasies of each tree. Wide discrepancies are reported for Spanish acorn production with a startling range of 0.5–147 kg/tree (Chap. 7). Averaging acorn yields over several years, 9–12 kg/tree is considered a typical mean value, although with a wide variation (Vázquez et al. 2002). The presence of *castizos*—trees that produce large acorn falls—can substantially increase acorn availability. Historical accounts describe a superlative Portuguese holm oak averaging 409 kg/acorns/year, and a single massive California valley oak (*Q. lobata*) yielding acorns at 900 kg/year (Smith 1916). On the dehesa an acorn production of more than 400 kg/ha is needed to maintain typical livestock densities—and pigs in particular—during autumn and winter

Fig. 10.6 A Retinto consumes some green holm oak foliage, supplementing the dry grass. Much of the dehesa, and California oak woodland, has an identifiable browse line where cattle, goats, and deer have browsed. (Photograph by P. F. Starrs)

(Vázquez et al. 1996). Pruning in winter occurs every 10 or 12 years in dehesa, removing a portion of the canopy to improve subsequent acorn production, although recent experiments have found this ineffective (Cañellas et al. 2006; Alejano et al. 2008).

In California, data on acorn production varies widely (Chap. 7). Acorns, buds, and oak leaves are a hazard for livestock if they become too high a proportion of the diet. Cattle diets containing more than 50 % oak for even a day or two can make animals ill, and under rare conditions when there is no fresh grass and only new oak growth for forage there are documented cattle die-offs. Cows in poor flesh that ingest large numbers of acorns in the second trimester may have deformed calves (Miller 2011). In Spain, these conditions are prevented by the rapid consumption of acorns and leaves by pigs, wildlife, and goats. There are also possible differences in the toxicity of Spanish and Californian acorns.

10.3 Spanish Livestock and the Dehesa: An Overview

The dehesa of western and southwestern Spain is a mix of agronomic, social, and economic traits that forms a time-tested agrosilvopastoral system involving about half the Spanish free-ranging livestock. A dehesa will often be used by a succession of livestock types (Gaspar et al. 2009). Cattle may share ground with pigs until December or January, and after the pigs are removed, sheep and goats can be brought onto the land. The aftermath from staple grains once so reliably planted under the oak canopy is less available now that understory grain growing is

Fig. 10.7 Rock walls reflect the passage of time, and ongoing maintenance, of a dehesa in the Sierra Norte de Sevilla. The long rock walls are a product of hundreds of years of rock removal from a hillside once used for grain production in the oak understory. The square in the center is a *colmenar*—an exclosure to keep out domesticated and wild animals—and inside are beehives, once made from cork oak bark. Oaks are carefully retained, and at the hilltop are dense brush stands that envelop the occasional oak. (Photograph by P. F. Starrs)

minimally profitable. Instead, livestock are grazed in large pastures that are fenced or surrounded by rock walls that may be hundreds of years old, as part of a system that is conservatively 2,000 years old (Fig. 10.6 and 10.7). But through the last 1,500 years, some dehesas were only a seasonal resource, as part of the cycle of transhumance that moved animals—traditionally sheep, but more often now cattle—from northern to southwestern Spain and then back (Ruiz and Ruiz 1986) .

Livestock that graze the dehesa increasingly are from breeds imported from elsewhere in Europe. These are often crossbred animals bred to adapt to local conditions, with genetic traits for enhanced weight gain and heat-tolerance. But it is common to see rustic autochthonous breeds that have a long history, and whose use reflects landowner preferences and selection by past breeders for desired attributes. A major advantage of native stock is the ability to make do with reduced forage during dry months, maintaining meat production. It is similarly important for range-based sheep, goats, and cattle used for dairying to maintain milk yields in summer. Sheep for milking are normally stabled and fed; sheep raised for meat remain on the dehesa for much of the year. When indigenous breeds are used, animals are selected for color, self-sufficiency, disposition, willingness to defend offspring, temperament, tradition, or simply because they are breeds of long standing, which appeals to owners. Idiosyncrasy and past history are important attributes influencing landowner preference. European Union agreements in the

Fig. 10.8 Landowners maintain the dehesa in a mosaic of tree and understory densities, with pruning of trees, removal of shrubs, and uplands minimally disturbed to provide big game habitat. (Photograph by P. Ovando, from historical photos at Dehesa Los Rasos, Córdoba)

last two decades particularly favor low-intensity agriculture and native breeds—as they also favor afforestation with native trees (Fig. 10.8).

A mix of livestock can achieve an optimal use of dehesa resources, taking advantage of dissimilar grazing behavior: goats browse; sheep prefer broadleaf plants; cattle concentrate on grass; pigs consume mast and grass, and grub in the soil for roots and tubers. Adding wildlife that are taken in organized hunts adds a further advantage to exploitation of dehesa natural resources, and so it has been for dozens of generations. Yet use of feed supplements to speed weight gain is increasingly common, especially for Iberian pigs and wild game. Specialization is on the upswing as managers eliminate mixed grazing. Settling on two or even a single species means there is a suboptimal use of dehesa and that elevates producer risk, although there can be advantages from simplified management and reduced labor and maintenance costs. The economic crisis of 2010 has driven some of this behavior as landowners seek to maximize short-term profit (Chap. 13) .

Although the use of some 3.6 million ha of oak woodlands in the dehesa of Spain is remarkably well covered in the literature (Parsons 1962; Díaz et al. 1997; Joffre et al. 1999; Eichhorn et al. 2006; Linares 2007; Gade 2010), no official count tallies livestock production unique to the dehesa landscape. Constructing a total livestock census as a result is a matter of estimates rather than exactitude. There are additional animal breeds whose main role is service-provision, including the fighting bull (*toro bravo, toro de lidia*) and the Spanish or Andalusian horse. And the Andalusian donkey, the white Cacereña cow, the Segureña ewe, the Retinta Extremeña goat, and the Andalusian blue hen are but a few cases of livestock being sustained as genetic reservoirs or to prevent the extinction of a unique strain of livestock (García Dory et al. 1990). Motivations for preserving races of livestock are no less varied than the breeds of the Spanish dehesa.

10.3.1 Sheep

Among all Spanish livestock, the Merino sheep is a signature breed with a uniquely long history. For nearly 600 years fine Merino wool was a key monopoly for the Kingdom of Castile, with sheep grazed in transhumant migrations from Extremadura to northern Spain. The guild of sheepherders, backed by the crown, had considerable power and was able to maintain grazing and trailing rights across Spain (Klein 1920; Zapata 1986). Other breeds were reared for meat, wool, and milk for cheese (Fig. 10.9). Meat production is now more important than wool, since declining fiber prices make wool breeds uneconomical. Quickly fattening meat breeds include the Merino Precoz, Île-de-France, and Berichon du Cher, even though these yield wool of lower quality. With globalization and demand from the European Union market, artisanal cheese from sheep milk (Queso de la Serena, Torta del Casar) is prized by consumers and brings premium retail prices. Some dehesa managers now focus on milk, with lamb meat, wool, and mutton as secondary products. Sheep operations are adopting ultra-specialized crossbreeds, such as Assaf and Awassi ewes that originate in Israel and the Middle East, and which, like the Lacaune, are dual-purpose meat-and-milk breeds.

Fig. 10.9 Segureña sheep are no longer a common breed in the dehesa, but here they travel in an area near the Córdoba-Jaen border. Famed for their ability to move swiftly, they were long a preferred variety of sheep for herders practicing transhumance. (Photograph by A. Caparrós)

A 2009 count estimated eight million sheep in the dehesa region, more than 40 % of the overall Spanish flock (MARM 2009a). The local breed of choice remains the Merino, followed by the Manchega, Castellana, and Talaverana, with the last an endangered breed (MARM 2008). A decrease in the national herd is notable: in 2000 there were 25 million sheep; a dozen years later, under 20 million. Milking sheep tend to be in Castilla León and Castilla-La Mancha, while the main meat herds are in Andalucía and Extremadura. Sheep for milking are normally stabled and fed; sheep raised for meat remain on the dehesa for much of the year. In either case, lambs (dropped once a year, or three times every two years) are pen-fattened to market weight. A special pasture with higher-quality feed, the *majada,* is reserved for ewes about to give birth, and once they do, they and their lambs are kept there so the ewes can gain weight and breed back. Sheep raised as denomination of origin stock—known among all classes of livestock as PGI (Protected Geographical Indication)—are subject to special rules. Extremaduran Lambs (the Cordero de Extremadura) are slaughtered before they reach 100 days of age, and sheep crossbred for meat and wool can be admitted to the PGI.

A broader look at Spanish sheep shows a pattern of recent neglect. The latest epizootics, decreasing lamb consumption, and low prices paid by commercial carcass buyers make public intervention desirable. A resolution of the European Parliament in 2008 noted that sheep herding plays a key role by making use of less fertile areas and preserving sensitive ecosystems and landscapes (DOUE 2009). How public investment in the dehesa will play out in the current crisis remains to be determined. There are multiple and overlapping EU funds that have supported native breeds, low-intensity agriculture, and cropping patterns deemed beneficial to wildlife. Whether these will continue given difficulties in supporting common agricultural policy subsidies from the EU remains to be seen.

10.3.2 Goats

Forty-five percent of nearly three million goats raised in Spain are in the dehesa, a count essentially unchanged over the last two decades (MARM 2009a). Dehesa goats fall into three classes: small herds used for meat production on large pastures; mixed use herds, where goats are milked but also graze on pasture; and intensive operations with goats kept penned and the kids weaned early so milk can be diverted to dairying. Milk goats outnumber meat goats in the dehesa by three to one, and constitute about half Spanish milk goats (Fig. 10.10). Shepherds of milk goats earn more pay, with better living conditions, than herders of meat goats (Sierra 2003). The goat census shows a marked shift toward intensive milk production.

Meat goat breeds formerly pastured on the most difficult areas of the dehesa include the Blanca Andaluza (Andalusian white goat), the Retinta, Verata, and Jurdana, all of which are currently considered at risk of disappearance. Low goat meat prices and a sizeable increase in pasture rental prices discourage the keeping

Fig. 10.10 A herd of dairy goats rest and graze near Grazalema, an area famed for Payoyo cheese that blends tart goat milk with sweet sheep milk. Goats are browsers, favoring shrubs and tree leaves as forage and will climb trees when possible. (Photograph by P. F. Starrs)

of meat goat herds (Sierra 2003). A sizable proportion of dehesa properties once used for goats are now used for big game hunting. This change, driven by economics, portends a likely disappearance of goat grazing from mountainous areas, and reinforces the growing risk of fire hazard (de Rancourt et al. 2006). A European Union report notes that goats, "thanks to their preference for eating in areas where browse or shrubs play an important role, contribute to the preservation of biodiversity in flora, protect the fauna, and clean dry plant material from natural spaces, essential for fire prevention in Mediterranean countries" (DOUE 2009).

10.3.3 Cattle

Most cattle reared in the dehesa are commercial crosses raised for meat, with the goal production of a calf to be weaned at six months at a live weight of 250 kg. More than half the mother cows in Spain are crosses; only in Andalucía does a single local breed stand out, the Retinto (Fig. 10.11). Proliferation of Retinto cattle (Retinta/Retinto are used interchangeably) is associated with the quality assurance "SAT Carne de Retinto" (Sociedad Agraria de Transformación; SAT 2010). In Extremadura and both Castiles (Castillay León and Castilla–La Mancha) purebred stock grazers prefer the Limousin. Herds of native breeds are popular among consumers who support local races of cattle, and these include the Avileña–Negra

Fig. 10.11 Retinto cattle maintain a wary presence in the municipal Montes Propios de Jerez, where they graze by permit. The Retinto is among the most successful native cattle breeds in Spain, and is favored by landowners. The Salamancan Morucha and the Avileña Negra are two additional breeds in the same sought-after category. (Photograph by P. F. Starrs)

Ibérica (PGI–Carne de Ávila), the Morucha (PGI–Carne de Morucha de Salamanca), and the Extremaduran cattle breed (PGI–Ternera de Extremadura). While regionally significant breeds include the Avileña–Negra Ibérica (Bociblanca variety), Berrenda en Colorado, Berrenda en Negro, Marismeña, Morucha (black variety) and Negra Andaluza, they are not widespread, and fewer than 1,000 exemplars persist of once prominent breeds such as Blanca Cacereña and Cárdena Andaluza (Fig. 10.12).

More than one million mother cows graze the dehesa, 60 % of the total Spanish cattle census. That count is on the upswing, nearly doubling in the 1990s and accelerating further since 2006 (MARM 2009a), reflecting hopes for profits based on encouragement provided by government subsidies, and a marked shortage of workers skilled in the care of sheep or goats. Range cattle require less vigilance than other livestock, but cattle operations are notably inefficient, and limited supervision reduces calf survival and profitability. Average herd size is 23 mother cows, mid-sized by Spanish standards. Dehesa cattle operations face not only high labor costs but low prices for weaned calves, making for tight profit margins or significant losses. A particular concern is herd health. On some dehesas bovine tuberculosis persists despite strict veterinary controls, which leads one research team to suspect that big game animals may infect livestock. Reducing the contact between wildlife and cattle might improve the success of bovine TB eradication programs (Castillo et al. 2010).

Fig. 10.12 The Blanca Cacereña, with a distinctive horn-spread, bright white coat, and white muzzle, is sustained by a few devoted breeders who work to overcome its low fertility, as with this herd on a dehesa near Monfragüe Natural Park in Extremadura. (Photograph by P. F. Starrs)

10.3.4 Fighting Bulls

Unique to the dehesa is the fighting bull ranch (Fig. 10.13). The workforce needed on these extensive livestock operations is large and highly specialized, with sizable capital investments required in holding facilities and fencing, management (horses

Fig. 10.13 A group of young fighting bulls maintains watch in southern Salamanca province at the edge of a deciduous dehesa bounded by a stand of *quejigo* (*Q. faginea*). (Photograph by P. F. Starrs)

are used to carry out many tasks), and breeding management, since only bulls of high quality are destined for bullrings. Payoff is in part through prestige, and the success or failure of a breeder is documented widely and much discussed in national newspapers and other fan media. The number of registered fighting bull brands and properties is on the rise, from 873 in 1990 to more than 1,200 in 2008 (MARM 2009a). There are about 135,000 breeding cows in 540,000 ha of dehesa (Boza 2007), and four overlapping associations govern fighting bull registries. For all that, a perceived slippage in the quality of fighting bulls sent to the bullring is lamented and denounced by bullfighting journalists and fans, although such complaints have a history of at least 80 years (Hemingway 1932).

From a socioeconomic and workforce point of view fighting bulls are of obvious importance, but ranches for *toros bravos* are no less crucial ecologically, since they restrict public access and protect flora and fauna. Raising fighting bulls has a strong emotional component, with bull raisers affording themselves what economist Pablo Campos describes as "self-consumed environmental income" (Campos 2005). About 30 % of the Spanish population affirms an interest in bullfighting—a proportion on the downswing. But enthusiasm is widespread for fighting bulls roaming the dehesa countryside. In fact, Spain's foremost bull-fighting journalist, Antonio Lorca, notes in a sidebar of *El Pais* in August 2012 (in translation) that "The fighting bull is a defender of the environment; a fundamental value in the maintenance of the dehesa, an ecosystem that is unique and exclusive to the Iberian Peninsula," and continues, quoting a Huelva–based breeder of fighting bulls who insists that "The dehesa is the most intelligent form of resource exploitation that humans have developed in nature: it is a sink for CO_2, it sustains the population of rural areas, and it exists, after all, because of a rusticity of fighting bulls that takes advantage of the dehesa's environments year-round" (Lorca 2012).

10.3.5 Horses, Donkeys, Mules

Equine production is important in the dehesa, although it is in official statistics a great unknown. The latest statistical yearbook identifies 221,000 registered horses in Spain, 30 % of them in the dehesa (MARM 2009a). "Official" in this case does not mean "accurate," and alternate figures suggest registered horses in Spain are at least double that number, with two-thirds in dehesa environments (MARM 2009b). A 2011 count claims a total of 680,000 head (MARM 2011).

Almost all equine holdings in dehesa areas are breeding operations for private or non-profit use, or for equestrian sports. Unsurprisingly, some 60 % of the 732 stud farms that claim to raise the Purebred Spanish Horse (Pura Raza Español, PRE) are in dehesas, with 186 additional studs of Anglo-Arabian horses, and 23 studs of the Caballo de Deporte Español (CDE). Horses used for managing and testing fighting bulls are often crossbreeds, described as "Tres Sangres" [three-way crosses] and draw on lines of Hispanic-Anglo-Arab origin, with "lusitanos" also a common parent-stock, valued for agility. The Andalusian donkey, once essential to farm tillage in Córdoba and adjoining provinces, is in marked decline and rarely seen outside of government-run Reproduction Centers where breeding populations are kept (García Dory et al. 1990).

If unusual, the Purebred Spanish Horse is a case worthy of note. Since 2000, horse breeding in the dehesa grew chic, and like raising fighting bulls, became a symbol of conspicuous wealth. This is reflected in prices and in breeders applying selection criteria that prized color above functionality. A breeding bubble now exists, with the registry unchecked and young studs and broodmares sold purely as brood animals. With investment and maintenance costs rising, and an economic crisis of the early 2010s in full force, many horses are simply sent to slaughter due to low market demand and high upkeep costs. Horse and donkey meat, once popular in northern Spain, is still sold, though less so thanks to EU regulations.

10.4 Iberian Pig Production and *Montanera* in the Dehesa

The Iberian pig is a native Spanish breed that avoided near extinction in the 1970s to become the foremost economic contributor to livestock production in the dehesa, with an expanded range in the last two decades to regions of Spain where the Iberian pig was never before reared (Fig. 10.14). Consumer interest in Iberian pork products has gone from inattention and resistance to a cult of global admiration, thanks in part to scientific research that finds dietary advantages in cured pork products from the dehesa where *jamón* and other pork products are a praised and a locally-sourced food (Campillo 2001).

10.4.1 Census and Breed Diversity

Classification of the nearly 3 million Iberian pigs registered in 2009 is complex indeed (MARM 2010), with 8 categories (Table 10.1) (BOE 2007). Only where the specification is "Pure Iberian" is there assurance that pigs issue from a pure ancestry; the rest may, in varying percentages, be crossed with Duroc or Jersey lines. In commercial branding, breeding is one constraint and the other is finishing. Just one-third of Iberian pigs mature on grass and acorns, almost all in Andalucía and Extremadura.

A 2007 census counted nearly 300,000 breeding sows that as brood stock do not feature in the "fattened pig" census. The number of Iberian sows and the total number of Iberian and crossed pigs fattened in intensive systems outside dehesa sites is unknown but considerable, with figures untabulated for Murcia and Cataluña. Iberian pigs being finished in Portuguese fattening pens and montados are also uncounted in statistics.

Fig. 10.14 Pigs like this are a paragon of Spanish culinary life, "jamón ibérico de bellota," or ham from acorn-finished black-footed Iberian pigs. Also produced are chorizo and other cured pork products, all at a premium if from the right stock. (Photograph by J. de D. Vargas)

The breakdown of breed origins within the Iberian pig category is also unknown for each autonomous region. Two varieties—Retinto and Entrepelado—are recognized as "locally expanding," although the Torbiscal, Lampiño, and Manchado de Jabugo breeds are believed at risk (BOE 2009). Other races such as Dorada Gaditana, Negra de los Pedroches, and Cana Campiñesa (believed extinct) distinguish the Iberian pig as a great genetic reservoir. Researchers have examined qualitative (Delgado et al. 2000), productive (Forero et al. 2000), reproductive (Suárez et al. 2002) and genetic (Martínez et al. 2000) aspects of these varieties, and see considerable differences—to a startling degree in some cases, as with the Manchado de Jabugo—that makes preservation of Iberian pig races a worthy effort.

10.4.2 Production Systems of the Iberian Pig

Production of the Iberian pig is unlike production of most other livestock, since the pig is single-stomached rather than a ruminant; fattening depends on high-quality feed. Intensive Spanish farming of non-native "white pig" breeds is vertically integrated and, in the pigs raised, marked by abject conformity: fattening males and females are pen-confined, unneutered and slaughtered young, weighing

Table 10.1 Fattened Iberian pig census in 2009, by category and autonomous community

Category/Autonomous community	Pure Iberian Bellota[a]	Pure Iberian Recebo[b]	Pure Iberian Cebo Campo[c]	Pure Iberian Cebo[d]	Iberian Bellota[a]	Iberian Recebo[b]	Iberian Cebo campo[c]	Iberian Cebo[d]	Total
Andalucía	170,041	13,947	11,991	58,810	295,026	11,196	19,022	284,190	864,223
Castillay León	4,899	426	480	16,683	44,055	4,128	9,069	677,525	757,265
Extremadura	77,293	6,123	2,744	15,069	252,240	13,314	3,358	664,980	1,035,121
Castile—La Mancha	1,757			3,106	7,482	1,180	205	220,586	234,316
Madrid	241				56				297
Cataluña								43,083	43,083
Murcia								13,914	13,914
Total	254,231	20,496	15,215	93,668	598,859	29,818	31,654	1,904,278	2,948,219

"Pure Iberian bellota" and "Iberian bellota" are the only two groups finished on montanera, and total 853,090 out of 2,948,219, or under 30 % of the total. All others are fed supplements to some degree. (Source MARM 2010)

[a] Bellota (meaning acorn): Free-range pigs fattened in dehesa, eating only acorns and other native feeds, in what is called montanera

[b] Recebo: Free-range pigs graze and eat acorns and other native feed during montanera, but are supplemented to reach optimal slaughter weight

[c] Cebo campo: Concentrated feed is provided in distant feeders. Pigs walk distances, with no access to natural feeds in a feedlot situation

[d] Cebo: Pigs are fed concentrated food in fattening pens

90–100 kg at about 6 months. Iberian pig production systems manifest great diversity in handling facilities, animals kept, purebred versus crossbred stock, and management. Iberian pigs are fattened to an age of 15–22 months, which makes castration of males obligatory, and an elevated weight at slaughter of 150–160 kg makes possible production of high-quality long-cured hams and sausages (Vargas and Aparicio 2000. Final slaughter weights must be relatively homogeneous to facilitate post-slaughter salting and curing.

10.4.3 Fattening During the Montanera

Montanera refers both to the acorn mast and other feeds that pigs eat while at large on the dehesa, and to the managed utilization of those resources. The relationship of these pigs to the holm and cork oak woodland is ancient: the Iberian pig is the sole member of its breed that actually peels acorns, removing shells rich in fiber and tannins that are indigestible to a single-stomached animal. In areas with abundant cork oaks, the montanera lasts longer since those acorns (*bellotas Palomeras*) ripen slowly (Aparicio 1987).

An expert resource appraiser, the *aforador*, will visit several times before the montanera begins to decide how many pigs a dehesa can accommodate (Chap. 7). When the montanera starts, other livestock are confined, with acorns reserved for the fattening of Iberian pigs (Fig. 10.15). The sites exploited first are the steepest, and those farthest away from the pens where pigs are confined at night. Flatlands and areas nearest farm buildings and the houses of workers are reserved for the final fattening phase when rotund pigs reach a weight where they can move only with difficulty. As pigs consume acorns, other livestock are gradually reintroduced. This allows the growing piglets that will be adult pigs for the following year's montanera to feed on the grass and acorn remnants, known as *retales*.

Aspects of traditional practices persist in the dehesa-swineherd culture, though with a minimal transmission of knowledge to new generations of workers, mainly due to a lack of interest. Supplanting *vareadores* and swineherds now are stout fenced enclosures. Municipalities still exist where a dehesa belongs to the town (*dehesas boyales*), and each village family will contribute a fixed number of pigs to the communally-grazed herd, tended by a hired worker. The tradition of a household pig slaughter continues, but an activity once essential for food preservation (*del cerdo, hasta los andares*, "of the pig, everything down to the way it walks") has turned into a folklore-rich family gathering.

The difference in prices for Iberian pigs fed exclusively on the dehesa during the montanera (*de bellota,* or acorn-finished pigs) and those only briefly fattened on those resources (*de recebo*), are covered by strict rules that govern how stockbreeders market their products, which in turn influences the carrying capacity and profitability of a dehesa. Although scientific systems for estimating acorn production exist (Chap. 7; Vázquez 1998; Gea-Izquierdo et al. 2006), aforadores are increasingly important. The profession recently gained stature with an upswing

Fig. 10.15 A traditional swineherd, or *vareador*, knocks acorns from an oak with a staff, or *vara*, while surrounded by cork oaks and a sizable group of Iberian pigs. Historically, a herder tended the swine (sometimes a child), a practice referred to in *Don Quixote* (1615) and documented in the *Très Riches Heures of the Duc of Barry* (1416). (Photograph by J. J. Parsons 1962)

in demand for leased dehesa grazing. Stratospheric prices for Iberian pork, and especially jamón, lured into the business national and international meat companies that count Iberian hams and sausages among their signature products (Navidul, El Pozo). These companies ensure production in part by prearranging their access to rented dehesa plots, securing an acorn supply and pasture access. Although it would be wise to adjust stocking density to the minimum estimated production of acorns in a dehesa, with prices high, overstocking is common, which is why supplement-fed Iberian pigs exist.

To be considered de bellota, quality regulations require that pigs graze free-range for a minimum of 60 days, gaining around 800 g/day and adding a minimum of 46 kg. Since a pig's ability to eat and digest increases with age, older pigs fatten faster than younger pigs, with the resulting meat products of better quality as more fat marbles the muscle fibers. Large dehesas with good infrastructure are returning to the traditional system: pigs, after spending a year and a half feeding on the grasses and other resources of the dehesa (with a slight supplement of concentrates) start the montanera.

During the montanera, pigs feed on grass and acorns, roots, tubers, seeds, insects, annelids, and small mammals. The exact proportion of each of these components in a pig's diet during the montanera is unknown, although recent studies provide interesting information (Rodríguez-Estévez et al. 2009). Pigs ingest around 10 kg/day of acorns and 6 kg/day of grass as they end their fattening (López-Bote et al. 2000) and increase in weight by 1 kg/day (Chap. 13). Although there exist areas with a tree density of more than 70 trees/ha, most areas have 20–45 trees/ha (Cañellas et al. 2007). Pigs move constantly in the dehesa with daily travel of up to 9 km documented (Aparicio et al. 2006). They are selective and prefer healthy acorns and search out the largest acorns possible (García et al. 2003; Rodríguez-Estévez et al. 2009). Pigs are nose-ringed to decrease rooting and consequent soil stirring.

10.4.4 An Increasing Sophistication in Marketing the Iberian Pig

Diet composition, continuous exercise, climatic conditions, genetics, and animal welfare are factors that contribute to a quality product. Hams sourced from Iberian pigs represent 25 % of all cured hams produced in Spain, up from 2005 when they were only 7.5 %. Contributing to reputation are designations of origin under European Union laws (Dehesa de Extremadura, Guijuelo, Jamón de Huelva, Los Pedroches, and Jamón de Teruel). Under the first four of those designations are 4,140 pig husbandries and 212 processing facilities (Rueda and Diéguez 2007), Jamón de Trevélez has the closely-related "protected geographical indication."

Thanks to profitability, jamón overproduction has triggered something of a crisis. Technology, facilities, and handling techniques earlier used only in "white" pig raising operations now dilute the Iberian pig population on the dehesa, creating a widespread fraud, with nonnative swine presented as Iberian pigs. The commercial feeds used to finish most Iberian pigs are composed of up to 10 % monounsaturated fats (Durán and Lizaso 1997), which is above the recommended maximum of 5–6 % (López-Bote et al. 2002). Some commercial feeds attempt to mimic the fatty acid balance of the montanera (Cava 2007). Distinguishing dehesa-finished pigs from supplement-fed pigs was once believed possible, but recent work indicates that gas chromatograph analysis—required by official registries—was unable to distinguish an actual montanera feeding regime from a surrogate (Arce et al. 2009). Recently imposed standards and regulations recognize this in a warning that "control of livestock … has been reinforced to verify the origin of feed supplied to the animals" (BOE 2007).

10.5 California Ranch Statistics

Livestock have been in California for a comparatively brief period, with dramatic shifts in numbers through time. A history of livestock in California shows huge increases following the Gold Rush of 1849 when prospectors and associated industries departed, reducing demand for the animals once brought into feed them (Fig. 10.16). After the mid-twentieth-century sheep were supplanted by cattle as

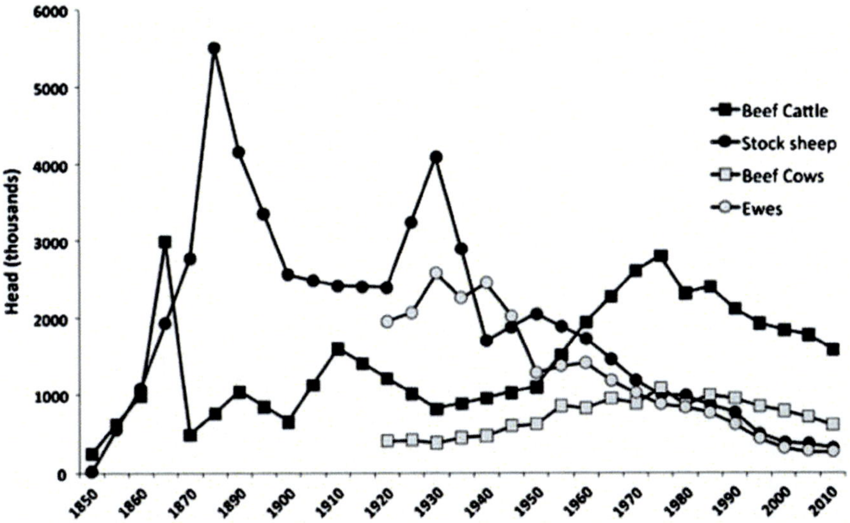

Fig. 10.16 Range cattle and sheep in California, 1850–2010. Beef cattle are all meat breed cattle in California; stock sheep are all meat breed sheep; beef cows are the meat breed brood cows of the kind most common on rangelands; ewes are meat breed brood ewes. Separate data for cows and ewes was not provided until 1920 (Data recompiled from Burcham 1982, USDA–NASS 2011b)

the dominant animal. Beef cattle totals include animals in feedlots, and stock sheep includes animals on feed and brought into the state for feeding or pasturing. Most beef cows and ewes, not counted separately until 1920, obtain a substantial amount of their forage from rangeland. The large number of cattle in feedlots and yearlings on pasture and range is revealed by the gap between beef cattle and beef cows.

Livestock producers who graze animals on the woodland are known as ranchers in California. Ranchers manage more California oak woodlands than any other type of landowner, and provide livestock that graze on millions of hectares of public and private leased land. More than three-quarters of oak woodland properties larger than 80 ha are grazed by livestock (Huntsinger et al. 2010a), better than 60 % are owned by those who produce livestock for sale, and another 10 % of owners produce livestock for their own use only. Another 10 % of oak woodland landowners graze livestock on their property by leasing land to ranchers (Huntsinger et al. 2010a).

More than three-fourths of oak woodland ranchers live on their properties and manage the land themselves, and have owned their properties for an average of 39 years. In 2005, 25 % of oak woodland ranchers in a statewide survey reported that the majority of household income came from ranching, while 10 % reported it was farming. About 22 % cited off-ranch wages as a major income source, and another 22 % earned most of their income from other forms of self-employment, including investments, pensions, and the like (Huntsinger et al. 2010a). Livestock producers may earn income through marketing oak trees as firewood and selling access to their land for hunting, but grazing is the major and usually the most common oak woodland land use in California.

A ranch relies on techniques that come to mind when we think of the American cowhand and extensive rangeland: corrals, stock chutes, cattle roping, horseback herding, trail drives, branding, and mobility. On sheep ranches, highly trained dogs replace the rope, animals are worked on foot, and herders that live with a flock are sometimes employed—though fences, rather than herders, are the norm in oak woodlands. What ranchers say makes ranching worthwhile is experiencing the lifestyle, raising a family on a ranch, working with livestock, and relishing the natural environment.

This attitude is common throughout the western United States on a wider variety of vegetation types than just oak woodlands (Martin and Jeffries 1966; Smith and Martin 1972; Bartlett et al. 1989; Starrs 1997; Rowe et al. 2001; Gentner and Tanaka 2002; Sulak and Huntsinger 2002; Torell et al. 2005). Many consider land appreciation an important, long-term financial asset (Ferranto et al. 2011), and plan retirement and inheritance accordingly. As a result they strongly defend their right to market their land at a good price when demand is there. And in comparison with Spain's strict zoning, there is often little to keep them from doing so, even when it means converting large properties into smaller parcels for development (Huntsinger et al. 2004).

Traditionally, cash-short ranchers sell parts of a property to raise capital. At the extreme, an entire ranch may be sold for development when the owner retires, when there is no heir interested in taking over the ranch or when there are conflicts among heirs, or when the family decides to move to a more rural area and buy another ranch at lower cost (Brunson and Huntsinger 2008). Few individuals can afford to buy a ranch and keep it intact with ranching as their sole income base. The constant attrition and fragmentation of the resource base undercuts the long-term sustainability of oak woodland ranching and feeds competition for grazing land (Fig. 10.17).

Eighteen percent of oak woodlands are in public ownership (CDF–FRAP 2003). A diverse array of public agencies—local, state, and federal—make available leases of public grassland to be used for grazing, and many ranches make use of such leases (Fig. 10.18) (Liffmann et al. 2000; Sulak and Huntsinger 2002). National forests are used when allowed as part of an annual cycle of transhumance, although these days few ranching families actually live in line camps in the mountains during the summer. Hired cowhands and shepherds, or frequent visits, suffice, which is another reason cattle able to defend themselves are chosen over sheep.

There is strong competition for available grazing leases (Sulak and Huntsinger 2002). Competition is augmented by the administrative withdrawal of millions of acres of federal lands from grazing, and the continued decline in numbers of stock allowed (CDF–FRAP 2003). Declining public forage supply puts stress on the industry, and on the private lands associated with public leases (Sulak 2007). A study in the Sierra Nevada foothills that included oak woodlands showed that ranchers with federal grazing permits have been ranching for a longer time, and are more affected by land use change, than ranchers not practicing transhumance

Fig. 10.17 A group of ranchers, government resource managers, and nonprofit conservationists look at a luxury home implanted on former ranch land in Alameda County, California. (Photograph by P. F. Starrs)

(Huntsinger et al. 2010b). Most have owned their land for a long time by U.S. standards: 63 % of transhumant foothill oak woodland ranchers reported that their families have owned their ranch for more than 100 years.

10.6 California Oak Woodland Livestock Production

California livestock production is not diverse, with the vast majority of ranchers producing cattle, usually owning a cow herd that produces calves for market (Table 10.2). In 2004, less than 20 % of woodland landowners grazed goats, sheep, or llamas (*Lama glama*), and most of those grazed cattle also (Huntsinger et al. 2010a). The changing profile of woodland owners, towards small properties used primarily for amenities, suggests that the livestock types grazing in oak woodlands are also changing.

In oak woodlands, ranchers are encouraged to manage to leave behind a certain amount of ungrazed plant material, or residual dry matter (RDM) at the end of the grazing season to protect the soils and encourage the abundant growth of useful forage species by influencing germination conditions. This type of management, called residue management, is designed for the annual grasses, such as *Avena* spp., that have long supplanted California natives. Recommendations for oak woodland call for leaving 110–960 kg/ha depending on canopy cover of oaks and slope

Table 10.2 Livestock owned by oak woodland owners in California, 2008, including alpaca (*Vicugna pacos*); ostrich (*Struthio camelus*), and emu (*Dromaius novaehollandiae*). (Unpublished data from Ferranto et al. 2011)

Percent of properties with the following livestock varieties	Properties of <10 ha (%)	Properties of ≥10 ha (%)
Horses	19	39
Cattle	11	42
Sheep	12	8
Goats	8	7
Llama or Alpaca	2	7
Poultry/birds	11	3
Ostrich or Emu	1	2
Pigs/hogs	0.5	5
Mules	0	0.04
Donkey	0	2

(Bartolome et al. 2002). Some ranchers and public owners manage for native species, wildlife habitat improvement, or intensive production, and practices vary accordingly, including efforts at rotational and targeted grazing.

10.6.1 Cattle

About 610,000 beef cows grazed California rangelands in 2010, down from a million in 1985. About 88 % are from counties in the Mediterranean region. One-half-million to a million yearling beef cattle, known as stockers, will graze rangelands, depending on rainfall, markets, and other factors (USDA NASS 2011b). Most oak woodland ranchers that graze stockers have a cow-calf herd, with less than 10 % of oak woodland cattle ranchers producing stockers only in 2004 (Huntsinger et al. 2010a). Stocker producers or operators, who may or may not have their own property, often lease land for grazing stockers in the woodland. Many stockers are shipped in from states where calves are weaned in the fall, to take advantage of California's winter growth and spring forage peak.

Oak woodland cow-calf producers traditionally time the cow-calf production cycle so that the time of year with the highest forage quality and quantity (mid-spring) is when a herd has the greatest nutrient and energy needs (Fig. 10.3). The typical herd calves in the fall, so the period of greatest need typically occurs in spring when the cows are nursing growing calves and are simultaneously pregnant. In late spring calves are weaned and usually sold. Forage demand for a herd drops sharply immediately before the natural forage available on rangelands senesces and loses nutrient quality. Weaned calves may be sold, fed supplemental hay or feed, or grazed on fallow crop land, dry grassland, or irrigated pasture. Some buyers ship the animals as stockers to different regions or higher elevations where there is green forage available. Cows are kept on dry grass, perhaps with some

Fig. 10.18 Cows rest under an oak at Castle Rock Regional Recreation Area, in the eastern San Francisco Bay Area. Such behavior has implications for soils, understory growth, and biodiversity (Chaps. 4, 6 and 8). (Photograph by S. Garcia)

nutrient supplements, or on irrigated pasture, or are sent to mountain meadows in a cycle of transhumance (Huntsinger et al. 2010b). Mountain meadows in California are frequently in federal ownership and ranchers are granted permits for grazing them in the summer. There are more than 40 cattle breeds produced in the state, but state's beef cattle industry is dominated by five to ten breeds that best fit California's climatic and forage conditions. The typical animal in the oak woodlands is a Hereford-Angus cross, with a small percentage of Charolais or Brahma. Most cattle produced in California are of mixed breeds reflecting generations of rancher efforts to develop a cow herd adapted to local environmental conditions that produces calves with characteristics in demand on the market.

The livestock system that has evolved in the twentieth century on California oak woodland range is to sell calves or stockers to feedlots where they are fed intensively on agricultural supplements, hay, grain, and other feed rations for 100–150 days and slaughtered at 18–24 months of age and 475–520 kg. The largest feedlots are near the U.S. grain production centers in the middle of the country, though there are a few in California. Cattle may be raised at different stages in places that are geographically quite distant. Animals are produced faster and in larger numbers than they can be on rangelands alone, with high energy feeds and trucks replacing sparse rangeland vegetation and herding at various points in the production chain. The system responds to national and global markets, including prices set by commodities markets and influenced by grain imports, the demand for forages from the dairy industry, and policies that require the use of corn to produce fuel, or restrict cattle imports from Canada in response to an outbreak of mad cow disease. Oak woodland producers get the calf and stocker prices that derive from all this, but also must respond to the availability of range and pasture, which is often driven by weather subject to high levels of uncertainty.

The last decade has seen a growing interest in niche-marketed beef. This includes grass-fed, organic, family farm, and local beef. A small but growing number of ranchers are marketing these products directly online and through farmers markets. Some animals never see feedlots and may go from birth to slaughter on a single ranch. Consumers are attracted by the marketing emphasis on healthy food and sustainable management, and the food-based press likes the change. Ranchers do face some formidable barriers: animals must be processed in a licensed and federally-approved slaughter facility, and these are disappearing fast in California. Some ranchers ship cattle to processors in Nevada or several states away, and this reduces if it does not eliminate profitability, especially with rising fuel prices.

10.6.2 Sheep

The sheep ranching industry in California is comparatively small, yet California and Texas have the highest numbers of sheep in the United States (USDA NASS 2011a). More than $67 million in receipts came from the marketing of sheep and lambs in 2010 in California (USDA NASS 2011b). The industry has seen dramatic declines. In 2010, there were 263,000 ewes, or breeding females, in California, down from 770,000 in 1985 (USDA-NASS 2011b). The drop is attributed to low profits stemming from competition from imported lamb, high labor costs, and a decline in consumer preference for lamb meat. In oak woodlands, with wildlife protections predators have increased and have had an impact (Conner et al. 1998; Neale et al. 1998), and with new neighbors, domestic dogs are a growing problem. Increasing immigration of people into the U.S. from non-Western cultures who bring with them a taste for lamb and goat are helping to stimulate the market again. The rising costs of hay and grains in the last few years has driven some expansion in sheep production, as rangeland forage can comprise the majority of feed. There is a interest in accelerated lambing in California, where ewes lamb more than once per year, however the sharp periodicity in forage production characteristic of Mediterranean grasslands is an obstacle making more supplemental feed or irrigated pastures necessary to support year-round lambing. This may be too costly to make it feasible.

The production cycle for sheep is similar to that of cows with some notable exceptions. Sheep may produce twins in a pregnancy and sheep ranchers favor twin-bearing ewes. Traditional range sheep production synchronizes lambing. As with cattle, fall-winter lambing is considered desirable, so growing lambs and lactating ewes can benefit from strong spring growth, but fattening on feed, if needed, is required only for a short period and lambs do not need to be carried through the dry season like calves. For protection, sheep are often kept in smaller improved pastures or, if on rangelands, they are turned out with a herder, or guard dogs, donkeys or llamas. Transhumance from oak woodlands is relatively rare, while in the desert regions of California it is common, using herders brought from other countries—once the Basque country or Italy, but now Peru, Chile, and even Mongolia.

10.6.3 Less-Common California Livestock: Goats, Pigs, Horses

About 140,000 goats resided in California in 2011, with 38,000 of them dairy animals (USDA NASS 2011b). Goats today offer three sources of rancher income: meat, milk/cheese, and vegetation management. The varieties of goats chosen are diverse, and depend on end goals; many of the meat goats are Boer goats, a meat breed of South African origin. Dairy goats are of remarkably diverse breeds, which tend to have fanatical advocates. And when employed for brush control or vegetation management, the goats are often tough hybrids, frequently selected for shorter stature to make them easier to keep fenced in. Goats are used to control weeds and reduce fire hazard, often on hills so steep that other brush control methods would be expensive and physically difficult. Typically, a site is surrounded by electric fence and the goats introduced for a week or so, and generally with a herder on-site or at least nearby to guard against predation (Fig. 10.19). In urban populations that would not countenance herbicide use or cattle grazing, goats find a much readier audience, public support, and goat owners receive a net revenue for bringing in and supervising the animals that provide vegetation management from grazing .

Free-ranging pigs are a rarity in California, except as wild boar when they are hunted as pests, plagues, and predators. Still, in an early twentieth-century article the geographer J. Russell Smith noted "The U.S. Forest Service annually admits two hundred thousand or more swine to the national forests for a consideration [a fee]. The acorn is one of the chief reasons why the owners of these animals pay for their admission" (Smith 1916). He does not specify where, but specifically notes the practice is similar to the montanera of Spain and Portugal. Older ranchers interviewed in 2004 near Red Bluff, California remarked that pigs on their ranch and those of neighbors grazed on acorns, and they recalled "pig drives" to move the mature pigs more than 100 miles to market near Sacramento. The practice has undergone a recent revival for the gourmet market, with at least two producers experimenting with raising acorn-fed pigs (Reed 2010; O'Rourke 2012). The increased interest in Spanish jamón, and sustainable agriculture, has stimulated these efforts (Fig. 10.20).

The count of horses in California, as in the Spanish dehesa, is impossible to determine with any certainty. The last census undertaken by the U.S. Department

Fig. 10.19 California has had an upsurge in its goat population, in part to sate demand for goat meat among the State's Hispanic, Islamic, and South Asian populations. Weed and grass management is another important use. This Boer goat herd grazes alongside a vineyard in Santa Margarita, California, watched over by several Great Pyrenees guard dogs. (Photograph by P. F. Starrs)

Fig. 10.20 While most raising of pigs in California follows the feedlot, grain-fed model so characteristic of the United States confined-feeding style, some California ranchers are even experimenting with acorn-fattening of hogs, as here advertised in a grocery store, all a part of producing value-added products. (Photograph by L. Huntsinger)

of Agriculture was in 1999, when California had 240,000 horses, with 15,000 of them sold for an income of $60 million. The American Horse Council estimates that there are more than 698,000 horses in California, producing $4.1 billion in goods and services, with 70 % used for showing and recreation (AHC 2012). In oak woodlands, horses may have a devastating impact when small property owners attempt to keep them in grazing areas that are too small. The changing landowner profile (Table 10.2) includes many who do not understand that a single horse requires at least 6–8 ha of rangeland for year round grazing, and still needs supplemental feed in the dry season. Erosion, tree loss, horses in poor condition, and foundering in the spring are the result. On the other hand, there is reduced fire hazard.

10.7 A Comparison of Livestock in California and Spain

A table can quickly sum up contrasts of livestock raising in the Spanish and California oak woodlands (Table 10.3). But a bit of discussion helps with context. In both Spain and California, a sizable majority of woodland properties are privately owned, in long-term ownership, and large. California ranchers favor cattle, with almost no diversity in economic products beyond some sales of firewood and a few operations that allow fee-hunting (Chap. 11). In California, most ranchers live on their properties, while in Spain, full-time residence by owners is rare. Even dehesas under long-time ownership are often visited by owners only on weekends or for recreational use, and a resident manager, a caretaking family, or a guard will live on the property. Nonetheless, owners place high amenity values on their property, and prices for ranches and dehesas easily exceed their value based on production alone (Chap. 13).

For livestock-producing operations, the situation in Spain is dramatically different than California, with many products to be gleaned from the land, and cash-earning outputs generally oriented around a mixed livestock operation. If cattle and Iberian pigs are viewed now as profit centers, even recently sheep and goats were prominent, and are still seen on many a dehesa—though rarely in the oak woodland ranches of California, where predation is seen as a severe impediment. Management is changing in Spain; less so in California, where perhaps the most prominent threat to a woodland ranch is conversion into either a residential subdivision of 2–10 ha ranchettes, or type-conversion that push vineyards and other intensively cultivated agricultural crops into former rangeland. Over the last 30 years the dehesa has seen a vast upswing of interest in fee-hunting and recreation, and a dehesa managed primarily for big game will see far less clearing of shrubs in the woodland understory, less investment in grain or small livestock production, and may be given over almost entirely to hunting. The balance of management for game and management for livestock is still being worked out. In California, hunting and game management is ultimately controlled by the state, and potential profits are lower and less direct (Chap. 11).

Table 10.3 Characteristics of oak woodland ranches and dehesas from multiple surveys and reports

	Californian oak woodland	Spanish wooded dehesa
Ownership of oak woodlands or dehesa properties	82+ % private land ownership[a]	75–80+ % private land ownership[b]
Percent properties where owner is resident manager	90 %[c]; 93 %[d]	<25 %[e]; 56 %[f]; 17 %[g]
Years of ownership	49[c]; 50[d]	73[e]
Average property size	927 ha[h]; 985 ha[c]	507 ha[e]; 710 ha[f]; 493 ha[g]
Commodity products typically produced	Beef, lamb, wool, firewood, game, grazing resources	Beef, Iberian pig, lamb, acorns, firewood, hay, cereal grain, grazing resources, wool, cabrito, goat milk, game, truffles, mushrooms, charcoal, cheese, fodder, honey, cork. fighting bulls
Mean number cattle on property	150[h]; 336[h]	102[e]
Stocking rate	5–10 ha/Animal Unit/year[i]	4 ha/Animal Unit/year[j]
Small stock status	Declining	Declining
Large stock status	Declining; 92 % of animal demand is for cattle[k]	Increasing; 42 % of animal demand is for cattle[j]
Percent properties rearing more than one kind of stock	6 %[c]; 22 %[d]	79 %[e]

[a] Statewide (CDF–FRAP 2003); [b] Representative area in Extremadura (Campos 1984); [c] Statewide survey oak woodland ranchers (Huntsinger et al. 2010); [d] Alameda, Contra Costa, and Tehama county ranchers (Liffmann 2000); [e] Caceres province (Plieninger et al. 2004); [f] Cadiz (Campos et al. 2009); [g] Central & Iberico (Campos et al. 2009); [h] Central Sierra Nevada foothills, California (Sulak and Huntsinger 2002); [i] Oak woodlands (CDF–FRAP 2003); [j] Extremadura (Campos 1997); [k] Statewide (USDA–NASS 2011b)

While land use change is a major threat to the traditional oak woodland grazing in California, in Spain, the declining economic value of some dehesa products has caused changes in the livestock grazed, with a shift to game production, and resultant declines in the diversity of products managed for. In Spain, markets for unique local livestock products are well developed, with denominations of origin and certification of livestock breeds and production systems. In California, attempts to develop niche markets for unique rangeland livestock products are in early development, but fit a growing interest in local, sustainable, forms of agricultural production, and demand for grass-fed, organic, or "natural" meats (with no hormones or sub-therapeutic antibiotics). A developing market involves the use of livestock (cattle, sheep, or goats) for grazing for fire hazard and invasive species reduction. It remains to be seen whether these changing markets, in Spain as in California, can make grazing in the woodlands an enterprise profitable enough to sustain itself in the decades ahead.

Acknowledgments The authors thank the reviewers of this material, the editors, and the many dehesa and oak ranchland owners who have shared their time and knowledge with us.

References

AHC [American Horse Council] (2012) A state breakout study, 2005: California. American Horse Council: your unified voice in Washinton. http://www.horsecouncil.org/state-breakout-studies-following-states. Accessed June 2012

Alejano R, Tapias R, Fernández M, Torres E, Alaejos J, Domingo J (2008) Influence of pruning and the climatic conditions on acorn production in holm oak (Quercus ilex L.) dehesas in SW Spain. Ann For Sci 65:1–9

Arce Jiménez L, Rodríguez Estévez V, Valcárcel Cases M (2009) Desarrollo y aplicación de estrategias analíticas de vanguardia para determinar el régimen de alimentación del cerdo ibérico. Eurocarne 174:84–92

Aparicio MJB (1987) El Cerdo Ibérico. Sánchez Romero Carvajal (ed) Jabugo S.A. Huelva. España

Aparicio MA, Vargas JD, Robledo J, Andrada JA, Atkinson A, San José JJ, Prieto L (2006) Nuevas tecnologías y cerdo ibérico. Solo Cerdo Ibérico 16:83–90

Bartlett ET, Taylor RG, McKean JR, Hof JG (1989) Motivation of colorado ranchers with federal grazing allotments. J Range Mgmt 42:454–457

Bartolome JW, Frost WE, McDougald NK, Connor JM (2002) California guidelines for residual dry matter (RDM) management on coastal and foothill annual rangelands. In: Rangeland Management Series. UC Division of Agriculture and Natural Resources, Oakland, CA

Beaufoy G (1995) Distribution of extensive farming systems in Spain. In: McCracken DI, Bignal EM, Wenlock SE (eds) Farming on the edge: the nature of traditional farmland in Europe. Joint Nature Conservation Committee, Peterborough, pp 46–49

Bentley JR, Talbot MW (1951) Efficient use of annual plants on cattle ranges in the California foothills. USDA Circular 870. USDA, Washington DC

BOE (2007) Real Decreto 1469/2007, de 2 de noviembre, por el que se aprueba la norma de calidad para la carne, el jamón, la paleta y la caña de lomo ibéricos

BOE (2009) Real Decreto 2129/2008, de 26 de diciembre, por el que se establece el Programa nacional de conservación, mejora y fomento de las razas ganaderas

Boza López J (2007) Recursos ganaderos de Andalucía. Anales Real Academia de Ciencias Veterinarias de Andalucía Oriental 20:217–242

Brunson M, Huntsinger L (2008) Ranching as a conservation strategy: can old ranchers save the new west? Rangeland Ecol Manag 61:137-141

Burcham LT (1982) [1857] California range land: an historico-ecological study of the range resource of California. Center for Archeological Research 7, University of California, Davis, CA

Campillo Álvarez JE (2001) Jamón Ibérico y Riesgo Cardiovascular: Mitos y Realidades. In: Proceedings of the 1st Congreso Mundial del Jamón. Córdoba, España 59–62

Campos P (1984) Economía y energía en la dehesa extremeña. Instituto de Estudios Agrarios, Pesqueros y Alimentarios/MAPA, Madrid

Campos P (1997) Análisis de la rentabilidad económica de la dehesa. Situación Serie estudios regionales: Extremadura 111–140

Campos P (2005) La renta ambiental en las dehesas de producción de ganado de lidia. R Inst Estud Económicos 22:141–161

Campos P, Oviedo J, Caparrós A, Huntsinger L, Coelho I (2009) Contingent valuation of woodland owners private amenities in Spain, Portugal and California. Rangel Ecol Mgmt 62:240–252

Cañellas I, Roig S, Montero G (2006) Pruning influence on acorn yield in cork oak open woodland. In: Mosquera-Losada MR, McAdam J, Rigueiro-Rodríguez A (eds) International congress on silvopastoralism and sustainable management, Lugo, España. CABI Publishing, Oxfordshire

Cañellas, I, Roig S, Poblaciones MJ, Gea-Izquierdo G, Olea L (2007) An approach to acorn production in Iberian dehesas. Agrofor Sys 70:3–9

Castillo L, Fernández-Llario P, Mateos C, Carranza J, Benítez-Medina JM, García-Jiménez W, Bermejo-Martín F, Hermoso de Mendoza F (2010) Management practices and their association with Mycobacterium tuberculosis complex prevalence in red deer populations in Southwestern Spain. Prev Vet Med 98:58–63

Cava López, R (2007) Calidad de carne y Cerdo ibérico Seminario de producción porcina en condiciones tropicales y mediterráneas. Facultad de Veterinaria de la UEX, Cáceres. Spain

CDF-FRAP [California Department of Forestry and Fire Protection-Fire and Resource Assessment Program] (2003) Changing California: forest and range 2003 assessment. State of California Resources Agency, Sacramento, CA. http://www.frap.cdf.ca.gov/assessment2003. Accessed June 2012

Conner MM, Jaeger MM, Weller TJ, McCullough DR (1998) Effect of coyote removal on Sheep depredation in Northern California. J Wildl Mgmt 69:690–699

Delgado JV, Barba C, Diéguez E, Cañuelo P (2000) Caracterización exteriorista de las variedades del cerdo ibérico basada en caracteres cualitativos. Arch Zootec 49:201–207

Díaz M, Campos P, Pulido FJ (1997) The spanish dehesa: a diversity in land use and wildlife. In: Pain DJ, Pienkowski MW (eds) Farming and birds in Europe: the common agricultural policy and its implications of bird conservation. Academic Press, London, pp 178–209

DOUE [Diario Oficial de la Unión Europea] (2009) Futuro del sector ovino y caprino en Europa. P6_TA(2008)0310. Resolución del Parlamento Europeo, de 19 de junio de 2008, sobre el futuro del sector ovino y caprino en Europa (2007/2192(INI)) (2009/C 286 E/10)

Durán R, Lizaso J (1997) Alimentación del cerdo ibérico. Anaporc 170:82–106

Eichhorn MP, Paris P, Herzog F, Incoll LD, Liagre F, Mantzanas K, Mayus M, Moreno G, Papanastasis, VP, Pilbeam DJ, Pisanelli A, Dupraz C (2006) Silvoarable systems in Europe—past, present, and future prospects. Agrofor Sys 67:29–50

Ferranto, S, Huntsinger, L, Getz, C, Nakamura, G, Stewart, W, Drill, S, Valachovic, Y, DeLasaux, M, Kelly, M (2011) Forest and rangeland owners value land for natural amenities and as financial investment. Cal Ag 65(4):184–191

Ferrer C, Broca A (2001) Pastos y Biodiversidad. Una revisión científica (1960–2000) de la producción bibliográfica de la SEEP En Biodiversidad en Pastos. Ponencias y

Comunicaciones de la XLI Reunión Científica de la SEEP y I Foro Iberoamericano de Pastos. 23–27 de Abril, Alicante. Centro Iberoamericano de Biodiversidad (CIBIO) y Generalitat Valenciana

Forero J, Cumbreras M, Venegas M, Ferrer N, Barba C, Delgado JV (2000) Contribución a la caracterización productiva del cerdo Manchado de Jabugo en el período predestete: resultados preliminares Arch Zootec 50:189–190

Fuentes SC (1991) La Encina. R Agropecuaria 706:434–437

Gade DW (2010) Parsons on pigs and acorns. Geogr Rev 100:598–606

García S, Ramos S, Josemaria A, Isabel B, Blanco J, Jucas AB, Aguilar S, Doncel E, Vázquez FM (2003) Consumo de bellotas por el cerdo ibérico durante la montanera. Solo Cerdo Ibérico 10:65–72

García Dory MA, Martínez Vicente SF, Orozco Piñán F (1990) Guia de campo de las razas autóctonas de España. Alianza Editorial, Madrid

Gaspar P, Mesías FJ, Escribano M, Pulido F (2009) Sustainabiity in spanish extensive farms (Dehesas): an economic and management indicator-based evaluation. Rangel Ecol Mgmt 62:153–162. doi: http://dx.doi.org/10.2111/07-135.1 Accessed July 2012

Gea-Izquierdo G, Cañellas I, Montero G (2006) Acorn production in Spanish holm oak woodlands. Inves Agrar Sist y Recur For 15:339–354

George M, Nadar G, McDougald N, Connor M, Frost W (2001) Forage quality, rangeland management series publication 8022. UC Division of Agriculture and Natural Resources, Oakland, CA

Gentner BJ, Tanaka JA (2002) Classifying federal public land grazing permittees. J Range Mgmt 55:2–11

González F, Schnabel S, Prieto PM, Pulido-Fernández M, Gragera-Facundo J (2012). Pasture productivity in dehesas and its relationship with rainfall and soil. In: Rosa Maria Canals Tresserras RM, Sanemeterio Garciandía L (eds) Proceedings of SEEP Nuevos retos de la ganadería extensiva: un agente de conservación en peligro de extinción. Sociedad Española para el estudio de los pastos

González Rebollar JL, Robles Cruz A, Boza López J (1998) Sistemas pastorales In: Jiménez Díaz RM, Lamo de Espinosa J (eds) Agricultura Sostenible. Ediciones Mundi Prensa

Hart GH, Guilbert HR, Goss H (1932) Seasonal changes in the chemical composition of range forage and their relation to the nutrition of animals. Bulletin 543, UC Agricultural Experiment Station, Berkeley

Hemingway E (1932) Death in the afternoon. Scribners, New York

Hernández Díaz-Ambrona CG (1998) Ecología y fisiología de la dehesa. In: Hernández Díaz-Ambrona CG (ed) Jornadas de Agronomía. La Dehesa. Aprovechamiento sostenible de los recursos naturales. Editorial Agrícola Española, S.A, Madrid

Huntsinger L, Starrs PF (2006) Grazing in arid North America: a biogeographical approach. Sécheresse 17:219–233

Huntsinger L, Sulak A, Gwin L, Plieninger T (2004) Oak woodland ranchers in California and Spain: conservation and diversification. Adv Geoecology 37:309–326

Huntsinger L, Johnson M, Stafford M, Fried J (2010a) Hardwood rangeland landowners in California from 1985 to 2004: production, ecosystem services, and permanence. Rangel Ecol Mgmt. 63:324–334

Huntsinger L, Forero L, Sulak A (2010b) Transhumance and pastoralist resilience in the western United States. Pastor Res Pol Practice 1:1–15

Jackson LE (1985) Ecological origins of California's mediterranean grasses. J Biogeog 12:349–361

Joffre R, Rambal S, Ratte JP (1999) The dehesa system of southern Spain and Portugal as a natural ecosystem mimic. Agrofor Sys 45:57–79

Klein J (1920) The mesta: a study in Spanish economic history, 1273–1836. Cambridge UP, Cambridge

Linares AM (2007) Forest planning and traditional knowledge in collective woodlands of Spain: the dehesa system. For Ecol Mgmt 249:71–79

López-Bote C, Fructuoso G, Mateos GG (2000) Sistemas de producción porcina y calidad de la carne. El cerdo ibérico XVI Curso de Especialización FEDNA: Avances en Nutrición y Alimentación Animal. Fundación Española para el Desarrollo de la Nutrición Animal. Eds.: P.G[a]. Rebollar, C. de Blas y G.G. Mateos. Fira de Barcelona, España

López-Bote C, Fructuoso G, Mateos GG (2002) Sistemas de producción porcina y calidad de la carne. El cerdo ibérico. Solo Cerdo Ibérico 9:59–85

Liffmann RH, Huntsinger L, Forero LC (2000) To ranch or not to ranch: home on the urban range? J Range Mgmt 53:362–370

Lorca A (2012) La fiesta del toro se desploma [and] ¿Un blen ecológico?. El Pais [Madrid] 07 August 2012. http://sociedad.elpais.com/sociedad/2012/08/06/actualidad/1344278036_271436.html (Accessed August 2012)

MARM [Ministerio de Medio Ambiente Rural y Marino] (2008) Diagnóstico de las Dehesas Ibéricas Mediterráneas. TRAGSATEC, Servicio de Publicaciones del Ministerio de Medio Ambiente, Rural y Marino http://www.marm.es/es/biodiversidad/temas/montes-y-politica-forestal/anexo_3_4_coruche_2010_tcm7-23749.pdf. Accessed June 2012

MARM [Ministerio de Medio Ambiente Rural y Marino] (2009a) Anuario de estadística 2009. Servicio de Publicaciones del Ministerio de Medio Ambiente, Rural y Marino

MARM [Ministerio de Medio Ambiente Rural y Marino] (2009b) El sector equino en cifras. Principales indicadores económicos en 2008. Servicio de Publicaciones del Ministerio de Medio Ambiente, Rural y Marino

MARM [Ministerio de Medio Ambiente Rural y Marino] (2010) Censo de animales y productos comercializados en 2009. http://www.marm.es/es/alimentacion/temas/calidad-agroalimentaria/calidad-comercial/mesa-del-iberico/riber-publico/censos-de-animales-y-productos-comercializados-2009/default.aspx. Accessed June 2012

MARM [Ministerio de Medio Ambiente Rural y Marino] (2011) El sector equino en cifras. Principales indicadores económicos en 2011. Servicio de Publicaciones del Ministerio de Medio Ambiente, Rural y Marino

Martin WE, Jeffries GL (1966) Relating ranch prices and grazing permit values to ranch productivity. J Farm Econ 48:233–342

Martínez AM, Rodero A, Vega-Pla JL (2000) Estudio con microsatélites de las principales variedades de ganado porcino del tronco ibérico Arch Zootec 49:49–52

Miller K (2011) Acorns: small but potentially deadly. Angus Beef Bulletin Extra October 20, 2011. http://www.angusbeefbulletin.com/extra/2011/10oct11/1011hn_acorns.html Accessed March 2013.

Montero G, San-Miguel A, Cañellas I (1998) Sistemas de selvicultura mediterránea: la dehesa. In: Jiménez Díaz R, Lamo de Espinosa J (eds) Agricultura sostenible. Ediciones Mundi-Prensa, Madrid

Montoya Oliver M (1993) Encinas y encinares. Agroguías Mundi-Prensa. Mundi-Prensa. Madrid

Neale JCC, Sacks BN, Jaeger MM, McCullough DR (1998) A comparison of bobcat and coyote predation on lams in north-coastal California. J Wildl Mgmt 62:700–706

Olea L, Paredes J, Verdasco MP (1991) Características y producción de los pastos de las dehesas del S.O. de la Península Ibérica. R Pastos 20–21:131–156

O'Rourke T (2012) Sustainable ranching renaissance takes hold in northern California. San Jose Mercury 02 May. http://www.mercurynews.com/food-wine/ci_20517168/sustainable-ranching-renaissance-takes-hold-northern-california. Accessed June 2012

Parsons JJ (1962) The Acorn-Hog Economy of the Oak Woodlands of Southwestern Spain. Geog Rev 52:211–235

Plieninger T, Pulido FJ, Schaich H (2004) Effects of land-use and landscape structure on holm oak recruitment and regeneration at farm level in Quercus ilex L. dehesas. J Arid Envi 57:345–364

Plieninger T (2006) Habitat loss, fragmentation, and alteration: quantifying the impact of land-use changes on a spanish dehesa landscape by use of aerial photography and GIS. Landsc Ecol 21:91–105 de Rancourt M, Fois N, Lavin MP, Tchakérian E, Vallerand F (2006) Mediterranean sheep and goats production: an uncertain future. Sm Rumin Res 62:167–179

Reed B (2010) Magruder Ranch Slideshow. The Ethical Butcher blog site. 07 January. http://ethicalbutcher.blogspot.com/2010/07/magruder-ranch-slideshow.html. Accessed June 2012

Rodríguez-Estévez V, García A, Perea J, Mata C, Gómez AG (2008) Dimensiones y características nutritivas de las bellotas de los Quercus de la dehesa. Arch Zootec 57(R):1–12

Rodríguez-Estévez V, García A, Peña F, Gómez AG (2009) Foraging of Iberian fattening pigs grazing natural pasture in the dehesa. Livest Sci 120:135–143

Rueda Sabater L, Diéguez Garbayo E (2007) Manual de Cerdo Ibérico. AECERIBER. Zafra (Badajoz) España

Rowe HI, Bartlett ET, Swanson LE Jr (2001) Ranching motivations in two Colorado counties. J Range Mgmt 54:314–321

Ruiz M, Ruiz JP (1986) Ecological history of transhumance in Spain. Biol Cons 37:73–86

SAT [SAT Carne de Retinto] (2010) Estatutos sociales de la Sociedad Agraria de Transformación "Carne de Retinto" N° 9954. Madrid Spain. http://www.retinta.es/images/stories/pdf/normas_sacrificio.pdf. (Accessed June 2012)

Sierra Alfranca I (coord.) (2003) Evolución y cambio en el sector ovino-caprino en España durante la última década. Dirección general de ganadería. Subdirección general de vacuno y ovino del MARM y la SEOC

Smith AH, Martin WE (1972) Socioeconomic behavior of cattle ranchers, with implications for rural community development in the west. Am J Agric Econ 54:217–225

Smith JR (1916) The oak tree and man's environment. Geogr Rev 1:3–19

Starrs PF (1997) Let the cowboy ride: cattle ranching in the American west. Johns Hopkins University Press, Baltimore

Suárez MV, Barba C, Forero J, Sereno JRB, Diéguez E, Delgado JV (2002) Análisis multivariante entre poblaciones porcinas de origen ibérico basado en parámetros reproductivos. Arch Zootec 51:249–252

Sulak, A (2007) Land conservation and environmental policy: Public land grazing for private land conservation? Dissertation, UC Berkeley

Sulak A, Huntsinger L (2002) Sierra Nevada grazing in transition: the role of forest service grazing in the foothill ranches of California. In: A report to Sierra Nevada alliance, California Cattlemen's Association, and California Rangeland Trust. Sierra Nevada Alliance, California. http://www.sierranevadaalliance.org/publications/publication.shtml?type=pgm02. Accessed June 2012

Sullins J, Maas, J (2011) Blue oak acorn toxicity risk in cattle. Oak Conservation information from the University of California, ANR blog. http://ucanr.edu/blogs/blogcore/postdetail.cfm?postnum=5987 Accessed March 2013.

Torell LA, Rimbey NR, Ramirez OA, McCollum DW (2005) Income earning potential versus consumptive amenities in determining ranchland values. J Ag Res Econ 30:537–560

USDA NASS [United States Department of Agriculture—National Agricultural Statistics Service] (2011a) Sheep and goat review report, 27 Jan 2012. https://docs.google.com/viewer?url=http%3A%2F%2Fwww.nass.usda.gov%2FPublications%2FTodays_Reports%2Freports%2Fshep0112.pdf. Accessed June 2012

USDA NASS [United States Department of Agriculture—National Agricultural Statistics Service] (2011b) California agricultural statistics, Crop Year 2010, Livestock and Dairy. Oct. 28, 2011. http://www.nass.usda.gov/Statistics_by_State/California/Publications/California_Ag_Statistics/2010cas-lvs.pdf. Accessed Jan 2012

Vargas Giraldo JD, Aparicio Tovar MA (2000) El cerdo ibérico en la dehesa extremeña. Análisis técnico y económico. Diputación de Badajoz y Caja Rural de Extremadura. Badajoz. España

Vázquez FM (1998) Producción de bellotas en quercus. I. Métodos de estimación. Solo Cerdo Ibérico 1:59–65

Vázquez FM, Doncel E, Pozo J, Ramos S, Lucas AB, Medo T (2002) Estimación de la producción de bellotas de los encinares extremeños en la campaña 2002–2003. Sólo Cerdo Ibérico 2:95–100

Vázquez FM, Montero G, Suárez MA, Baselga P, Torres E (1996) Estructura de una masa mixta de frondosas (Quercus rotundifolia Lam. y Q. suber L.) I Densidad de arbolado Actas de la reunión de Córdoba. Grupo de trabajo selvicultura mediterránea. Cuadernos de la Sociedad española de estudios forestales 3. Sociedad Española de Ciencias Forestales. Cádiz. España

Zapata S (1986) La producción agraria en Extremadura y Andalucía Occidental. Universidad Complutense de Madrid

Chapter 11
Hunting in Managed Oak Woodlands: Contrasts Among Similarities

Luke T. Macaulay, Paul F. Starrs and Juan Carranza

Frontispiece Chapter 11. The California tule elk (*Cervus canadensis* ssp. *nannodes*) is a subspecies of elk once almost lost to commercial overhunting. Successful reintroduction efforts have allowed renewed sport hunting opportunities of great value to landowners. (Photograph by B. Voelker)

L. T. Macaulay (✉)
Department of Environmental Science, Policy, and Management, University of California, 130 Mulford Hall, Berkeley, CA 94720-3114, USA
e-mail: luke.macaulay@berkeley.edu

P. Campos et al. (eds.), *Mediterranean Oak Woodland Working Landscapes*, Landscape Series 16, DOI: 10.1007/978-94-007-6707-2_11,
© Springer Science+Business Media Dordrecht 2013

Abstract Distinct cultural and legal histories governing the property rights that regulate wildlife and land tenure in California and Spain have created dissimilar hunting systems. The differences that are manifest in the methods of hunting, the economic return to landowners, the actions taken to manage game species, and the accompanying environmental effects. Private landowners in Spain retain greater control of game species, while in California, the state and federal government exerts greater authority. After providing background on the game species and systems of hunting in California and Spain, a review of the legal and cultural history illustrates how distinct systems evolved in places that are similar in many other ways. In terms of economics, hunting revenue in Spain is often greater than in California, due to higher hunter participation rates, fewer governmental restrictions that limit the commercialization of hunting, and greater liberties in hunting methods and game management practices. As such, income from hunting provides a greater incentive for Spanish landowners to maintain areas of habitat for game species. Some of the greatest contrasts between these places are illustrated in wildlife management practices, where Spanish landowners can implement far more intensive practices to manipulate populations of game species. Numerous environmental effects can result from these management practices, which include changes to vegetation, erosion, genetic impacts, invasive species introductions, and impacts to non-game species.

Keywords Hunting · California · Spain · Game · Wildlife management · Property rights · Predator control

11.1 The Basics

The oak woodlands of California and Spain share commonalities of climate, vegetation, livestock production, and biodiversity of global importance. Yet distinct cultural and legal histories governing property rights over wildlife and land tenure have created dissimilar hunting traditions. The geography, game species, hunting participation rates, land tenure, and other factors have all shaped these unique systems, and the legal and cultural history of hunting illustrates how these hunting systems evolved. This historical discussion elucidates the economic, legal,

P. F. Starrs
Department of Geography, University of Nevada, MS 0154 Reno, NV 89557-0048, USA
e-mail: starrs@unr.edu

J. Carranza
Department of Zoology, University of Córdoba, Campus de Rabanales, Colonia San José Ctra. Nacional IV-A, Km 396 14071 Córdoba, Spain
e-mail: jcarranza@uco.es

and social roots of present-day hunting, while the final sections dwell on wildlife management practices, and the associated environmental impacts of hunting in Spain and California oak woodlands.

11.1.1 Geography and Game Species

Three regional *autonomías,* or autonomous regions—Castilla-La Mancha, Extremadura, and Andalucía—contain a sizable part of the Spanish dehesa (Chap. 1) and include what historically and today are Spain's most productive hunting lands. Forty-two percent of all Spanish hunting areas are contained in these autonomías, where two-thirds of all big game harvests take place (Metra and Seish 1985; López Ontiveros and Verdugo 1991). The dehesa, in association with surrounding and understory shrublands, provides quality food and cover for popular game species including Iberian red deer (*Cervus elaphus hispanicus*), wild pig (*Sus scrofa*), and occasionally mouflon (*Ovis orientalis musimon*) or fallow deer (*Dama dama*) (Blas Aritio 1974). High quality hunting in the early 1900s made the Spanish dehesa a destination for big game hunters from Europe and lands farther afield (Delgado and Muñoz 1982). Iberian red deer and wild pigs make up the largest share of Spain's annual harvest, reaching averages of 100,000 Iberian red deer and 180,000 wild boar (Carranza 2010; Garrido Martín 2011) (Tables 11.1 and 11.2).

The hunting of small game in Spain is a secondary activity, although preferred by some hunting classicists because it permits solitary and contemplative travel in a natural realm (Delibes 1982). Small game species predominate in more agriculturally developed areas. Popular small game species include thrushes (*Turdus* spp.), the European rabbit (*Oryctolagus cuniculus*) and the red-legged partridge (*Alectoris rufa*). While over 3 million partridges are harvested every year, few are from natural populations, with most birds farm-raised and released onto dehesa sites before the start of each year's hunting season (Garrido Martín 2011) (Figs. 11.1 and 11.2).

The oak woodlands of California are located in foothill and coastal regions from sea level to an elevation of 1300 m in the southern Sierra Nevada. The main oak habitat is in California's Central to North Coast region and in a vast ring around the Central Valley. These offer food and cover for big game animals,

Table 11.1 Predominant game species in the oak woodlands of California and Spain

California	Spain
Predominant big game species	*Predominant big game species*
Black-tailed/Mule deer (*Odocoileus hemionus* ssp.)	Iberian red deer (*Cervus elaphus hispanicus*)
Wild pig (*Sus scrofa*)	Wild pig (*Sus scrofa*)
Predominant small game/upland game	*Predominant small game/upland game*
Turkey (*Meleagris gallopavo*)	European rabbit (*Oryctolagus cuniculus*)
Quail (*Callipepla californica*)	Red-legged partridge (*Alectoris rufa*)

Table 11.2 2005–2009 averaged annual harvest of game species in Spain (Garrido Martín 2011)

Species	Average harvest
Big game	
Wild boar (*Sus scrofa*)	183,739
Red deer (*Cervus elaphus*)	102,220
Roe deer (*Capreolus capreolus*)	22,591
Fallow deer (*Dama dama*)	8,805
Mouflon (*Ovis musimon*)	6,150
Spanish ibex (*Capra pyrenaica*)	2,722
Barbary sheep (*Ammotragus lervia*)	601
Chamois (*Rupicapra rupicapra*)	485
Iberian wolf (*Canis lupus signatus*)	80
Small game	
Thrush (*Turdus* spp.)	6,120,587
European rabbit (*Oryctolagus cuniculus*)	5,628,208
Red-legged partridge (*Alectoris rufa*)	3,803,460
Common quail (*Coturnix coturnix*)	1,287,014
Granada hare (*Lepus granatensis*)	1,202,869
European turtle dove (*Streptopelia turtur*)	840,888
Eurasian woodcock (*Scolopax rusticola*)	75,611

especially in terms of acorns that yield significant nutrition for wildlife in the fall when herbaceous forage is desiccated and of minimal value. Migratory deer in many parts of California descend in the fall from montane environments to lower elevation oak woodlands where they over-winter. Big game species commonly hunted include the native mule deer (*Odocoileus hemionus*, classified into as many as six subspecies) and non-native wild pigs. California quail, mourning dove, wild turkey, pheasant, rabbit, jackrabbit, and squirrels are the small game and upland game commonly hunted in oak woodlands and on adjoining cropland (Figs. 11.3 and 11.4) (Table 11.3).

Fig. 11.1 Game birds: Spain's red-legged partridge (*Alectoris rufa*).

Fig. 11.2 Changes in legislation, including the hunting law of 1970, means that an overwhelming proportion of Spanish woodland properties are posted as some form of reserved-access land (*cotos*). (Photograph by P. F. Starrs)

Fig. 11.3 Game birds: the wild turkey (*Meleagris gallopavo*) in California. (Photograph by L. T. Macaulay)

While waterfowl is hunted on water impoundments or rivers in the oak woodlands of California and Spain, most duck and goose hunting occurs in wetland habitats or flooded agricultural fields, areas outside the focus of this chapter.

Fig. 11.4 Game birds: the California quail (*Callipepla californica*). (Photograph by L. T. Macaulay)

Table 11.3 2004–2008 averaged annual harvest of game species in California (CDFG)

Species	Average harvest
Big game	
Mule deer (*Odocoileus hemionus*)	40,470
Wild boar (*Sus scrofa*)	19,864
Small game	
Dove (*Columbidae* spp.)	1,759,337
Quail (*Callipepla* and *Oreortyx* spp.)	626,970
Farmed ring-necked pheasant (*Phasianus colchicus*)	342,710
Wild ring-necked pheasant (*Phasianus colchicus*)	107,082
Black-tailed jackrabbit (*Lepus californicus*)	63,137
Cottontail rabbit (*Sylvilagus* spp.)	55,977
Western gray squirrel (*Sciurus griseus*)	50,310
Wild turkey (*Meleagris gallopavo*)	26,202

11.1.2 Hunting Seasons

California deer hunting seasons close prior to the breeding season or rut. The seasons open at varying times across the state, with archery-only hunting commencing in August–September, and general gun hunting from September to early November (CDFG 2012b). Wild pigs, a non-native and invasive species in California, are hunted year-round. Bears are occasionally found in California's oak woodlands, and the bear season often coincides with the deer season (Fig. 11.5).

In Spain, the general season for hunting runs from early October to late February, closing just before the onset of spring breeding for most game species. Hunting after February may adversely impact pregnant ungulate females, since they are easily captured by hunting dogs even when not a target of the hunt. Red deer are hunted in the Spanish *montería*—an organized hunt with dogs driving big

Fig. 11.5 Wild pigs harvested in California. (Photograph from the collection of R. Barrett)

game toward waiting hunters—within the general hunting period from early October to mid-February, and by stalking during the rutting season (from September to opening of the general season in October) (Aguayo 1986). One notable exception is the roe deer season, which generally runs from mid-April to the end of July with additional hunting periods in September and October.

11.1.3 Structural Organization of Hunting Operations

In Spain, two main hunting venues exist, the *coto social* (or *coto deportivo*) and the *coto privado* (Barbosa et al. 2004). The coto social is organized by a community and maintained for recreational use. Hunts historically occur on community hunting grounds and are generally oriented towards small game without a profit motive (Chap. 2). In modern times, these clubs may secure the use of a private property, or in some cases, a community actually owns a dehesa where a productive use of the land is hunting. The coto privado involves hunts on private lands oftentimes for profit. Viability and profitability of these properties has increased since 1970 due to four significant changes: (1) landowners can enclose wildlife behind high game fences, which makes wildlife management much easier; (2) there is more interest in big game hunting in Spain; (3) small game is increasingly scarce on community lands due to disease outbreaks that have significantly reduced European rabbit populations; and (4) the Hunting Law of 1970 has made it easier for associations to own an estate, lowered taxation on hunting lands, and allowed landowners to prevent trespassing on marked hunting reserves (López Ontiveros and Verdugo 1991; Barbosa et al. 2004; Martínez Garrido 2009).

Significant areas of public land (most of which is owned by the U.S. Forest Service) in California are open to free hunting without any reservation or membership. Some public lands require reservations or special tags drawn by lottery to hunt. Hunting operations on private lands in California generally fall into two main types: (1) year-long leases, where hunters pay for access to the property year-round with restrictions on game taken, and (2) day hunts, where hunters pay for access to a property for a given number of days with a specified amount of game to be taken. Although there is a wide array of hunting styles in California, there is no significant cultural and historical distinction between the social hunt and private hunting as in Spain; however, there are social and cultural differences between what one researcher deems "sport hunters," "meat hunters," and "ecologistic hunters," with varying priorities and motivations among the types (Kellert 1980). Small game hunting is often practiced by youth or as a secondary quarry for many hunters in California (Shaw 2008).

11.1.4 Trends in Hunting Participation

A feature shared by hunters in California and Spain is their enthusiasm for the rhythms and associations of hunting. Nonetheless, a demographic shift in hunters appears to be underway, with hunting license sales decreasing in California and Spain. As one California journalist reported in 2003, after discussions with Department of Fish and Wildlife legislative analysts, "the growing urban populations in California are 'less supportive of hunting'" (Stienstra 2003). Yet the overall populations of California and of Spain are increasing, if only gradually since 2008, so while the total count of licensed hunters decreases, the percentage of hunters with licenses in the population drops still more (Fig. 11.6). Of late the pace of decline in hunter interest has slowed slightly, with increasing sales of tags to hunt wild pigs and bear. In California, a decreasing overall percentage of hunting license holders tracks with a general decline in the state's most popular game species, the mule deer and black-tailed deer, which are particularly effected by increasing numbers of predators and a decrease in readily available habitat.

Curiously, while urbanized populations may be less familiar with hunting's traditions, schedules, and attractions, there is rising interest in hunting from a new

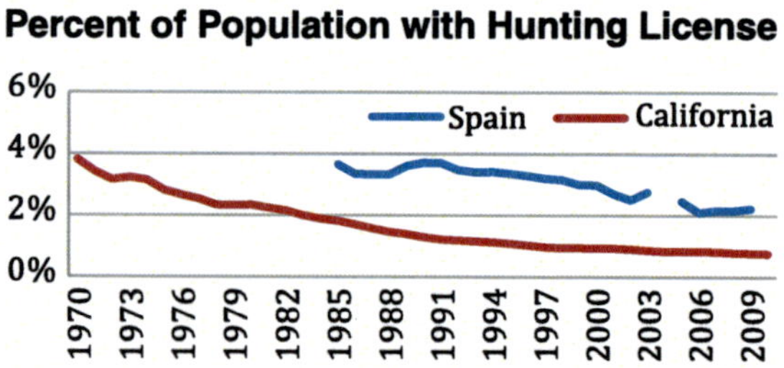

Fig. 11.6 In both Spain and California the percentage of population that hunts is decreasing. (CDFG 2012c; MARM 2012)

generation seeking to reconnect with food sources (Marris 2012). In the U.S., the most recent national survey on hunting participation has revealed an increase in the number of hunters of 9 % from 2006 to 2011, which could be indicative of a renewed desire to learn how to hunt, clean, and cook one's food from sources in the wild (Pollan 2006; Goldstein 2010; Shaw 2011; USDOI and USDOC 2011; Heyser 2012).

11.1.5 Land Ownership

The majority of dehesa habitat in Spain is under private ownership and that is not government-owned. The properties where hunting occurs reflect three distinct histories of land ownership: one in aristocratic proprietorship; another with Church, or ecclesiastical, control of property; and finally, an ownership by village communities of land that is hunted. This history has opened up a variety of ways in which hunters and landowners access and manage game species (Gómez Mendoza 1992; López Ontiveros 1993).

Approximately 50 % of California is public land, and much of that is open to free hunting access. However, more than 80 % of oak woodlands in California are privately owned (CDF-FRAP 2003). Hunter success rates on private holdings are generally much higher than on public lands because private lands generally contain higher quality habitat, food, and water resources and because private landowners can regulate hunting pressure (Scott et al. 2001; CDFG 2012a).

11.2 Brief History of Hunting

11.2.1 Spain

As Upper Paleolithic and Magdalenian cave art bears witness, interest in the taking of game in Spain was widespread even in prehistoric and ancient times. Hunters in more recent years derive from two main types, the first of aristocratic background who traditionally sought game on the properties of friends or relations, and the other residents of towns and of communities owning land suitable for hunting (Chap. 2). Wealthy landowners would hunt big game or small game, maintaining the right to exclude others, and sustaining a privileged land use with the force of law by employing hired gamekeepers or guards. Hunting by rural residents often focused on small game, taken by snare, shotgun, rifle, or net. Historically, these village hunters—almost exclusively male—hunted for recreation and for food.

During the nineteenth and first half of the twentieth century, hunting in Spain was a popular activity, though one restricted in practice. The divide between the wealthy landowning gentry, who hunted socially with friends, and the rural village

Fig. 11.7 From 1940s issues of *Caza y Pesca,* a monthly Spanish sports magazine, came two visions of "hunters" in the Spanish countryside: at *right*, a group of loden-garbed gentry; at *left*, a poacher, whose work has been discovered by horse-mounted Guardia Civil, the Spanish national police force that was especially vigilant in rural areas. (Photographs composited by P.F. Starrs)

residents and poachers who hunted for food or recreation was distinctive and large. This is much-studied for parts of Andalucía, with enduring accounts issuing especially from researchers and historians working in Córdoba (Chap. 2). Hunting and the *montes* of Spain are remarkably linked, in topics laced with political and social significance (Gómez Mendoza and López Ontiveros 2001) (Fig. 11.7).

Beyond historical accounts, there is a vast literature on twentieth-century hunting in Spain that extends even into cinema, and unease about invidious distinctions of class and access are a feature of many narratives (Almazán 1934). While residents of villages might hunt community properties, some hunted game illegally, trespassing onto the private land of others. These hunters were *furtivos,* whose actions were, as the name suggests, furtive in their pursuits. Poaching was not only an act of villager disregard and disdain for upper class norms, it was an act of defiance toward any ruling authority, including the long-lived Franco government, as depicted in literature and films such as Carlos Saura's *La caza* (1965), Borau's *Furtivos* (1975), Berlanga's *La escopeta nacional* (1977), and Brasó's *El mundo de Juan Lobón* (1989; TV miniseries). In latter-day Spain, however, fenced properties enclose valuable game animals and exclude rural hunters who once regarded taking game by poaching as an adventurous sport and a birthright. Little wonder that many properties now managed principally for game hire gamekeepers who reside on the property to protect against illicit access.

Well into the 1970s, two distinct groups participated in a montería. The shooters were one group and the *rehaleros* or *podenqueros,* dog fanciers and handlers, made up the other. In today's commercially-organized hunts, however, the rehaleros are itinerant professionals hired for wages by hunt organizers—a marked departure from older montería traditions, and a change spoken of with regret by long-time "monteros." Hunting dog owners until the last 30 years or so were prized and fully-fledged members of the hunting party. Traditionally, demonstrating the prowess of a *rehala* (the dog pack) earned the rehalero prestige at least equal to any vested in the successful shooter (Gibert Buch 1975).

11.2.2 California

California's history of hunting exploitation began at the arrival of Native Americans approximately 14,000 years ago, with early arrivals practicing subsistence hunting, which continued among some Native American groups into the early 1900s. While Spanish missions marked the first European settlements in California, their impact on wildlife resources was relatively minor compared to the influx of immigrants during the Gold Rush of 1849, when the population ballooned from 15,000 non-Native Americans to hundreds of thousands (Starr and Orsi 2000).

The immense wildlife resources present at the time of European arrival made it seem impossible that hunting could be a threat to their sustainability, but with settlers increasing in numbers, market hunting, and improved firearm technology many species began to decline and reach extinction levels. Wildlife were harvested as a source of food without any limit or regulation (CDFG 2002). Grizzly bears and wolves were extirpated from California in 1922 and 1924, respectively, and once-immense numbers of tule elk, a California subspecies of *Cervus canadensis*, were reduced to a herd of 28 animals protected by a single rancher near Bakersfield (McCullough et al. 1996). Populations of many wildlife species declined precipitously in the late 1800s, some due to overhunting, others because of habitat loss and human encroachment. Federal and state regulations were enacted in the early 1900s to create hunting seasons and bag limits, with the support of sportsmen hunters, and against the protests of some market hunters. These regulations have led to the recovery of many game species and are the foundation of the current system of state regulation of wildlife in North America—known as the North American Model of Wildlife Conservation (Tober 1981; Geist 1988) (Fig. 11.8).

Fig. 11.8 Animals taken, whether in Spain and California, are always an object of hunter pride, as here at a property in California. (Photograph from the collection of L.T. Macaulay)

Table 11.4 Wildlife practices in Spain and California

Wildlife practice	California	Spain
Shrub management	X	X
Predator control	X	X
Food planting	X	X
Wildlife fencing (2 m+)		X
Sale of meat from hunting		X
Supplemental feeding		X
Restocking and transporting native big game		X
Both sex harvest on native deer		X
Landowners set bag limit for big game		X
Hunting over bait		X
Hunting at night		X
Driving wildlife with dogs		X

X denotes commonly and legally practiced

11.3 Economics of Hunting

11.3.1 Governance, Property Rights, and Wildlife Management

Although a number of Spanish and Mexican legal principles were borrowed in the early state history of California, divergent legal histories governing wildlife since colonial times have resulted in distinct property rights over wildlife that have led to readily apparent differences in the wildlife management regimes of each place (Table 11.4). In Spain, landowners have much greater ability to manipulate and control the movement and management of wildlife, while in California regulations prohibit landowners from fencing, relocating, and feeding wildlife. If wildlife is nominally considered *res nullius*[1] in both locations, Spanish landowners have much greater rights to control and prevent the escape of game animals, capturing *de facto* ownership, even if not *de jure* ownership. Although game species are managed in the public interest by the state and federal government in California, private landowners do control access to property where the game resides, and can charge a fee for hunters to hunt on their property.

In the United States there have been three major historical influences on the development of wildlife governance (Huffman and Wallace 2011). The first was rebellion against oppressive English policies, including the restriction of hunting to those with sufficient wealth or status. The second was the need of colonists to hunt for food and clothing. The third was the belief that given the huge expanse of America and abundant wildlife, restrictions on hunting would hinder economic growth. Eventually, laws and policies developed granting states and the federal government some aspects of managing wildlife in the public interest. There remains a strong sentiment among hunters in the United States that hunting is a time to experience nature, share adventure with friends and family, and to acquire food, and that it should be open to all regardless of wealth or status (IAFWA 2002). There is a tension between this widespread view of the role of hunting and the practice of trophy hunting for high fees, which is often seen as the domain of wealthier individuals. While there is historic opposition to paying for access to hunting opportunities, in California the deer herd has been in decline for the past several decades and the success rate of public land hunting has dropped significantly. As a result, individuals may stop hunting, or become willing to pay for access to higher quality hunting opportunities that occur on private lands.

In many rural areas, it was once and to some extent still is a common courtesy to allow friends and community members, or even polite visitors, to hunt on one's land for free or in exchange for other favors. One study found that even in 2004 nearly two-thirds of ranches on oak woodlands had hunting on their property, but fewer than ten percent charged fees for hunting access (Huntsinger et al. 1997).

The most visible manifestation of the differences between the two countries in property rights over wildlife is the enclosure of game species behind ~2 m high game-fencing (Spanish: *malla* or *valla cinegética*), a common practice in Spain but illegal in California.

An extensive literature discusses the intricacies of property rights of fugitive resources, which Ostrom (1990) terms "common-pool resources" (Ostrom 1990; Lueck 1995). An essential quandary for these resources is that they are often

[1] *Res nullius* is a Latin term borrowed from Roman law. It holds that when an object (which conspicuously includes wildlife) is unowned by any particular person or entity it is ownerless property and free to be owned by anyone. When such an object is found, however, it may in some circumstances belong to the first person who takes it (as in beachcombing).

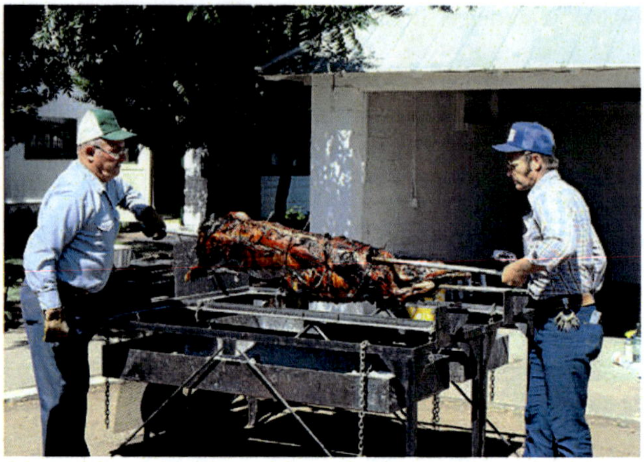

Fig. 11.9 The preparation, cooking and consumption of meat after a hunt in California is an important part of the California hunting experience: here a wild boar is barbecued. (Photograph from the collection of R. Barrett)

jointly produced by more than one person's property, yet these resources, when harvested, are generally not jointly used. For example, deer may live on, reproduce on, and utilize the habitat on several properties that are owned by different individuals, but when a deer is hunted and harvested, it is usually harvested and consumed by one person and is no longer available for someone else to harvest.

The lack of game-fencing in California and in most areas of Spain means game animals are a fugitive resource because they can easily move off a landowner's property and are a type of common-pool resource. Organizing the harvest of common-pool resources is "usually an uncertain and complex undertaking" and if the right conditions are not present, many of these resources may be utilized in a way that reduces the productivity and quality of the resource (Ostrom 1990). The limitations of institutional management and challenges to cooperative efforts such as the presence of free riders can result in difficulties in managing the fugitive resource. For example, male deer in many areas of California are often harvested at two years of age—the first opportunity that they may be legally harvested—due to the possibility that one's neighbor may harvest it first. As such, deer in California and in unfenced areas of Spain often exhibit an age structure that is skewed to younger males and very low proportion of male animals, which can lead to diminished overall harvest and biological problems (Ostrom 1990; Clark 2010; McCullough 2001; CDFG 2002; Milner et al. 2007; Pérez-Gonzalez and Carranza 2009). Game-fencing in Spain, while a costly investment, allows those landowners in Spain to control the fugitive wildlife resource and thereby avoid having to cooperate with others in management and harvest the resource—though not without creating distinct management problems and biological quandaries (Díaz et al. 2009).

Other important property rights differences between California and Spain is the ability of Spanish landowners to provide supplemental feed to wildlife, set bag limits on big game, and sell live game animals to other landowners. Feeding allows landowners to maintain larger game populations, and owners of Spanish hunting dehesa properties can retain meat from the animals taken in a hunt, which yields an additional income source. The situation is significantly different in California, where wildlife may not be fed, sold (dead or alive), or transported, and bag limits are largely state-controlled. While hunters may self-consume or donate meat from the hunt, they may not sell it on the open market. Of note, California law requires possession of the meat, prohibiting hunters from allowing an animal's meat to go "needlessly to waste" (CFGC 2012) (Fig. 11.9).

11.3.2 Commercialization

Hunting in California is far less commercialized than in Spain due to four main factors: (1) cultural opposition to charging fees for hunting, (2) fewer hunters in California as a percentage of the population (Fig. 11.6); (3) the extensive public lands that are open to free hunting, and (4) the vastly different regulatory structures and property rights regimes that govern wildlife. The more state-oriented property rights regime in California and the rest of the U.S. developed in large part as a reaction to overexploitation of wildlife by profit-motivated market hunting (Stine 1980; Tober 1981). As such, there is deep-seated suspicion and long-standing opposition to programs that would allow landowners to profit from wildlife (Fitzhugh 1989).

Greater freedoms to manipulate and control wildlife in Spain makes for a more intensively managed and significantly commercialized hunting industry. A significant shift has occurred in the role of hunting in Spain from 1970 to today. Before 1970, hunting was in the main a leisure activity for socializing, and lacked an overt focus on generating cash income, whether in the realm of large landholders or village hunting associations of small means. The Hunting Law of 1970 modernized the previous Hunting Law from 1902, and made possible the expansion of commercialized hunting and intensified management as discussed above (Structural Organization of Hunting Operations). Since the incorporation of the law in 1970, hunting increasingly is managed as a for-profit commercialized business, because hunting resource rent can surpass returns from grazing or field crops (López Ontiveros and Verdugo 1991; Vargas et al. 1995; Martínez Garrido 2009). While this increases motivations to maintain better habitat for game species, increased fencing and feeding in Spain has yielded high ungulate densities on hunting properties and can lead to deleterious environmental effects as discussed below (Olea and San Miguel-Ayanz 2006; Díaz et al. 2009).

California has developed a pair of programs to help landowners earn income from recreational hunting of wildlife on their land in exchange for habitat improvement or expanded public access. The Private Lands Management program in California has a contentious history but mimics a Texas program that provides landowners greater harvest flexibility over game species in exchange for improving habitat. Landowners must submit a lengthy wildlife management plan to the California Department of Fish and Wildlife and must obey requirements such as informing neighbors of their intent to enroll in the program. The SHARE program provides private landowners with a liability waiver and a cash payment in exchange for allowing public access to a property for hunting or other forms of outdoor recreation (CDFG 2012d).

11.3.3 Costs, Revenue, and Net Profit

Costs and revenues are known to be highly variable between woodland properties and from year to year (Loomis and Fitzhugh 1989; Peiró and Seva 1996). This chapter draws on data from a 1987 survey of 55 ranches with hunting in California and from three case studies in Spain on properties of 658, 4000, and 7000 ha (Loomis and Fitzhugh 1989; Campos et al. 1995; Lenzano and Zamora 1999). Cost and revenue data for hunting properties is relatively limited, but new economic research on hunting is underway in each location to better understand profitability in this industry (Macaulay in prep, RECAMAN in Chap. 13). In general the economic activity, in terms of average costs, revenues, and profit per hectare, is several times greater in Spain than in California. All cost and revenue figures from this section have been converted to 2011 US dollars.

11.3.3.1 Costs

Tables 11.5 and 11.6 illustrate a sample of the differences in costs between a case study of a fenced Spanish hunting property and an average of a subset of 41 ranches in California with >10 % oak cover. These tables should only be compared generally and with caution as different accounting methods were used.

Personnel wages and vegetation management expenses account for a significant portion of total costs on fenced Spanish properties, while administrative and infrastructure costs are significantly higher in California. Economic data from unfenced properties in Spain is unavailable, but would presumably be considerably lower, especially in personnel costs, as there is less need for hired hands to census and cull the red deer population and less need of a guard to protect game animals, as there are fewer valuable trophies present (Campos et al. 1995) (Tables 11.5 and 11.6).

11.3.3.2 Revenue

Lenzano and Zamora (1999) estimate income from the sale of 76 positions at about $700 per person on an unfenced montería hunt, yielding $53,200 in revenue for a

Table 11.5 Costs of managing a hunting property in Spain. The Lenzano and Zamora case study of a 658 ha dehesa is one of the few in Spain that includes specific costs of operating a hunting estate, summarized in the table below; dollars normalized to 2011 (Lenzano and Zamora 1999; BLS 2012)

Costs	2011 dollars/ha	Percent of total (%)
Guard	26.5	50.3
Dog packs—Rehalas	6.5	12.4
Bidders	1.4	2.8
Food	1.4	2.6
Horses/Mules/Oxen	1.3	2.5
Vehicles	1.0	1.9
Veterinarian	0.9	1.7
Guides	0.6	1.1
Personnel cost subtotal	39.6	75.3
Brush clearing	8.3	15.7
Pruning trees	0.4	0.7
Enhancing oak stands	0.2	0.3
Vegetation management cost subtotal	8.8	16.7
Housing (each year)	1.4	2.8
Roads	0.8	1.4
Ponds	0.5	1.0
Water wells (each year)	0.3	0.5
Fences (each year)	0.2	0.3
Infrastructure costs subtotal	3.2	6.1
Taxes on Luxuries	0.6	1.1
Renewal of Registration	0.4	0.8
Montería Hunting Permit	0.0	0.1
Administrative costs subtotal	1.0	1.9
Total Costs/ha	52.6	100.0

red deer montería. Added income of $5,000 can come from on-site wild boar, with $5,000 from the sale of venison and boar meat after the hunt (Lenzano and Zamora 1999). This leads to a total revenue stream of $63,200 resulting in approximately $96/ha in revenue (income before costs). The value of a montería position can vary from $500 to $6,000, with most montería positions costing $800–1,200 (Campos et al. 1995). The sale of live deer for restocking dehesa properties can offer an additional source of income. If the value of live animals exceeds the meat value, landowners may find it more profitable to capture and sell live animals instead of culling their deer herd (Vargas et al. 1995).

In California, research has shown that hunter success rate, percentage of trophies, and the wealth of hunting participants added to revenues (Loomis and Fitzhugh 1989). Additional analysis of the data shows that proportion of a ranch with oak habitat significantly improved revenue in California hunting operations. Ranches in the study on average earned $19.54/ha, while those with 10 % or more

Table 11.6 Average costs of 41 Ranches in California with hunting and ≥ 10 % oak woodland cover [adapted from data from (Loomis and Fitzhugh 1989; Standiford 1989)]

Cost	2011 dollars/ha (BLS)	Percent of total (%)
Non-professional wages: cooks, operators, etc.	1.5	22.7
Guides	0.7	9.7
Accountants	0.1	1.9
Attorney and consultants	0.1	1.5
Personnel wages subtotal	2.4	35.8
Vegetation management	0.1	2.0
Vegetation management cost subtotal	0.1	2.0
Vehicles: maintenance	0.3	4.5
Vehicles: depreciation	0.3	4.5
Irrigation water	0.2	2.9
Improvements (roads fences and water): maintenance	0.2	2.6
Phone	0.2	2.4
Road construction	0.1	2.0
Ranch equipment: depreciation	0.1	1.8
Other annual costs	0.1	1.7
Ranch equipment: maintenance	0.1	1.2
Supplemental feeding	0.1	1.0
Improvements (roads fences and water): depreciation	0.1	1.0
Predator control	0.0	0.7
Other costs (gas, electric, structure maintenance)	0.4	5.3
Infrastructure/Maintenance costs subtotal	2.3	32.3
Liability insurance	1.0	15.0
Lease of hunting rights on other lands	0.5	7.0
Advertising	0.5	6.9
License, permit, and legal fees	0.1	1.7
Administrative costs subtotal	2.1	30.5
Total costs	6.8	100

oak woodland cover received \$21.09, and those with 15 % cover or more earned \$22.47/ha on average (calculated from data in Standiford 1989).

11.3.3.3 Net Profit

A significant issue to consider is how costs factor into final profitability calculations. Some costs, as with water provisioning, are undertaken for livestock, yet benefit wildlife. Comparisons of net profit among hunting operations is difficult. While accounting differences make a detailed comparison impossible, the profitability of Spanish dehesa hunting properties appears significantly higher than an equivalent California site.

Spanish studies show net profit (owner hunting net operating margin) of $23.80–72.47/ha (Campos Palacín et al. 2001), while other studies tally net losses when costs such as capital depreciation are incorporated. Lenzano and Zamora (1999) suggested that by using a third party intermediary a landowner could receive a return of $6/ha, but under self-management, profitability increased to $37/ha in an unfenced property, with increases up to $42–63/ha for fenced properties while amortizing fencing investments. Net profit in California is estimated at $5.68–7.31/ha. There is significant variation in profitability levels, with 30 out of 41 ranches operating profitably, and 11 ranches taking losses (calculated from data in Standiford 1989).

While data suggests Spanish dehesas may earn 10 times as much from hunting than California ranches, these figures should be compared with caution given differences in accounting for costs and revenues (Standiford and Howitt 1993).

11.4 Wildlife Management

Wildlife management can generally be divided into two major types:

1. Top-down population regulation: actions taken by humans that directly impact the movement and population composition of game species and predators, such as harvest, culling, fencing, predator control and restocking; and
2. Bottom-up management practices: actions taken by humans that influence or improve resources needed by wildlife. These are generally improvements to food, cover, and water resources for the benefit of wildlife species.

11.4.1 Top-Down Population Regulation: Harvest, Fencing, Restocking, and Predator Control

11.4.1.1 Harvest Methods and Regulations

A major component of any wildlife management program is managing the harvest of animals by hunting. Harvest rates are impacted by the method of hunting, bag limits, and the age and sex of animals harvested. Distinct regulations in Spain and California govern each of these aspects of hunting. California's hunting methods are far more restrictive (Table 11.7).

Table 11.7 Hunting methods in Spain and California

Hunting method	Spain	California
Big game		
Spot and stalk, i.e. rececho	X	X
Stand hunting, i.e. aguardo, espera	X	X
Montería	X	
Gancho	X	
Hunting over bait	X	
Hunting at night	X	
Small game		
Opportunistic, i.e. la caza al salto	X	X
Walking lines, i.e. ojeo	X	X
Driven shooting	X	
Live decoy hunting	X	
Use of sighthounds (galgos)	X	

Big Game

Common to Spain and California are the spot and stalk (*rececho*) and stand-hunting methods. In spot and stalk hunting, hunters use binoculars to spot game animals from a vantage point where a large area of terrain can be surveyed. They then stalk the animal to get close enough to take a shot. In Spain, rececho is the most expensive hunting method for a hunter to undertake, and is usually used to harvest a single trophy animal. In California it is one of the primary ways to hunt, and generally costs no more than hunting from a blind or stand. Stand hunting involves placing hunters in a concealed or elevated position (the stand) where they wait for an animal to approach. Stand hunting (*aguardo* or *espera*) is often done over bait in Spain, may be performed at night, and is generally used for selective wild boar hunting. In California, stand hunting is used by hunters who wait in stations above locations frequented by game, or they may wait over grain fields, since hunting over bait is illegal.

Unique to Spain is the montería, where 50–75 hunters take up shooting positions in fixed locations on a property for a single day of the year. The *rehaleros* or beaters move into the lower reaches of the property with one or more *rehalas* or *recovas*, a group of 20–25 hounds, and chase animals from hiding in the often-dense brush and into the open where hunters shoot the game, often on the move. The gancho is a small-scale montería, usually focused on wild boar. In California dogs can be used to hunt non-native wild pigs, but the method usually involves following dogs to the game rather than dogs driving animals to hunters waiting in designated locations.

Small Game

Small game methods of hunting are generally similar in Spain and California, although Spanish hunters may use packs of dogs, as in big game hunting, to drive

Fig. 11.10 Dogs are an essential element in the drive hunt, or montería, of Spain. (Photograph by P. F. Starrs)

Fig. 11.11 Dogs await transport and release with eager anticipation before a hunt in the Sierra Morena. (Photograph by L. T. Macaulay)

small game toward hunters (Vargas et al. 2006). With rabbits, hunters can utilize methods of "la caza al salto," which roughly translates as "hunting on the jump," where a solitary hunter with or without dogs walks an area and flushes rabbits. Additionally, rabbit hunting can be performed with sighthounds that rely on vision and speed to hunt. Spanish hunters usually use the *galgo*, a breed similar in appearance to the greyhound (Figs. 11.10 (at left) and 11.11).

California considers rabbits, jackrabbits, and squirrels as small game, and generally classifies game birds including quail, dove, and wild turkey as upland game. In California, small game mammal hunting is practiced with and without dogs, but lacks the widespread popularity and cultural significance of hunting rabbits and hares in Spain. Blogger Hank Shaw describes a prevalent attitude toward jackrabbits: "Most "normal" hunters wouldn't waste a shell on hares…. [T]hese folks view jackrabbits as beneath them" (Shaw 2008). Small game mammal hunting is generally practiced by youth or opportunistically as a secondary quarry, when hunting for other game.

Upland game bird hunting of quail and turkey generates more hunter interest than small mammal hunting (Table 11.3). Live decoys are illegal in upland game bird hunting in California, although artificial decoys are utilized, particularly for turkey hunting. Quail in California are generally hunted by individuals or small groups of hunters. The practice of walking lines is less productive in California quail hunting, given the prey's preference for thicker habitat cover and the pattern in which they flush. Hunters will use dogs to locate and retrieve birds, but rarely to drive the animals. Smaller bag limits for wild turkey result in lower harvest numbers, which belie the significant interest in wild turkey hunting in California, particularly during the spring mating season, when male turkeys are lured within shotgun range by mimicking the call of a female hen.

Bag Limits and Antlerless Harvest

California's Department of Fish and Wildlife generally imposes a bag limit of two deer, and various limits on small game. Because wild pigs are nonnative in California and have high fecundity, there is neither a closed season or bag limit. In Spain, each regional government, or autonomía, sets its hunting regulations. Generally, there is no limit on the big game species, but small game species have various limits.

While Spanish landowners generally control the age and sex of harvest on big game on their properties, the California Department of Fish and Wildlife retains control over bag limits and the age and sex of harvests. Due to political controversy dating back to the state's first antlerless harvest in 1956, regulatory authorities prohibit female and juvenile male harvest of deer in most parts of the state (Fitzhugh 1989). This has led to high hunting pressure on male deer and sex ratios that, while variable, are oftentimes far less than 0.5 bucks-to-doe ratio, with some estimates as low as 0.05 bucks per doe (CDFG 2012e). Despite research that shows density-dependent responses in California deer and increased buck harvests as a result of removing female deer, political opposition persists, and today the California Department of Fish and Wildlife (prior to 2013, the California Department of Fish and Game, or CDFG) describes California as having the "dubious distinction" of being the only state in the United States where doe hunts cannot be carried out even when such hunts are biologically justified (McCullough 2001; CDFG 2002).

A similar tradition of not harvesting female red deer historically existed in Spain. This was rooted in a belief that hunting females would drive populations to local extinction, which was occurring just after the Civil War in the late 1930s and 1940s. However, after the hunting law in 1970 with populations increasing for most Spanish big game species, management policies embraced the need to regulate population size and the ratio of males to females to improve game animal condition, increase the value of trophies, and reduce the impact on vegetation and the risk of traffic collisions (Carranza 2010).

11.4.1.2 Fencing

The most significant wildlife management consideration on Spanish hunting lands is whether a property is fenced. On unfenced (open) properties, traditional agricultural practices and livestock are usually the main productive activities and big game hunting is of secondary importance (Vargas et al. 1995). Given a propensity of wildlife to stray onto adjacent properties, unfenced lands generally have low intensity management, since only very large properties or those in cooperative agreements with neighbors can reap the benefits of targeted harvest and habitat improvement strategies. These smaller unfenced properties tend to manage for a greater number of wildlife than for high quality trophies. Unfenced hunting lands have been documented to have a 0.25 male-to-female sex ratio, and an age structure where over 45 % of the males are juveniles (Pérez-González and Carranza 2009; Torres-Porras et al. 2009).

Fenced properties require significantly different management practices than open properties because the wildlife cannot escape the property, allowing managers to more accurately census the population of game species and control the harvest. This includes knowing the sex ratio and age structure of the population, which allows a landowner to practice much more effective population management through selective hunting, allowing for more stags to mature to trophy quality and increasing the overall number of valuable males in the population.

Although fenced properties may entail significant benefits for a landowner in terms of quality and quantity of trophies, and can result in a wildlife population with more natural characteristics and age structure, their management requires resources to maintain healthy wildlife populations and habitats. This is particularly true in the Spanish Mediterranean ecosystem known for high variability in rainfall and in resource production. A landowner must maintain the fence itself, ensure a water supply within the fence, and provide adequate forage for the animals within the property to prevent excessive grazing and habitat degradation in times of drought. Many fenced properties cull to reduce population size, particularly of females and of smaller, non-trophy males to counteract selective pressures of trophy hunting for large males (Martínez et al. 2005; Mysterud and Bischof 2010).

Fig. 11.12 Game fences represent a significant act of landowner control over a dehesa property. Fences allow game species containment and increased profits by intensified management. The problems associated with fenced properties made this particularly controversial when game fences first arrived, but the fences are now very common. (Photograph by P. F. Starrs)

Dehesa properties may also need to provide supplemental summer feed, especially in times of drought, to maintain a wildlife population that cannot migrate off the property to find food. Furthermore, gamekeepers and guards are required to police a property and prevent poaching of valuable trophy animals.

As with other aspects of Spanish hunting management, fencing regulations are controlled by regional governments. Most set a minimum size and enforce permeability restrictions on fencing. In southwestern Spain a big game hunting operation has a minimum area requirement of 700 ha—also the minimum size for a fenced estate. The tendency, however, is to enclose larger areas. Experts in Andalucía recommend that properties smaller than 2,000 ha obtain a certificate of quality as a prerequisite for fencing (Carranza and Vargas 2007; AAMAA 2008). Fences must have regular openings to allow the movement of non-game species such as the protected Iberian lynx (*Lynx pardina*). Fences retain deer species better than wild boar, which easily burrow under a fence. Enclosures for wild boar (*cercones*) are in general prohibited for their association with unnatural, high-density management, which is linked to an increase of diseases. However, these do occur and their numbers are increasing (Fig. 11.12).

11.4.1.3 Restocking and Farmed Game Species

Restocking and farming of game species plays a major role in both small and big game hunting in Spain (Barbosa et al. 2004), but in California the practice is utilized almost exclusively with exotic game birds, notably the ring-necked pheasant. In Spain, red deer, red-legged partridges, rabbits, roe deer and wild boar may be restocked, with red-legged partridges the most common among farmed game species. Vargas et al. (2006) estimated that one-third of red-legged partridges harvested in Andalucía were farmed birds, usually released into the field prior to a hunt. For other species, restocking is less common, but can involve animals from game farms, or from live captures in other areas, which are generally used to re-establish or reinforce natural populations.

For stocking game species in Spain, a permit is needed to move game animals and the approval depends on regional policy. Generally, only native species or subspecies can be transported, although enforcement of this regulation varies. There is no current regulation against the release of animals from game farms, provided that they belong to native varieties of the region where they are going to be released, although effective enforcement is not common. There is a growing tendency toward limiting import, export, transfer, or release of live game.

In California, three-quarters of exotic ring-necked pheasant harvested are from stocked game bird farms (CDFG). For native big game species, restocking is illegal unless undertaken by California Department of Fish and Wildlife to re-establish game species (CFGC). The state does license a number of game bird farms that exclusively utilize farmed birds.

11.4.1.4 Predator Control

California permits predator control on a number of species. Steel-jaw leg traps and poison have been banned, but box traps, padded leg traps, and snares are legal subject to their being checked every 24 h. Mountain lions (*Puma concolor*) are legally protected by a statewide voter referendum in 1979, and predator control of these populations is illegal, except when lions are killed for public safety concerns or with depredation permits granted by the state. Many landowners and hunters blame this law for a decline in deer populations in California, although research suggests that a combination of factors, including habitat change, is affecting deer populations (Longhurst et al. 1976; Kucera and Mayer 1999). Bobcats (*Lynx rufus*) can be harvested with the purchase of a special tag and a hunting license. There is no legal season on red foxes (*Vulpes vulpes*). Gray foxes (*Urocyon cinereoargenteus*) are considered a furbearer species with a separate season (Fig. 11.13).

In Spain, landowners practice predator control most intensively for predators of small game such as the red fox and the common magpie (*Pica pica*). While there is intensive effort to control red fox populations, studies have found that red fox populations are rarely controlled by these measures, and that oftentimes other non-target predator species may be most greatly impacted (Virgós and Travaini 2005).

Fig. 11.13 Not all hunters are older, or male. This wild boar eventually gained the young huntress an award for taking a silver medal specimen, in a hunt north of Córdoba. It was not her first hunt. (Photograph by P. F. Starrs)

The Iberian wolf (*Canis lupus signatus*) is the only predator that can significantly impact red deer populations. Wolf populations are increasing in central and northern Spain, while in southern Spain they appear to be declining. In Extremadura, wolves disappeared in the early 1990s (Rico et al. 2000), and in Andalucía there is only a small breeding population in the Sierra Morena. Authors describe an antagonistic relationship of rural landowners with wolves and some illegal killing may occur, but disease and traffic fatalities likely account for more deaths than illegal hunting or predator control (Rico et al. 2000). Ultimately, the wolf's range is so limited today that wolves are rarely encountered on dehesas (Blanco and Cortés 2002).

11.4.2 Bottom-Up Practices: Shrub Management, Food Plots, Supplemental Feed, and Water

In California and Spain, landowners practice a variety of habitat management actions to improve the food, cover and water needs for game species, which we will refer to as bottom-up management practices. In both California and Spanish oak woodlands this generally involves clearing understory brush, planting food plots, and providing water sources. In Spain, supplemental feeding is common (Fig. 11.14).

Fig. 11.14 Much of the dehesa is given to occupation by brush, if not regularly cleared of its aromatic understory vegetation. Sometimes, however, to give access to hunters, strips (or *manchas*) will be cut into the understory, as in this aerial photograph near Hornachuelos, Córdoba. Oaks were carefully left behind, in keeping with legal strictures. (Photograph by P. F. Starrs)

11.4.2.1 Shrub Management

While the level of shrub, or brush, presence and growth in a given area is dependent on a multitude of factors such as grazing, soil characteristics, topography, fire history, and climate, many landowners will clear brush to improve herbaceous production or to stimulate fresh and more palatable brush growth. While some brush species provide important cover and forage for ungulates and other wildlife, others provide little nutritional value for game species and can reduce the overall forage production for game and livestock. California deer prefer foraging on brush during the spring when new growth appears (Evans et al. 1976). Palatable species include manzanita (*Arctostaphylos* spp.), chamise (*Adenostoma fasciculatum),* ceanothus (*Ceanothus* spp.), and oaks (*Quercus* spp.) (Sampson 1963). A less palatable species rarely eaten by deer in California is coyote brush (*Baccharis* spp.). Palatable brush species in Spain include species of phyllirea (*Phyllirea* spp.), madrone (*Arbutus* spp.) and mytle (*Myrtus* spp.), while less palatable species include rockrose (*Cistus* spp.), heather (*Erica* spp.), and mastic (*Pistacia lentiscus)* (Rodríguez Berrocal 1993; Bugalho and Milne 2003). Many of these brush species are highly resilient and persistent in oak woodlands, and often return at various time intervals after clearing. Rockrose, *Jara pringosa* (*Cistus ladanifer*), will return aggressively 5–8 years after it is cleared. Even brush species that provide important forage for wildlife grow into tall thickets beyond the reach of deer, so some managers clear brush to encourage herbaceous production and new brush growth (Sampson 1963). After clearing, California landowners may retain brush piles as habitat, in particular for quail and a variety of other game and nongame wildlife (Gorenzel et al. 1995). In Spain, brush piles may shelter partridges and rabbits, but creating brush piles is not a widespread practice.

In California, ranchers will sometimes use tracked vehicles to drag a large chain with 60-pound links across brush areas to remove brush and enhance habitat for deer. Brush-clearing costs can be substantial. One study in Andalucía found that 17 % of expenses on a hunting property were devoted to clearing brush (Lenzano and Zamora 1999), while marginal expenses were devoted to pruning and enhancing oak cover. Discing every 4–6 years is the most commonly used practice to limit brush encroachment with hand-grubbing of brush and grazing utilized to reduce the frequency of follow-up discing treatments (Huntsinger et al. 1991).

Traditional management of the dehesa involves clearing all brush species, leaving older oaks interspersed with herbaceous vegetation. While this improves livestock feed and some agricultural returns, recent research suggests that maintaining a mosaic of brush species in dehesa habitat is ideal for game species (Carranza 2010). Leaving a mosaic of brush in the dehesa contributes to natural regeneration of oak trees (Chaps. 5, 8), adds to biodiversity and provides an increasingly important component of the diet of big game species during the summer drought period (Bugalho and Milne 2003; Plieninger et al. 2003; Díaz 2009; Carranza 2010).

While controlled burning can produce similar benefits, it is rarely used in Spain because escaped fires could harm valuable cork oak bark, and because arson historically was a means of expressing dissent with political authority by vandalizing the property of landowners who fence land (Huntsinger et al. 1991). In California prescribed burns are made difficult by air pollution regulations, which limit the days that a rancher can conduct a burn, and it is logistically difficult to coordinate fire-fighting personnel on short notice to reduce the liability risk of the fire escaping a property.

11.4.2.2 Food Plots

Ranchers in both Spain and California plant improved pastures for wildlife. In some places in California, this includes planting fields of barley or other grains to attract wild pigs. In Spain, land managers will sometimes sow fields to supplement the dietary needs of wildlife on the property, especially in closed or fenced properties.

11.4.2.3 Supplemental Feeding

Direct feeding of wildlife is generally illegal in California, although difficult to enforce. Several decades ago in Spain, supplemental feeding was a practice seen as beneficial to game and wildlife. Today, ecological knowledge and management policy generally discourages interfering with natural processes, although feeding threatened species is often justified and practiced. Recent decreases in population sizes of rabbits, for example, has resulted in supplementary feeding, although feeding to promote concentration of animals for hunting purposes is discouraged. For big game, supplementary food and making salt available is only allowed during extreme drought months in late summer. Many private land managers nonetheless provide food without control by authorities. For example, a wildlife study in Spain describes 23 t of feed being placed out for pig hunting on a 920 ha property (Braga et al. 2010).

11.4.2.4 Water Provisioning

Spanish and California landowners provide wildlife with supplemental water sources. While this historically was to serve livestock, wildlife species benefit. Water sources can range from stock ponds that collect and store surface water runoff to troughs that are filled with groundwater pumped from aquifers. Ranchers in California equip "wildlife friendly" water troughs with ramps to allow the escape of birds, bats, and other animals that may fall into the trough. Water troughs may be inset at near ground level for ease of access of game species. Ranchers in drier areas of California's oak woodlands may construct "guzzlers" designed to provide water for small or big game species (Bleich et al. 2005). In Spain, the most common and traditional practice is to build stock ponds that fill with water pumped from a shallow aquifer or from surface runoff. These are designed to maintain water storage until the end of the summer dry period.

11.5 Environmental Effects

There are many positive environmental effects from management for hunting. Most notably, wildlife habitat is conserved in a relatively undisturbed state, especially when compared to conversion of oak woodlands to intensive agriculture or residential development. Because habitat loss is the greatest threat to biodiversity worldwide, this conservation of open areas provides critical habitat for many species in biodiversity hotspots (Brooks et al. 2002). Standiford and Howitt (1992, 1993) have demonstrated that incorporation of hunting revenue provides an incentive to retain oaks on woodland properties. The economic return from hunting ventures reduces the economic incentive to use the oak woodlands in less sustainable ways, resulting in greater habitat conservation and increased provisioning of ecosystem services including the following:

- Habitat conservation
- Reduce opportunity cost for habitat conservation

Table 11.8 A generalized overview of the complex mix of environmental effects of wildlife management actions

Environmental Effect	Wildlife Management Action
Animal Density	Fencing
Overgrazing and Vegetation Impacts	Feeding
Disease Impacts	Game Farming/Translocations/Harvest
Fragmentation	Fencing
Genetic Impacts	Fencing
Domestication	Feeding
Changed mating behaviors	Harvest
Hybridization	Game Farming/Translocations
Inbreeding	
Genetic drift	
Artificial selection and breeding	
Loss of local genetic adaptations by introgression	
Population Structure	Harvest
Age structure	
Sex ratios	
Altered dispersal patterns	
Erosion and water quality	Fencing
	Feeding
	Brush management
Invasive species	Brush management
	Discing
	Food plots
	Game farming/Translocations
Impacts on protected or non-target species	Predator control

- Nutrient cycling
- Provisioning of food (game, livestock, crops)
- Water filtration
- Air filtration
- Carbon sequestration
- Aesthetic value
- Recreational experiences
- Cultural significance
- Spiritual inspiration

Management actions for hunting have particular impacts on the environment, and our focus in this section will evaluate those effects. The impacts can be described by two mechanisms: (1) the effects of people on an environment through wildlife management actions, and (2) the effects of game animals on their habitat, which is largely determined by the species present and animal density. These are inextricably linked and will be discussed in an integrated way.

A brief and generalized overview of environmental impacts of management actions is classed by management action and environmental effect in Table 11.8.

Much of the discussion will focus on environmental effects in Spain due to the greater range of wildlife management practices legally available.

11.5.1 Animal Density: Overgrazing and Disease/Parasite Concerns

Management actions of fencing, feeding, game farming/restocking, under-harvest and excessive predator control can lead to high animal densities and increased risks of vegetation degradation and disease outbreak. Overgrazing can result in vegetation community simplification with a loss of palatable species, detrimental effects on biodiversity and ecosystem function, and erosion (Murden and Risenhoover 1993; Mysterud 2006; Acevedo et al. 2008). Risks of disease and parasitic outbreaks also increase with high animal density (Andrews 2002; Vicente et al. 2007; Castillo et al. 2011)

Disease emergence or increased disease risk is a frequent consequence of ungulate overabundance (Acevedo et al. 2008). Several studies show a prevalence of tuberculosis-like lesions in southern Spain, which is noted for its intensively managed and fenced hunting properties (Vicente et al. 2007). Recent research increasingly posits that although overabundance plays a significant role in disease and parasite transmission, management practices that promote the clumping of animals, even of different species, at water sites or supplementary feeding sites can exacerbate disease even at low animal densities (Vicente et al. 2007; Castillo et al. 2011).

11.5.2 Erosion and Water Quality

Erosion is a concern on oak woodlands because the loss of fertile topsoil can reduce the vegetative capacity of a property. High animal density resulting from fencing or feeding can lead to over-grazing, and loss of many plants that anchor soils into the ground increasing erosion risks and reducing water quality (Kauffman and Krueger 1984). Brush clearing practices can leave large areas of land without vegetative cover, increasing erosion risks and impacting water quality in riparian zones (Sampson 1963).

11.5.3 Fragmentation

Fences increase fragmentation of a landscape. While fences generally are designed only to prevent the movement of certain game species, they can make migration of other species more difficult and produce unwanted genetic impacts by separating animal populations. Furthermore, fences can increase predation at crossing points, increase mortality of birds and other animals that get caught in the fence, and can lead to animals exceeding carrying capacity of a landscape (Andrews 2002; Hayward and Kerley 2009).

11.5.4 Genetic Impacts of Fencing, Feeding, Harvest, and Restocking

Genetic impacts that result from management activities can include the domestication of wild species, changed mating behaviors, inbreeding, genetic drift, hybridization, artificial selection, and loss of local genetic adaptations.

Fencing isolates populations, encouraging both genetic drift and inbreeding. For red deer populations in southwestern Spain, fences encourage genetic differentiation between neighboring properties, accounting for some level of inbreeding (Martinez et al. 2002). But the full role of fences on inbreeding is complex. In the dehesa, fenced lands are some of the only areas that sustain a natural age structure of red deer populations. In unfenced areas, overhunting of males leads to female-biased populations with mostly young males (Pérez-González and Carranza 2009). Research has shown that transmission of genetic variability is compromised in the paternal lineage in open lands compared to fenced ones (Pérez-González and Carranza 2009). The result is that inbreeding does not differ on average between open and fenced properties (Martinez et al. 2002) and cases of extreme inbreeding are found even more easily in open lands (Pérez-González et al. 2010a, b). Inbreeding can lead to decreased survival, increased vulnerability to disease, reduced fitness, and decreased lifetime breeding success. Biologists recommend prioritizing strategies to minimize the effects of inbreeding (Mysterud and Bischof 2010).

Feeding tends to increase the gathering of animals and have potential effects on natural mating behaviors. In Spanish hunting properties with supplemental feeding, the placement of feed can be more important in female red deer aggregation during the breeding season than the distribution of natural resources (Pérez-González et al. 2010a, b). Males then shift their strategy from harem defense to territoriality around the presence of supplemental feed (Carranza et al. 1995). Supplemental feeding also tends to increase harem size, but brings a reduction in sexual harassment of females by males (Sánchez-Prieto et al. 2004). The evolutionary consequence benefits males who control the larger female harem groups (Carranza et al. 1995). In red deer populations with mature males (mostly fenced populations), supplementary feeding may further increase the degree of polygyny and affect male mating success (Pérez-González and Carranza 2011).

Harvest has a major impact on population density, age structure, and sex ratio, leading to selective pressures that influence the genetic makeup of game animals. Several studies discuss how harvesting trophy-quality game selects for smaller and potentially less fit animals (Martínez et al. 2005; Mysterud 2010). Compensatory culling targeting low-quality yearlings may successfully counter the selective effects of trophy hunting, although combinations of human mediated management practices may lead to semi-domestication (Torres-Porras et al. 2009; Mysterud 2010; Mysterud and Bischof 2010).

Game farming and restocking practices mix genetic lines and leads to hybridizing among game animals that used to be separated by significant geographic distance. These practices increase the risk of losing local adaptations due to genetic introgression from non-local populations. Indiscriminate translocation practices throughout Europe introduced genetic matrilines of Scottish red deer and Eastern European red deer in some Spanish red deer populations (Martinez et al. 2002; Fernández-García et al. 2006). Recent regulations have attempted to reduce this practice, requiring genetic tests of trophies prior to their entry into Spanish records to ensure they are not hybrids. Some regions of Spain have begun to require these genetic tests prior to authorizing translocations, and some landowners utilize these genetic tests to purge hybridization from their herds (Carranza et al. 2003; Carranza 2010).

Genetic impacts are not limited to big game species. The red-legged partridge faces threats from farm-raised animals, including parasite and disease transmission, and genetic dilution through introgression of farmed bird genes (Vargas et al. 2006; Blanco-Aguiar et al. 2008).

11.5.5 Population Structure

In California, buck-only harvest regulations result in female-biased sex ratios and the selection for smaller-antlered deer because larger-antlered males are often preferentially removed from the population. California has not implemented any compensatory culling scheme to reduce this effect. Bucks that lack forked antlers (with at least two points on each side) are illegal to harvest. While this is designed to prevent the harvest of one-year-old males, there are examples of 2–3 year old bucks that do not grow branched antlers being selectively protected and presumably allowed to breed. This artificially selects for deer whose antlers never mature beyond the size of a one-year-old buck, which runs counter to expected breeding fitness characteristics.

Harvest practices can impact the age structure and sex ratios of game species, leading to altered dispersal patterns (Pérez-González and Carranza 2009). The phenomenon of artificially skewed age and sex ratios is captured effectively by Pérez-González and Carranza (2009) who found that red deer sex ratios in fenced properties was approximately 0.79 males to females, while unfenced populations exhibited a 0.25 male to female ratio. They also found lower percentages of adult males in unfenced properties than fenced ones, at 54 versus 74 %.

The population characteristics of the unfenced properties described above are often the result of overharvest of adult males, which leads to reduced male mating competition. This altered population structure can affect genetic exchange between populations in unfenced areas. Pérez-González and Carranza (2009) found that low male mating competition changed natural dispersal patterns to female-biased dispersal instead of the expected male-biased dispersal for polygynous species such as the red deer.

A study comparing two harvest methods of wild boars, espera (stand hunting) vs. montería (driven hunts by dogs), showed that the montería was much less selective, yielding harvest of animals across all age classes and genders, while espera hunting was much more selective for mature males (Braga et al. 2010).

11.5.6 Invasive Species

Management actions that clear vegetation can create openings for invasive species. In California, perennial native grasslands are usually extirpated from an area after plowing, discing, or planting of food plots. Invasive weeds, such as yellow star thistle (*Centaurea solstitialis*), and non-native grasses are usually the dominant species that re-vegetate plowed areas.

Translocations of non-native game animals can lead to invasions by the game animals that outcompete native species or disrupt ecosystem function. This can spread exotic diseases or parasites to native species. Wild boars have spread throughout many oak woodlands in California, causing changes in the ecosystem (Wilcox and Van Vuren 2009). In Spain, the Barbary Sheep (*Ammotragus laervia*) has expanded in certain areas and can compete with native ungulates, in particular the Spanish ibex (*Capra pyrenaica*) (Carranza 2010). Translocation of rabbits has been implicated to some extent in the spread of both myxamatosis and rabbit hemorrhagic disease (RHD), which have decimated rabbit populations throughout

Fig. 11.15 So significant is hunter success in wild boar hunting in Spain that proud landowners will often create a trophy plaque showing the "defenses" of the boar, especially for a mature animal, as seen here at a dehesa in northern Córdoba. (Photograph by A. Caparrós)

Spain (Delibes-Mateos et al. 2008). Since the rabbit is the primary prey species for at least 29 top predators in Spain, its decline has led to declines in predator species, including the Spanish imperial eagle (*Aquila adalberti*) and the Iberian lynx, which are highly dependent on rabbits (Delibes-Mateos et al. 2008) (Fig. 11.15).

11.5.7 Protected and Non-Target Species

Predator control in California's oak woodlands does not currently threaten any endangered species, although historical hunting and culling of predators such as the gray wolf (*Canis lupus*) and the grizzly bear (*Ursus arctos horribilis*) extirpated those species from the state. While intensive predator control was historically a part of California's management for hunting and livestock, recent regulations banning poisons and limiting methods of trapping have decreased the impact of predator control. Mountain lions are protected, and while ranchers can obtain a permit to kill a mountain lion that has preyed on livestock, the administrative process may take several days, during which time the mountain lion often moves far from a kill site.

Spain's predator control practices appear somewhat detrimental to non-target species. While the target of predator control tends to be the red fox or the common magpie, many other species are affected by illegal or non-selective predator control. Studies have found declines in populations of the common kestrel (*Falco tinnunculus*) and the common genet (*Genetta genetta*) on hunting properties where intensive predator control is practiced (Beja et al. 2009). Additional research shows that non-selective control measures of box traps, snares, and illegal poison that is intended for red foxes may not significantly impact fox populations, but in fact adversely affect species such as the badger (*Meles meles*), wildcat (*Felis silvestris*), and stone marten (*Martes foina*), which are unable to cope with intensive predator control (Virgós and Travaini 2005).

11.6 Conclusions

The oak woodlands of Spain and California are ecologically similar, but significant disparities in history and governance yield dramatically different hunting systems on the two sites. These distinctions are manifest in the methods of hunting, the economic return to landowners, game management practices, and the environmental impacts of such management.

Income from hunting can provide an economic incentive to maintain areas of relatively undisturbed wildlife habitat. Properties that earn revenue from hunting are more common in Spain and often earn more than those in California. Current law and economic conditions in Spain make it favorable for landowners to continue to utilize areas of dehesa for recreational hunting, and reinforces a trend away from traditional livestock operations, brush clearing, and agricultural use that were the norm even a decade or two ago. However, while properties managed for income as hunting reserves may be profitable (Chap. 13), they are not lucrative enough to entirely replace the traditional multiple uses of dehesa habitat. In California, historical precedent, widespread public lands, lower hunting participation rates, limitations on commercialization of wildlife, and increasing restrictions on hunting and wildlife management practices all serve to reduce the amount ranch owners earn from hunting. Nonetheless, the local food movement and the current interest in wilderness experiences are shaping a new narrative about hunting as an ecologically sound way to connect to one's food source while also obtaining organic meat. Even though hunting can provide sustainable economic return on dehesas and wooded oak ranchlands, uninformed wildlife management practices that are too narrowly focused on game animals can cause environmental degradation (Díaz et al. 2009). However, thoughtful management practices that seek to improve habitat for a variety of species can not only maintain but improve environmental values on these properties.

References

AAMAA (Asociación de Agentes de Medio Ambiente de Andalucía) (2008) Blog oficial de la Asociación de Agentes de Medio Ambiente de Andalucía (AAMAA): descargas. http://www.aamaa.info/p/descargas.html#caza. Accessed Feb 2012

Acevedo P, Ruiz-Fons F, Vicente J, Reyes-García A, Alzaga V, Gortázar C (2008) Estimating red deer abundance in a wide range of management situations in Mediterranean habitats. J Zool 276:37–47

Aguayo M (1986) Relatos de Caza. Anonymous Publicaciónes del Monte de Piedad y Caja de Ahorros de Córdoba. Córdoba, Spain

Almazán DD (1934) Historia de la Montería en España. Instituto Gráfico Oliva de Vilanova, Madrid

Andrews L (2002) Biological and social issues related to confinement of wild ungulates. Wildlife Society, Bethesda, Maryland

Barbosa AM, Vargas J, Farfán MA, Real R, Guerrero JC (2004) Caracterización del aprovechamiento cinegético de los mamíferos en Andalucía. Galemys 16:41–59

Beja P, Gordinho L, Reino L, Loureiro F, Santos-Reis M, Borralho R (2009) Predator abundance in relation to small game management in southern Portugal: conservation implications. Eur J Wildl Res 55:227–238

Blanco JC, Cortés Y (2002) Ecología, censos, percepción y evolución del lobo en España: análisis de un conflicto. Sociedad Española para la Conservación y Estudio de los Mamíferos (SECEM), Spain

Blanco-Aguiar JA, González-Jara P, Ferrero M, Sánchez-Barbudo I, Virgós E, Villafuerte R, Dávila J (2008) Assessment of game restocking contributions to anthropogenic hybridization: the case of the Iberian red-legged partridge. Anim Conserv 11:535–545

Blas Aritio L (1974) Guia de Campo de los Mamiferos Espanoles (Para Cazadores y Amantes de la Naturaleza). Omega, Barcelona

Bleich VC, Kie JG, Loft ER, Stephenson TR, Oehler Sr MW, Medina AL (2005) Managing rangelands for wildlife. Tech Wildl Investig Manag:873–897

BLS (Bureau of Labor Statistics) (2012) CPI Inflation Calculator. http://data.bls.gov/cgi-bin/cpicalc.pl?cost1=1&year1=1995&year2=2011. Accessed Feb 2012

Braga C, Alexandre N, Fernández-Llario P, Santos P (2010) Wild boar (*Sus scrofa*) harvesting using the espera hunting method: side effects and management implications. Eur J Wildl Res 56:465–469

Brooks TM, Mittermeier RA, Mittermeier CG, Da Fonseca GAB, Rylands AB, Konstant WR, Flick P, Pilgrim J, Oldfield S, Magin G (2002) Habitat loss and extinction in the hotspots of biodiversity. Conserv Biol 16:909–923

Bugalho MN, Milne JA (2003) The composition of the diet of red deer (*Cervus elaphus*) in a Mediterranean environment: a case of summer nutritional constraint? For Ecol Manage 181:23–29

Campos Palacín P, Caparrós Gass A, Rodríguez Jiménez Y (2001) Towards the dehesa total income accounting: theory and operative Monfragüe study cases. Sist Recur For Fuera de Serie n, Invest Agr, pp 1–2001

Campos P, Vargas J, Calvo J (1995) Gestión económica del ciervo en ambiente mediterráneo. Trabajo integrado en el Proyecto CAMAR: CT90–28 Análisis técnico y económico de sistemas de dehesas y de montados

Carranza J (2010) Wild ungulates in Spain. In: Apollonio M, Andersen R, Putman R (eds) European ungulates and their management in the 21st century. Cambridge University Press, UK

Carranza J, Vargas JM (2007) Criterios para la Certificación de la Calidad Cinegética en España. Universidad de Extremadura, Spain

Carranza J, Garcia-Muñoz AJ, de Dios Vargas J (1995) Experimental shifting from harem defence to territoriality in rutting red deer. Anim Behav 49:551–554

Carranza J, Martinez J, Sanchez–Prieto C, Fernandez-Garcia J, Sanchez-Fernandez B, Alvarez–Alvarez R, Valencia J, Alarcos S (2003) Game species: extinction hidden by census numbers. Anim Biodivers Conserv 26:81–84

Castillo L, Fernández-Llario P, Mateos C, Carranza J, Benítez-Medina J, García-Jiménez W, Bermejo-Martín F, Hermoso de Mendoza J (2011) Management practices and their association with Mycobacterium tuberculosis complex prevalence in red deer populations in Southwestern Spain. Prev Vet Med 98:58–63

CDF-FRAP (California Department of Forestry and Fire Protection-Fire and Resource Assessment Program) (2003) The changing California forest and range 2003 Assessment. http://frap.cdf.ca.gov/assessment2003/. Accessed Feb 2012

CDFG (California Department of Fish and Game) (2002) Guide to hunting deer in California. http://www.dfg.ca.gov/wildlife/hunting/deer/publications.html. Accessed Feb 2012

CDFG (California Department of Fish and Game) (2012) Game take hunter survey reports. http://www.dfg.ca.gov/wildlife/hunting/uplandgame/reports/surveys.html. Accessed Apr 2012

CDFG (California Department of Fish and Game) (2012a) Wild pig management program—take index. http://www.dfg.ca.gov/wildlife/hunting/pig/takeindex.html. Accessed Feb 2012

CDFG (California Department of Fish and Game) (2012b) Tag and season date information—hunting—deer management program. http://www.dfg.ca.gov/wildlife/hunting/deer/tags/index.html. Accessed Feb 2012

CDFG (California Department of Fish and Game) (2012c) License statistics. http://www.dfg.ca.gov/licensing/statistics/. Accessed Feb 2012

CDFG (California Department of Fish and Game) (2012d) SHARE program. http://www.dfg.ca.gov/wildlife/hunting/share/. Accessed Feb 2012

CDFG (California Department of Fish and Game) (2012e) Deer management program—zone specific information. http://www.dfg.ca.gov/wildlife/hunting/deer/zoneinfo.html. Accessed Feb 2012

CFGC (California Fish and Game Commission) (2012) Regulations. http://www.fgc.ca.gov/regulations/current/. Accessed Feb 2012

Clark CW (2010) Mathematical bioeconomics: the mathematics of conservation. Wiley, Hoboken, NJ

Delgado F, Muñoz R (1982) Los Libros de Caza de la Biblioteca del Palacio de Viana: Estudio Bibliográfico. Publicaciónes de la Obra Cultural de la Caja Provincial de Ahorros de Córdoba. Córdoba, Spain

Delibes M (1982) El Libro de Caza Menor. Ediciónes Destino, Barcelona

Delibes-Mateos M, Ferreras P, Villafuerte R (2008) Rabbit populations and game management: the situation after 15 years of rabbit haemorrhagic disease in central-southern Spain. Biodivers Conserv 17:559–574

Díaz M (2009) Biodiversity in the dehesa. In: Mosquera MR, Rigueiro A (eds) Agroforestry systems as a technique for sustainable land management pp 209–225. Programme Azahar, AECID

Díaz M, Campos P, Pulido FJ (2009) Importancia de la caza en el desarrollo sustentable y en la conservación de la biodiversidad. In: Sáez de Buruaga, M, Carranza J (coords). Gestión Cinegética en Ecosistemas Mediterráneos, 21–33. Consejería de Medio Ambiente de la Junta de Andalucía, Sevilla

Evans CJ, Haines RD, Chesemore DL (1976) Winter range food habits of the north kings deer herd. Cal-Neva Wildl Trans 1976:25–37

Fernández-García JL, Martínez JG, Castillo L, Carranza J (2006) Phylogeography of Iberian red deer populations and their relationships with main European lineages. In: Proceedings of 6th international deer biology congress

Fitzhugh EL (1989) Innovation of the private lands wildlife management program: a history of fee hunting in California. Trans West Sect Wildl Soc 25:49–59

Garrido Martín JL (2011) Estimación de aprovechamientos cinegéticos en España por especies y comunidades autónomas. Caza menor y caza mayor. Real Federación Española de Caza. http://www.fecaza.com/. Accessed Feb 2012

Geist V (1988) How markets in wildlife meat and parts, and the sale of hunting privileges, jeopardize wildlife conservation. Conserv Biol 2:15–26

Gibert Buch J (1975) Perros de Caza en Espana: Razas, adiestramiento y aplicación, Pulide

Goldstein J (2010) The new age cavemen and the city. New York Times, 8 Jan 2010, ST1

Gómez Mendoza J (1992) Ciencia y política de los montes españoles (1848–1936). Icona, Madrid

Gómez Mendoza J, López Ontiveros A (2001) Montes y caza. In: Gil Olcina A, Gómez Mendoza J (coords) Geografía de España, pp 405–422. Ariel, Barcelona

Gorenzel WP, Mastrup SA, Fitzhugh EL (1995) Characteristics of brushpiles used by birds in northern California. Southwest Nat 40:86–93

Hayward MW, Kerley GIH (2009) Fencing for conservation: restriction of evolutionary potential or a riposte to threatening processes? Biol Conserv 142:1–13

Heyser H (2012) Norcal cazadora. http://norcalcazadora.blogspot.com/#0. Accessed 22 Feb 2012

Huffman JE, Wallace JR (2011) Wildlife forensics: methods and applications. Wiley, Oxford

Huntsinger L, Bartolome JW, Starrs PF (1991) A comparison of management strategies in the oak woodlands of Spain and California, pp 300–306. USDA For Serv Gen Tech Rep PSW–126

Huntsinger L, Buttolph L, Hopkinson P (1997) Ownership and management changes on California hardwood rangelands: 1985–1992. J Range Manage 50:423–430

IAFWA (International Association of Fish and Wildlife Agencies) (2002) The economics of hunting in America. IAFWA, Washington, DC

Kauffman JB, Krueger WC (1984) Livestock impacts on riparian ecosystems and streamside management implications… a review. J Range Manage 37:430–438

Kellert SR (1980) Attitudes and characteristics of hunters and antihunters. Trans 43rd NA Wildl Conf 1:87–119

Kucera TE, Mayer KE (1999) California. Dept of Fish and Game, Mule Deer Foundation (Reno, Nev.) University of California, Berkeley. Division of Ecosystem Sciences A sportsman's guide to improving deer habitat in California. Calif Dept Fish Game, California

Lenzano R, Zamora R (1999) Census methods and big-game optimization in Sierra Morena (Cordoba, Spain). Invest Agr: Sist Recur For 8:241–262

Longhurst WM, Garton EO, Heady HF, Connolly GE (1976) The California deer decline and possibilities for restoration. Cal-Neva Wildl Trans 1976:74–103

Loomis JB, Fitzhugh L (1989) Financial returns to California landowners for providing hunting access: analysis and determinants of returns and implications to wildlife management. Trans N Am Wildl Natl Resour Conf 54:196–201

López Ontiveros A (1993) Caza, ecología y ética. Revista de Occidente 149:90–108

López Ontiveros A, Verdugo FJG (1991) Geografía de la caza en España. Agric Soc 58:81–112

Lueck D (1995) Property rights and the economic logic of wildlife institutions. Nat Resources J 35:625-670

MARM (Ministerio de Agricultura, Alimentación y Medio Ambiente) (2012) Producción-Estadísticas Forestales- Montes y Política Forestal- Temas- Biodiversidad. http://www.magrama.es/es/biodiversidad/temas/montes-y-politica-forestal/estadisticas-forestales/produccion_2006.aspx#para2. Accessed Feb 2012

Marris E (2012) Hipsters who hunt. Slate magazine. December 2012.

Martínez Garrido E (2009) Visiones territoriales del boom cinegético español, 1970–1989. Bol Asoc Geogr Esp 51:325–351

Martinez JG, Carranza J, Fernández-García J, Sánchez-Prieto C (2002) Genetic variation of red deer populations under hunting exploitation in southwestern Spain. J Wildl Manag 66:1273–1282

Martínez M, Rodríguez–Vigal C, Jones OR, Coulson T, Miguel AS (2005) Different hunting strategies select for different weights in red deer. Biol Lett 1:353–356

McCullough DR (2001) Male harvest in relation to female removals in a black-tailed deer population. J Wildl Manag 65:46–58

McCullough DR, Fischer JK, Ballou JD (1996) From bottleneck to metapopulation: recovery of the tule elk in California. In: McCullough DR (ed) Metapopulations and wildlife conservation pp 375–403. Island Pr, Washington DC

Metra Seis (1985) Turismo cinegético en España. Secretaría General de Turismo, Madrid

Milner J, Nilsen EB, Andreassen HP (2007) Demographic side effects of selective hunting in ungulates and carnivores. Conserv Biol 21:36–47

Murden SB, Risenhoover KL (1993) Effects of habitat enrichment on patterns of diet selection. Ecol Appl 3:497–505

Mysterud A (2006) The concept of overgrazing and its role in management of large herbivores. Wildl Biol 12:129–141

Mysterud A (2010) Still walking on the wild side? Management actions as steps towards "semi-domestication" of hunted ungulates. J Appl Ecol 47:920–925

Mysterud A, Bischof R (2010) Can compensatory culling offset undesirable evolutionary consequences of trophy hunting? J Anim Ecol 79:148–160

Olea L, San Miguel-Ayanz A (2006) The Spanish dehesa. A traditional Mediterranean silvopastoral system linking production and nature conservation. Grassl Sci Eur 11:3–13

Ostrom E (1990) Governing the commons: the evolution of institutions for collective action. Cambridge University Press, New York, NY

Peiró V, Seva E (1996) Conservación y explotación de la fauna en ecosistemas mediterráneos. Instituto de Cultura Juan Gil-Albert, Diputación de Alicante, Spain

Pérez-González J, Carranza J (2009) Female-biased dispersal under conditions of low male mating competition in a polygynous mammal. Mol Ecol 18:4617–4630

Pérez-González J, Carranza J (2011) Female aggregation interacts with population structure to influence the degree of polygyny in red deer. Anim Behav 82:957–970

Pérez-González J, Carranza J, Torres-Porras J, Fernández-García JL (2010a) Low heterozygosity at microsatellite markers in Iberian red deer with small antlers. J Hered 101:553–561

Pérez-González J, Marcia Barbosa A, Carranza J, Torres-Porras J (2010b) Relative effect of food supplementation and natural resources on female red deer distribution in a Mediterranean ecosystem. J Wildl Manage 74:1701–1708

Plieninger T, Pulido FJ, Konold W (2003) Effects of land-use history on size structure of holm oak stands in Spanish dehesas: implications for conservation and restoration. Environ Conserv 30:61–70

Pollan M (2006) The omnivore's dilemma: a natural history of four meals. Penguin Press, New York

Rico M, Llaneza L, Fernández–Llario P, Carranza J (2000) Datos sobre el lobo ibérico (Canis lupus signatus Cabrera, 1907) en Extremadura. Galemys 12:103–111

Rodríguez Berrocal J (1993) Utilización de los recursos alimenticios naturales. Nutrición y alimentación de rumiantes silvestres. Centro de Cálculo, Facultad de Veterinaria, Spain

Sampson AW (1963) California range brushlands and browse plants. Agriculture and Natural Resources, California

Sánchez–Prieto CB, Carranza J, Pulido FJ (2004) Reproductive behavior in female Iberian red deer: effects of aggregation and dispersion of food. J Mammal 85:761–767

Scott JM, Davis FW, McGhie RG, Wright RG, Groves C, Estes J (2001) Nature reserves: do they capture the full range of America's biological diversity? Ecol Appl 11:999–1007

Shaw H (2008) Hunting jackrabbits in California. http://honest-food.net/2008/07/09/stalking-the-elusive-hare/. Accessed Feb 2012

Shaw H (2011) Hunt, gather, cook: finding the forgotten feast. Rodale Pr, New York

Standford RB, Howitt RE (1992) Solving empirical bioeconomic models: a rangeland management application. Am J Agr Econ 74:421–433

Standford RB, Howitt RE (1993) Multiple use management of California's hardwood rangelands. J Range Mgmt 46:176–181

Standiford RB (1989) A bioeconomic model of California's hardwood rangelands. Dissertation, University of California, Berkeley

Starr K, Orsi RJ (2000) Rooted in barbarous soil: people, culture, and community in gold rush California. Univ of California Pr, Berkeley

Stienstra T (2003) DFG (Department of Fish and Game) in tug-of-war between hunters, anti-hunters. San Fran Chronicle, 27 Mar. http://www.sfgate.com/default/article/DFG-in-tug-of-war-between-hunters-anti-hunters-2625693.php Accessed Aug 2012

Stine SW (1980) Hunting and the faunal landscape: subsistence and commercial venery in early California. Dissertation, University of California, Berkeley

Tober JA (1981) Who owns the wildlife? The political economy of conservation in nineteenth-century America. Greenwood Press, Westport

Torres-Porras J, Carranza J, Pérez-González J (2009) Selective culling of Iberian red deer stags (*Cervus elaphus hispanicus*) by selective montería in Spain. Eur J Wildl Res 55:117–123

USDOI & USDOC (U.S. Department of the Interior, Fish and Wildlife Service, and U.S. Department of Commerce, U.S. Census Bureau) (2011) National Survey of Fishing, Hunting, and Wildlife-Associated Recreation

Vargas J, Calvo J, Aparicio M (1995) Red deer (Cervus elaphus hispanicus) management in the dehesa system in central extremadura, Spain. Agrofor Syst 29:77–89

Vargas J, Guerrero J, Farfán M, Barbosa A, Real R (2006) Land use and environmental factors affecting red–legged partridge (*Alectoris rufa*) hunting yields in southern Spain. Eur J Wildl Res 52:188–195

Vicente J, Höfle U, Garrido JM, Fernández de Mera IG, Acevedo P, Juste RA, Barral M, Gortazar C (2007) Risk factors associated with the prevalence of tuberculosis-like lesions in fenced wild boar and red deer in south central Spain. Vet Res 38:451–464

Virgós E, Travaini A (2005) Relationship between small-game hunting and carnivore diversity in central Spain. Biodivers Conserv 14:3475–3486

Wilcox JT, Van Vuren DH (2009) Wild pigs as predators in oak woodlands of California. J Mammal 90:114–118

Part IV
Economics

Chapter 12
Economics of Ecosystem Services

Alejandro Caparrós, Lynn Huntsinger, José L. Oviedo, Tobias Plieninger and Pablo Campos

Frontispiece Chapter 12. People in oak woodland communities enjoy a rural lifestyle, as here, with onlookers and participants taking in the scene of a small town parade. Oak woodlands provide many, often non-market, ecosystem services. (Photograph by L. Huntsinger)

A. Caparrós (✉) · J. L. Oviedo · P. Campos
Institute of Public Goods and Policies (IPP), Spanish National Research Council (CSIC), Albasanz 26–28 28037 Madrid, Spain
e-mail: alejandro.caparros@csic.es

J. L. Oviedo
e-mail: jose.oviedo@csic.es

P. Campos
e-mail: pablo.campos@csic.es

P. Campos et al. (eds.), *Mediterranean Oak Woodland Working Landscapes*, Landscape Series 16, DOI: 10.1007/978-94-007-6707-2_12,
© Springer Science+Business Media Dordrecht 2013

Abstract A better appreciation of the value of ecosystem services produced on private lands opens the door to programs that offer incentives to landowners and managers for specific conservation and production practices. This chapter reviews studies of ecosystem services provided by oak woodlands in California and Spain, focusing on those that may be difficult to quantify and value, and therefore are often undervalued in decision-making processes drawing on economic analysis. We first examine how ecosystem services are defined and valued, and then review research done from an economic perspective in California ranch and Spanish dehesa oak woodlands. We conclude with a brief exploration of differences in institutions and policies that bear on oak woodland ecosystem services in these two regions. The next step in ecosystem service valuation and use in policy is to extend case studies and to undertake analyses at the regional, state, and nation-wide scales. Despite scientific advances, the need for preservation of the natural capital of oak woodlands and the many ecosystem services the woodlands provide is far from fully recognized by society. An important future policy task will be incorporating payments for provision of biodiversity and ecosystem services into agricultural, water, energy, and other policies.

Keywords Ecosystem services · Oak woodlands · Ranches · Dehesas · Spain · California

12.1 Introduction

While the term has been defined a number of ways, a straightforward definition of ecosystem services is "the benefits people obtain from ecosystems" (MEA 2003). Oak woodlands in California and Spain provide rich flows of these benefits. Landowners in each location place a high value on living and working among oaks and their natural beauty; the public enjoys oaks and grasslands; water cycles through the system supply water to towns and villages; wildlife flourishes in oak-dominated habitat, supported by the annual acorn crops and the mosaic of vegetation; crops and commodities including meat, cheese, firewood, honey, and hunting are produced for the market from dehesas and ranches.

This chapter focuses on ecosystem services that are not traditional market commodities, since they are discussed later (Chap. 13). We explore efforts to identify oak woodland ecosystem services that may be difficult to quantify and value, and are

L. Huntsinger
Department of Environmental Science, Policy, and Management, University of California, Berkeley, 130 Mulford Hall MC 3110 CA 94720, USA
e-mail: huntsinger@berkeley.edu

T. Plieninger
Ecosystem Services Research Group, Berlin-Brandenburg Academy of Sciences and Humanities, Jägerstr 22/23, Berlin 10117, Germany
e-mail: plienint@geo.hu-berlin.de

therefore often undervalued in economic decision-making and in terms of their potential role in policy and management. A better understanding of the value of ecosystem services produced on private lands opens the door to programs that offer incentives to landowners and managers for conservation and production practices.

We review how ecosystem services are defined and valued, and the research that has been done from an economic perspective on ecosystem services in California oak woodland ranches and Spanish dehesas. We conclude with a brief exploration of differences in institutions and policies that bear on oak woodland ecosystem services in these two regions.

12.2 The Ecosystem Services Concept

Ecosystem services is a term generally used when there is a need to value the spectrum of societal benefits from ecosystems. Ecosystem services can include specific goods like mushrooms, or services such as water cycling. In recent years, the need to better understand the value of benefits deriving from the environment has been the subject of much discussion, and with publication of the Millennium Ecosystem Assessment (MEA 2005) initiated by the United Nations, ecosystem services garnered widespread attention from ecologists and economists. As a concept ecosystem services can be traced to the 1970s when a classification of nature's services was undertaken in the U.S. *Study of Critical Environmental Problems* (SCEP 1970). In 1981 the conservation biologists Paul and Anne Ehrlich coined the term "ecosystem services" (Ehrlich and Ehrlich 1981; Ehrlich and Mooney 1983). The concept was popularized through a controversial paper published in the journal *Nature* (Costanza et al. 1997) that attempted to respond to an increasingly evident gap in our ability to measure the economic value of beneficial outputs from the environment because such benefits lacked a market price. In an economic sense, "ecosystem services" are distinguished from other ecosystem functions in that there are beneficiaries willing to pay for the use or preservation of those scarce services (Chan et al. 2006).

The provisional Common International Classification of Ecosystem Goods and Services (CICES) distinguishes three main categories of ecosystem services: Provisioning (grass, acorns, meat, cork, firewood), regulating (climate, floods, pollination and pest control for food production), and cultural (serenity, identity, inspiration). Table 12.1 uses this provisional proposal to classify several of the studies discussed below (EEA 2011).

12.3 Valuing Ecosystem Services

Few universally accepted methods exist for setting a value on nonmarket ecosystem goods and services. All seek to estimate what a person or household would willingly pay for a good or a service, or to set a value for damages from losses or

Table 12.1 Ecosystem services valuation studies in Spanish and Californian oak woodlands reviewed in this chapter classified in accordance with the *common international classification of ecosystem goods and services* [adapted from Haines-Young and Potschin (2010) and EEA (2011)]

Ecosystem services from oak woodlands	Spanish studies	Californian studies
Provisioning services (resource function)	Livestock, crops, cork, firewood, grazing resources, hunting and mushroom gathering (Campos and Riera 1996; Campos et al. 1996, 2001 , 2007a, 2008; Campos 1997, 1999, and 2002; Caparrós et al. 2003; Campos and Caparrós 2006)	Firewood, forage, and wildlife (Standiford and Howitt 1992) livestock, firewood, hunting, crops (Chap. 13) and ecosystem functioning (Chan et al. 2006) Wildlife habitat (Kroeger et al. 2010)
Regulating and maintenance services (service function)	Carbon sequestration (Caparrós et al. 2010 and 2011; Joffre et al. 2003; Pereira et al. 2007; Li et al. 2008)	Pollinator habitat (Chaplin-Kramer et al. 2011) Carbon sequestration (Kroeger et al. 2010) Water and nutrient cycling (O'Geen et al. 2010) Sudden oak death (Kovacs et al. 2011)
Cultural services (service function)	Recreation (Ruiz-Avilés et al. 2001; Arriaza Balmón et al. 2002; Oviedo et al. 2005; Campos et al. 2007a) Landscape conservation (option values) (Campos 1998; Oviedo et al. 2005; Caparrós et al. 2010, 2011) Landowner consumption of private amenities (Campos and Mariscal 2003; Campos et al. 2009)	Landscape enjoyment values (Standiford and Scott 2001; Thompson et al. 2001; CA-LAO 2004; Newburn et al. 2005) Endangered and threatened species preservation (Loomis and Gonzalez-Caban 1996) Retaining oaks for property values (Diamond et al. 1987) Easement value (Rilla and Sokolow 2000) Landowner consumption of private amenities (Martin and Jeffries 1966; Smith and Martin 1972; Huntsinger et al. 2010; Oviedo et al. 2012)

costs avoided (Olewiler 2004; Tanaka et al. 2011). When market prices are available, this is used to value outputs. Common methods for valuing ecosystem services without a set market price include stated preferences (SP) and revealed preferences (RP). Contingent valuation surveys, the most common stated preference technique, estimates how much individuals would be willing to pay to maintain the continuity of an environmental feature, such as biodiversity. This is a "stated preference" approach because it requires a prospective buyer to estimate an economic value or, more commonly, to state whether or not they would be willing to pay a given amount. The "stated choice model" approach, a variation on the stated preference method, offers alternatives with multiple attributes for the respondent to choose from. A revealed preference approach associates the existing price of a good or service with some attribute of interest—for example: Are ranches with a secure and plentiful water supply worth more than ranches with poor water supplies? As water sources vary, how do prices increase or decrease and by how much? The relationship between ranch price and distance to water supplies is a "hedonic" regression or relationship. As it is based on existing ranch prices, the value of water to ranch buyers is considered to be "revealed" through changes in the actual value of ranches rather than "stated."

When valuing ecosystem services, distinguishing between final and intermediate goods and services is important to avoid double-counting of values (Campos et al. 2001; Boyd and Banzhaf 2007). Final outputs are traded in markets or consumed by society as they are, and are usually the focus of economic analysis, while intermediate outputs are used to create final outputs. For example, if a "grass-fed" steer is sold, the final output is the steer, while the grass the animal ate is an intermediate output from the ecosystem. The grass is an important intermediate output, especially if we are looking at the value of services from the ecosystem. Grass clearly has a value that can be quantified because it can be sold directly to someone else. Nevertheless, to avoid double-counting it is best not to include this value twice, in other words once as the ecosystem service "grazing" and a second time as a part of the commodity, "the steer" (Campos 1999; Caparrós et al. 2003; Boyd 2007; Campos et al. 2008). Defining an economic value and establishing the methods for valuing ecosystem services requires a precise definition of those services as final or intermediate outputs.

There is debate about the best means of classifying and valuing ecosystem services, but there is a widespread consensus among environmental economists that total economic value (TEV) is the appropriate framework. TEV includes all the reasons individuals are motivated to attribute economic value to scarce goods and services and classifies them into current, option and existence values (Table 12.2). The easiest motivations to ascertain are those leading to the current active use, such as public recreation. Another motivation is to ensure the option for a future use. Option value emerges when the current generation accepts an additional management cost to preserve the option to use the service in the future. People can give economic value to passive uses (existence value), such as efforts to preserve wildland or avoid extinction of threatened species. This value is based on the observation that humans spend economic resources to prevent non-replaceable ecosystems, biological varieties, and unique cultural values from disappearing forever. This behavior occurs even in situations where the passive user only knows these unique assets from reading, conversation or audiovisual mediums, and without the requirement of anticipating a future active use (Krutilla 1967). The concept of existence value of an ecosystem has led to a lively controversy, which is not yet fully resolved, over the difficulty of valuing it (Fig. 12.1).

Table 12.2 Total economic value of oak woodland ecosystems

Active use		Passive use	
Current value		Option value	Existence value
Final	Intermediate	Final	Final
Exchange value of ecosystem goods and services that are consumed or invested in during the accounting year	Exchange value of ecosystem goods and services used as an intermediate input into another output in the same accounting year	Consumer or institutional willingness to pay a premium in addition to the ordinary price of an ecosystem good or service to ensure its conservation and future use	Consumer or institutional willingness to pay with the exclusive purpose of preventing the future extinction of a unique feature in the oak woodland ecosystem studied
Examples: – Commercial goods and services from forestry, livestock, or game – Public recreational services – Collection of mushrooms, plants and wildlife –Carbon sequestration	Examples: – Natural grass and fruits consumed by livestock and wildlife – Old trees kept to favor biodiversity – Crop plants and seeds for planting and sowing	Examples: – Conservation of biological resources for research into new drugs and for the biological control of pests – Continuity of the future supply of goods and services from traditional activities in oak woodland ecosystems	Examples: – Preservation of a forest ecosystem or a threatened species – Preservation of architectural heritage or the cultural institutions of the ecosystems that are threatened

12.4 Ecosystem Services Valuation in Spain and California Oak Woodlands

Over the last decade, the value of ecosystem services from oak woodlands has become more widely appreciated in Spain and California, including services associated with recreation, landscape enjoyment, carbon sequestration, biodiversity, and watershed maintenance. Markets and incentives for the production of ecosystem services can make profound changes in oak woodland management and conservation and contribute to the quality of life for all residents.

The status of ecosystem services from the Spanish dehesa have been assessed by expert opinion in the National Millennium Ecosystem Assessment (EEME 2011), though no economic valuation was performed. The Assessment highlighted production of food of extraordinary quality, water regulation, minimizing soil erosion, and the provision of recreation as the most important ecosystem services of the dehesa. It concluded that 5 of the 19 evaluated ecosystem services were

Fig. 12.1 Wildlife is believed to have important option and existence values for residents in the area and for urban dwellers. Here, a California tule elk stands in oak woodland. (Photograph by R. Keiffer)

degraded or had been used unsustainably over the past 30 years. Land use change, in particular rural land abandonment and agricultural intensification, and changes in global biogeochemical cycles were identified as the most influential drivers behind this exploitation. Land use change in California in the form of urban sprawl, fragmentation of large properties into smaller "ranchettes," and property development are widely recognized as a broad-scale threat to the continued production of ecosystem services from oak woodland ranches (CDF-FRAP 2003).

Research on non-market benefits is on the upswing in agroforestry research (Montambault and Alavalapati 2005). Studies focus on improving economic analysis for public policy, and on the real and potential influence of ecosystem services on landowner decision-making. The economic valuation of individual ecosystem services provided by California oak woodland ranches has been addressed in a number of studies as detailed below. In Spain, in addition to studies focusing on individual ecosystem services, a set of studies has tried to integrate several ecosystems services into a common framework based on national accounting concepts. The ultimate aim is to estimate the Hicksian total income, which is is the monetary flow generated in a given period that, totally spent within the period, leaves the same stock at the end of the period as there was at the beginning (Caparrós et al. 2003) (Chap. 13). Research toward this goal in Spanish oak woodlands has been intense (Campos and Caparrós 2006) and these efforts have stimulated such research in California (Chap. 13).

12.4.1 Woodland Ecosystem Services at the Landscape Scale

The "wide open spaces" and aesthetics of ranch and dehesa country are cherished elements of the quality of life in California and Spain. In California, many ecosystem services depend upon the extensive and undeveloped character of oak woodland ranches. Services at the landscape scale are universally appreciated by landowners and the public. Oak woodlands and ranch lands can provide a buffer around parks and natural reserves (Talbert et al. 2007), reducing conflicts between urban and natural areas. Thompson et al. (2001) did a contingent valuation study to estimate the values attached to agricultural land and oak woodlands in San Luis Obispo County (California). They found a one-time willingness to pay per voter ranging from $75 to $83 for avoiding conversion from extensive to intensive agriculture, or from agriculture to residential-commercial development. If all voters had contributed this much at the time of the study in 1997, it would have totaled $12 million, which could potentially have been used for purchasing development rights for preserving oak woodlands on some key properties.

Kovacs et al. (2011) analyzed the impacts of the spread of invasive oak pathogens, and the subsequent loss of oaks, on home values, showing implicit positive values to conserving ecosystem services by avoiding oak mortality. They simulated the spread of Sudden Oak Death (*Phytophthora ramorum)* in the 21 California counties predicted to have pathogen-related oak mortality from 2010 to 2020. The simulation predicted that more than 10,000 oak trees would need to be replaced and that there would be housing development property value losses of $135 million in the studied communities. Extrapolated beyond the studied communities to all oak woodland housing developments, the predicted losses increase to $350 million.

Oaks on the landscape provide dehesa and ranch owners with marketable ecosystem services. Oaks directly improve property values, through on-site effects, in California. On 2 ha lots, land with at least 100 oaks per ha was worth 27 % more than land without oaks (Diamond et al. 1987). The proximity of oak woodland open space influences property values through off-site effects. Standiford and Scott found in 2001 that house and land prices increase when they are closer to oak stands. In particular, undeveloped land was worth 19 % more when it was adjacent to an oak woodland open space area than when the oak woodland open space was 1,000 ft or more away. In an analysis of house prices, a 10 % reduction in distance to the nearest oak stand increased the price by 3 %.

12.4.2 Biodiversity, Recreation, and Watershed as Ecosystem Services

Research on the existence values associated with biodiversity in oak woodland and dehesa ecosystems is scarce in Spain and California and employs diverse methods.

Loomis and Gonzalez-Caban (1996) estimated the willingness to pay of Californian residents for the conservation of habitat for northern spotted owls, found in dense, old-growth oak woodlands. These authors found that the median willingness to pay per household was $56 per year for preventing a loss of 1,028 ha of habitat for the endangered northern spotted owl (*Strix occidentalis caurina*). Chaplin-Kramer et al. (2011) found that the value of pollination services provided to California agriculture by the pollinators that are part of the high biodiversity in the oak woodlands and grasslands surrounding the state's agricultural valleys is more than $2 billion.

Valuing recreational services offers fewer methodological challenges than calculating existence values since they are based on a direct use value (Caparrós et al. 2003). In Spain, Campos et al. (2007a) calculated the value of recreational visits to Alcornocales Natural Park, an extensive area in Andalucía with cork oak and holm oak dehesas, and found the mean willingness to pay in 2002 reached €21 per visit. That is not immediately comparable to California figures that are based on consumer surplus, or how much each buyer might be willing to pay. The values for Spain estimate potential market prices as discussed in Sect. 12.3. For California oak woodlands, the only study in the literature that performs a direct calculation of contingent valuation for recreational ecosystem services is by Thompson et al. (2001), and focuses on landscape valuation.

Water availability is one of the most serious natural resource issues facing California, especially given projections for future climate change. Oak woodlands play a critical role in California's water supply system, providing runoff primarily from winter rainfall events and hosting two-thirds of the state's drinking water reservoirs (O'Geen et al. 2010). The value of reducing erosion from runoff in California was estimated at $9.00 per ton/year in 2011, and the value of reducing wind erosion at $1.25 per ton/year (Tanaka et al. 2011). In the dehesas of the Sierra Morena (central-west Andalucía, Spain) a hedonic analysis of irrigated land in the Guadalquivir Basin offers a regulated average water natural asset value of €3.46/m^3 for 2005 (Berbel and Mesa 2007). This water natural asset value, discounted at a real rate of 5 %, gives an annual average rainwater resource valuation of €0.17/m^3. Dehesa and ranch woodlands with medium to low oak canopy cover consume significantly less rainwater (green water), producing more runoff to reservoirs than heavily forested or dense shrubby vegetation types. However, dense woody vegetation in dehesa stores more above-ground carbon (unpublished data, RECAMAN[1] project).

[1] The RECAMAN project (Valoración de la Renta y el Capital de los Montes de Andalucía) of the Junta de Andalucía, initiated in 2008, is ongoing and applies the Agroforestry Accounting System at the regional scale to measure total income and capital from the montes of Andalucía in Spain.

12.4.3 Carbon Sequestration

Carbon sequestration is an ecosystem service that has received attention in Spanish oak woodlands and has been valued and analyzed both from a national accounting perspective (Caparrós et al. 2003; Campos and Caparrós 2006) and by comparing the outcome of alternative land uses. The latter approach is followed in Caparrós et al. (2010) where the potential of reforestation with cork oaks for carbon sequestration is assessed and compared with the alternative option of planting fast-growing exotic or non-native species. If payments reward growth of trees with fast cycles of harvest and regrowth, then species such as eucalyptus are favored. On the other hand, if payments come in increments based on enduring biomass and secondary products created by trees, the slower-growing and seldom harvested oaks are favored. The oaks provide more biodiversity than the non-natives, and are generally preferred as a part of the landscape by the public (Caparrós et al. 2010). The study pointed out tradeoffs like this among ecosystem services that need to be considered in policy development (Fig. 12.2).

In California, Kroeger et al. (2010) used a 100-year per acre carbon gain scenario to estimate that rangeland oak afforestation or reforestation projects would have earned Chicago Climate Exchange carbon credits worth between $118 and $568 per ha, while returns from livestock production alone could not justify

Fig. 12.2 Tradeoffs are a feature of management for ecosystem services. For example, small wetlands in oak woodlands are home to a rare bird known as the California black rail (*Laterallus jamaicensis coturniculus*). Planting woody riparian species in the wetlands to stabilize the soil, increase habitat for other woodland birds, and sequester carbon makes the habitat unusable for the rail, as on this California site where willows were restored along a creek that was formerly only grassland, and which was then mostly abandoned by rails. (Photograph by L. Huntsinger)

oak tree plantings by private landowners. Typical costs for rangeland oak plantings were estimated at $7,500 to $15,000 per ha (Kroeger et al. 2010). Despite this, a 2005 survey showed that about a third of oak woodland landowners do plant oaks, mostly for amenity, browse, and real estate values (Huntsinger et al. 2010).

The initial focus on reforestation, both in research and international negotiations, has shifted to "avoided deforestation and degradation" by reducing the emissions caused by current deforestation and reductions in tree and shrub cover. Caparrós et al. (2011) extended their analysis of reforestation to include programs to avoid dehesa degradation. Taking into account commercial values, carbon sequestration values, and public landscape preferences, cork oaks are shown to make more sense for planting than non-native species. Avoiding dehesa degradation is preferable to reforestation and the only reason reforestation is more popular in Spain is that public subsidies encourage it. European subsidies might better be used to promote traditional and natural regeneration, avoiding dehesa deterioration in the first place. Other potential benefits of favoring avoided degradation in terms of biodiversity and landscape were not fully valued, but doing so would probably only reinforce an argument for this policy shift (Caparrós et al. 2011).

12.4.4 Integrating Multiple Sources of Income

In Spain, broadening the spectrum of oak woodland ecosystem services accepted into the economic analyses employed in forging public policy has led to creation of an Agroforestry Accounting System that integrates the commercial and non-commercial values of agroforestry ecosystems. This accounting framework was initially developed for commercial benefits in the dehesa (Campos 1984), and later extended to take private amenities (Campos and Riera 1996), public recreation, and conservation values into account (Campos et al. 1996; Campos 1998). The Agroforestry Accounting System was expanded beyond dehesa ecosystems to the timber, recreation, and conservation values of Sierra de Guadarrama pine forests by Caparrós et al. (2003). They introduced the Simulated Exchange Value method to model a market for integrating commercial and non-commercial values. This, together with the Agroforestry Accounting System, was applied to an oak woodland setting in the Monfragüe shire in Spain. Table 12.3 shows the methodology at work (Campos and Caparrós 2006), highlighting a significant share of total income (the last column) that is unrecorded in current accounts (the first column).

Simulated markets are quite commonly used in national accounting to obtain prices for goods without a market price, as when market prices for berries are used to estimate the value of berries gathered by recreational visitors to a forest. This methodology can be extended to include cases where no similar market prices exist, as in open access recreational services in Spain (Caparrós et al. 2003).

Simulation of a market for services lacking a fixed price can employ stated preference environmental valuation techniques. But the direct use of a value set by either contingent valuation or choice models assumes that every visitor pays the

Table 12.3 Value of ecosystem services produced from Monfragüe Shire (1998 Euros per hectare), assuming sustainable levels of production (Campos and Caparrós 2006)

Class	Monfragüe cork and holm oaks			
	European accounting system, 1995		Visitor environmental values	Agroforestry accounting system
	European accounting system for forestry, 1997	Omitted		
	(1)	(2)	(3)	(1) + (2) + (3)
Outputs				
Intermediate output:				
Livestock-grazing		16		16
Final outputs				
Timber				
Cork	462			462
Firewood	76			76
Hunting		37		37
Mushrooms				
Owner amenity values				
Min		n.a.		
Max		85		85
Public access recreation				
Min			[3]	[3]
Max			8	8
Conservation, for visitors			9	9
Total outputs	**538**	**138**	**17**	**693**
Costs				
Production expenses				
Private	48			48
Governmental		6		6
Labor				
Private	142			142
Governmental		n.a.		
Fixed capital consumption/ Depreciation	2			2
Total costs	192	6		198
Net operating margin/Net benefit (NOM)	**346**	**132**	**17**	**495**
Net value added at market prices (NVA = NOM + Labor)	**488**	**132**	**17**	**637**
Gross value added at market prices (depreciation not subtracted from NVA)	**490**	**132**	**17**	**639**

Source Campos and Caparrós (2006)

maximum price they are willing to pay ("consumer surplus"). That assumption is unrealistic if the objective is to simulate a real-world market. Not all customers, after all, pay the maximum amount they would be willing to pay when going to a grocery store to buy potatoes. The farmer-retailer who sells potatoes sets the price beforehand, since it is not feasible to charge each customer a different price. Consequently, to simulate a market price, Caparrós et al. (2003) assumed that the owner can only choose one price for access to a property, given demand for recreational access estimated using a contingent valuation. The amount a landowner could obtain was modest in these oak woodlands (Table 12.3), although for other regions and ecosystems the value can be high (Caparrós et al. 2003).

A recent research project (RECAMAN) in Andalucía, an autonomous region in southern Spain with significant oak woodlands, is taking first steps toward a bottom-up national accounting system for natural resources. The system includes ecosystem services and commodity goods, and spatially tracks where these outputs are produced. Although the project targets more than oak woodlands, it provides valuable information on oak woodlands at the scales of both individual property and landscapes, since holm and cork oaks are by far the most abundant tree species in Andalucía. Capturing data for an area of about four and a half million hectares, the project basically follows the methodology described above for the accounting framework and the consistent aggregation of environmental values with commercial values produced by ecosystem goods and services.

Ecosystem services are important in regional models that prioritize lands for conservation and set-aside payments from government entities and private conservation organizations. The price of properties acquired for conservation purposes reflects ecosystem service values. In an analysis aimed at prioritizing land purchase and set asides, a model developed by Newburn and others takes into account environmental benefits, land costs, and likelihood of land use conversion to determine optimal sites for conservation investment (Newburn et al. 2005, 2006). The method was applied to developable parcels in the oak woodlands of unincorporated Sonoma County in California that represented some 94 % of the county area ($\sim$4,000 km^2). The researchers concluded that only considering environmental characteristics and ignoring land costs would bias the strategy toward urban fringe parcels with a high likelihood of conversion, and that ignoring the likelihood of land use conversion would bias the strategy toward protecting low-cost sites in the hinterlands.

Chan et al. (2006) explored the potential trade-offs among ecosystem services in California's Central Coast ecoregion using a spatial analysis methodology. Much of the area studied was oak woodland, although that was not an explicit focus. They worked with six ecosystem services: carbon storage, crop pollination, flood control, forage production, outdoor recreation, and water provisioning. They did not attempt to measure all the ecosystem services in the same unit (such as a dollar value). Instead, they estimated the physical quantities of each of the services provided and analyzed possible conflicts and synergies between them. The study demonstrated that carbon storage, flood control, forage production, and outdoor recreation are positively correlated with high biodiversity, while there are trade-offs with other agriculture-focused services.

12.5 Ecosystem Service Policies and Programs

Policies, programs and regulations have evolved to protect and encourage ecosystem services that can have far-reaching effects on human welfare and quality of life. Government intervention plays a large role. A 2011 nationwide survey sponsored by the Packard Foundation (The David and Lucile Packard Foundation 2011) shows that most Americans favor public investment in agricultural conservation efforts because of their environmental and social benefits, and support programs that provide environmental benefits over direct commodity subsidies. There are three basic types of policies used in Spain and California: regulatory, market, and payments for ecosystem services.

The first and conventional method is regulatory. This can take the form of rules that require, or prevent, specific practices or land use. While such policies may be effective, enforcement costs and unintended consequences can constrain their usefulness. In addition, a purely regulatory approach offers no incentives to landowners who create or manage habitat above the regulatory standard (Bean and Wilcove 1997).

A second type of policy creates markets for ecosystem services. Willing buyers purchase ecosystem services from willing sellers, ideally allowing competition and supply and demand to set prices and foster innovation. Markets for ecosystem services are created by government-sponsored or private labeling and certification programs that can add value to products and brands. In the United States, government-sponsored labeling—for example for organic products—tends to be broad, general, inclusive, and often highly contentious. Government controlled origin and quality labels are important in Spain.

A third kind of program simply pays individuals or groups to produce ecosystem services—a direct contract between an entity, governmental or private, and the landowner. Such Payment for Ecosystem Services (PES) programs are active in the United States and in Spain. In California, the federal Environmental Quality Incentives Program (EQIP) offers cost sharing for landowners who carry out improvements or maintenance that protects or enhances wildlife habitat, as when dredging stock ponds or providing wildlife escape ramps in livestock watering troughs (USDA-NRCS-EQIP 2011). Subsidies that support increased landowner production of ecosystem benefits are a form of PES, as are "avoided degradation" programs, that pay landowners to protect carbon storage or other services, or programs that reduce property taxes for those who use or manage their lands in certain beneficial ways.

12.5.1 Regulation and Zoning to Maintain Ecosystem Services

In Spain, for oak woodland classified as protected or not, private management is subjected to public regulation that prevents specific management practices and development. For example, dehesa landowners cannot cut oaks without government permission (BOE 2003). As a result, no significant impact on dehesas are expected from urban development since they affect mainly urban buffer zones (BOE 2007a, b), but public transport infrastructure development has created landscape fragmentation that damages habitat.

The main laws on environmental protection in Spain are the Natural Patrimony Law (BOE 2007a) and the National Parks Law (BOE 2007b). Dehesas in national parks are rare and in most cases protected dehesas are in natural parks, which are the autonomous region equivalent of national parks and bear a lower level of protection. Nevertheless, except for strict preservation areas, even Spanish national parks continue to be grazed by private livestock owners. Regional governments promulgate the regulations that govern natural parks. Their restrictions generally focus on the habitat and conservation needs of wild fauna and flora, actions that rarely impose large economic costs on private landowners (Fig. 12.3).

The main impact on dehesas derives from non-environmental regulations such as livestock subsidies (in the past) and reforestation programs for agricultural land, discussed below. Livestock subsidies did help maintain dehesas as working landscapes, but they have incentivized overstocking. The European Common

Fig. 12.3 Cattle grazing in a dehesa property located in the Natural Park Sierra Norte de Sevilla. Regulation for conservation and protection is compatible with management in dehesa oak woodland ecosystems. (Photograph by S. García)

Agricultural Policy (CAP), which began in the late 1950s, aims now, through its Pillar 1, to maintain active use of farmland in good agricultural and environmental condition. However, in part due to the overstocking problem, there is a significant change taking place from the old mechanism that paid a subsidy for livestock by the head to a new formula based on hectares of land in use for agriculture. As a consequence, large parts of the area covered by dehesa will no longer be eligible for subsidies under Pillar 1, since they are covered by shrublands and trees at a density of more than 50 trees per hectare and do not qualify as eligible agricultural land use. Marginal pastures face likely abandonment.

Zoning should act to reduce the asset value of a property to be more in line with the use it is zoned for. In California, it is common for zoning limiting development to be flexible in response to the financial power of development interests, making reduction in asset value questionable (Fig. 12.4). Additionally, there is a perception of injustice on the part of landowners when land on one side of a zoning line is worth a great deal more than similar land on the other side of the line, just because of a zoning decision. This discontent threatens sustainability in the long run. Citizens of the neighboring state of Oregon, for example, passed Measure 37 in 2004 (later reversed), requiring government compensation when zoning reduced property values, making such protections almost impossible. The measure was widely viewed as a backlash against zoning efforts to conserve open space and agricultural land (Berger 2009). Finally, without a viable agricultural operation on the property, zoning that restricts land use to agriculture can in some cases be

Fig. 12.4 In California, oak woodlands are attractive areas for suburban development. Ironically the oaks themselves are considered a valuable attractant for prospective buyers. Here the development itself is titled "Live Oak Ranch," even though the cows (and agricultural fields) are long gone. (Photograph by P. F. Starrs)

argued to be a constitutionally defined "taking", or an unfair governmental expropriation or confiscation of property value.

In California, another effort to conserve oak woodland ecosystem services at the landscape scale is the California Land Conservation Act (the Williamson Act) of 1965, a form of payment for ecosystem services through tax relief. Landowners can enroll land actively used for grazing or other agriculture, committing to a contract stipulating they will keep their land in this use for 10 years at least. The contract is renewed each year, and enrolled landowners pay an agriculture-based property tax that is less than they would pay if the land was valued for development. If they do not wait 10 years to develop their land after de-enrolling, then they must repay the taxes on the full development value. Annual cost of the Williamson Act to the State, when fully funded, was approximately $80 million (CA-LAO 2004). About two-thirds of oak woodland landowners had their land registered in the Williamson Act in 2005 (Huntsinger et al. 2010). The program was cut back by the elimination of most state funding starting in 2009.

12.5.2 Carbon Markets and Credits

Carbon trading markets allow landowners to market carbon sequestration credits earned by managing their land for increased carbon flux to plants and soil. The additional carbon sequestered as a result of landowner action is a "credit" that can be sold on the market to industries or other entities needing to emit carbon. The Chicago Climate Exchange (CCX 2009) operated from 2003 to 2010 as a trading platform and a registry for more than 400 U.S. corporations and governmental entities that were interested in purchasing independently verified greenhouse gas emission allowances. Between 2008 and 2010, 1,000 U.S. ranchers participated in the exchange (Gosnell et al. 2011). However, prices for CCX credits were consistently less than those in government-sanctioned exchanges in Europe and most of the smaller global voluntary markets. Without the prospect of CCX credits being grandfathered into an official U.S. trading system, the demand and price of credits dropped to levels that did not justify the costs of continuing to run the exchange, which closed down in 2010. In 2006, the Global Warming Solutions Act (Assembly Bill 32) was passed in California, setting limits on greenhouse gas (GHG) emissions (ARB 2006), reducing them to 1990 levels by 2020—a reduction of 30 %—and another 80 % reduction by 2050 (see Chap. 9 for details). In the European Union, an active mandatory market exists for carbon emissions from industries, but carbon sequestration in terrestrial ecosystems is not included. Nevertheless, under the Kyoto Protocol, Spain can use the net emissions reductions achieved by increasing carbon sequestration in oak woodlands to meet its international commitments (Caparrós et al. 2011).

12.5.3 Conservation and Mitigation Easements as Examples of Payments for Ecosystem Services in California

Although unknown in Spain, probably because of already strict land use controls (Huntsinger et al. 2004), conservation easements are now the most widely used private sector land conservation method in the United States (Gustanski and Squires 2000). Adapted from programs designed for protecting rights of way for public utilities, the amount of California land under conservation easements increased by 34 % since 2005, and doubled since 2000 (Land Trust Alliance 2010). In exchange for tax benefits or outright payment, a landowner voluntarily agrees to a permanent deed restriction on the property title that prohibits development. This right is then held by a third party, sometimes a public agency, but often a non-governmental organization known as a land trust. Although far from perfect as a conservation strategy (Merenlender et al. 2004; Reiner and Craig 2011), and not specifically oriented to oak woodland conservation, easements allow ranchers to continue consuming the ecosystem services from the property, while extracting some of the capital value of the land by donating or selling the right to develop (Sulak et al. 2004).

The placement of a conservation easement on a property is strictly voluntary for the landowner—the easement is sold or donated (Sulak et al. 2004), creating a market for the "easement value". This value is the difference between the development price and the value of the oak woodland for ranching (including landowner private amenity values), and is what a land trust or other entity pays for the development rights from the property. Appraisals of the value of the easement, and financial arrangements between the parties (land owner and land trust), generally are kept private, despite the fact that substantial amounts of the funding used are often from government sources (Merenlender et al. 2004). A 2005 survey of oak woodland landowners found that approximately 6 % of the properties had a conservation easement on them (Huntsinger et al. 2010). Ferranto et al. (2011) found that 6 % of California forest and rangeland owners in 10 representative California counties had a conservation easement in place in 2008 (Fig. 12.5).

Mitigation easements are similar, but are purchased using the funds of property developers to preserve specific types of habitat that will be lost as a result of the development. A landowner might have one part of a property designated as a "mitigation easement" for a particular threatened or endangered species, for example. An average "easement value" for California has been estimated at approximately $5,000 per ha, with costs ranging from less than $1,000 per ha to more than $100,000 per ha in urban-fringe areas. California programs spent approximately $103 million in cash on easements for conservation or mitigation in 2002, much of it public funds, and mostly on oak woodland grazing lands (Rilla and Sokolow 2000).

The most visible conservation easement negotiation in California's oak woodlands recently was on the Hearst Ranch, where some $80 million in public funds and $15 million in tax credits went to purchase conservation easements on

Fig. 12.5 Found near a busy road, the sign informs visitors that a conservation easement is in place on this ranch property. The development rights are held by the Marin Agricultural Land Trust while the owner continues to produce livestock and manage the property. The sign emphasizes the collaborative nature of the relationship. (Photograph by L. Huntsinger)

32,000 ha of the ranch surrounding Hearst Castle and state ownership of 600 ha for public access in 2005 (California Resources Agency 2004). Because the easement was purchased with public funds, the price paid and the stipulations of the easement are known. The Hearst Ranch woodlands rise from sea level to peaks of the Santa Lucia Mountains, and provide habitat for nearly 1,000 plant and animal species, including many rare, threatened, or endangered species. It is the largest privately-owned working cattle ranch remaining on the California Coast. In addition, about one million people a year view the ranch scenery on the long bus ride up to the castle, and all who drive up California's Highway 101 see an undeveloped coast and hills. The people of California, via State government, have valued these services and a host of others that come from the land at approximately $95 million. The ranch is now niche marketing value-added natural beef, which could be considered an enhancement of ecosystem services from the property.

12.5.4 Labeling and Landowner Marketing of Ecosystem Services

Landowners use value-added products to market ecosystem services valued by consumers. While the main motivation for purchasers of such products probably lies in the quality and characteristics of the product itself, there is no doubt that

Fig. 12.6 This cattle producer sells grass-fed beef directly to consumers, who come to the ranch to pick up their meat. The consumer meets the rancher and sees the operation, and is assured that the meat is produced sustainably and in a way that the consumer values enough to drive out to the ranch and to pay a higher price for the meat. (Photograph by L. Huntsinger)

heritage values, belief in "sustainable" uses of land, and the appeal of woodland and local landscapes has an as yet un-quantified role. In Spain, there are government certified regional and local appellations to add value to culturally significant products such as acorn-fed ham and other meats and cheeses (Chap. 10). In California, labeling programs are far less developed, although consumer interest in various kinds of designations is on the increase. Some producers market to the public by stating that they manage for ecosystem services and "sustainability," incorporating this into the price. Non-governmental certification programs play a growing role in informing consumers of the ecosystem services associated with buying various products or brands (Fig. 12.6).

Origin and quality labels are important in Spain, with about 10 % of farm products carrying such a label (Chap. 10), although in dehesa-dominated areas the proportion is likely larger. Qualifications may be a geographical origin, a special livestock breed, or the guarantee of a certain level of quality (Bartolomé-García 1994). The premium prices charged can help increase dehesa profits. For example, ham certified to be of the pure Iberian breed, certified as raised in the proper manner, and certified again as finished on acorns in the dehesa in Extremadura, brings an extremely high price as Jamón Ibérico Dehesa de Extremadura (López-Bote 1998; Chap. 10). Another successful label is Corderex, for lamb produced from sheep breeds with a defined quality and origin. In 2011, 792 Extremaduran sheep operations with a stock of 559,000 ewes were registered members. The use of cork is a conservation activity as it directly improves the economic situation of

Fig. 12.7 In spite of an increasing aggressiveness in marketing synthetic cork stoppers and screw caps, there are wine producers who defend the advantages of real cork in wine and there are many consumers who prefer corks. Sometimes the reasons go beyond a belief that cork is an essential part of the culture of wine, to an environmental commitment to maintaining a market that is a critical source of support for many dehesa woodlands. (Photograph by P. F. Starrs)

cork oak woodlands (Campos 1999). Unfortunately, the conservation argument has been insufficiently communicated to wine consumers, and as a consequence there is a limited awareness of the conservation benefits of cork use. This communication is even more important as synthetic corks and screw caps are marketed with increasing aggressiveness (Fig. 12.7).

There are markets for some ecosystem services provided by livestock in California. One prominent example is grazing for control of fire and invasive weeds. Companies have sprung up offering to provide goats specifically for vegetation management, and they charge up to $1,300 per ha for this service. Goats will consume some invasive species that are hard to get other livestock to consume, and they like brush, which is invasive on some rangelands. Ironically, this has caused some ranchers who have traditionally only produced cattle to acquire a herd of goats to rent out. Cattle can be used for fire hazard management, and in fact this is one rationale for grazing on some public lands (Byrd et al. 2009). However, cattle owners most often pay for the privilege of grazing grass even if reduction in fire hazard is a recognized service. The emergence of detailed and highly constrained grazing prescriptions for improving biodiversity and creating specific habitat characteristics is leading to reductions in rancher payments, and may eventually lead to payments for cattle grazing as well as the costs of compliance increase (Germano et al. 2010) (Fig. 12.8).

Fig. 12.8 The use of livestock in the provision of ecosystem services is evident in this California Grazing website, offering "holistic land management and brush & weed control through grazing." The firms in California now making offers of such services are numerous, and the literature suggests that goats are far preferred to cattle or sheep, and that the public finds active pleasure in seeing goats at work in "fuel load reduction" (Brown 2001; Wood 2006) (Courtesy of California Grazing, at http://www.californiagrazing.com/)

12.5.5 Incentivizing Management Practices to Increase Ecosystem Services

Within the oak woodland, the priorities, practices, and tradeoffs among ecosystem services become more complex. A closer look reveals that while large-scale ecosystem function is less altered by ranching than many other forms of agriculture, livestock production has short and long term impacts on the land. Grass and water is consumed, and trails and fences are created. Some types of vegetation and wildlife may flourish with the short and long term changes caused by management practices, while some may decline. Soils and water quality may be affected. Livestock management can be maximized for the production of a single product, meat, or goals can be diversified. Livestock grazing can be used as a tool to create vegetation and soil conditions that favor the co-production of various ecosystem services, or it can be seen solely as a means to an end: animal gain. Programs that motivate landowners to manage for increased and multiple ecosystem services can be a powerful tool for the conservation of oak woodland ecosystem services.

There is good reason to keep oak woodland in production—the provision of some ecosystem services depends on the owner's agricultural and management activities. In Spain, several rare species are dependent on the anthropogenic environment of the dehesa (Díaz and Pulido 1995; Díaz et al. 1997, 2006; Chap. 8). In California, the role of agricultural producers as providers of ecosystem services is increasingly being recognized. In the San Francisco Bay region, half of the available habitat for the endangered California tiger salamander (*Ambystoma californiense*) and some other wildlife and rare species is provided by stockponds created and managed by oak woodland ranchers. In this case too, grazing seems to benefit the target species (DiDonato 2007). In a more complex case, more than half of the habitat for the state-threatened California black rail comes from the leaky irrigation works associated with ranching (Richmond et al. 2010), yet at the pasture-scale grazing should be managed in the small seeps and spring habitats so as not to change the structure of the vegetation at the wrong time of year and hinder bird use. Tradeoffs among ecosystem services are typical: grazing reduced methane emissions from oak woodland seeps and springs, but was associated with a decline in insect species richness (Allen-Diaz et al. 2004). Other examples of habitat improvement with grazing include lowering grass height for burrowing owls (*Athene cunicularia*) (Nuzum 2005) and endangered Stephen's kangaroo rats (*Dipodomys stephensi*) (Kelt et al. 2005); and increasing the abundance of flowers (Barry 2011) and of broadleaved plants for rare butterflies (Weiss 1999). There have been notable cases where grazing exclusion has caused the species being "protected" by the exclusion to leave or disappear (Weiss 1999). In dehesas, maintenance of productive open woodlands with grazing increases biodiversity but hampers tree recruitment, whereas shrub encroachment increases carbon storage and tree regeneration but decreases aesthetic values, biodiversity, and commercial production (Díaz 2009; Chap. 8).

12.5.6 Payments for Ecosystem Services in Practice

More than two-thirds of ranchers surveyed in California were receptive to the idea of being rewarded monetarily "to improve the quantity and/or quality of environmental benefits that their land provides to society," even though many were unfamiliar with the specific term "ecosystem services" (Cheatum et al. 2011). The duration of the commitment they would need to make, and the amount of payment, were important factors in rancher willingness to participate in such ecosystem services production programs, with preference for shorter contracts and higher payments (Cheatum et al. 2011). The kind of entity that would offer the payments was important to prospective sellers, with non-profit organizations or private firms strongly preferred over federal agencies, and state agencies the least preferred.

Fig. 12.9 Unsurprisingly, when asked, ranchers and dehesa owners definitely prefer incentive-based programs to regulatory approaches. Learning about the motives, practices, and goals of woodland owners is essential to designing effective programs. (Photograph by L. Huntsinger)

This probably reflects the regulatory and enforcement roles of governmental agencies (Ferranto et al. 2012) (Fig. 12.9).

Provision of wildlife habitat was the service that ranchers in California would prefer to market or be rewarded for producing, but there was considerable willingness to restore native plants, improve water quality, and increase carbon storage (Cheatum et al. 2011). Ranchers were slightly less interested in increasing oak numbers, perhaps because most are familiar with the difficulties involved and may feel too many oaks will interfere with forage production (Huntsinger et al. 2010). In fact, "improving wildlife habitat" was found to already be a management goal for the majority of California forest and rangeland owners with more than 20 ha of property (Ferranto et al. 2011). Additionally, more than half of forest and rangeland landowners with more than 200 ha stated they currently manage for water quality, build erosion control structures, and manage streams and ponds for wildlife.

In Spain (as in the European Union as a whole), payments for ecosystem services to farmers have been widespread since 1992 under Pillar 2 of CAP. These commonly termed "agro-environmental schemes" aim to promote environmentally compatible production processes and have often been implemented as part of national or regional rural development plans (Oñate et al. 1998). Each region in Spain has designed agro-environmental schemes, some "horizontal," meaning applicable region-wide, and others "zonal," meaning they pertain only to specific areas, usually around protected zones such as natural parks. Horizontal schemes include extensification, organic farming, preservation of indigenous breeds, and agro-environmental training. Zonal schemes may promote livestock stocking rate reduction, reducted of fertilizer use, or the conversion of tilled cropland into extensive grassland. These programs encourage traditional and low-intensity agricultural practices, but not specifically dehesa management or oak protection (the programs are not aimed, in name, at dehesa conservation). Farmers participate voluntarily, with a five-year commitment co-financed by the EU for up to 75 % of the costs. Current proposals for CAP after 2013 indicate the "greening" of European farm policies is likely to continute into the future (BOE 2010, 2011).

Nevertheless, dehesas receive the highest priority for afforestation subsidies. European Union subsidies cover planting and maintenance costs through the first 5 years, and provide a premium to cover the loss of income resulting from afforestation—grazing is generally not allowed in plantations for 20 years after planting (15 years in the last regulation). In the period 1994–2000, reforestation subsidies supported the planting of 197,600 ha of holm oak and 83,435 ha of cork oak (Ovando et al. 2007).

Although there are local ordinances protecting certain trees, in general California has approached conserving trees in oak woodlands mostly through education of landowners and managers. As part of the Integrated Hardwood Range Management Program (IHRMP; Chap. 2), outreach focused on the values of oaks. These included the contribution of oaks to maintaining property values, increasing wildlife habitat, and in some cases, extending the green forage season for grazing. Over the period of the program's duration, 1985 – 2010, there was an increase in oak planting by landowners and a reduction in cutting (Huntsinger et al. 2010). Efforts were linked to an understanding of rancher needs and values derived from survey research.

California ranchers have already captured some of the value of some ranch ecosystem services through cost-share programs for habitat improvement and environmental quality improvements. The Wildlife Habitat Incentives Program (WHIP), the Environmental Quality Incentives Program of the Natural Resources Conservation Service (EQIP), and the Grassland Reserve Program (GRP), are three examples of payment for ecosystem services programs available to oak woodland landowners in California. In 2011, EQIP provided $74 million to California farmers and ranchers for carrying out projects to improve "environmental quality" while WHIP paid out $3.6 million (USDA-NRCS-EQIP 2011; USDA-NRCS-WHIP 2011). Cost-shares are based in the idea that public benefits make public investment in these projects worthwhile. Altogether the USDA Natural Resources Conservation Service spent another $5.4 million on conservation practices in California between 2005 and 2009, including brush management, prescribed grazing, and upland wildlife habitat (Tanaka et al. 2011).

12.6 Ecosystem Services and Landowner Decisions

Ecosystem services have major impacts on landowner behavior and on land markets. They affect the prices landowners are willing to pay for land, shape management decisions, and offer opportunities to diversify and expand income streams. When private ecosystem service values are not included in models, erroneous conclusions are reached about likely management behavior and appropriate public policies. The applications of the conventional model to cattle ranchers in the American West once led agricultural economists to think that either they were wrongly evaluating the models or that the landowners were behaving "irrationally." As stated in an Economic Research Service report in 1972, ranchers, "when contrasted to more progressive agriculturalists, seem to make irrational economic decisions and continue to employ economically unproductive managerial strategies" (Schultz 1972). Smith and Martin (1972) attributed ranch prices that were higher than could be justified by beef production to "ranch fundamentalism" (an attachment to the land and lifestyle) and "conspicuous consumption" (investment and enjoyment of owning a significant property). These concepts can be seen as a way of describing the non-commodity ecosystem services that ranchers and dehesa landowners personally consume from their land (Starrs 1997). When they are valued and included in an analysis of ranching's "bottom line," rancher choices are much easier to understand, and outreach and research programs can be developed that better recognize their motivations.

In California, the IHRMP, initiated in 1985, was perhaps one of the first large scale research and management programs that funded research into landowner attitudes and values as part of developing a statewide outreach program (Huntsinger and Fortmann 1990). Research indicated that landowners with smaller properties were growing in number, and had different goals and motives for living in the oak woodlands than did owners of large properties. Owners of large properties were more likely to produce livestock, while owners of small properties focused on amenity and investment values. These findings were used to design outreach approaches attractive to each type of owner.

Ranchers in California oak woodlands readily agree that "income maximization" in the conventional sense is not their goal (Liffmann et al. 2000). Instead, a financially sustainable operation that maximizes landowner autonomy in decision-making, provides a good place to raise a family, and is based on enjoyable work is more important to most ranchers (Liffmann et al. 2000; Huntsinger et al. 2010). In both the U.S. and Spain, ranch prices are consistently above those that can be justified by commercial production value alone, indicating substantial landowner consumption of non-market benefits from the land (Campos and Riera 1996; Torell et al. 2005; Campos et al. 2009). In the United States, the majority of ranch owners work off-ranch to support the property (Smith and Martin 1972; Torell and Kincade 1996; Huntsinger et al. 2010). In fact, in their attitudes toward conservation of ecosystem services, Californian and Spanish oak woodland landowners have much in common (Huntsinger et al. 2004). For the most part, they are enthusiastic about the amenities produced from the management of their properties (Campos et al. 2009), and aware that society values them, but at the same time, they strongly seek to maintain control over management decisions and practices on their land.

The ecosystem services consumed as amenities by the oak woodland owner have been compared in Spain and in California using the contingent valuation technique (Campos et al. 2009; Sect. 12.3). These benefits include those of using the land for private recreation, having the opportunity to leave the land to heirs, to welcome friends and visitors, and to enjoy a "country way of life". In both Spain and California, landowners reported they enjoyed the beauty of the woodlands, hunting and fishing, legacy values, and other lifestyle benefits (Liffmann et al. 2000; Campos et al. 2009; Huntsinger et al. 2010). Spanish landowners tended to be more focused on recreational values, while in California, landowners reported a greater focus on lifestyle and legacy values in general (Campos et al. 2009). Part of the reason for this difference is probably the fact that dehesa owners tend to live off-site in the city and only visit their properties when they are not working (Fig. 12.10), while oak woodland ranches and their families generally live and work on-site (Campos et al. 2009). The difference in income between keeping the land and making an alternative, more profitable investment can be considered what landowners actually pay for the ecosystem services they enjoy from their land. Campos et al. (2009) analyzed responses of landowners from a sample of oak woodlands in Spain and in California who were asked the maximum amount they were willing to lose before selling their property. The authors found a high willingness to pay by landowners for keeping their property and enjoying ecosystem services from oak woodlands: \$135 per ha in California and up to \$213 per ha in south Spain in 2002.

Fig. 12.10 The dehesa provides a beautiful setting for this traditional, if palatial, estate. The status of having such a home in the dehesa is one of the ecosystem services consumed by landowners. (Photograph by P. F. Starrs)

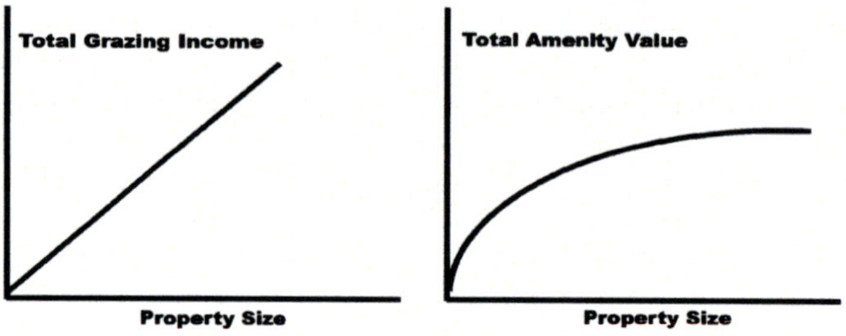

Fig. 12.11 The relationship between grazing income and property size, and amenity values and property size. For grazing, commercial income increases with each additional hectare as property size increases. The value to the landowner of amenities does not increase much after a certain property size is reached, shown by the flattened curve. This demonstrates that the amenity values of landownership for most owners is satisfied when they have acquired a property of a certain size, and having an even larger property does not increase their enjoyment much (Oviedo et al. 2012). This helps to explain the popularity of 10–30 acre "ranchettes"

Oviedo et al. (2012) offers a detailed description for the California situation, showing that private amenity values are an important influence on landowners. However, the authors show that there remains the risk of subdivision and habitat fragmentation if commercial operations disappear, because the ecosystem service value per acre to landowners levels off when properties get large (Fig. 12.11). Ovando et al. (2010), in an extended analysis of these values in Spanish forest investment scenarios, showed that the increasing value of private ecosystem services on a per hectare basis as properties increased in size also leveled off. Many non-market ecosystem service values can be satisfied with a property of a few hectares. On the other hand, commercial values from livestock production and other natural resource products continue to increase with property size, and some minimum, yet unknown, property size would be needed to maintain biodiversity, populations of endangered species, and tree recruitment in dehesas (Díaz 2009; Chap. 8). Combining these two "valuations" is the basis of the "working landscapes" effort in California to encourage joint production of commercial and non-commercial ecosystem goods and services in order to create sustainable rangeland enterprises.

Standiford and Howitt (1992) developed a model to evaluate optimum management strategies for privately owned oak woodland ranches producing cattle, firewood and hunting in California. This model was calibrated for the early 1990s and the researchers concluded that markets at that time would lead landowners to clear their oaks to increase forage yield for livestock production. Instead, clearing oaks was actually rare in the 1990s compared to the 1940s to 1970s (Standiford et al. 1996), contradicting the prediction of the model. The model's shortcomings were due to the failure to accurately account for the landowner's desire to keep oaks for their amenity value. Later, using an alternative approach that took into account the ecosystem service value of the oak to landowners, the decisions landowners made to retain oaks were explained (Howitt 1995; Standiford et al. 1996). This model is detailed in Chap. 9.

Fig. 12.12 Madrid's traditional Fiesta de la Trashumancia (Trashumance Festival) is celebrated yearly in October. This celebration evokes the traditional movement of livestock from the high mountains of northern Spain to the more temperate dehesa regions of south-west Spain where the livestock are wintered. Owners and herders of livestock move in a procession along the Calle de Alcala, which has its origins as a *cañada real* used by practitioners of transhumance in the distant past to bring animals to market, but is still recognized in Spanish law as a stock driveway. Celebration of the annual return to the Puerta de Alcala began anew in 1994, in part as activists insisted on passage of a new law of *Vias Pecuarias* in Spain. Few scenes are more colorful than seeing sheep, cows, and herders walking to the Puerta, one of the main symbols of the city of Madrid. (Photograph by J. Garzón)

12.7 Conclusions

The reviewed studies show societal benefits from oak woodland ecosystem services. Quantifying such values can contribute to policy and management, but it is at the same time all but impossible to capture the entire social-economic value of an ecosystem. Ecosystem services have unquantifiable cultural and spiritual meanings, and we have at best only limited knowledge of the ecological relationships that make an ecosystem function and that increase human well-being. Natural resources including space, scenery, water, skies, have profound and diverse meanings to people. However, the lack of quantified information has often led to undervaluing and neglect of ecosystem services in public and private decision-making. Analyses of ecosystem service values attempt to prevent undervaluation, not to capture the entire value of a service to society (Fig. 12.12). Ecosystem services can make a powerful argument for improved management and greater conservation of oak woodlands and dehesa.

Opponents of the concept object that such framing includes only services delivering economic benefits for society today or, hypothetically, in the future, based on the present generation's needs. Therefore, the concept neglects all ecosystem services whose use and value are still unclear (Pistorius et al. 2012). Concerns are voiced that poorly designed instruments may inflate the value of single ecosystem services (for example carbon sequestration) at the risk of other services that are less marketable (Redford and Adams 2009; Caparrós et al. 2010). Enhancing ecosystem services in a "bundle" is an exception to policy so far, unless one considers the conservation easement a form of bundling services. One avenue to resolving this is to evaluate ecosystem services at a large scale and within an accounting system that incorporates competing ecosystem services in a consistent manner (Campos et al. 2001, 2007b; Caparrós et al. 2003; Campos and Caparrós 2006).

While it is important to keep conceptual shortcomings in mind, more information about the value of ecosystem services from oak woodlands could improve the quality of decision-making and policy development. The value of ecosystem services that are lost through rangeland fragmentation, for example, is not clearly known. In much of California, costs would include increased wildfire suppression and firefighting, loss of wildlife habitat on lands directly affected, loss of wildlife on nearby lands, reductions in water quality due to pollutants and urban runoff from developed lands, loss of air quality due to auto exhaust (especially when long commutes, as is typical from the wildland fringe, are factored in), loss of range-based livestock products, and so forth. There are of course many compelling reasons for allowing land development, but analysis of trade-offs is hindered by limited knowledge of the environmental costs.

Research effort in this direction for oak woodlands, in Spain as in California, is expanding. The next step is to extend case studies and to undertake analyses at the regional, state, and nation-wide scales. Despite scientific advances, the need for preservation of the natural capital of oak woodlands and the many ecosystem services the woodlands provide is far from fully recognized by society. An important future policy task will be incorporating payments for provision of biodiversity and ecosystem services into agricultural, water, energy, and other policies.

Acknowledgments The authors are grateful to two anonymous referees for constructive criticisms on an earlier draft.

References

Allen-Diaz B, Jackson RD, Bartolome JW, Tate KW, Oates LG (2004) Long-term grazing study in spring-fed wetlands reveals management tradeoffs. Cal Agric 58(3):144–148

ARB [Air Resources Board] (2006) California Global Warming Solutions Act of 2006, California State Assembly Bill 32. http://www.arb.ca.gov/cc/docs/ab32text.pdf

Arriaza Balmón M, González Arenas J, Ruiz Avilés P, Cañas Madueño JA (2002) Determinación del valor de uso de cinco espacios protegidos de Córdoba y Jaén. Rev Esp Estud Agrosoc y Pesq 196:153–172
Barry S (2011) Current findings on grazing impacts: California's special status species benefit from grazing. Cal Cattlemen June:18–20
Bartolomé-García TJ (1994) Denominaciones de origen y calidad. In: La agricultura y la ganadería extremeñas en 1993: Caja de Extremadura
Bean MJ, Wilcove DS (1997) The private-land problem. Cons Biol 11(1):1–2
Berbel J, Mesa P (2007) Valoración del agua de riego por el método de precios quasi-hedónicos: aplicación al Guadalquivir. Econ Agrar y Recurs Natur 7(14):127–144
Berger B (2009) What owners want and governments do—evidence from the oregon experiment. Fordham Law Rev 78:1281–1331
BOE [Boletín Oficial del Estado] (2003). Ley 43/2003, de 21 de noviembre, de Montes. BOE núm. 280, de 22-11-2003, 41422–41442
BOE [Boletín Oficial del Estado] (2007a). Ley 5/2007, de 3 de abril, de la Red de Parques Nacionales. BOE num. 81, de 4 de abril de 2007:14639–14649
BOE [Boletín Oficial del Estado] (2007b). Ley 42/2007, de 13 de diciembre, del Patrimonio Natural y de la Biodiversidad. BOE num. 299, de 14 de diciembre de 2007: 51275–51327
BOE [Boletín Oficial del Estado] (2010). Real Decreto 1852/2009, de 4 de diciembre, por el que se establecen los criterios para subvencionar los gastos en el marco de los Programas de Desarrollo Rural cofinanciados por el Fondo Europeo Agrícola de Desarrollo Rural (FEADER). BOE num. 1, de 1 de enero de 2010: 3–9
BOE [Boletín Oficial del Estado] (2011). Real Decreto 1336/2011, de 3 de octubre, por el que se regula el contrato territorial como instrumento para promover el desarrollo sostenible del medio rural. BOE num. 239, de 4 de octubre de 2011: 104199–104206
Boyd J (2007) Nonmarket benefits of nature: What should be counted in green GDP? Ecol Econ 61:716–723
Boyd J, Banzhaf S (2007) What are ecosystem services? The need for standardized environmental accounting units. Ecol Econ 63:616–626
Brown PL (2001) Goats used in California to prevent brush fires. New York Times 14 Oct, http://www.nytimes.com/2001/10/14/us/goats-used-in-california-to-prevent-brush-fires.html. Accessed Sep 2012
Byrd KB, Rissman AR, Merenlender AM (2009) Impacts of conservation easements for threat abatement and fire management in a rural oak woodland landscape. Landscape and Urban Plann 92(2):106–116
CA-LAO [California's Legislative Analyst's Office] (2004) Williamson Act—Subventions for Open Space. Analysis of the Budget Bill–General Government Feb 18. http://www.lao.ca.gov/laoapp/pubdetails.aspx?id=1147 Accessed Dec 2011
California Resources Agency (2004) Hearst ranch conservation transaction overall summary, prepared by California State Resources Agency Staff, July 12. http://resources.ca.gov/hearst_docs/overall-transaction-summary-1.pdf Accessed Sep 2012
CCX [Chicago Climate Exchange] (2009) Chicago climate exchange sustainably managed rangeland soil carbon sequestration offset project protocol. http://theccx.com/docs/offsets/CCX_Sustainably_Managed_Rangeland_Soil_Carbon_Sequestration_Final.pdf. Accessed Feb 2012
Campos P (1984) Economía y energía en la dehesa extremeña. Instituto de Estudios Agrarios, Pesqueros y Alimentarios, MAPA, Madrid: 336
Campos P (1997) Análisis de la rentabilidad económica de la dehesa. Situación. Serie de Estudios Regionales Extremadura, 111–121
Campos P (1998) Contribución de los visitantes a la conservación de Monfragüe. Bienes públicos, mercado y gestión de los recursos naturales. In: Hernández CG (ed) La dehesa: aprovechamiento sostenible de los recursos naturales. Fundación Pedro Arce y Editorial Agrícola Española, Madrid, pp 241–263

Campos P (1999) Alcornocales del suroeste ibérico. In: Marín F, Domingo J, Calzado A (eds), Los montes y su historia. Una perspectiva política, económica y social. Universidad de Huelva, Huelva:245–285
Campos P (2002) Economía del uso múltiple de los Montes Propios de Jerez de la Frontera (1991–1993). Rev Esp Estud Agrosoc y Pesq 195:147–186
Campos P, Caparrós A (2006) Social and private total Hicksian incomes of multiple use forests in Spain. Ecol Econ 57:545–557
Campos P, Mariscal P (2003) Preferencias de los propietarios e intervención pública: el caso del las dehesas de la comarca de Monfragüe. Invest Agrar Sist Recu For 12(3):87–102
Campos P, Riera P (1996) Rentabilidad social de los bosques. Análisis aplicado a las dehesas y los montados ibéricos. Información Comercial Española 751:47–62
Campos P, Andrés RD, Urzainqui E, Riera P (1996) Valor económico total de un espacio de interés natural. La dehesa del área de Monfragüe. In: Azqueta D, Pérez L (eds), Gestión de espacios naturales. La demanda de servicios recreativos. McGraw-Hill, Madrid, pp 192–215
Campos P, Rodríguez Y, Caparrós A (2001) Towards the Dehesa total income accounting: theory and operative Monfragüe study cases. Investigación Agraria: Sistemas y Recursos Forestales Monográfico fuera de serie New forestlands economic accounting. Theories Appl 1:45–69
Campos P, Bonnieux F, Caparrós A, Paoli JC (2007a) Measuring total sustainable incomes from multifunctional management of corsican maritime pine and Andalusian cork oak mediterranean forests. J Envi Plan Manag 50(1):65–85
Campos P, Caparrós A, Oviedo JL (2007b) Comparing payment-vehicle effects in contingent valuation studies for recreational use in two Spanish protected forests. J Leis Res 39(1):60–85
Campos P, Daly H, Oviedo JL, Ovando P, Chebil A (2008) Accounting for single and aggregated forest incomes: application to public cork oak forests of Jerez in Spain and Iteimia in Tunisia. Ecol Econ 65:76–86
Campos P, Oviedo JL, Caparros A, Huntsinger L, Coelho I (2009) Contingent valuation of woodland-owner private amenities in Spain, Portugal, and California. Rangel Ecol Mgmt 62(3):240–252
Caparrós A, Campos P, Montero G (2003) An operative framework for total hicksian income measurement: application to a multiple use forest. Envi Res Econ 26:173–198
Caparrós A, Cerdá E, Ovando P, Campos P (2010) Carbon sequestration with reforestations and biodiversity-scenic values. Envi Res Econ 45:49–72
Caparrós A, Ovando P, Oviedo JL, Campos P (2011) Accounting for carbon in avoided degradation and reforestation programmes in Mediterranean forests. Envi Devel Econ 16(4):405–428
CDF-FRAP [California Department of Forestry and Fire Protection-Fire and Resource Assessment Program, CalFire] (2003) Changing California: forest and range 2003 assessment. Sacramento, State of California Resources Agency, CA. http://www.frap.cdf.ca.gov/assessment2003. Accessed June 2012
Chan KMA, Shaw MR, Cameron DR, Underwood EC, Daily GC (2006) Conservation planning for ecosystem services. PLoS Biol 4(11):1–15
Chaplin-Kramer R, Tuxen-Bettman K, Kremin C (2011) Supplying pollination services to California agriculture. Rangel 33(3):33–41
Cheatum M, Casey F, Alvarez P, Parkhurst B (2011) Payments for ecosystem services: a California rancher perspective. Nicholas Institute for Environmental Policy Solutions, Duke University, Washington D.C
Costanza R, Ralph d'Arge R, de Groot R, Farberk S, Grasso M, Hannon B, Limburg K, Shahid Naeem S, O'Neill RV, Paruelo J, Raskin RG, Suttonkk P, van den Belt M (1997) The value of the world's ecosystem services and natural capital. Nature 387:253–260
Diamond NK, Standiford RB, Passof PC, LeBlanc J (1987) Oak trees have varied effect on land values. Cal Ag 41:4–6
Díaz M (2009) Biodiversity in the dehesa. In: Mosquera MR, Rigueiro A (eds) Agroforestry systems as a technique for sustainable land management. AECID, Madrid, pp 209–225

Díaz M, Pulido FJ (1995) Wildlife-habitat relationships in the Spanish dehesa. In: McCracken DI, Bignal EM, Wenlock SE (eds) Farming on the edge: the nature of traditional farmland in Europe. Joint Nature Conservation Committee, Peterborough, pp 103–111

Díaz M, Campos-Palacín P, Pulido FJ (1997) The Spanish dehesas: a diversity in land use and wildlife. (Chapter 7) In: Pain DJ, Pienkowski M (eds) Farming and birds in Europe: the common agricultural policy and its implications for bird conservation. Academic Press, London, p 436

Díaz M, Baquero RA, Carricondo A, Fernández F, García J, Yela JL (2006) Bases ecológicas para la definición de las prácticas agrarias compatibles con las Directivas de Aves y de Hábitats. Ministerio de Medio Ambiente-Universidad de Castilla-La Mancha, Toledo. http://www.mma.es/portal/secciones/biodiversidad/desarrollo_rural_paisaje/naturaleza_rural/estudios_bases_ecologicas.htm

DiDonato J (2007) Endangered amphibian research within grazed grasslands. Keeping landscapes working, University California Cooperative Extension Newsletter for Rangeland Managers Winter: 4–6. http://cesantaclara.ucdavis.edu/newsletterfiles/Keeping_Landscapes_Working10641.pdf

EEA [European Environment Agency] (2011) An experimental framework for ecosystem capital accounting in Europe. EEA. Copenhagen

Ehrlich PR, Ehrlich AH (1981) Extinction: the causes and consequences of the disappearance of species, 1st edn. Random House, New York

Ehrlich PR, Mooney HA (1983) Extinction, substitution, and ecosystem services. BioSci 33:248–254

EEME [La Evaluación de los Ecosistemas del Milenio de España] (2011) La Evaluación de los Ecosistemas del Milenio de España. Síntesis de resultados. Fundación Biodiversidad. Ministerio de Medio Ambiente, y Medio Rural y Marino

Ferranto S, Huntsinger L, Getz C, Nakamura G, Stewart W, Drill S, Valachovic Y, DeLasaux M, Kelly M (2011) Forest and rangeland owners value land for natural amenities and as financial investment. Cal Ag 65(4):185–191

Ferranto S, Huntsinger L, Stewart B, Getz C, Nakamura G, Kelly M (2012) Consider the source: the impact of media and authority in outreach to California's forest and rangeland owners. J Environ Manage 97:131–140

Germano DJ, Rathbun GB, Saslaw LR (2010) Effects of grazing and invasive grasses on desert vertebrates in California. J Wildl Mgmt 9999:1–13

Gosnell H, Robinson-Maness N, Charnley S (2011) Engaging ranchers in market-based approaches to climate change mitigation: opportunities, challenges, and policy implications. Rangel 64:20–24

Gustankii JA, Squires RH (2000). Protecting the land: conservation easements past, present and future. Island Press, Washington D.C.

Haines-Young R, Potschin M (2010) Proposal for a Common International Classification of Ecosystem Goods and Services (CICES) for Integrated Environmental and Economic Accounting (V1). Fifth Meeting of the UN Committee of Experts on Environmental-Economic Accounting. Department of Economic and Social Affairs, Statistics Division, United Nations. New York, 23–25 June 2010. Available online

Howitt R (1995) Positive mathematical programming. Am J Ag Econ 77:329–342

Huntsinger L, Fortmann LP (1990) California privately owned oak woodlands—owners, use, and management. J Range Man 43(2):147–152

Huntsinger L, Sulak A, Gwin L, Plieninger T (2004) Oak woodland ranchers in California and Spain: conservation and diversification. In: Schnabel S and Ferreira A (eds) Sustainability of Agrosilvopastoral Systems: Dehesas, Montados. Chapter 6. Adv Geoecology 37:309–326

Huntsinger L, Johnson M, Stafford M, Fried J (2010) Hardwood rangeland landowners in California from 1985 to 2004: production, ecosystem services, and permanence. Rangel Ecol Manage 63(3):324–334

Joffre R, Ourcival J, Rambal S, Rocheteau A (2003) The key role of topsoil moisture on CO2 efflux from a mediterranean quercus ilex forest. Ann For Sci 60:519–526

Kelt DA, Konno ES, Wilson JA (2005) Habitat management for the endangered Stephens' kangaroo rat: the effect of mowing and grazing. J Wildl Manag 69(1):424–429

Kovacs K, Václavík T, Haight RG, Pang A, Cunniffe NJ, Gilligan CA, Meentemeyer RK (2011) Predicting the economic costs and property value losses attributed to sudden oak death damage in California (2010–2020). J Environ Manage 92:1292–1302

Kroeger T, Casey F, Alvarez P, Cheatum M, Tavassol L (2010) An economic analysis of the benefits of habitat conservation on California rangelands, a conservation economics white paper. Defenders of Wildlife, Washington D.C

Krutilla JV (1967) Conservation reconsidered. Am Econ Rev 57(4):777–786

Land Trust Alliance (2010) 2010 National land trust census report. A Look at voluntary land conservation in America. Available at http://www.landtrustalliance.org/land-trusts/land-trust-census/national-land-trust-census-2010/2010-final-report. Accessed Sep 2012

Li YL, Tenhunen J, Mirzaei H, Hussain MZ, Siebicke L, Foken T, Otieno D, Schmidt M, Ribeiro N, Aires L, Pio N, Banza J, Pereira J (2008) Assessment and up-scaling of CO_2 exchange by patches of the herbaceous vegetation mosaic in a Portuguese cork oak woodland. Agric For Meteorol 148(8–9):1318–1331

Liffmann RH, Huntsinger L, Forero LC (2000) To ranch or not to ranch: Home on the urban range? J Range Man 53(4):362–370

Loomis J, Gonzalez-Caban A (1996) A willingness to pay function for protecting acres of spotted owl habitat from fire. Ecol Econ 25:315–322

López-Bote CJ (1998) Sustained utilization of the Iberian pig breed. Meat Sci 49:17–27

Martin WE, Jeffries GL (1966) Relating ranch prices and grazing permit values to ranch productivity. J Farm Econ 48:233–242

Merenlender AM, Huntsinger L, Guthey G, Fairfax SK (2004) Land trusts and conservation easements: Who is conserving what for whom? Conser Biol 18(1):65–75

Millennium Ecosystem Assessment (2003) Ecosystems and human well-being: a framework for assessment. Island Press, Washington, DC

Millennium Ecosystem Assessment (2005) Ecosystems and human well-being: synthesis. Island Press, Washington, DC

Montambault JR, Alavalapatti JRR (2005) Socioeconomic research in agroforestry: a decade in review. Agrofor Sys 65:151–161

Newburn D, Reed S, Berck P, Merenlender A (2005) Economics and land-use change in prioritizing private land conservation. Cons Biol 19(5):1411–1420

Newburn DA, Berck P, Merenlender AM (2006) Habitat and open space at risk of land-use conversion: targeting strategies for land conservation. Am J Ag Econ 88(1):28–42

Nuzum RC (2005) Report: Using livestock grazing as a resource management tool in California. Contra Costa Water District report. www.ccwater.com/files/LivestockGrazingFinal72005.pdf. Accessed Aug 2006

O'Geen AT, Dahlgren RA, Swarowsky A, Tate KW, Lewis DJ, Singer MJ (2010) Research connects soil hydrology and stream water chemistry in California oak woodlands. Cal Ag 64(2):78–84

Olewiler N (2004). The value of natural capital in settled areas of Canada. Ducks Unlimited Canada and the Nature Conservancy of Canada, Canada, p 36 http://www.ducks.ca/aboutduc/news/archives/pdf/ncapital.pdf

Oñate JJ, Malo JE, Suárez F, Peco B (1998) Regional and environmental aspects in the implementation of Spanish agri-environmental schemes. J Envir Mgmt 52:227–240

Ovando P, Campos P, Montero G (2007) Forestaciones con encina y alcornoque en el área de la dehesa en el marco del Reglamento (CE) 2080/92 (1993–2000). Rev Esp Estud Agrosoc y Pesq 214:173–186

Ovando P, Campos P, Oviedo JL, Montero G (2010) Private net benefits from afforesting marginal crop and shrublands with cork oaks in Spain. For Sci 56:567–577

Oviedo JL, Caparrós A, Campos P (2005) Valoración contingente del uso recreativo y de conservación de los visitantes del parque natural los Alcornocales. Rev Esp Estud Agrosoc Pesq 208:115–140

Oviedo JL, Huntsinger L, Campos P, Caparrós A (2012) Income value of private amenities assessed in California oak woodlands. Cal Agric 66(3):91–96

Pereira JS, Mateus JA, Aires LM, Pita G, Pio C, David JS, Andrade V, Banza J, David TS, Paco TA, Rodrigues A (2007) Net ecosystem carbon exchange in three contrasting mediterranean ecosystems—the effect of drought. Biogeosci 4(5):791–802

Pistorius T, Schaich H, Winkel G, Plieninger T, Bieling C, Konold W, Volz K-R (2012) Lessons for REDDplus: a comparative analysis of the German discourse on forest functions and the global ecosystem services debate. For Pol Econ 18:4–12

Redford KH, Adams WM (2009) Payment for ecosystem services and the challenge of saving nature. Conser Biol 23(4):785–787

Reiner R, Craig A (2011) Conservation easements in California blue oak woodlands: testing the assumption of livestock grazing as a compatible use. Nat Areas J 31(4):408–413

Richmond OMW, Chen SK, Risk BB, Tecklin J, Beissinger SR (2010) California black rails depend on irrigation-fed wetlands in the Sierra Nevada foothills. Cal Ag 64(2):85–93

Rilla E, Sokolow AD (2000) California farmers and conservation easements: motivations, experiences and perceptions in three counties. In California Farmland and Open Space Policy Series: U of California Agricultural Issues Center

Ruiz-Avilés P, Madueño C, Arenas JG, Antonio J (2001) Economía ambiental de los parques naturales de Córdoba. Universidad de Córdoba, Córdoba

SCEP [Study of Critical Environmental Problems] (1970) Man's impact on the global environment; assessment and recommendations for action. MIT Press, Cambridge

Schultz JL (1972) Sociocultural factors in financial management strategies of western livestock producers. Final report. Economic Research Service Farm Production Economics Division, U.S. Department of Agriculture. J., Washington, D.C. http://www.sbcouncil.org/Ranching-History-and-Trends. Accessed July 2012

Smith AH, Martin WE (1972) Socioeconomic behavior of cattle ranchers, with implications for rural community development in the west. Am J Ag Econ 54(2):217–225

Standiford RB, Howitt RE (1992) Solving empirical bioeconomic models: a rangeland management application. Am J Ag Econ 74:421–433

Standiford RB, Scott T (2001) Value of oak woodlands and open space on private property values in southern California. Inv Agrar Sist y Recurs, special issue 1:137–152

Standiford RB, McCreary D, Gaertner S, Forero L (1996) Impact of firewood harvesting on hardwood rangelands varies with region. Cal Ag 50:7–12

Starrs PF (1997) Let the cowboy ride: cattle ranching in the American west. Johns Hopkins University Press, Baltimore

Sulak A, Huntsinger L, Standiford R, Merenlender A, Fairfax SK (2004) A strategy for oak woodland conservation: the conservation easement in California. In: Schnabel S and Ferreira A (eds) Sustainability of Agrosilvopastoral Systems: Dehesas, Montados. Chapter 6. Adv Geoecology 37:353–364

Talbert CB, Knight RL, Mitchell JE (2007) Private ranchlands and public land grazing in the southern rocky mountains. Rangel 29(3):5–8

Tanaka JA, Brunson M, Torrell A (2011) Chapter 9: a social and economic assessment of rangeland conservation practices, pp 373–422. In: Briske DD (ed) Conservation benefits of rangeland practices: assessment, recommendations, and knowledge gaps. United States Department of Agriculture, Natural Resources Conservation Service. Lawrence, Allen Press, KS, p 429

The David and Lucile Packard Foundation (2011) Agriculture Survey. Conservation highlights. Available at: http://www.iaenvironment.org/documents/Packard%20Farm%20Bill%20Conservation%20FINAL.pdf. Accessed Aug 2012

Thompson RP, Noel JE, Cross SP (2001) Oak woodlands economics: a contingent valuation of conversion alternatives. In Proceedings of the fifth symposium on oak woodlands: Oaks in California's changing landscape, San Diego, California, 22–25 Oct 2001

Torell LA, Kincaid ME (1996) Public land policy and the market value of New Mexico ranches, 1979–94. J Range Man 49(3):270–276

Torell LA, Rimbey NR, Ramirez OA, McCollum DW (2005) Income earning potential versus consumptive amenities in determining ranchland values. J Agric Resour Econ 30(3):537–560
USDA-NRCS-EQIP [United States Department of Agriculture Natural Resources Conservation Service-Environmental Quality Incentives Program] (2011) EQIP 2011 contracts and funding data. http://www.nrcs.usda.gov/wps/portal/nrcs/detailfull/national/programs/financial/eqip/?&cid=nrcs143_008294. Accessed Dec 2011
USDA-NRCS-WHIP [United States Department of Agriculture Natural Resources Conservation Service-Wildlife Habitat Improvement Program] (2011) WHIP 2011 contracts and funding. http://www.nrcs.usda.gov/wps/portal/nrcs/detail/national/programs/financial/whip/?&cid=stelprdb1041997. Accessed Dec 2011
Weiss SB (1999) Cars, cows, and checkerspot butterflies: nitrogen deposition and management of nutrient-poor grasslands for a threatened species. Conser Biol 13(6):1476–1486
Wood DV (2006) How to keep fires down in California scrub: chew it. The Christian Science Monitor, 18 Sept

Chapter 13
The Private Economy of Dehesas and Ranches: Case Studies

José L. Oviedo, Paola Ovando, Larry Forero, Lynn Huntsinger, Alejandro Álvarez, Bruno Mesa and Pablo Campos

Frontispiece Chapter 13. Since the 1273 implementation of the Crown's Law of the Mesta—a regulation protecting the producers of Merino sheep and preventing the loss of dehesa grasslands—amenity consumption has been a common characteristic of dehesa landownership. This palatial dehesa estate in Navalvillar de Pela, Extremadura, dates from 1778 and belonged to the Orellanas. This family had a prominent role in colonization of the Americas, and can count the conquerors Hernán Cortés, Francisco Pizarro, and Francisco de Orellana among its ancestral members. (Photograph by A. Adámez)

J. L. Oviedo (✉) · P. Ovando · A. Álvarez · B. Mesa · P. Campos
Institute of Public Goods and Policies (IPP), Spanish National Research Council (CSIC), Albasanz 26-28 28037 Madrid, Spain
e-mail: jose.oviedo@csic.es

P. Ovando
e-mail: paola.ovando@cchs.csic.es

P. Campos et al. (eds.), *Mediterranean Oak Woodland Working Landscapes*, Landscape Series 16, DOI: 10.1007/978-94-007-6707-2_13,

© Springer Science+Business Media Dordrecht 2013

Abstract This chapter's objective is to measure and analyze total private income and profitability for five case study privately-owned dehesas and oak woodland ranches. The *Agroforestry Accounting System* is applied at the farm scale. Results are estimated for individual forestry, game, livestock, crop, and service activities, and for activities aggregated as a whole. The case study application incorporates landowner consumption of private amenities as part of the total income from the dehesa or ranch, showing that these private amenities are the most important contributor to total income, while the contribution from livestock production is low or even negative. Hunting activities show low revenues. Dehesas with a high stocking rate are significantly supported by European Union livestock subsidies, while livestock production and other activities on California ranches are more sensitive to market conditions. Both in Spain and California, real profitability is competitive with alternative non-agricultural investments when amenity consumption and increases in land value are considered. These results are relevant to understanding current and future trends in landowner motivations for land and enterprise investment, and should be considered in conservation policy development.

Keywords Total income · Private amenity · Livestock production · Land appreciation · Oak woodland economics

13.1 Introduction

The land use and vegetation mosaic characteristic of dehesas in Spain and oak woodland ranches in California are shaped by human management practices and activities (Bolsinger 1988; Díaz et al. 1997; FRAP 2003; MARM 2008). Dehesas and ranches are typically managed for multiple uses, combining production of

---

A. Álvarez
e-mail: alejandro.alvarez@cchs.csic.es

B. Mesa
e-mail: bruno.mesa@cchs.csic.es

P. Campos
e-mail: pablo.campos@csic.es

L. Forero
Cooperative Extension, Shasta County, University of California Cooperative Extension, 1851 Hartnell Avenue, Redding, CA 96002-2217, USA
e-mail: lcforero@ucanr.edu

L. Huntsinger
Department of Environmental Science, Policy, and Management, University of California, Berkeley, 130 Mulford Hall , Berkeley, CA, MC 3110 94720-3110, USA
e-mail: huntsinger@berkeley.edu

livestock, game, wood products, and crops. Crops are often used as supplemental fodder for livestock and game. The particular combination of uses depends on the unique setting of each dehesa or ranch, the kinds of trees, the fertility of the soil, local rainfall and topography, adjacent land uses, understory species, and markets. Analysis and comparison of the economies of these working landscapes must be framed by differences and similarities in institutional settings, property rights, national and regional economies, management practices, and landowner objectives (Huntsinger et al. 2004).

Economic activities in dehesas and ranches have evolved through time, departing from an almost exclusive reliance on grazing-based livestock operations to developing new practices, capital investments, and products in response to changing landowner and social demands (Díaz et al. 1997; Huntsinger and Sayre 2007; Campos et al. 2001, 2009a) (Fig. 13.1). There is increasing economic value attributed to the public benefits of ranch and dehesa ecosystem services. This chapter, however, focuses solely on the private economy of dehesas and ranches, excluding the value of the public ecosystem services analyzed in Chap. 12.

In this chapter private total income (hereafter total income) and profitability rates (hereafter profitability) are estimated and measured for five case studies of privately-owned dehesas and ranches using the Agroforestry Accounting System (Campos 2000; Campos et al. 2001, 2007a, b, 2008a, b, 2009a, b). This approach extends the governmental standard System of National Accounts by incorporating private amenities, land revaluation (which can be any change in land value), annual natural growth of timber, cork and firewood, and growth from the previous accounting period that is used during the current accounting period. Improvement in standardized physical statistics and economic accounting tools at micro and macro scales is critical for a better understanding of landowner decisions and response to natural resource conservation policy. Results are estimated for individual forestry, game, livestock, crop, and service activities, and for the aggregated activities as a whole.

The case study approach provides a comprehensive description of landowner management and motivations within the context of a specific political, social, and ecological setting. The economic analysis of these case studies considers commercial and private amenity benefits (hereafter "amenities"), because landowner management and investment decisions are motivated by a mix of both (Torell et al. 2001, 2005; Campos et al. 2009b; Oviedo et al. 2012). In analyzing the multiple uses of dehesas and ranches, few researchers have integrated market and non-market landowner amenity economic values consistently in a standardized total income system of accounts.

First we present the Agroforestry Accounting System (AAS) and the case studies. Next the physical indicators used in the analysis are laid out, followed by analysis of the economic results from the case studies. Findings from a comparison of the case studies are then summarized and conclusions presented.

Fig. 13.1 Castilian Mastiff dogs are still put to work in the dehesa region north-west of Castilla-León, where wolves are a continuing threat to Merino sheep in the melojo oak (*Quercus pyrenaica*) dehesas. In this view, the distinguished Spanish economist J.M. Naredo is hiking a medieval path (*camino de herradura*) in El Cardoso de la Sierra, 100 km north of Madrid, in the company of Pinto, a mastiff. Naredo's pioneering work integrating nature conservation and human needs has notably influenced conservationists and academics. (Photograph by P. Campos)

13.2 The Agroforestry Accounting System Methodology

13.2.1 In Relation to the System of National Accounts

The United Nations-based System of National Accounts (SNA) is the internationally agreed-upon standard set of recommendations on how to compile measures of current production (EUROSTAT 2000; European Communities et al. 2009; BEA 2010). The SNA lays out a coherent, consistent, and integrated set of macroeconomic accounts drawing on internationally agreed upon concepts, definitions, classifications and accounting rules. However, the SNA has shortcomings for measuring total income and capital from dehesas and ranches. On the output side, the SNA does not measure the value of forage grazed[1] by livestock in the accounting period and of the annual natural growth of tree and shrub products that will be harvested in a future accounting period (e.g.: firewood and cork annual natural growth[2]) (Fig. 13.2). On the cost side, the SNA omits the value of tree and shrub products grown in previous years that are harvested, or contribute to

[1] Forage is considered an "intermediate output" on the output side and an "own intermediate consumption" on the cost side. An intermediate output is one produced in the accounting period and used in the production of other things rather than marketed directly. "Own intermediate consumption" means "on-farm produced" intermediate consumption.

[2] Known in SNA terminology as "final gross work in progress formation" (GWPF).

Fig. 13.2 Cork stripping is unique to oaks of the Mediterranean forest. Alcornocales Natural Park in Spain is the largest cork oak forest in the world. Cork stripping is a specialized task and workers need to be trained to become harvesters of cork, locally known as "hachas" (literally, "hatchets"). Such highly trained workers receive a wage several times higher than that of non-specialized workers. (Photograph by P. Campos)

products harvested, during the analyzed accounting period.[3] This is the case of shrubs browsed by game and of wood products from trees. These omissions are relevant when measuring the contribution of single and aggregated activities to dehesa and ranch total income. Furthermore, private amenity consumption, considered an important form of income to landowners (Campos et al. 2009b), and capital gain, are not part of total income in the SNA. The system also does not offer an individualized criterion for measuring the value of self-employed labor, but it does offer a measure of mixed income that includes jointly the value of self-employed labor and operating capital benefit.

The goal of the Agroforestry Accounting System (AAS) is to extend the criteria of the SNA in order to include omitted values when estimating the total income and capital from dehesas and ranches. The AAS criteria that were applied for valuing forage grazed by livestock, the annual natural growth of trees and shrubs

[3] Referred to in the SNA as "work in progress used" (WPu). These are the result of natural growth from previous accounting years. When these stored goods are harvested or used, they are accounted for as a cost to the "forestry" activity. This allows integrating the physical use of natural resources as withdrawals from the production and capital accounts, since economic goods already produced that are harvested or used in the accounting period have a price higher than zero before entering as work in progress used in the current production.

Fig. 13.3 Cattle can be fixed capital if they are brood cows that remain on the ranch from year to year, products if they are sold within the year as with calves, or work in progress if they are young female cattle being raised to replace a brood cow in future years. In all three cases, the California rancher asserts his or her ownership with a brand, a permanent reminder of the origin and owner of a cow or calf. A cow may accumulate more than one brand if she is sold. (Photograph by L. Huntsinger)

that are not harvested or used in production by the end of the accounting period, and previous natural growth that has already accumulated ("stored") at the beginning of the accounting period, are detailed in Campos et al. (2007a, 2008a, b, 2009a). The application of the AAS methodology at the farm scale generates a comprehensive set of income, capital and profitability measurements for the analyzed case studies for individual and aggregated dehesa and ranch activities.

The AAS organizes economic information into two accounts. The production account measures output and cost flows from current production, including investments made by the landowner in the woodland,[4] final work in progress formation, and work in progress used. The capital balance account incorporates entries, withdrawals, and revaluation of fixed capital and stored work in progress, like annual wood growth that is not harvested in the accounting period (Fig. 13.3).

Outputs and fixed good (durable goods) sales are valued at producer prices in the AAS. These are prices received by the producer excluding operating subsidies and any value added tax or similar deductible tax (European Communities et al. 2009: para. 6.51b, 101). Indicators at producer prices have interest both for society as a whole and landowners since these are prices before government intervention via taxes and subsidies, which are transfers between sectors of the economy and do not generate new income. We also present indicators at basic prices, the price received by the producer including operating subsidies on production minus any tax payable.

[4] Described in accounting terminology as gross fixed capital formation (GFCF). This is a finished final output that is produced in the dehesa or ranch in the current year with the aim of being a fixed investment contributing to the production of goods and services in upcoming years.

This is the most relevant price for producer decision-making (European Communities et al. 2009: paras 6.51a and 6.52, 101). Raw materials, services, and fixed goods bought are valued at purchase prices, which are prices paid by the landowner at farm gate, excluding operating subsidies and any value added tax or similar deductible tax (European Communities et al. 2009, para. 6.64, 102). Operating subsides minus any payable tax will be referred hereafter as net operating subsidies.

13.2.2 Total Income

Total income has been defined "as that which can be consumed while keeping real wealth intact, saving [=investment] is the difference between this measure of income and actual consumption. Both income and saving will then include real capital gains. To preserve the saving-investment identity, investment would also have to include these capital gains. Failure to include them causes a disparity between income statements and balance sheets that reflect market values" (Eisner 1989: 17). If we follow the SNA criteria, we find that the production boundary is limited to traded products and to a reduced number of produced outputs supplied free by the government and other nonprofit economic units.[5] The missing measurement of environmental amenities in the SNA is a relevant gap because these are "a significant part of the real income of many individuals" (Krutilla 1967). The AAS incorporates both capital gain and amenities in the total income measurement to overcome these omissions.

The two components of total income measured by the AAS are the net value added and capital gain. The net value added at producer prices has two components, the net operating margin and labor. The latter includes income to both employed and self-employed labor.[6] Net operating margin plus capital gain constitutes the capital income, which is the total benefit that landowners obtain annually from dehesas and ranches both as actual monetary and expected benefit. These benefits pay both for land resource rent[7] and manufactured capital. The net

---

[5] In the NIPAs [National Income and Product Accounts], "the definition of income is narrower [as a satellite system of the SNA], reflecting the goal of measuring [net value added from] current production" (European Communities et al. 2009: paras. 1.46, p. 7 and 6.27, p. 98; BEA 2010: 18).

[6] Self-employed persons "are persons who are the sole or joint owners of the unincorporated enterprises in which they work" (European Communities et al. 2009: para. 19.25, p. 407). Although not all self-employed persons are necessarily part of the landowner family or all family members are necessarily self-employed, in our case studies they match up. For simplicity, we will refer to self-employed labor as family labor throughout the paper.

[7] Resource rent is the income receivable by the owner of a natural resource (the lessor or landlord) for putting the natural resource at the disposal of another institutional unit (a lessee or tenant) for use of the natural resource in production (European Communities et al. 2009: para. 7.154, 156). The resource rent does not include returns from any manufactured capital involved in the current production.

operating margin is the difference between outputs and costs at producer prices. Capital gain is measured from capital revaluation less capital destruction[8] plus normal depreciation during the accounting period. Land and livestock revaluation are important sources of capital gain. Other asset revaluations come from equipment and buildings.

In our case studies, we present the results at producer prices for the different components of the total income: net operating margin, labor, and capital gain. The net value added at basic prices (net operating surplus plus labor) and the operating profitability at basic prices are also presented, since these indicators have interest from the landowner perspective. We also present results of net cash flow estimated as revenues less expenditures in the accounting year, since this gives an indication of the landowner liquidity when setting management goals and deciding on investment.

The AAS application in this chapter extends the income measurement to identify the value of family labor and of the non-market amenity output flow to the landowner during the year. It also distinguishes the share of land revaluation attributed to commercial and amenity land activities when estimating capital gains as part of total income.

When family labor is involved in dehesa and ranch operating activities, the family commercial net value added, measured after subtracting employee compensation, is a mixed commercial operating income which is integrated by resource rent, manufactured net operating margin and family labor. Since from the production account we obtain the value of this family mixed commercial net value added (European Communities et al. 2009: paras. 20.49 and 20.50, p. 421), we are able to estimate a joint value for the manufactured net operating margin and the family labor, given that the resource rent value is assumed to be known from the local market. When this joint value is negative, we attribute all of it to the manufactured net operating margin since we assume that family labor services will never have a negative value but that these labor services are unpaid (hereafter unpaid family labor). When the above joint value is positive we set a criteria so that part of the value is attributed to family labor and part to the manufactured net operating margin (RECAMAN [9] project)[10].

The amenities consumed by landowners are the only annual economic output without a market price in the case studies. For the dehesa case studies, this non-

---

[8] This a fixed good withdrawal from the capital balance account during the accounting period that results in zero revenue for the landowner (e.g., the death of reproductive or draft livestock).

[9] The RECAMAN project *(Valoración de la Renta y el Capital de los Montes de Andalucía)*) of the Junta de Andalucía , initiated in 2008, is ongoing and applies the Agroforestry Accounting System at the regional scale to measure total income and capital from the montes of Andalucía in Spain.

[10] When this value is positive and on a per hour basis is lower or equal than 80% of employee wages per hour in the area, we assume that all the mixed income value is attributed to family labor and the manufactured family net operating margin is zero. When this value is positive and on a per hour basis is higher than 80% of employee wages per hour in the area, we assume that the family labor value corresponds to this 80% of employee wages per hour, and the remaining value is attributed to the manufactured family net operating margin.

market amenity output is estimated from a 2010 contingent valuation study of 765 privately-owned dehesa and forest landowners in Andalucía (RECAMAN project). For the ranch case studies, this amenity output is estimated by Campos et al. (2009b) from a contingent valuation survey to landowners in California in 2004. This value is updated to 2010 dollars for our application to ranch case studies.

Since the land price of dehesas and ranches are explained by landowner commercial and amenity activities, we offer income and capital estimates distinguishing between these activities, since they affect capital gain estimates. To that end, we assume that annual land commercial revaluation follows the historical annual nominal cumulative rate of commercial operating benefit. As the total land revaluation in the analyzed year is known, the land amenity revaluation is measured by subtracting land commercial revaluation from total land revaluation considering the percentage of land price which corresponds to each of these two components (RECAMAN project).

13.2.3 *Profitability Measure*

The estimation of profitability refers to net operating margin or surplus, capital gain and capital income. It is measured on the basis of immobilized capital in the accounting period in order to be consistently compared with alternative investments (Campos et al. 2001). The immobilized capital is the average annual landowner economic investment in a dehesa or ranch operation.

The ratios of net operating margin and net operating surplus to immobilized capital (IMC) provide the operating profitabilities earned from the enterprise at producer and at basic prices, respectively. The capital gain profitability is measured at producer prices as the ratio of capital gain to immobilized capital. These profitabilities are expressed in nominal terms. Total profitability, estimated as the ratio of capital income to immobilized capital, is presented both at nominal and real terms, being the latter a nominal rate deflated by consumer price index.

13.3 Case Study Typologies

13.3.1 *Spanish Dehesas*

Dehesa natural resource management is typically dominated by livestock production. Holm oak is the most widespread tree species, but there is often a mixture of other oaks and pine species. In the past, a significant number of dehesa owners have converted the shrublands that are part of the mosaic of vegetation into plantations of native or exotic pine and eucalyptus species in response to government incentives to increase domestic timber supply. Dehesa case studies A and

Fig. 13.4 Family residential houses in dehesas persist along with livestock production, cork stripping, hunting and small areas of crops such as olives or other fruit and vegetable crops for landowner consumption. Other buildings and structures are common, including worker housing, warehouses and livestock *corrales*. This family house is next to the livestock production infrastructures in a dehesa in the Sierra Norte de Sevilla. (Photograph by S. García)

C (see below) did not use silvicultural practices within the analyzed year for regenerating native oak species and improving grass, acorn and pine nut productivity. Dehesa case study B is illustrative of holm and cork oak plantations and natural regeneration stands. Dehesa landowners are often motivated by lifestyle benefits and they do not always need to achieve a positive net cash flow (Fig. 13.4).

Dehesa A—Holm oak woodland/crossbred cattle/Iberian pig. This is a flat property of 179 ha, located in the municipality of Pozoblanco in Córdoba province. Holm oak is the only tree species and occupies almost the entire property (Table 13.1), with an understory of 92 % grasses and 7 % shrubs. Overstory canopy cover is 55–60 %. Firewood is harvested for landowner use and silvopastoral practices are carried out for improving pasture productivity. Cattle are crossbred

Table 13.1 Land use and vegetation for the dehesa and oak woodland ranch case studies

Vegetation type	Dehesa A	Dehesa B	Dehesa C	Ranch A	Ranch B
Woodland (%)	99.6	89.0	89.3	33.6	65.3
Oak (%)	99.6	86.7	43.2	33.6	65.3
Pine (%)			37.2		
Other (%)		2.3	8.9		
Shrubland (%)		6.2	4.9		14.9
Grassland (%)		3.0		54.7	8.9
Cropland (%)			5.5	11.7	9.7
Unproductive land (%)	0.4	1.8	0.2		1.2
Total land (%)	100.0	100.0	100.0	100.0	100.0
Total land (ha)	178.8	1,336.4	1,260.4	2,671.3	1,358.1
Unproductive land (ha)	0.8	24.4	2.8	0.0	16.2
Useful agrarian land (ha)	178.0	1,312.0	1,257.6	2,671.3	1,341.9

meat breeds, and purebred Iberian pigs graze on acorns and grass during the *montanera* from October to January (Chap. 10). Wild boar is occasionally present and small game is hunted at a low rate by the landowner and friends. Income from this and other dehesas owned by him are the landowner's main income source. Most labor is from hired workers, although family labor is also used.

Dehesa B—Holm oak-cork oak woodland/retinta cattle/big game. Dehesa B is a moderately sloped *sierra* property of 1,336 ha, located in the municipality of Constantina in Sevilla province. Holm and cork oaks cover 87 % of the property (Table 13.1), with a mostly grass understory and, to a lesser extent, mixed grass-shrub understory. The estimated canopy cover is 25–30 %. Cork and firewood harvesting is carried out, and silvopastoral management aims to improve grass and acorn yields. Purebred Andalusian beef cattle are the only livestock breed raised. The property includes an enclosed commercial hunting reserve (*coto*) for big and small game, with high red deer and wild boar stocking rates. Landowner family income from offsite activities is significantly greater than family livelihood needs. This landowner is primarily motivated by the amenity benefits of owning a dehesa. All labor is from employees and the property is professionally managed.

Dehesa C—Holm oak-stone pine woodland/segureña sheep/big game. Dehesa C has the greatest variety of native trees and highest wildlife biodiversity of the three Spanish cases. It is located in the municipality of Montoro in Córdoba province. It is a property of 1,260 ha on a moderate to highly sloped *sierra*. Pure stands of holm oak cover 43 % of the property, with an estimated canopy cover of 35–40 %, and pure stands of stone pine extend over 37 % of the property with an estimated canopy cover of 50 % (Table 13.1). There is no timber or firewood harvest and no silvopastoral interventions are carried out. Iberian pigs and *Merino* sheep are reared (Fig. 13.5), and *segureña* sheep graze on the property, but the latter do not belong to the landowner and they are moved to another property from May to October (*transtermitancia*). This dehesa has a partially enclosed commercial hunting *coto* for big and small game, with the hunter success rate for red deer and wild boar, partridge, rabbit, hare, turtle dove, dove and thrush considered average for the area. Family labor is used in this dehesa. The landowner is highly motivated by amenity benefits, but with net cash flow restrictions. The main family income comes from offsite sources.

13.3.2 California Ranches

Oak woodland ranches often have woodlands that are naturally open (Chaps. 5 and 8) and grazed by livestock and game. Cattle are the main—and usually the sole—livestock. There is some cropping, mainly to grow supplemental feed for years of low forage productivity or to attract game, although animals also graze croplands during months of low forage production. Blue oak (*Q. douglasii*) is the most common oak, but most woodlands include a mix of oak and conifer species (Chaps. 1, 6). The main silvicultural practice is firewood harvesting, but this is

Fig. 13.5 The black *merina* (Merino) sheep is an endangered native breed. Government compensation programs aim to increase the production of black Merino in the dehesas to avoid their extinction. This young herd is being introduced into a Dehesa Boyal (Chap. 1) belonging to the municipality of Arroyo de la Luz in Cáceres province. The landowner is changing from traditional feeding of draft *bueyes* (oxen) to managing for domestic animal preservation and public recreation. (Photograph by F. Pulido)

increasingly rare as owners have become aware of the value of retaining oaks, with less than 10 % of ranchers reporting the sale of fuelwood in 2005 (Huntsinger et al. 2010). Firewood is typically sold standing, at the stumpage price, with the landowner being paid by a timber operator for the value of the firewood prior to harvest by the operator. Since California oaks grow slowly, firewood is not cut every year, but when market conditions and cash flow needs dictate. Firewood sales are often considered a by-product of oak thinning to increase forage production. Family ownership and family labor is common, and ranchers often live on the property year-round (Huntsinger et al. 2010).

Ranch A—Blue oak woodland/English crossbred cattle. This 2,671 ha ranch is located in Shasta County in the Northern Cascade region of California. It has an open oak canopy with gentle slopes and deep soils. Oak woodland and annual grassland cover 34 and 55 % of the property, respectively (Table 13.1). The ranch also has irrigated pasture, which contributes substantially to livestock feeding activities (Fig. 13.6). Blue oak is dominant with an estimated canopy cover of 20–30 %. Firewood is sold by the ranch owner as stumpage. No other silvicultural treatments besides tree harvest are used. A fee is collected annually from hunters who are allowed access to hunt in compliance with California Department of Fish and Wildlife regulations. From May through October, livestock primarily graze the irrigated pastures. There is a stocker operation, where weaned calves are grazed through the green forage season from November to June (Chap. 10). Additional pasture is leased to several cow-calf operators both seasonally and on a year-round basis. The ranch is grazed by English crossbreeds. Horses are used to manage the cattle. The ranch goals are to maximize commercial cattle value and amenities, and only employee labor is used.

Ranch B—Blue oak woodland/big game. This 1,358 ha ranch is located in Tehama County in the Northern Sacramento Valley. The ranch is steeper, has poorer soils, and is less open than Ranch A, and has no irrigated land. Oak woodlands occupy 65 % of the property, while 15, 9 and 10 % is chaparral shrubland, cropland and annual grassland, respectively (Table 13.1). The oak woodland has blue oaks and scattered grey pines (*Pinus sabiana*), with an estimated canopy cover of 20–30 %. Hunting is the main commercial activity, including black-tailed deer, dove, wild boar, quail and squirrel. As in Ranch A, firewood is harvested. The landowner maintains a small herd of English crossbred cattle that are used to help meet vegetation management goals. The rancher also leases winter pasture (grassland and oak woodland) to a local livestock producer. Wheat, alfalfa and perennial grasses are cultivated for cattle and big game, and for direct sale in the case of alfalfa. Most ranch labor is supplied by the landowners, who live on the ranch year-round, but they also have a part-time employee to help out. The household obtains off-ranch income from a retirement pension, which supports the enjoyment of ranching amenities and lifestyle.

13.4 Results: Physical Indicators

Physical indicators can reveal how much natural resources, manufactured resources and labor contribute to extensive livestock grazing and the production of other tree and crop products. We focus on a group of output and input indicators (Tables 13.2, 13.3, 13.4, 13.5). Outputs include firewood and cork, grass and acorns, and animal products, both from livestock and game. Inputs include animal feed consumption, fossil fuel, and labor (employee and self-employed labor).

13.4.1 Spanish Dehesas

It is not common to fatten livestock in the dehesa on feedlots, though supplemental feeding while the animals are on the woodland is more common than in California (Chap. 10). For the Iberian pig supplemental feeding may take place before and after the *montanera*, which is the period when pigs consume the available acorns and grass in the fall. Tables 13.2 and 13.3 show livestock and big game forage and supplemental feed consumption. This illustrates differences among the dehesa case studies and the decline of livestock grazing in favor of big game species in those case studies with enclosed hunting cotos (dehesas B and C) (Fig. 13.7).

Fossil fuel consumption in Dehesa management is lower than usage on croplands. The infrastructure (livestock shelter, warehouses, roads, fences, water

Table 13.2 Forage and supplemental feeding for the dehesa case studies (FU/ha of UAL, year 2010)

Class	Forage			Supplemental feeding	Total	Forage			Supplemental feeding	Total	Forage			Supplemental feeding	Total
	Acorns	Grass	Total			Acorns	Grass	Total			Acorns	Grass	Total		
	Dehesa A					Dehesa B					Dehesa C				
1. Livestock	45.98	978.26	1,024.24	704.09	1,728.33		142.23	142.23	70.08	212.31	6.82	58.06	64.88	29.70	94.57
1.1 Cattle		978.26	978.26	643.13	1,621.38		142.23	142.23	70.08	212.31					
1.2 Sheep											5.12	58.06	63.17	20.52	83.70
1.3 Pig	45.98		45.98	60.96	106.95						1.70		1.70	9.17	10.88
2. Big game						89.95	484.70	574.65	140.37	715.01	28.59	180.21	208.80		208.80
2.1 Red deer						77.34	416.82	494.16	18.36	512.52	24.15	130.16	154.31		154.31
2.1.1 Priced						77.34	171.84	249.17	18.36	267.54	15.35		15.35		15.35
2.1.2 Unpriced							244.99	244.99		244.99	8.80	130.16	138.97		138.97
2.2 Wild boar						12.61	67.88	80.49	118.53	199.02	4.25	22.86	27.10		27.10
2.2.1 Priced						12.61	27.98	40.59	118.53	159.12	2.70		2.70		2.70
2.2.2 Unpriced							39.90	39.90		39.90	1.55	22.86	24.41		24.41
2.3 Others									3.47	3.47	0.19	27.19	27.38		27.38
2.3.1 Priced									3.47	3.47	0.19	2.53	2.72		2.72
2.3.2 Unpriced												24.66	24.66		24.66
Total	45.98	978.26	1,024.24	704.09	1,728.33	89.95	626.94	716.88	210.44	927.33	35.41	238.26	273.67	29.70	303.37

FU forage unit
UAL useful agrarian land (Table 13.1)

Table 13.3 Forage and supplemental feeding on the ranch case studies (FU/ha of UAL, year 2010)

Class	Forage			Supplemental feeding	Total	Forage			Supplemental feeding	Total
	Acorns	Grass	Total			Acorns	Grass	Total		
	Ranch A					Ranch B				
1. Livestock		560.50	560.50	2.31	562.81		176.18	176.18	1.16	177.34
1.1 Cattle		560.50	560.50	0.00	560.50		176.18	176.18	1.16	177.34
1.2 Horses		0.00	0.00	2.31	2.31					
2. Big game		n. a.	n. a.	n. a.	n. a.		28.71	28.71	0.00	28.71
2.1 Deer		n. a.	n. a.	n. a.	n. a.		9.85	9.85	0.00	9.85
2.1.1 Priced		n. a.	n. a.	n. a.	n. a.		9.85	9.85	0.00	9.85
2.1.2 Unpriced		n. a.	n. a.	n. a.	n. a.		n. a.	n. a.	n. a.	n. a.
2.2 Wild boar		n. a.	n. a.	n. a.	n. a.		18.86	18.86	0.00	18.86
2.2.1 Priced		n. a.	n. a.	n. a.	n. a.		18.86	18.86	0.00	18.86
2.2.2 Unpriced		n. a.	n. a.	n. a.	n. a.		n. a.	n. a.	n. a.	n. a.
Total		560.50	560.50	2.31	562.81		204.89	204.89	1.16	206.05

FU forage unit; *n. a.* not available
UAL useful agrarian land (Table 13.1)

Table 13.4 Selected input and output indicators for the dehesa case studies ($/ha of UAL, year 2010)

Class	Unit (u)	Relevant area (ha)	Dehesa A			Dehesa B			Dehesa C		
			Quantity (u/ha)	Price ($/u)	Value ($/ha)	Quantity (u/ha)	Price ($/u)	Value ($/ha)	Quantity (u/ha)	Price ($/u)	Value ($/ha)
1. Raw materials											
Total forage units	FU	UAL	1,728.34	0.17	285.40	927.33	0.08	70.06	303.37	0.05	16.22
Fossil fuels	Liters	UAL	16.36	1.20	19.56	11.91	1.20	14.34	2.71	1.30	3.53
2. Labor	Hour	UAL	16.05	11.18	179.38	13.94	15.00	209.12	7.50	8.65	64.81
Employees	Hour	UAL	15.42	11.64	179.38	13.94	15.00	209.12	5.16	12.57	64.81
Self-employed	Hour	UAL	0.63	0.00	0.00				2.34	0.00	0.00
3. Forestry outputs											
Cork	kg	UAL_{Qs}				154.80	1.70	263.21			
Firewood	t	UAL_{Qsp}				0.02	98.10	2.16			
4. Grazing			1,024.25	0.06	62.36	716.88	0.05	32.28	273.67	0.02	5.99
Grass	FU	UAL	978.26	0.05	46.68	626.94	0.04	25.56	238.26	0.02	3.87
Acorns	FU	UAL_{Qi}	45.99	0.34	15.68	89.95	0.07	6.72	35.41	0.06	2.12
5. Livestock sales											
Calves	Head	UAL				0.02	561.10	9.35			
Lambs	Head	UAL							0.07	55.68	4.07
Lechones	Head	UAL							0.02	121.09	2.89
Piglets	Head	UAL	0.81	321.64	260.17				0.05	188.27	9.43
Montanera piglets	kg	UAL	41.49	2.32	96.25						
6. Red deer						14.22	10.26	145.83	0.48	52.54	25.32
Hunting fees	Positions	UAL				0.04	2,839.83	108.23	0.03	761.42	24.22
Meat	Head	UAL				14.18	2.65	37.60	0.45	2.44	1.10

FU forage unit
UAL useful agrarian land (Table 13.1)
UAL_{Qs} useful agrarian land of *Q. suber*
UAL_{Qsp} useful agrarian land of *Q. spp*
UAL_{Qi} useful agrarian land of *Q. ilex*

Table 13.5 Input and output indicators for the ranch case studies ($/ha of UAL, year 2010)

Class	Unit (u)	Relevant area	Ranch A			Ranch B		
			Quantity (u/ha)	Price ($/u)	Value ($/ha)	Quantity (u/ha)	Price ($/u)	Value ($/ha)
1. Raw materials								
Total forage units	FU	UAL	562.81	0.11	63.11	206.05	0.10	20.23
Fossil fuels	Liters	UAL	n.a.	n.a.	n.a.	n.a.	n.a.	n.a.
2. Labor	Hours	UAL	2.02	16.65	33.67	5.00	5.37	26.83
Employees	Hours	UAL	2.02	16.65	33.67	0.64	25.05	15.91
Self-employed	Hours	UAL				4.36	0.84	3.67
3. Forestry output								
Firewood	t	UAL_{Qsp}	0.03	90.61	2.73	0.04	90.61	3.22
4. Grazing								
Grass	FU	UAL	560.50	0.11	62.28	204.89	0.10	19.80
5. Livestock sales								
Calves	Head	UAL				0.00[a]	549.86	2.43
Yearlings	Head	UAL	0.98	578.88	564.72			
6. Deer								
Hunting fees	Fees	UAL	0.00[a]	8,028.46	3.01	0.04	537.91	20.99

n.a. not available; *FU* forage unit
UAL useful agrarian land (Table 13.1)
UAL_{Qs} useful agrarian land of *Q. suber*
UAL_{Qsp} useful agrarian land of *Q. spp*
[a] Value lower than 0.01

Fig. 13.6 Irrigated pasture can provide nutritious green forage during the dry months on oak woodland ranches, although it is relatively rare today. Pastures are flood irrigated or sprinklers are used. (Photograph by L. Huntsinger)

Fig. 13.7 The Morucha cow is one of the pure cattle breeds whose preservation is being pursued in Spanish dehesas. Livestock owners often express pride in their native livestock breeds, and competitions showcasing the autochthonous breeds are often held at livestock fairs. (Photograph by P. Campos)

facilities) tends to not take up a large land area. Family housing is a key investment that helps the household capture amenity benefits.

Dehesa labor in these cases ranges from 7.5 to 16 h per hectare per year (Table 13.4), which is within the range of other research results (Campos and Riera 1996; Campos et al. 2001; RECAMAN project). In dehesas A and C all family labor is unpaid. This can be considered labor that seeks to enhance current and future expected amenity consumption by the household.

13.4.2 California Ranches

In the ranch case studies, all grazing is attributed to income from grass forage (Table 13.3). In both ranches, forage is the main feed for cattle, with no supplementation in Ranch A and less than 1 % supplementation in Ranch B. A significant contribution to total forage needs is provided by the irrigated pasture on Ranch A. Horses are present only on Ranch A, and rely on supplemental feeding

Fig. 13.8 In both California and Spain, rock walls demonstrate a traditional commitment to long-term stability in property boundaries and modes of production. Within the walls, the landowner is a consumer as well as producer of ecosystem services. (Photograph by L. Huntsinger)

(Table 13.3). The difference in total forage consumption between the two ranches highlights the different landowner objectives. Big game forage consumption from the oak woodland could not be estimated due to a lack of wildlife population models, but for Ranch B, we show wildlife forage provided by cropland since the landowner cultivates crops for wildlife feeding (Table 13.3). On Ranch B, 66 % of cropland forage is consumed by wild boar and the remaining by black-tailed deer. Livestock and big game are partial competitors for the available grazing resources on Ranch B, but the lack of data on oak woodland grazing by black-tailed deer and wild boar makes it infeasible to draw conclusions.

Labor input intensity is low on the two ranches, at 2 and 5 h per hectare per year (Table 13.5). For the total hours of family labor in Ranch B, 4 % is paid, while the remaining 96 % is unpaid. The divergences in livestock grazing and sales highlight the different goals of the owners for their livestock management, as well as the different characteristics of each ranch (Fig. 13.8).

13.5 Results: Economic Analysis

The production accounts of the AAS (Tables 13.6, 13.7) applied to the dehesa and ranch case studies distinguishes five different activities: forestry, hunting, livestock, cropping, and services. Services include both commercial services (housing and machinery services) and amenities. Income, capital and profitability indicators (Tables 13.8 and 13.9) are used to differentiate between commercial activities and amenities. We do not present the capital balance account, although Table 13.8 shows the two key indicators derived from this account: land price at the beginning of the year and capital gains. Data was collected from account books, in-depth interviews and field data in 2010 for the dehesa case studies and in 2007 for the ranch case studies. All monetary data presented are in 2010 US dollars per hectare, and the economic results are expressed in nominal terms, except when stated otherwise.

13.5.1 Spanish Dehesas

13.5.1.1 Income Results

In dehesas with hunting in enclosed cotos, there is forage rent for game grazing from the oak woodland only if game replaces livestock and there is an effective local market demand for livestock grazing. Thus, in the enclosed cotos of dehesas B and C part of the grazing consumption is attributed as an imputed payment to game-related activity (Table 13.4).

The net operating margin is positive for forestry activities in dehesas A and C, and negative in Dehesa B, where the owner has invested in oak regeneration and grass productivity improvement. Table 13.6 shows how in Dehesa B oak regeneration increases depreciation costs (CFC) over the natural growth of trees (GWPF), and the oak growth negative net operating margin induces, additionally, an expected capital loss from tree growth at the end of the year. Hunting in dehesas B and C generates negative net operating margins. This is due to the low hunting market prices observed in the year and, specifically in dehesa B, to the landowner's choice of managing for unprofitable trophy deer for his own hunting enjoyment. Cattle and sheep production have a negative net operating margin in all case studies. The *montanera* for Iberian pig is the only profitable livestock production activity in Dehesa A (Table 13.6). Amenity services have a high positive net operating margin, making it possible to reach positive total net operating margins in the three dehesa case studies (Table 13.6).

Dehesa amenities make the greatest contribution to net value added at producer prices, while commercial activities make a significantly lower contribution. The livestock stocking rate is low in dehesas B and C, which receive eight times less from net operating subsidies than does the cattle-oriented Dehesa A (Table 13.8)

Table 13.6 Production accounts for the dehesa case studies. Data are at producer prices which exclude net operating subsidies ($/ha of UAL, year 2010)

Class	Dehesa A						Dehesa B						Dehesa C					
	FOR	GAM	LIV	CRO	SER	TOT	FOR	GAM	LIV	CRO	SER	TOT	FOR	GAM	LIV	CRO	SER	TOT
1. Total output	62.62		901.57		666.65	1,630.85	218.11	303.79	40.99		843.59	1,406.49	24.08	68.49	21.72	10.29	636.27	760.83
1.1 Intermediate output	62.36				38.16	100.52	108.33	91.40			41.04	240.77	5.99	27.63			15.91	49.52
1.1.1 Raw materials	62.36					62.36	102.23	91.40				193.63	5.99	27.63				33.62
1.1.2 Services					38.16	38.16	6.10				41.04	47.14					15.91	15.91
1.2 Final output	0.26		901.57		628.49	1,530.33	109.79	212.39	40.99		802.54	1,165.72	18.09	40.86	21.72	10.29	620.36	711.31
1.2.1 Own fixed investment (GFCF)			122.24			122.24		3.71	21.17			24.88		0.98	1.24		16.94	19.16
1.2.2 Sales			399.55			399.55	102.48	158.66	10.60		4.16	275.91	5.69	26.49	16.51	10.29		58.98
1.2.3 Work in progress (GWPF)	0.26		379.79			380.05	5.70	22.01	9.22			36.92	12.39	5.79	2.12			20.30
1.2.4 Owner consumed					628.49	628.49	1.61	4.05			795.09	800.75		7.09	1.84		594.47	603.40
1.2.5 Other final output								23.96			3.29	27.25		0.52			8.95	9.47
2. Total cost	30.92		1,079.64		38.16	1,148.72	258.30	315.47	186.92		92.78	853.46	0.21	91.64	57.32	9.82	37.69	196.68
2.1 Intermediate consumption	9.59		882.39		38.16	930.14	147.98	247.98	119.31		41.04	556.32		56.51	33.28	2.30	24.60	116.68
2.1.1 Raw materials			316.45			316.45	75.98	156.12	68.92			301.02		30.83	16.24	1.33		48.41
2.1.2 Services	9.59		36.82		38.16	84.56	6.65	39.13	34.14		41.04	120.97		9.74	2.12	0.96	24.60	37.43
2.1.3 Work in progress used (WPu)			529.13			529.13	65.35	52.73	16.26			134.34		15.94	14.91			30.85
2.2 Labor cost	19.43		159.94			179.37	101.00	34.44	50.67		23.01	209.12	0.19	31.16	20.02	5.18	8.25	64.81
2.2.1 Employees	19.43		159.94			179.37	101.00	34.44	50.67		23.01	209.12	0.19	31.16	20.02	5.18	8.25	64.81
2.2.2 Self-employed																		
2.3 Consumption of fixed capital (CFC)	1.90		37.31			39.21	9.32	33.04	16.94		28.72	88.02	0.01	3.97	4.02	2.34	4.84	15.19
3. Net operating margin	31.70		−178.07		628.49	482.13	−40.18	−11.68	−145.92		750.81	553.02	23.87	−23.15	−35.60	0.46	598.58	564.15

FOR Forestry; *GAM* Game; *LIV* Livestock; *CROP* Crop; *SER* Services; and *TOT* Total; *UAL* useful agrarian land (Table 13.1)

Table 13.7 Production accounts for the ranch case studies. Data are at producer prices which exclude net operating subsidies ($/ha of UAL, year 2010)

Class	Ranch A						Ranch B					
	FOR	GAM	LIV	CRO	SER	TOT	FOR	GAM	LIV	CRO	SER	TOT
1. Total output	42.06	3.01	877.41	45.99	120.30	1,088.78	15.55	21.34	12.16	18.30	211.93	279.28
1.1 Intermediate output	36.87		6.84	25.41		69.12	1.61			7.42	16.59	25.62
1.1.1 Raw materials	36.87			25.41		62.28	1.61			7.42		9.03
1.1.2 Services			6.84			6.84					16.59	16.59
1.2 Final output	5.19	3.01	870.57	20.58	120.30	1,019.65	13.94	21.34	12.16	10.88	195.34	253.66
1.2.1 Own fixed investment (GFCF)	2.46		8.90	20.58		31.95	3.52	0.35	6.32	1.05	0.35	11.59
1.2.2 Sales	2.73	3.01	564.72			570.45	10.43	20.99	2.43	9.83	3.58	47.26
1.2.3 Work in progress (GWPF)			296.95			296.95			3.41			3.41
1.2.4 Owner consumed					115.62	115.62					187.56	187.56
1.2.5 Other final output					4.69	4.69					3.84	3.84
2. Total cost	3.87		833.33	32.14	0.31	869.65	9.12	24.88	25.21	17.14	23.19	99.54
2.1 Intermediate consumption	0.81		808.32	17.74		826.86	2.83	13.36	10.77	7.58	17.54	52.09
2.1.1 Raw materials	0.09		65.73	10.51		76.33	0.78	8.40	5.90	3.88	0.57	19.53
2.1.2 Services	0.72		30.55	7.23		38.50	2.05	4.96	1.69	3.70	16.97	29.37
2.1.3 Work in progress used (WPu)			712.03			712.03			3.18			3.18
2.2 Labor cost	2.89		20.76	11.93		35.58	2.30	4.33	5.76	2.49	4.57	19.45
2.2.1 Employees	2.89		20.76	11.93		35.58	2.30	4.33	5.76	2.49	0.89	15.78
2.2.2 Self-employed							0.00	0.00	0.00	0.00	3.67	3.67
2.3 Consumption of fixed capital (CFC)	0.17		4.26	2.46	0.31	7.20	3.98	7.19	8.68	7.06	1.09	28.00
3. Net operating margin	38.19	3.01	44.09	13.86	119.99	219.13	7.35	–3.54	–13.05	1.16	188.74	180.66

FOR Forestry; *GAM* Game; *LIV* Livestock; *CROP* Crop; *SER* Services; and *TOT* Total; *UAL* useful agrarian land (Table 13.1)

(Fig. 13.9). These net operating subsidies contribute to the commercial net value added at basic prices to a significant amount in Dehesa A but to a lesser extent in dehesas B and C. Net cash flow is positive for Dehesa A and negative for dehesas B and C (Table 13.8).

The case studies show that up to 42 % of the land's market price -including bare land, trees, shrubs and game- can be attributed to commercial benefits (Table 13.8), while amenities account for 58 % of the land price (RECAMAN project, unpublished data). In 2010, dehesa land prices decreased at a nominal rate of −3.4 % (MARM 2011). Land revaluation distribution between commercial and amenity benefits for Spanish dehesas is estimated according to total land revaluation observed in 2010 and to an estimation of the cumulative commercial operating profitability from dehesas in the period 1994–2010. Available research allows estimating a negative cumulative rate of commercial operating profitability for dehesas of −5.3 % for this period, resulting from comparing 10 dehesas from the Extremadura region in 1994 (Campos and Riera 1996) and 6 from the Sierra Morena in the Andalucía region in 2010 (RECAMAN project). Thus, the revaluation of the share of land price attributed to amenity benefits is estimated so that total land revaluation rate is −3.4 % and the commercial land revaluation is −5.3 %. The total capital gain figures agree with these figures as well as with the other commercial capital revaluation components from the dehesa case studies. All dehesa case studies offered negative capital gains in 2010 (Table 13.8).

The major contribution to total income comes from the net operating margin in the three dehesas. In all cases amenities account for more than 80 % of the total income, reaching up to 96 % in Dehesa A (Table 13.8).

13.5.1.2 Profitability Measure

The commercial operating profitability at producer prices is negative in all dehesa case studies. The total profitability is between 2.5 and 3.6 %, and the commercial total profitability is negative in dehesas A and B and nearly zero in dehesa C. The amenity total profitability ranges from 2.9 to 3.9 % (Table 13.9). At basic prices, the only positive commercial operating profitability is in dehesa A, reaching 0.5 % (Table 13.9). However, analyzed on a one-year basis, dehesa annual total profitability does not capture the longer trend in land market prices that explains commercial and amenity benefits expectations.

These one-year results are also affected by the Spanish economic recession that drove land price losses in 2010. If we consider the nominal cumulative land revaluation rate for Spanish dry natural grassland for the period 1994–2010, which is 6.7 % (MARM 2011), and the inflation rate (consumer price index) in that same period, which is 3.1 %, we obtain on average for the three dehesa case studies a real land revaluation rate of 3.6 %. Assuming that the operating profitability at producer prices estimated in 2010 for the case studies is a lower bound in the period 1994–2010, we estimate a lower bound for the real total profitability at producer prices of 7.3, 7.3 and 7.2 % for Dehesas A, B and C, respectively. This

Table 13.8 Income and capital indicators in dehesa and ranch case studies ($/ha of UAL; year 2010)

Class	Dehesa A			Dehesa B			Dehesa C			Ranch A			Ranch B		
	COM	PA	TOT	COM	PA	TOT	COM	PA	TOT	COM	PA	TOT	COM	PA	TOT
1. Net value added at producer prices	71.17	590.34	661.50	8.10	754.05	762.14	50.40	578.56	628.96	139.40	115.32	254.71	29.96	170.15	200.11
2. Capital gain at producer prices	−49.87	−105.63	−155.50	144.91	−160.26	−15.35	27.56	−122.53	−94.97	−45.06	886.80	841.74	−10.95	729.74	718.79
3. Capital income at producer prices	−158.08	484.71	326.62	−56.11	593.78	537.67	13.15	456.03	469.18	58.76	1,002.12	1,060.88	−0.44	899.89	899.45
4. Total income at producer prices	21.29	484.71	506.00	153.01	593.78	746.79	77.96	456.03	533.99	94.34	1,002.12	1,096.46	19.01	899.89	918.90
5. Land price	3,667.51	5,053.24	8,720.75	3,049.22	6,361.67	9,410.90	4,254.12	5,861.49	10,115.61	1,832.74	2,429.44	4,262.18	1,506.62	1,997.14	3,503.76
6. Immobilized capital (IMC)	7,635.63	5,053.24	12,688.87	8,215.09	6,361.67	14,576.76	9,246.11	5,861.49	15,107.60	2,317.96	2,431.30	4,749.25	2,429.40	1,993.69	4,423.09
7. Net operating surplus	64.48	590.34	654.82	−176.54	754.05	577.50	−1.47	578.56	577.09	97.20	115.32	212.52	7.40	170.15	177.55
8. Net value added at basic prices	243.86	590.34	834.20	32.58	754.05	786.63	63.34	578.56	641.90	132.78	115.32	248.10	26.85	170.15	197.00
9. Net cash flow	55.56		55.56	−184.07		−184.07	−44.39		−44.39	484.11		484.11	5.86		5.86

COM Commercial; *PA* Private amenity; and *TOT* Total; *UAL* useful agrarian land (Table 13.1)

Table 13.9 Profitability rates in dehesa and ranch case studies (%; and year 2010)

Class	Dehesa A			Dehesa B			Dehesa C			Ranch A			Ranch B		
	COM	PA	TOT	COM	PA	TOT	COM	PA	TOT	COM	PA	TOT	COM	PA	TOT
1. Operating profitability at producer prices	−0.83	4.53	3.70	−1.35	5.06	3.71	−0.09	3.73	3.63	2.19	2.43	4.61	0.24	3.85	4.08
2. Gain profitability at producer prices	−0.38	−0.81	−1.19	0.97	−1.08	−0.10	0.18	−0.79	−0.61	−0.95	18.67	17.72	−0.25	16.50	16.25
3. Total profitability at producer prices	−1.21	3.72	2.50	−0.38	3.99	3.61	0.08	2.94	3.02	1.24	21.10	22.34	−0.01	20.35	20.34
4. Operating profitability at basic prices	0.49	4.53	5.02	−1.19	5.06	3.88	−0.01	3.73	3.72	2.05	2.43	4.47	0.17	3.85	4.01

COM Commercial; *PA* Private amenity; and *TOT* Total

Fig. 13.9 Protection of a monopoly on wool production from the Spanish transhumant Merino sheep was the origin of the Castilian Crown's Law of the Mesta in 1273. Today, cheese and lambs from the breed bring top prices and awards for quality at national markets. As is common in native livestock races, the Merino is characterized by high natural fertility and does well on dehesa grasses of sometimes low palatability, but Merino milk and meat production is low and labor costs can be high compared to improved races. (Photograph by P. Campos)

profitability is likely to determine landowner decisions more than the profitability observed in a single year.

These real profitabilities are competitive with alternative investments of similar risks and time horizon. However, they likely do not consider oak natural regeneration and the loss of aging oak trees. Amenity, commercial benefits, and public economic values are directly linked with dehesa forestry management practices for maintaining and improving grazing productivity, but the current trend of ceasing livestock grazing and its associated management activities, and the natural expansion of shrubs into grasslands and woodlands whether due to abandonment, oak regeneration needs, or game habitat, could put at risk this current profitability. It has been shown that oak silvicultural investments are not profitable on a market basis and their abandonment could induce future land capital losses that are not currently reflected in land market prices. European Union Common Agricultural Policy (CAP) subsidies have a high to moderate effect in our case studies, but subsidies are likely to be more demanded by landowners with limited willingness to accept short term net cash flow losses to invest in land and vegetation conservation practices (Campos and Mariscal 2003).

Maintaining the current production of both livestock and big game—mainly red deer and wild boar—will play a crucial role in the conservation of dehesa as an open oak woodland, and as an important source of employee labor demand. On Dehesa A and B, hired work comes mainly for livestock (cattle and *montanera* pig), and on Dehesa C from big game activities. Commercial activities need to be maintained on an annual basis to produce this dehesa working landscape but a negative net cash flow in our 2010 case studies indicates that the market for commercial products alone is not able to self-finance dehesa commercial management. Our case studies show that landowners substitute hired labor with family labor hours, and that they will even work without remuneration to maintain their commercial activities.

Franco et al. (2012) estimated a higher average commercial operating profitability than our dehesa case studies and others in the literature (Campos and Riera 1996; Campos et al. 2001). This discrepancy in dehesa commercial operating profitability results from the fact that Franco et al. (2012) define an inconsistent "profitability rate" at basic prices as the ratio between landowner net value added (mixed income) and the mean annual total fixed capital, with mixed income defined as "the yield on land, [manufactured] capital and self-employed labour". Franco et al. (2012) find, for a sample of 69 dehesas, a mean commercial operating "profitability" of 4 % for 2003–2004, but their mixed income estimate includes paid self-employed labor. The dehesas from this sample use high level of family labor. In other words, the Franco et al. (2012) dehesa average commercial operating profitability estimate is not grounded in standard landowner profitability theory.

13.5.2 California Ranches

13.5.2.1 Income Results

Forestry activity generates positive net operating margins in the two ranch case studies, with woodland grazing an important source of resource income in ranch A (Table 13.7). The two ranches have low firewood sales per hectare per year. Game-related activity generates only a small hunting fee operation in ranch A while it supports an active hunting operation on ranch B. Ranch B incorporates unpaid family labor and has a negative net operating margin for the game operation (Table 13.7). This may be due to a poor market for hunting in the region or the steep topography of the property. The landowner's personal recreation and lifestyle preferences may explain why this activity is continued as part of the ranch B operation.

On Ranch A, where livestock are only grazed during the months of high forage productivity, cattle rearing generates a positive net operating margin using a low level of hired labor (Table 13.7). Livestock activity in ranch B uses unpaid family labor and results in a negative net operating margin. Crops play an important role as fodder for livestock on both ranches, generating positive net operating margins (Table 13.7). Services are an important source of net operating margin in both cases, coming from amenities in Ranch A, and from varied sources, such as rancher housing, amenities, and tractor services on Ranch B. The Ranch B net operating margin for service activities is about 1.5 times higher than that of Ranch A.

Aggregated commercial activities offer a low net operating margin on Ranch A and a negative one on Ranch B. Amenity values make the total net operating margin from Ranch B positive (Table 13.7). Since there are minimal net operating

Fig. 13.10 The retention of oaks on ranches is affected by land management goals such as maintaining pasture productivity, avoiding soil erosion, and preserving landscape views. Ironically, one reason ranchers maintain oaks is to maintain land values for eventual sale to residential developers. Prospective buyers like the look of an oak woodland beyond the deck-patios of their houses. (Photograph by L. Huntsinger)

subsidies for either ranch, the commercial net value added at basic prices is lower than at producer prices because of taxes. For both ranches, amenities contribute the highest percentage of the net value added (Table 13.8). Subsidies are present only in Ranch A for the subsidized irrigation system (from EQIP; Chap. 12), considered a capital subsidy at $10.9 per hectare (not shown in Table 13.8).

Based on the estimation of Campos et al. (2009b) from a sample of California oak woodlands, 43 % of the land price of the ranch case studies may be attributed to commercial benefits and 57 % to amenities (Table 13.8). During 2007, ranch-land revaluation in California was 19.6 %, according to the rangeland prices published in CASFMRA (2012) for several counties with oak woodland ranches (Fig 13.10). Similar to the dehesa case studies, land revaluation distribution between commercial and amenity benefits for the California ranches is estimated according to the land revaluation observed in 2007 and to an estimation of the past trend in the commercial operating profitability from ranches. In California and western US ranches, commercial operating profitability from the 80s and 90s (Torell et al. 2001) compared to the results from the case studies analyzed in this chapter suggests a negative cumulative commercial operating profitability rate of −2.8 % for the period 1990–2007. Thus, the revaluation of the share of land price attributed to amenity benefits is estimated so that total land revaluation is 19.6 % and commercial land revaluation is −2.8 %. This, along with the other capital gain components, offers a negative commercial capital gain and a positive amenity capital gain for the ranch case studies (Table 13.8). The total capital gain figures for ranches A and B are, respectively, 3.8 and 4.0 times more than their net operating margins.

The commercial and amenity total incomes are positive in both ranches (Table 13.8). Amenity total income represents more than 91 % of total income in both cases, as much as 98 % on Ranch B.

13.5.2.2 Profitability Measure

The owner of Ranch B accepts a commercial operating profitability figure near zero in exchange for high amenity profitability and a low net cash flow. The commercial operating profitability is positive for Ranch A. When including amenities, the operating profitability at producer prices reaches 4.6 and 4.1 % for Ranches A and B respectively. If we add capital gain, we obtain a total profitability of 22.3 and 20.3 % respectively (Table 13.9). This unusually high total profitability is explained by the high land appreciation obtained for ranchlands in 2007 (CASFMRA 2012). As in the dehesa case studies, an analysis including land appreciation over a longer period is needed.

For this reason, we extend our profitability analysis to consider recent historical cumulative land appreciation. Based also on the data provided by CASFMRA (2012), we estimate a land appreciation for the period 1999–2010 of 7.9 % for California ranchlands. Using this ranchland appreciation rate, and applying the same criteria as in dehesas, total profitability for Ranches A and B is 12.5 and 12.0 % respectively. If we adjust for inflation in California over the period of 1999–2010 (California Department of Finance 2012), we obtain a real total profitability of 9.8 and 9.3 % for Ranches A and B, respectively. However, the period for which we have available data is short and the contribution of land appreciation to total profitability figures could be still overestimated (prices have been falling since 2010). The expectation of future land price increases is affecting the actual land price such that the real capital gain is an expected income that cannot be made liquid without selling the ranch. However, the expected income exists as potential income available for consumption, and it is reflected at the end of the accounting period.

13.6 Comparative Analysis: Spain and California

A first difference is that supplemental livestock feeding is not significant in the ranch case studies but is common in the dehesas (Fig. 13.11). Forage consumption per unit area is higher in the dehesas (Tables 13.2 and 13.3), but livestock grazing-

Fig. 13.11 This view of the town of Arroyo de la Luz is from the Dehesa Boyal. The Dehesa is about 2 km from the town, which makes it possible for the traditional common livestock herds to be herded to the Dehesa in the morning, and in the evening. (Photograph by F. Pulido)

related profitability rates are higher on the ranches. This might indicate that other benefits make a higher contribution to dehesa land prices than to ranches. Cattle products are usually the only goods sold from ranches, compared to the multiple livestock products sold from dehesas (Tables 13.4, 13.5, 13.6, 13.7; Chap. 1). A potential market distortion of grazing fees in dehesas comes from government livestock subsidies, in contrast with the market-driven grazing rental prices on woodland ranches, where net subsidies are less important than in dehesas.

In the dehesas, livestock stocking rates are higher, with more diverse livestock breeds. The case study results from livestock activities (Tables 13.6, 13.7) indicate that livestock operations are commercially unprofitable when they are not subsidized, except for the Iberian pig montanera. In three of the case studies, family labor is unpaid, which would make these enterprises even less profitable if family labor were replaced by employee labor. If we can attribute the willingness of these families to maintain their operations with unpaid family labor to their desire to enjoy the lifestyle, environment, and livestock husbandry, and on future land appreciation based on the expectations of an increasing demand for such amenities, then such amenities could play a crucial role in sustaining dehesas and ranches (Fig. 13.12).

Other commercial products do not seem to adequately compensate for the low or negative profitability from livestock. For both the dehesa and ranch case studies, hunting has negative net operating margin, except for Ranch A, while other activities do not add much to commercial benefits (Table 13.6, 13.7). Emerging enclosed hunting enterprises may be a component of amenities or perhaps may prove to be profitable in the longer term. Hunting in open dehesas, which is not

Fig. 13.12 Part of the emergence of new products from California is sheep cheese, shown here during ripening. This high value product is enjoying considerable popularity with gourmand buyers. (Photograph by L. Huntsinger)

represented in our case studies, is profitable based on data from in-progress research (RECAMAN project).

Overall, commercial net operating margins are negative in dehesas and positive on ranches, but when basic prices are considered the commercial net operating surplus becomes positive for Dehesas A and C (Table 13.8). Examining amenity profitability for each case study, values are higher for dehesas than ranches. The results are not directly comparable, however, because of the different format of the contingent valuation questionnaire. The California cases determined value from an open-ended question, while the dehesa case used a closed-ended question. The contingent valuation literature shows that open-ended questions usually give lower values (Desvousges et al. 1993; Kealy and Turner 1993). Another explanation derives from the larger size of the case study ranches, which makes the amenity value per hectare lower, but the profitability from commercial enterprises probably

Fig. 13.13 Ranchettes are a type of exurban development becoming increasingly common in oak woodlands. Owning less land but obtaining the same amenity enjoyment is possible if commercial operations disappear from ranching. Amenity values to landowners are important to avoid land use change in oak woodland but commercial values are required to avoid fragmentation (Chap. 12). (Photograph by R. B. Standiford)

higher due to economies of scale. The implications from the Oviedo et al. (2012) results (Chap. 12) is that after a certain point, increases in ranch size no longer increase amenity values very much.

Land revaluation (appreciation) has an important influence on total profitability, but the land market situation was different during the two periods of the case studies. Nevertheless, the analysis of real land appreciation over a longer period shows that capital gain adds over 4 % to the profitability in both dehesas and ranches in an average year. Dehesa and ranch historical operating and capital gain profitability in real terms, accounts for around 7 and 9 %, respectively. These estimated rates, which integrate historical land appreciation trends and the value of private amenities that are usually not accounted for as income by the System of National Accounts (SNA), justify the observed land market price in Spanish dehesas and California ranches.

13.7 Concluding Remarks

The application of the *Agroforestry Accounting System* to these case studies allows an in-depth analysis of the private economies of dehesas and ranches. The applied methodology allows estimating physical, income and capital indicators in a standardized framework that includes market and non-market amenity values.

The contemporary decline in livestock production could have important impacts given the role of natural resource management to the sustainability of mixed commercial-amenity consumption enterprises. These are critical in maintaining the structure and function of dehesa (Chap. 6), conserving land (Chap. 8), maintaining grazing-related habitat characteristics, and supporting the flow of ecosystem services (Chap. 12). With the decline of commercial enterprises, the abandonment of forestry practices in dehesas will likely produce losses in natural and manufactured capital in the long term and it is uncertain how the current amenity values will persist. There is a potential risk of continued natural resource depletion if the market is not able to internalize natural capital losses, and government regulation and subsidies are not sufficient to avoid biodiversity and cultural losses from dehesa natural capital.

In California, threats come from property subdivision and subsequent habitat and landscape fragmentation (Oviedo et al. 2012; Chaps. 8 and 12) (Fig. 13.13). Paradoxically, the two components that contribute most to the total profitability of the ranch case studies, land appreciation and amenities, can undermine woodland conservation. Amenity values contribute to oak woodland profitability, but values drop on a per hectare basis for large properties. This could accelerate subdivision. High rates of land appreciation like those so prevalent in the early years of the twenty-first century could enhance the trend of ranch landowners selling off small parcels to meet cash flow shortages. At present, escalating prices for corn and other livestock feeds in response to growing energy markets, and drought in US grain producing areas, is resulting in livestock being sold off at low prices to avoid feed costs, depressing livestock prices nationwide. California oak woodland ranches, with the majority of livestock feed coming from natural grasslands, may benefit from subsequent high values for grass and livestock products in the coming years, although in the meantime many ranchers may have given up commercial operations. Dehesa owners, with a greater reliance on supplemental feeds, may not experience these same benefits.

These case study results, together with the literature on land prices and commercial profitability, show a cumulative increase in landowner amenity consumption and a decline in commercial operating profitability. With a persistent decline in commercial operating profitability and an uncertain future for landowner amenity preferences, markets and subsidies for ecosystem services, and for unpaid or low paid family labor, become important factors in the sustainable management of these oak woodlands. Amenity preferences and public non-market services turn out to be important factors and arguments for policy-makers to support landowners so that they would be able to maintain active ownerships, keeping the window open to finding further ways to conserve dehesa and ranch working landscapes.

Our case study approach has strengths, but also weaknesses. Case studies provide an outstanding laboratory to develop innovations in extending income measurement and illustrating new findings in the economies of these working landscapes. The application of the Agroforestry Accounting System contributes to a more complete accounting of the economics of dehesas and ranches and to understanding the economic forces that drive management, which is crucial for research and policy decisions in the near future. These applied case studies offer a detailed analysis at the farm scale. Although the case study approach is not a statistical approach and is year-specific in our application, we provide results that can be compared with the ranges of other statistically representative studies. Moreover, we extend our profitability analysis using historical land appreciation data, to smooth annual price fluctuations that would obscure the long term trend, and compare it with the available results in the literature of dehesa and ranch economies. Our results have confirmed the trend shown in the results of other studies. Although the case study approach is not enough to make broader policy recommendations and to draw representative conclusions, it is crucial to understanding how these multiple-use complex systems work and to highlight the need for improved accounting and economic valuation tools for applied research and analysis.

Acknowledgments The authors thank Kayje Booker, Samuel Gómez, Luis Guzmán, Casimiro Herruzo, Soledad Letón, María Martinez, Gregorio Montero, María Pasalodos, Dionisio Pérez, and Ana Torres for their technical support in data collection and treatment, and to Bill Stewart and Rick Standiford for their suggestions from an earlier version of the manuscript. This paper has been funded by and is a contribution to the projects *Valoración de la Renta y el Capital de los Montes de Andalucía* (RECAMAN) of the Junta de Andalucía, Assessing the Non-Market Values of California Ranches of the Division of Agriculture and Natural Resources of the University of California (CIG05-178), and Intramural Grant 200910I130 of the Spanish National Research Council (CSIC). The usual disclaimer applies.

References

Bolsinger CL (1988) The hardwoods of California's timberlands, woodlands, and savannas. Resource Bulletin PNW-RB-148. Pacific Northwest Research Station, USDA Forest Service, p 148

Bureau of Economic Analysis [BEA] (2010) Concepts and methods of the US national income and product accounts. Bureau of Economic Analysis. US Department of Commerce. Available at http://www.bea.gov/national/pdf/NIPAhandbookch1-4.pdf. Accesed June 2012

California Chapter, American Society of Farm Managers and Rural Appraisers [CASFMRA] (2012) Trends in agricultural land and lease values reports. Woodbridge. Available at http://www.calasfmra.com/trends.php. Accessed July 2012

California Department of Finance (2012) Consumer price index. Calendar year averages. Available at http://www.dof.ca.gov/html/fs_data/latestEconData/fs_price.htm. Accessed June 2012

Campos P (2000) An agroforestry account system. In: Joebstl H, Merlo M, Venzi L (eds) Institutional aspects of managerial and accounting in forestry. IUFRO and U of Viterbo, Viterbo, pp 9–19

Campos P, Mariscal P (2003) Preferencias de los propietarios e intervención pública: El caso de las dehesas de la comarca de Monfragüe. Investig Agrar Sis y Recursos Fores 12(3):87–102

Campos P, Riera P (1996) Rentabilidad social de los bosques. Análisis aplicado a las dehesas y los montados ibéricos. Información Comercial Española 751:47–62

Campos P, Rodríguez Y, Caparrós A (2001) Towards the Dehesa total income accounting: theory and operative Monfragüe study cases. Investig Agrar: Sist y Recurs Fores: Special issue on New Forestlands Economic Accounting: Theories and Applications 1:45–69

Campos P, Bonieux F, Caparrós A, Paoli JC (2007a) Measuring total sustainable incomes from multifunctional management of Corsican maritime pine and Andalusian cork oak Mediterranean forests. J Envi Plan Mgmt 50(1):65–85

Campos P, Daly H, Ovando P (2007b) Cork oak forest management in Spain and Tunisia: two case studies of conflicts between sustainability and private income. Int For 9(2):610–626

Campos P, Daly H, Oviedo JL, Ovando P, Chebil A (2008a) Accounting for single and aggregated forest incomes: application to public cork oak forests of Jerez in Spain and Iteimia in Tunisia. Ecol Econ 65:76–86

Campos P, Ovando P, Montero G (2008b) Does private income support sustainable agroforestry in Spanish dehesa? Land Use Pol 25:51–522

Campos P, Daly H, Ovando P, Oviedo JL, Chebil A (2009a) Economics of cork oak forest multiple use: application to Jerez and Iteimia agroforestry systems study cases. In: Mosquera-Losada MR, Rigueiro-Rodríguez A, McAdam J (eds) Agroforestry in Europe. Series Advances in Agroforestry 6. Springer, Dordrecht, pp 269–295

Campos P, Oviedo JL, Caparrós A, Huntsinger L, Coelho I (2009b) Contingent valuation of private Amenities from Oak Woodlands in Spain, Portugal, and California. Rangel Ecol Mgmt 62:240–252

Desvousges WH, Johnson FR, Dunford RW, Boyle KJ, Hudson SP, Wilson N (1993) Measuring natural resource damages with contingent valuation: tests of validity and reliability. In: Hausman JA (ed) Contingent valuation: a critical assessment. Amsterdam, pp 91–159

Díaz M, Campos P, Pulido FJ (1997) The Spanish dehesas: a diversity in land use and wildlife. In: Pain DJ, Pienkowski M (eds) Farming and birds in Europe: the common agricultural policy and its implications for bird conservation. Academic Press, London, p 436

Eisner R (1989) The total incomes system of accounts. The U of Chicago Press, Chicago, p 416

European Communities, International Monetary Fund, Organisation for Economic Co-operation and Development, United Nations and World Bank (2009) System of national accounts 2008 (SNA 2008). New York, p 662. Available at http://unstats.un.org/unsd/nationalaccount/docs/SNA2008.pdf. Accessed July 2012

EUROSTAT (2000) Manual on the economic accounts for agriculture and forestry EEA/EAF 97 (Rev. 1.1). European Communities, Luxembourg, p 172

Forest and Resource Assessment Program [FRAP] (2003) Department of forestry and fire protection. Available at http://www.frap.edf.ca.gov.assesment2003. Accessed March 2012

Franco JA, Gaspar P, Mesias FJ (2012) Economic analysis of scenarios for the sustainability of extensive livestock farming in Spain under the CAP. Ecol Econ 74:120–129

Huntsinger L, Sayre N (2007) Introduction: the working landscapes special issue. Rangel 29(3):3–4

Huntsinger L, Sulak A, Gwin L, Plieninger T (2004) Oak woodland ranchers in California and Spain: conservation and diversification. In: Schnabel S, Ferreira A (eds) Sustainability of agrosilvopastoral systems: dehesas, montados. Advances in Geoecology, vol 37, pp 309–326

Huntsinger L, Johnson M, Stafford M, Fried J (2010) Hardwood Rangeland Landowners in California from 1985 to 2004: production, ecosystem services, and permanence. Rangel Ecol Manage 63(3):324–334

Kealy MJ, Turner RW (1993) A test of the equality of closed-ended and open-ended contingent valuations. Am J Ag Econ 75:321–331

Krutilla JV (1967) Conservation reconsidered. Am Econ Rev 57(4):777–786

Ministerio de Medio Ambiente y Medio Rural y Marino [MARM] (2008) Diagnóstico de las Dehesa Ibéricas Mediterráneas. Tomo I. Available at http://www.magrama.gob.es/es/biodiversidad/temas/montes-y-politica-forestal/anexo_3_4_coruche_2010_tcm7-23749.pdf. Accessed July 2012

Ministerio de Medio Ambiente y Medio Rural y Marino [MARM] (2011) Encuesta de precios de la tierra 2010. Available at http://www.magrama.gob.es/es/estadistica/temas/encuesta-de-precios-de-la-tierra/. Accessed July 2012

Oviedo JL, Huntsinger L, Campos P, Caparrós A (2012) Income value of private amenities assessed in California Oak Woodlands. Cal Agric 66(3):91–96

Torell LA, Rimbey NR, Tanaka JA, Bailey S (2001) The lack of profit motive for ranching: implications for policy analysis. In Torell LA, Bartlett ET, Larrañaga R (eds) Proceedings of annual meeting society for range management, February 17–23. Hawaii, pp 47–58
Torell LA, Rimbey NR, Ramirez OA, McCollum DW (2005) Income earning potential versus consumptive amenities in determining ranchland values. J Agri Res Econ 30(3):537–560

Part V
Landscape

Chapter 14
Recent Oak Woodland Dynamics: A Comparative Ecological Study at the Landscape Scale

Ramón Elena-Rosselló, Maggi Kelly, Sergio González-Avila, Alexandra Martín, David Sánchez de Ron and José M. García del Barrio

Frontispiece Chapter 14. California and Spain have influenced one another since colonization in 1769. Ferreras de Abajo, in central Zamora, could easily be in California. Spanish and Californian oak woodland landscapes are a mosaic of ecosystems with oak woodlands as a central component, though the mosaic and the patterns diverge somewhat. This landscape shows spatial patterns somewhere in between those typical Spain and California. In the lowlands, there is a high proportion of cultivated land that includes oak trees, typical of Spain but unusual in California where the trees are usually eliminated. At the same time, the upper elevations are forested, which is more typical in California. The gentle hilly landform has allowed the development of windmill parks for energy production, appearing in California during the eighties and now is also common in Spain. (Photograph by 4Ullas)

R. Elena-Rosselló (✉) · S. González-Avila · A. Martín · J. M. García del Barrio
Department of Silvopascicultura, Universidad Politécnica de Madrid, UPM, Ciudad Universitaria s/n 28035 Madrid, Spain
e-mail: ramon.elena.rossello@upm.es

P. Campos et al. (eds.), *Mediterranean Oak Woodland Working Landscapes*, Landscape Series 16, DOI: 10.1007/978-94-007-6707-2_14,

© Springer Science+Business Media Dordrecht 2013

Abstract The tools of landscape ecology are used to compare oak woodland landscapes in California and Spain. The linkages between spatial patterns and functional processes are explored using Geographic Information System and remote sensing technologies, drawing on available databases in each country. The Spanish Rural Landscape Monitoring System (SISPARES) methodology is applied and tested in an attempt to answer questions about the comparative structure, patterns, and changes in oak woodlands in Spain and California. Landform at the macro scale was found to be the main driver for the mid and long-term for oak woodland pattern and distribution. Spanish woodlands are most frequently located in peneplains with acid bedrock, creating large contiguous areas of woodland, likely managed as dehesa. Californian oak woodland patches are smaller, highly dispersed, and intermixed with chaparral and dense forest in foothills with acid bedrock. Agro-silvo-pastoral practices are also important in shaping oak woodlands, especially in Spain. New land uses are increasing in both countries, such as agro-tourism, game hunting, golf courses, and suburban development, but their impacts are much greater in California. Oak woodlands in California are located in areas of much higher fire vulnerability because of the intermix with flammable brush, forest, and urban sprawl. The oak woodlands that are most similar in Spain and California, according to the TWINSPAN classification, are those in Spain close to the mountains, and in California located near the coast.

Keywords California · Change · CLATERES · Dehesa · Drivers · Functional Processes · Mediterranean · Monitoring · Ranch · SISPARES · Spain · Spatial Pattern · Structure · Tourism · Trends · Wildfire Vulnerability

S. González-Avila
e-mail: sergio.gonzalez@upm.es

A. Martín
e-mail: alexandra.martinfer@gmail.com

M. Kelly
Department of Environmental Science, Policy, and Management, University of California, 130 Mulford Hall , Berkeley, CA, MC 3110 94720, USA
e-mail: maggi@berkeley.edu

D. Sánchez de Ron · J. M. García del Barrio
Forest Research Centre, National Institute for Agriculture and Food Research and Technology, Crta La Coruña km 7.5 28040 Madrid, Spain
e-mail: dsanchez@inia.es

J. M. García del Barrio
e-mail: jmgarcia@inia.es

14.1 Introduction

Despite being an old term with subjective and aesthetic meanings, *landscape* has only recently come into use as a specifically defined ecological concept. Retasked in scholarly nomenclature in the mid-1980s as the biological level of organization above the ecosystem, landscape is the focus of landscape ecology, a branch of the ecological sciences that has grown significantly over the last 25 years (Forman and Godron 1986).

The aim of landscape ecology is to understand the linkages between spatial patterns and functional processes in complex systems that often include several land-cover types and land uses. Spatial and temporal dimensions are investigated by comparing landscape patterns (the composition and configuration of a landscape) and landscape processes (the landscape changes through time that involves functional processes). Landscape ecology has been the beneficiary of recent scientific advances in Geographic Information System (GIS) and Remote Sensing (RS) technologies, which allow complex spatial analysis of landscape pattern and process. Landscape ecology often relies on GIS and RS techniques to interpret landscape patterns in ecological and socioeconomic terms, identifying influential factors, and detecting processes at work in cultural landscapes such as dehesas and ranches.

The Spanish dehesa is considered an outstanding example of a managed, anthropogenic, ecosystem that offers challenges and opportunities for landscape ecologists. Dehesas are agro-silvopastoral farms that look a lot like savanna vegetation systems (Marañón 1986; Joffre and Lacaze 1993; Joffre et al. 1999; Valladares 2004; Marañón et al. 2009). Fully understanding the landscape spatial patterns of dehesa requires knowing the practices and history of dehesa management, and the motives and decisions of landowners.

Management is partly detectible in dehesa regions by the remote sensing (RS) technologies used by landscape ecologists. Traditional dehesa management has long included cycles of shrub removal and regrowth, beginning with the interrupting of shrub return in the understory through grazing, hand-removal, or mechanical plowing, followed by several years of regrowth and initiation of another removal cycle. This management changed drastically after the crisis in the 1950s, but it has had a strong influence on current dehesa landscapes due to the long life cycle of oak trees (Díaz et al. 1997; Moreno and Pulido 2009). Further, open oak woodlands that are not managed as dehesa are rare in Spain because abandonment of traditional uses causes tree and shrub regrowth that changes woodland structure (Chap. 6).

In California, today's oak woodlands are the legacy of successive historical activities, including indigenous management, Spanish colonization, Mexican governance, and Gold Rush exploitation, rather than a result of the kind of ongoing, deliberate, and complex management that shapes the Spanish dehesa. Spain and California do share common influences: Spanish settlers introduced livestock, including horses, cattle, and sheep, and a variety of Mediterranean plant species to California that spread rapidly through the native grasslands and woodlands

Fig. 14.1 Oak tree distribution in California's Sierra Nevada foothills follows natural gradients in soil and geological characteristics, but also has been shaped by past human impacts, for example, a power line corridor in the upper left, deliberate burning, and livestock grazing. The management regime—ranches, parks, or some other type—is much more difficult to detect from the air. Interpretation of remotely-sensed images can provide more reliable information about land cover than about management. Consequently, landscape ecological analysis requires complementary information on land management to better model landscape structure and functional processes. (Photograph by M. McClaran)

(Chaps. 2 and 6). These impacts are still not fully understood. Californian oak woodlands are relatively stable in the absence of management, with an estimated 85 % of woodland understory persisting as grassland without intervention (Allen-Diaz et al. 1999). When shrub encroachment occurs, it tends to happen slowly due to the exclusion of grazing and fire. Whether a Californian oak woodland patch is unmanaged or managed as part of a ranch is difficult if not impossible to discern using RS technologies. However, these oak woodlands are visually a lot like the dehesa of Spain, although with a distribution of oaks that follows natural ecological gradients and is generally less regular than in Spain (Fig. 14.1).

Similarities between Californian and Spanish oak woodland landscapes have been noted by many observers. The techniques of landscape ecology can be used to detect functional linkages between pattern and process. Two structurally similar landscapes may have very different underlying functional processes. If structurally similar, historical conditions and management are distinct: What is the impact on pattern? How can resemblance and differences between landscapes be compared in a quantifiable manner using the techniques of landscape ecology? This chapter evaluates techniques and data that are available for comparing landscapes by applying a similar sampling and analysis approach in Spain and California to detect change and to attempt to determine the drivers of each site's structure and pattern. The goal is to further research providing recommendations for multi-functional and sustainable management in both study sites.

14.2 Terminological Conventions

We rely on remote sensing, either aerial photography or satellite imagery, to conduct the analysis in this chapter. Dehesas and ranches are economic enterprises whose boundaries are not clearly defined from the air. To develop a study area that can be consistently defined with remote sensing in California and Spain, so an identical methodology can be applied to both places, some distance is needed from the definitions used in other chapters. *Oak woodland* is the term that we would like to use for denoting, whether in Spain or California, an ecosystem with scattered trees (canopy cover under 70 %) and an herbaceous layer that is detectible using aerial photography. Unless stated otherwise, this chapter will consider landscape as the study scale, where oak woodland patches cover more than 30 % of the total area, regardless of whether they are part of a dehesa agroforestry system or an oak woodland ranch. The area covered is necessarily much larger than that referred to as dehesa and oak woodland ranch in Chap. 1, and includes a variety of vegetation types beyond a pure oak woodland and savanna. This enables analysis of the distribution of oak woodlands within the matrix of vegetation in their potential habitat.

14.3 Oak Woodlands and Dehesa in Spain from a Landscape Perspective

Oak woodlands dominate the landscape of large areas of the south-west Iberian Peninsula, occurring in a broad gradient from north to south and a narrower one from east to west. They are almost always managed as dehesa (Fig. 14.2), and it is commonly inferred that the presence of open oak woodlands as detected using remote sensing indicates a dehesa-type management regime, with the characteristic physiognomy of an open oak canopy coexisting with crops and pastures. At least since the sixteenth century (Linares-Luján and Zapata-Blanco 2003), changes from dense forest to open woodland have been promoted and maintained in hilly to flat, but never rough, topographies, over soils predominantly developed on acid

Fig. 14.2 Twin aerial photographs from Spain and California show reservoirs, frequent components in the landscapes of each region. Both were built since the 1950s as key infrastructure for irrigation and for urban water supply. *Left* Spanish dehesa landscape at Orellana Municipality in Eastern Extremadura. There are clearly distinguished *yellow cereal fields*, *brownish pastures*, and *dense green irrigated crop* fields. Reservoir construction included reforestation to prevent sedimentation from eroded soils and newly developed croplands. *Right* A California oak woodland landscape in San Leandro (Alameda County) in the eastern San Francisco Bay Area. In the vicinity of a vast metropolitan area, the oak woodland landscape shows marked urban elements, a reservoir for urban water supply, and the dense forest that is more common in California oak woodland landscapes. It is becoming a forest-urban interface with high wildfire vulnerability. (Photographs by 4Ullas)

bedrocks with low soil fertility, particularly those deficient in phosphorus and calcium (Pérez Soba et al. 2007).

Sanchez de Ron et al. (2007) distinguished three main climatic typologies for dehesas (Fig. 14.3; Chap. 3). The main differences among these typologies are related to precipitation (types 1 and 2) and temperature (type 3) (Chap. 3). This climate gradient is reflected in patterns of distribution of the main tree species and in shrub and herb biodiversity. Although holm oak (*Quercus ilex)* is the dominant species in 80 % of Spain's dehesas, other species are found with them, including the cork oak (*Q. suber*) in warm and humid environments, the Pyrenean oak (*Q. pyrenaica)* in acidic and colder environments (Fig. 14.4), the Lusitanian oak (*Q. faginea)* on less acidic soils, and the narrow-leafed ash (*Fraxinus angustifolia)* on humid soils close to rivers. The shrub layer is deliberately restricted to enable production of forage and crops, and as a result is absent or scarce in traditionally-managed dehesa. It appears just before shrub control techniques are applied, or when dehesa management has shifted to big game activities (Chaps. 6 and 11). Patterns in herbaceous communities are highly dependent on local conditions such water availability, proximity to tree trunk, slope, and aspect (Chap. 6). In general, annual species are more abundant than perennials but in any case, vegetation strata contribute significantly to dehesa species richness, already among the highest in temperate and Mediterranean vegetation types (González Bernáldez et al. 1969; Naveh and Whittaker 1979; Marañón 1986; Díaz and Pulido 2009).

Management of a dehesa influences its structure in ways that can be detected using landscape ecology techniques. Diverse products are routine, and while some exclude other uses, many can be produced on the same property. Pasture is a

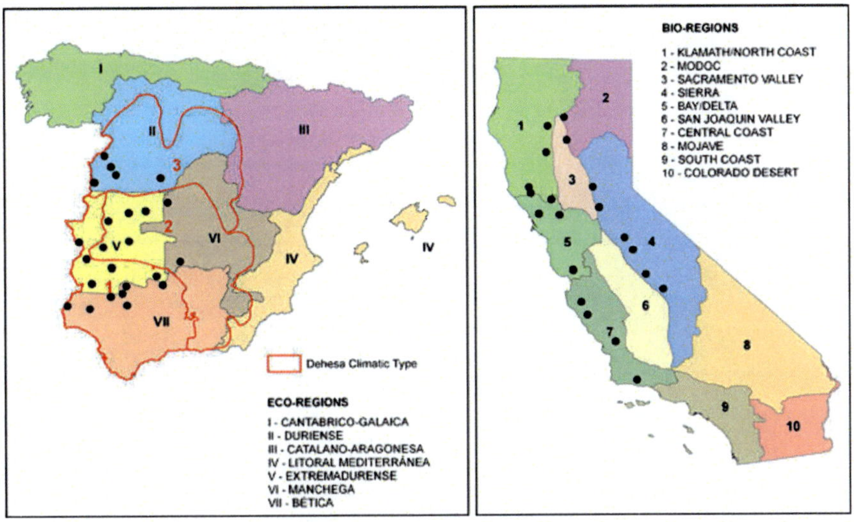

Fig. 14.3 Distribution of oak woodland dehesa plots in Spanish eco-regions and California oak woodland plots in state bioregions (CDF-FRAP 2002). Red lines on the Spanish map at left delineate three climate-based typologies for dehesa (Chap. 3). The dots represent the 4 km × 4 km plots sampled for this study in both Spain and California

Fig. 14.4 A dehesa with a mixed tree layer combining evergreen holm oaks together with semi-deciduous Luisitanean oaks and the marcescent Pyrenean oak. The brownish trees are holm oaks, and the lighter greens are Pyrenean and Luisitanean oaks. The site is an uncommon case of a dehesa of all three. It is found in colder areas on the border between Zamora and Salamanca provinces. (Digital rendition by 4Ullas)

priority in dehesa silvopastoral systems, as are acorns, bark tannins, cork, charcoal, and sometimes cereals as complementary products. The livestock species that graze a property play a decisive role in shaping dehesa ecological structure, by influencing the density of oaks maintained by the landowner (Chap. 10). For

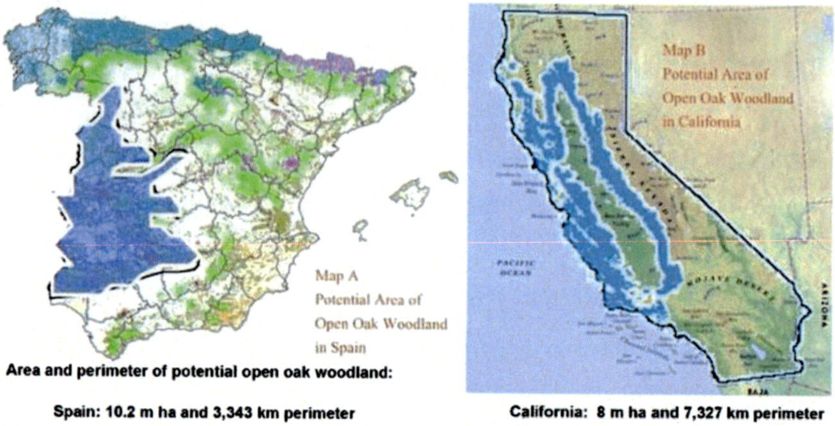

Fig. 14.5 The potential distribution of open oak woodland landscape in Spain and California. The area is defined using the Area Shape Index (ASI) by Ruiz de la Torre (2002). The area of oak woodland dehesas and ranches is much smaller and is located within each area (Chap. 3)

example, acorns from oaks are essential for producing Iberian pigs, and are an important food for game, but are less important for cattle and sheep. In the Mapa Forestal de España, the potential area of open oak woodland landscape in Spain encompasses 10.2 million ha (Ruiz de la Torre 2002; Fig. 14.5). The area of oak woodland dehesa is considerably smaller and is located within this area (Chap. 3), and as defined in this book (Chap. 1) there are 3–4 million ha currently managed as dehesa.

14.4 Oak Woodlands in California

On deep level soils where irrigation water is available, much of the original oak woodland of California has been converted to tree and field crops—eliminating perhaps as much as half of the original woodland (Fig. 14.6). Oak woodlands today are distributed through 6 out of the 10 bioregions defined for the state by CalFire, the agency responsible for statewide natural resource assessments (CDF-FRAP 2002; CDF-FRAP 2010c) (Fig. 14.3). In California, oak woodlands with an herbaceous understory are stable even without active management. Management is not required to prevent shrub encroachment, although in some areas shrubs will slowly encroach if fire and grazing are excluded, notably in coastal areas, at higher elevations, and on fertile volcanic soils. It is difficult if not impossible to infer a land use or management regime based on the presence of open woodland with an herbaceous understory. Two-thirds of oak woodlands in ownerships larger than 10 ha are owned by livestock producers, with some 75 % of the area used by ranchers for grazing (Huntsinger et al. 2010).

Fig. 14.6 The view east from the Sacramento Valley of California shows the occupation of lowland areas—once the domain of the valley oak (*Q. lobata*) and extensive marshes and grasslands—by field crops. Ranching remains the most widespread use in the hills, where cultivation and irrigation is impractical. Vineyards are now encroaching on some of this land, as is residential development. (Photo by Paul F. Starrs)

California's remaining oak woodland and savanna occurs on what is generally a rolling foothill topography (Ewing et al. 1988), including land along edges of the Sacramento and San Joaquin valleys and rising into the foothills of the Sierra Nevada. Oak woodlands include blue oak (*Q. douglasii*), valley oak (*Q. lobata*), Engelmann oak (*Q. engelmannii*), coast live oak (*Q. agrifolia*), interior live oak (*Q. wislizenii*), canyon live oak (*Q. chrysolepis)*, California black oak (*Q. kelloggii*), and Oregon white oak (*Q. garryana*) (Chaps. 5 and 6). The most common oak in the California oak woodland is blue oak, with coast live oak a common dominant or co-dominant in coastal foothills.

Ranch management can have impacts on woodland structure. The emphasis is on forage for cattle, with cover and acorns for game a secondary factor (Chap. 10). Owners usually think of managing oak density to prevent suppression of understory forage production, and most are satisfied with an oak canopy of 50 % or less (Huntsinger et al. 2010) (Fig. 14.7). Economic studies emphasize managing oaks to balance livestock and game production (Standiford and Howitt 1992). Ranch owners in the Sierra Nevada foothills pointed out in one recent study that fire suppression and the exclusion of grazing on public lands creates in some areas a difference in vegetation between private and public lands, with tree and shrub cover on public lands becoming more dense (Sulak and Huntsinger 2007). This trend might be detectable aerially.

Although native grasslands commonly show a higher proportion of perennials versus annuals, the characteristic oak woodland understory today is annual grasses and forbs (Fig. 14.8), most of which have emigrated and continue to emigrate from other Mediterranean regions (Huntsinger and Bartolome 1992). The distribution of these plants is shaped by factors similar to those in Spain (Chap. 6). But in a departure form the Spanish case, establishment of woody plants in the understory

Fig. 14.7 This woodland at the University of California's Sierra Foothill Research and Experiment Station was thinned in the 1970s, while the foreground hillside was cleared, to demonstrate the potential increase in forage production to landowners. The University has since stopped encouraging thinning and clearing of oaks. (Photograph by P. Gil)

Fig. 14.8 About 75 % of California oak woodlands are grazed by livestock, and some 67 % are owned by ranchers. At this site in a drier area of the Central Coast foothills, oaks and shrubs seem to follow patterns related to geological layering. The valley bottom was likely cultivated in the mid-nineteenth century. (Photograph by L. Huntsinger)

is uncommon (Bartolome et al. 1988) although it does occur on some ecological sites. Map B in Fig. 14.5 shows the potential distribution of open oak woodland ecosystems in California. Oak woodlands are usually grazed by cattle, though there are some sheep (Chap. 10), and mixed grazing is not usual. The potential area of open oak woodland landscape is a little over 8 million ha, although only some 2–3 million ha are currently managed as oak woodland ranches (Chap. 1).

14.5 Spanish Versus Californian Oak Woodland Landscapes: Hypotheses to be Tested

In Spain SISPARES—the Spanish Rural Landscape Monitoring System (*Sistema para seguimiento de paisajes rurales españoles*) was first used in 1993 to examine land use change in oak woodland landscapes (www.sispares.com). Here the

SISPARES methodology is applied to California to better answer questions about the comparative structure, patterns, and changes in oak woodlands in Spain and California.

14.5.1 Introduction to SISPARES

Two recent methodological additions to landscape ecology inquiry are study at short and medium time scales, and monitoring changes in structure and function in landscapes across spatial scales, potentially including local, regional, national and even continental scales. Analysis of changes in landscape over time would not be possible without the technological developments that have occurred in geographic information systems and remote sensing since the early 1990s.

SISPARES has evolved over 20 years from simple to complex, from a focus on landscape structure toward an inclusion of landscape function, and from the static toward the dynamic. Early on, the main output of SISPARES was static structural models. Currently, it is able to generate dynamic multifunctional models.

SISPARES is based on the data recorded from 215 environmental stratified samples using aerial photography. The required environmental stratification was provided by a land classification (LC) system, considered an integral part of the monitoring system (Bunce et al. 1996). SISPARES used the CLATERES, Spanish Territorial Classification (*Clasificación Territorial de España*) of Iberian Spain and the Balearic Islands (Elena-Rosselló et al. 1997). The design of the landscape sampling protocol (size, shape, area, and data to be recorded), was initially based on a pioneering study of landscape change in the state of Georgia in the U.S. (Turner and Ruscher 1988). Accordingly, in the REDPARES Spanish Rural Landscapes Network (*Red de Paisajes Rurales Españoles*) plots of 4 km × 4 km were selected using CLATERES Land Classes as the sampling strata. Each sample is 16 km^2 in extent, a common dimension in biodiversity and landscape monitoring studies (McCollin 1993; Honnay et al. 2003; Angelstam 2004), with a minimum mapped patch size of 1 ha.

Each sample unit was located at random within its land classification stratum. The spatial information describing landscapes entered into the database was recorded through interpretation of aerial photographs taken simultaneously all over Spain. Since SISPARES techniques began, the time period covered has increased with the completion of new surveys. Currently, in 2012, the period covered is 52 years with surveys in 1956, 1984, 1998, and 2008.

Changes in the Spanish dehesa were analyzed using the SISPARES Network for two time intervals, 1956–1984 (Regato et al. 2004) and 1956–1998 (García del Barrio et al. 2004). In each study the number of samples (15 in the first and 22 in the second) and the conclusions reached were similar.

The oak woodland landscape of Spain was relatively stable during the second half of the twentieth century with low rates of change in land use composition and configuration. A management intensification process that consisted of cutting trees and cultivating the land was detected in 10 % of cases. Increases in tree vegetation cover, tree colonization, and shrub encroachment were detected in 5 % of cases. These changes were related to abandonment of traditional land uses such as that of the dehesa system.

14.5.2 Applying SISPARES to California

SISPARES was applied to open oak woodland landscapes in California where adequate remotely-sensed images were available for two recent dates: 1993 and 2008. Sampling parameters were selected for conducting robust statistical analysis and estimations and to test statistically the similarities between Californian and Spanish oak woodland landscapes. The database obtained from California was compared to the Spanish database to address the following three main questions:

1. How similar are land use composition and landscape configuration in Spanish and Californian oak woodlands? What can be learned about functional similarities and differences? What explains differences?
2. Which factors are responsible for Spanish and Californian oak woodland landscape structure and function? Are the drivers of change the same? If yes, do they operate at the same temporal and spatial scales?
3. Are oak woodland landscapes changing or have they been stable in composition and configuration over the past fifteen years?

14.5.3 Protocol for Comparing Spanish and Californian Oak Woodland Landscapes

In Spain and California, a similar sampling protocol was used. A standardized SISPARES procedure was applied:

- Sample unit: Plots of 4 km × 4 km.
- 24 samples from the Spanish SISPARES network surveyed both in 1998 and 2008 (Fig. 14.3).
- 20 samples selected from California using the bioclimatic regions as environmental criteria, and the availability of remote images for the two studied dates: 1993 and 2006 (Fig. 14.3).
- Landscape composition requirement for eligibility: In 1993 more than 30 % covered by open oak woodland with less than 70 % covered by tree canopy.

The 24 Spanish samples are taken from almost the full distribution of dehesa ecosystems (Corine Land cover Map 2000) across western Spain. The range extends from a little less than 400 km along a north to south axis to 250 km from east to west (Fig. 14.5: Map A). The sampling intensity is 1.0 %, equivalent to 1 sample per 170,500 ha.

The 20 Californian samples are well distributed within the area occupied by oak woodlands according to the state's natural resources assessment program (Fig. 1.5 in CDF-FRAP 2010b) (Fig 14.3), from the north (Shasta County in the Modoc Bioregion) to the south (Santa Barbara County in the Central Coast Bioregion) extending almost 700 km and extending less than 400 km east to west. The sampling intensity is around 0.9 %, equivalent to 1 sample per 180,290 ha.

The location of sample units (latitude and longitude of the plot), climatic variables, mean annual temperature, summer precipitation, and mean maximum temperature of the warmest month were all derived from the WorldClim model (Hijmans et al. 2004). These and the topographic variables of average altitude and slope were derived from digital elevation models (DEMs) with 30 m × 30 m resolution for the 44 plots surveyed are shown in Table 14.1.

Ten land uses were defined for assessing the landscape composition of the 44 sample units: forests, oak woodlands, crops, grasslands, shrublands, gallery forests, plantation forests, artificial use, water bodies, and bare soils. An additional land use, mosaic, was defined in areas where a very fine-grained mixture of land uses was detected (e.g. patch size under 1 ha). The selection of dates for the landscape analysis was conditioned by the availability of remote images. In Spain we used a time range of 10 years (1998–2008). This time frame completes the intervals analyzed before, 1956–1984 (Regato-Pajares et al. 2004) and 1984–1998 (García del Barrio et al. 2004). For Californian samples, the timespan was 15 years (1993–2008).

Photo-interpretation and delineation of oak woodlands versus other landscapes is not an easy task, especially when it comes to distinguishing among the different typologies of the dehesa (Fig. 14.9). Pastures or arable land with a few scattered trees, less or more dense brush, and open woodlands with evenly spaced trees can be seen as different stages of the dehesa rotation system. It is in this type of context that we must rely on fuzzy classification systems (Haynes-Young 2005), when we talk about dehesa landscapes (Van Doorn and Pinto-Correia 2007). For avoiding this kind of vagueness we used a reduced number of typologies that can be interpreted from remote images.

14.5.4 GIS Methods

Drawing on the Patch Analyst extension in ArcGIS v.9.3.1 (ESRI 2009), configuration indices were calculated incorporating total landscape area, number of patches, mean patch size, largest patch index, edge density, total edge, mean size index, mean size index weighted by area, mean patch fractal dimension, mean patch fractal dimension weighted by area, landscape shape index, patch richness, patch richness density, Shannon's diversity index, Shannon's evenness index,

Table 14.1 Location of Spanish (D) and Californian (R) oak woodland landscapes, and the main abiotic parameters of each plot. Abreviations and unit of variables. AT Mean annual temperature (°C), MTWM Mean maximum temperature of warmest month (°C), AP Mean annual rainfall (mm), SP Summer rainfall (mm), ALT Altitude (m asl), Slo Slope (°)

	Plot location		Climatic variables				Topographic variables	
Plot	Latitude	Longitude	AT	MTWM	AP	SP	Alt	Slo
D033	41° 11' 46.386" N	6° 24' 43.544" W	11.9	28.5	612.6	58.2	752.6	2.7
D034	40° 34' 16.324" N	6° 41' 12.642" W	13.0	29.1	744.0	56.5	663.3	3.7
D035	40° 56' 49.877" N	6° 12' 0.324" W	12.1	28.9	548.3	55.1	787.5	3.6
D036	40° 45' 46.727" N	6° 2' 31.442" W	11.8	29.0	535.2	56.1	873.3	2.6
D055	40° 43' 8.969" N	4° 40' 9.722" W	10.7	27.7	408.8	68.2	1097.5	4.4
D127	38° 46' 48.011" N	6° 50' 18.049" W	16.4	33.7	525.6	34.2	242.9	2.2
D128	38° 46' 24.082" N	4° 2' 2.092" W	14.6	34.4	477.1	33.6	695.5	1.7
D129	38° 35' 50.689" N	6° 4' 58.046" W	15.6	34.1	579.3	33.0	508.9	8.9
D130	38° 26' 13.986" N	4° 44' 33.221" W	15.5	35.2	581.8	28.8	584.9	4.1
D131	39° 56' 27.683" N	5° 6' 2.444" W	15.5	33.5	367.0	34.9	364.6	1.8
D132	40° 8' 54.238" N	4° 26' 15.411" W	15.0	32.8	357.0	40.5	434.5	3.7
D133	39° 3' 58.417" N	6° 21' 12.853" W	16.5	34.4	501.4	31.6	289.6	4.1
D134	39° 13' 38.096" N	5° 35' 12.615" W	16.1	34.8	461.1	30.4	367.1	3.3
D135	39° 53' 19.680" N	5° 37' 0.778" W	16.0	33.9	394.6	33.0	282.0	1.2
D138	39° 41' 20.473" N	6° 13' 27.630" W	15.6	33.2	507.7	36.5	414.3	3.7
D139	38° 0' 49.228" N	5° 44' 50.269" W	15.5	34.4	661.3	30.9	581.7	10.9
D140	38° 13' 16.303" N	6° 39' 28.505" W	15.1	32.3	636.2	36.5	571.3	5.0
D143	37° 55' 55.258" N	6° 4' 56.846" W	15.6	33.8	650.5	33.0	551.0	12.1
D144	39° 10' 18.990" N	7° 5' 0.480" W	16.4	33.0	622.8	41.4	309.1	6.7
D180	37° 41' 6.424" N	7° 21' 2.618" W	17.0	31.1	511.2	25.0	132.4	6.3
D187	38° 11' 8.398" N	5° 38' 4.907" W	16.1	35.2	614.0	26.5	460.3	11.1
D189	37° 38' 18.347" N	6° 41' 16.329" W	16.5	31.6	569.0	28.5	314.9	9.6
D190	37° 44' 39.789" N	5° 35' 54.304" W	16.6	35.2	644.3	27.8	358.8	6.9

(continued)

Table 14.1 (continued)

Plot	Plot location		Climatic variables				Topographic variables	
	Latitude	Longitude	AT	MTWM	AP	SP	Alt	Slo
D191	38° 13' 59.841" N	4° 33' 25.374" W	15.5	35.0	604.0	28.3	626.8	5.6
R01	39° 0' 28.958" N	123° 5' 33.948" W	13.7	31.9	1025.6	22.2	317.5	11.5
R02	38° 44' 5.827" N	122° 24' 43.013" W	14.8	34.0	809.7	9.0	250.0	8.6
R03	37° 6' 18.142" N	121° 43' 53.630" W	14.5	29.2	631.5	12.5	238.6	12.0
R04	36° 22' 39.610" N	121° 28' 46.246" W	11.2	27.1	766.3	17.4	1092.4	18.5
R05	38° 52' 24.058" N	123° 1' 41.889" W	13.5	31.3	1065.6	20.8	387.5	13.4
R06	39° 49' 44.664" N	122° 35' 39.774" W	15.4	35.5	626.5	16.9	299.2	9.2
R07	40° 26' 33.217" N	122° 34' 9.351" W	16.1	36.2	1061.6	29.0	256.5	7.7
R08	40° 39' 19.452" N	122° 4' 32.303" W	16.1	36.1	1243.5	37.2	292.9	8.3
R09	40° 7' 31.528" N	121° 58' 41.746" W	15.9	34.9	813.0	24.9	374.4	8.4
R10	39° 2' 34.343" N	121° 10' 18.197" W	15.4	33.8	919.0	15.5	364.6	8.2
R11	38° 34' 0.844" N	120° 58' 21.617" W	15.6	34.3	694.0	14.8	275.9	5.7
R12	37° 1' 11.024" N	119° 35' 32.870" W	16.1	35.9	526.0	6.0	401.0	11.6
R13	34° 33' 56.339" N	119° 52' 28.423" W	15.4	28.1	513.2	8.3	302.9	8.8
R14	36° 40' 36.715" N	119° 5' 36.745" W	14.5	33.4	723.8	22.3	920.6	11.1
R15	37° 34' 25.062" N	119° 58' 14.949" W	13.8	33.5	874.5	19.0	789.4	14.2
R16	37° 52' 8.018" N	120° 13' 0.805" W	13.7	33.4	921.8	18.7	782.0	12.4
R17	35° 27' 22.124" N	120° 28' 45.106" W	14.2	32.2	527.9	13.0	435.0	7.1
R18	36° 4' 41.255" N	121° 17' 39.917" W	13.2	27.9	435.9	9.9	456.9	11.7
R19	38° 23' 58.172" N	122° 46' 6.632" W	13.8	27.5	867.3	15.4	26.0	1.7
R20	38° 22' 36.020" N	122° 9' 44.203" W	15.0	31.9	728.7	12.6	230.8	10.3

Simpson's evenness index, mean shape index, modified Simpson's evenness index, and an interspersion juxtaposition index.

14.5.5 Statistical Methods

TWINSPAN analysis (Hill 1979) allowed the building of a joint landscape classification model of open oak woodlands for Spain and California. This analysis provides a dendrogram with an attached taxonomy key. Landscape composition variables were used for ordering and grouping landscape samples according to their spatial pattern. TWINSPAN is especially designed to analyze flora inventory data, the same type of landscape composition data that SISPARES develops based on a hierarchical classification that generates disjoint classes. Each step of the classification is a new dichotomy based on simple indicators or attributes.

To verify the differences of landscape composition and configuration among the landscape types, an Analysis of Variance was implemented using the SPSS statistics package (IBM-SPSS 2008). We also completed an analysis of variance test to understand the similarities and differences between oak woodland landscapes in relation to physiographic and climatic variables.

14.6 Differences in Oak Woodland Distribution and Change

Just a glance at the land cover maps of Spain (SP) and California (CA) allows the reader to observe clear differences, not just in size but in the spatial dispersion of oak woodlands in California and Spain (Fig. 14.5). The analysis of the broadest potential spatial distribution at both sites shows the following results:

- Total broadest area of potential oak woodland landscape in SP: 10,200,456 ha.
- Total broadest area of potential oak woodland landscape in CA: 8,659,890 ha.
- Total perimeter length in SP: 3,543 km.
- Total perimeter length in CA: 7,367 km.
- Area/perimeter rate: 28.79 km in SP versus 11.77 km in CA.

These two simple geomorphic indexes (assessing patch shape and contagion) have high discriminative power for California and Spain. Those differences can easily be related to macro relief as the main driver responsible for the variation of other important ecological factors, such as climate and soil. The geomorphology of California is structured in successive north- to south-oriented mountain ranges with interspersed valleys. Such a tectonic landform constrains the ecological conditions required for oak woodlands to certain elevations. Consequently, native oak woodland ecosystems cover a widespread area elongated by narrow strips less

Fig. 14.9 Aspect and water availability also affect the distribution of oaks in the dehesa, here shown in the outskirts of Madrid. Oaks are more dense along the ravines and more sparse in the fields, on both cultivated and pasture lands. A history of silviculture, and a high value for the oaks is revealed in subtle dehesa patterns. Some researchers have suggested that oaks are managed at densities that minimize competition for soil water (Joffre et al. 1999). (Photograph by 4Ullas)

than 20 km wide. Mainly in the foothills of Sierra Nevada, steep slopes restrict the dispersal of oak woodland types upward into the Sierra Nevada and downward into the Central Valley.

In contrast, the tectonic structure of Spain creates a completely different pattern. Mountain ranges are oriented east–west, with high plateaus in between. Despite an average elevation higher than that of California, the mountains are less steep. As a consequence, the potential natural oak woodland distribution is widespread on vast peneplains (Fig. 14.10; Chap. 3).

14.6.1 Composition of Spanish and Californian Oak Woodland Landscapes

Similarities and differences in land use patterns and change exist for Spain and California (Table 14.2). Spanish oak woodland landscapes are more intensively managed than California's because of the higher proportion of crops (14.1 % vs. 3.47 % of the area considered "oak woodland landscape" in this study) and conifer forest plantations (4.3 % vs. 1.1 %). There are more open oak woodlands

Fig. 14.10 Dehesas tend to be on gentler topography as shown in this photograph of Cabañeros National Park, in the provinces of Toledo and Ciudad Real. A geographical curiosity, Cabañeros is the only National Park that has its antipode also in a National Park: Tongariro N.P. on the North Island of New Zealand. Like most of the dehesa, Cabañeros is on chain of low hills in the middle of the vast peneplain of the western regions of Spain. All in all, actual oak woodland landscapes in Spain are 15 % larger than in California, and the spatial pattern is much more contiguous. (Photograph by 4Ullas)

Table 14.2 Landscape composition, in percentage, of Californian and Spanish oak woodland landscapes, and percentage of change between the two dates

Land use	California (%)		Spain (%)		Change (%)	
	1993	2009	1998	2008	California	Spain
Forest	25.93	25.94	5.73	5.49	0.03	−4.20
Crops	2.94	3.47	14.01	14.06	18.18	0.35
Oak woodland	52.94	52.28	60.96	60.97	−1.24	0.01
Gallery forest	0.54	0.52	0.48	0.50	−2.61	2.94
Water bodies	0.59	0.62	0.59	0.59	4.55	0.00
Bare soil	0.49	0.40	0.12	0.19	−17.90	60.93
Shrubland	12.92	13.66	7.92	7.78	5.72	−1.84
Grassland	1.99	1.08	5.39	5.63	−45.67	4.31
Plantation forest	0.12	0.10	4.31	4.31	−20.33	0.12
Artificial	1.54	1.92	0.45	0.46	24.86	2.13
Mosaic	0.00	0.00	0.04	0.03	0.00	−10.59
Total	100.00	100.00	100.00	100.00		
Number of plots	20	20	24	24		

in Spain (61 % vs. 52.3 %) in the study area, and most of these result from dehesa management. In contrast, California has a higher proportion of dense forests (25.9 % vs. 5.5 %) and shrubland (13.7 % vs. 7.8 %) within the areas we define as oak woodland landscapes.

Another clear difference between California and Spain is non-agricultural land use at rates almost four times higher in California than in Spain's oak woodland landscapes (1.9 % vs. 0.5 %)—a figure on the increase. Rural houses and residential and small urban developments are mostly absent in Spanish oak woodlands. In fact, houses not used for agricultural activities but for residential primary and secondary houses and recreation are three times more common in Californian oak woodland landscapes than in Spanish ones. The presence of golf courses in Californian plots is more frequent than in Spain. Moreover, Californias oak woodland landscapes are more densely populated than Spain's and the distance to larger towns (measured as the average of the minimum distance from each landscape plot to the nearest town with more than 250,000 inhabitants) is significantly smaller in California than in Spain (26.8 km vs. 108.3 km) (Fig. 14.11). The distance to metropolitan nuclei is a very good index of urban demand for recreational, rural tourism, and second-home residential uses. These threaten the rural landscape with urbanization (Chap. 8), and a larger wildland-urban interface. A smaller distance to towns implies a greater risk of wildfire that causes extensive property damage, and of ignition from anthropogenic sources.

Finally, if we take into account the structure of Californian landscapes, where the fuel loads in forests and shrublands are greater than in Spain's dehesas, we can conclude that there are major differences in the likely causes of wildfire in the two regions. That result is of a great importance because wildfire is one of the main drivers of vegetation change in Mediterranean landscapes.

Oak woodlands have been stable over the study period with no changes in Spain (0.01 %) and a slight reduction in California (−1.24 %). Other land uses have changed more in the Californian setting, for example in grasslands (reduction of 45.7 %) and bare soils (reduction of 17.9 %). These reductions were mainly counterbalanced by an increase in croplands (18.2 %) and in residential use (24.9 %). We note again the "exurban development" that affects oak woodland landscapes in the California sample area, but not in Spain.

The average total landscape area of the Californian and Spanish samples is not as similar as we would expect given the sampling design (Table 14.3). The main differences between the two landscapes groups relate to patch size, which is greater in Spain than in California (64.3 ha vs. 43.8 ha in 2008). The opposite is true for other metrics: number of patches, edge density, size index, and shape index, all are greater in California than in Spain (53 %, 49.2 %, 32.8 %, and 56.1 %, respectively). By land use diversity indices, Californian landscapes reached higher measures of diversity than in Spain, which suggests a more heterogeneous land cover. Only interspersion and juxtaposition indices had higher values in Spain than in California, arguing for more intermixed land uses in Spain than in Californian landscapes. In relation to changes in configuration indices between the two dates, the percentages of change were never above 10 %. A slight trend related in fragmentation could be detected in both landscapes, as we can infer from the change in number of patches (6.37 % and 3.67 % in California and Spain, respectively) and the changes in mean patch size (−7.6 % and −6.42 %

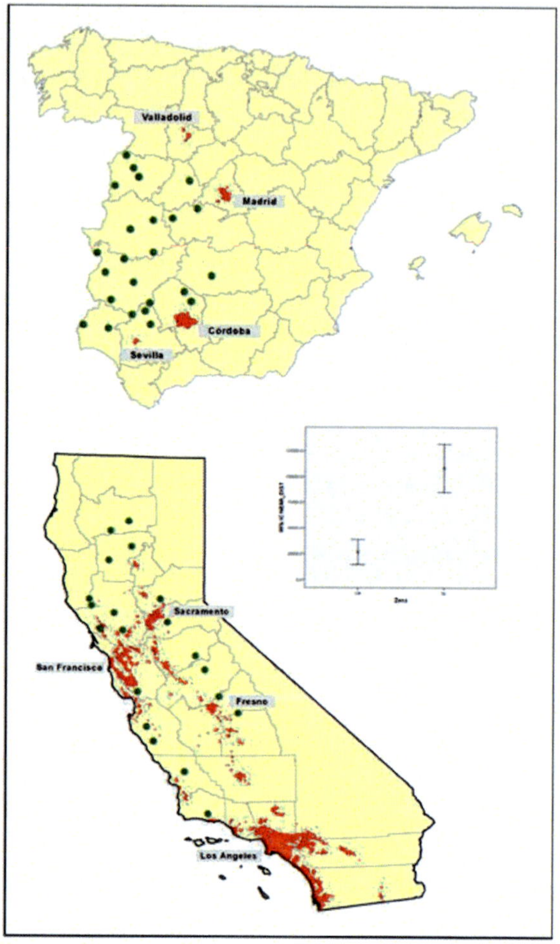

Fig. 14.11 Human population density in Spain and California. The maps show the larger towns (red) and the sampled areas (green dots). The chart shows the results of an analysis of variance for the distance to larger towns

respectively) indices. No other change trends were been detected that could be considered significant.

14.6.2 Ordination Analysis of Spanish and Californian Oak Woodland Landscape Samples

After exploring similarities and differences in composition and configuration, the next question was how to group these landscapes in terms of landscape

Table 14.3 Landscape configuration indices of oak woodlands in California and Spain and percent change from 1993 to 2009

Index	California		Spain		Change (%)	
	1993	2009	1998	2008	California	Spain
Total landscape area (ha)	1605.17	1605.17	1472.66	1472.66	0.00	0.00
Number of patches	40.80	43.40	27.28	28.28	6.37	3.67
Mean patch size (ha)	47.41	43.81	68.77	64.35	−7.60	−6.42
Largest patch size index (ha)	48.64	46.67	66.33	66.02	−4.06	−0.47
Edge density (m/ha)	66.57	68.09	45.43	45.64	2.29	0.45
Total edge (km)	106.84	109.28	66.73	66.96	2.28	0.35
Mean size index	2.15	2.13	2.02	1.99	−1.05	−1.27
Size index	4.16	4.13	3.11	3.11	−0.66	0.16
Mean patch fractal dimension	1.117	1.116	1.109	1.106	−0.13	-0.30
Mean patch fractal dimension Wtd. by area	1.176	1.178	1.140	1.140	0.13	0.00
Shape index	6.67	6.82	4.35	4.37	2.29	0.40
Patch richness	6.20	6.40	6.00	6.17	3.23	2.78
Patch richness density	0.39	0.40	0.41	0.42	3.36	2.56
Shannon's diversity index	0.98	1.01	0.85	0.86	3.06	1.58
Shannon's evenness index	0.53	0.54	0.48	0.48	2.16	0.23
Simpson's evenness index	0.64	0.65	0.51	0.51	1.41	0.77
Mean shape index	2.15	2.13	2.02	1.99	−1.05	−1.27
Modified Simpson's diversity index	0.79	0.82	0.61	0.61	3.36	1.47
Modified Simpson's evenness index	0.43	0.44	0.34	0.34	2.08	−0.16
Interspersion juxtaposition index	44.19	45.60	56.69	57.11	−1.40	−0.43

composition and configuration. Figure 14.12 shows the summarized results of landscape similarity using TWINSPAN analysis. At the first level two main groups can be distinguished, a Spanish group, group 1 (18 out of 19 components of the group were located in Spain), and a partly mixed group, group 0 (19 out of 25 components were located in California). At the second level were three "pure" groups corresponding to California, group 00 (all 13 landscapes were Californian), and in Spain, groups 10 (all 10 landscapes were Spanish) and 11 (8 out 9 landscapes were in Spain). The fourth group, group 01, is made up of the same number of Californian and Spanish landscapes, 6 each for a total of 12. Hereafter we designate group 00 as completely Californian landscapes, group 01 as a mixed group, group 10 as Spanish landscapes that are less intensively managed and group 11 as Spanish landscapes that are more intensively managed.

Analysis of variance results are analyzed for the four main TWINSPAN groups (00, 01, 10 and 11) and the two dates (Fig. 14.13). Variables that were comparatively discriminative for the groups were the percentage of the landscape covered by forests (higher in California's group 00, and in the mixed group 01, and lower in the two Spanish groups 10 and 11), the percentage of the landscape covered by crops (greater in the more intensively managed dehesa group 11, and lesser in the

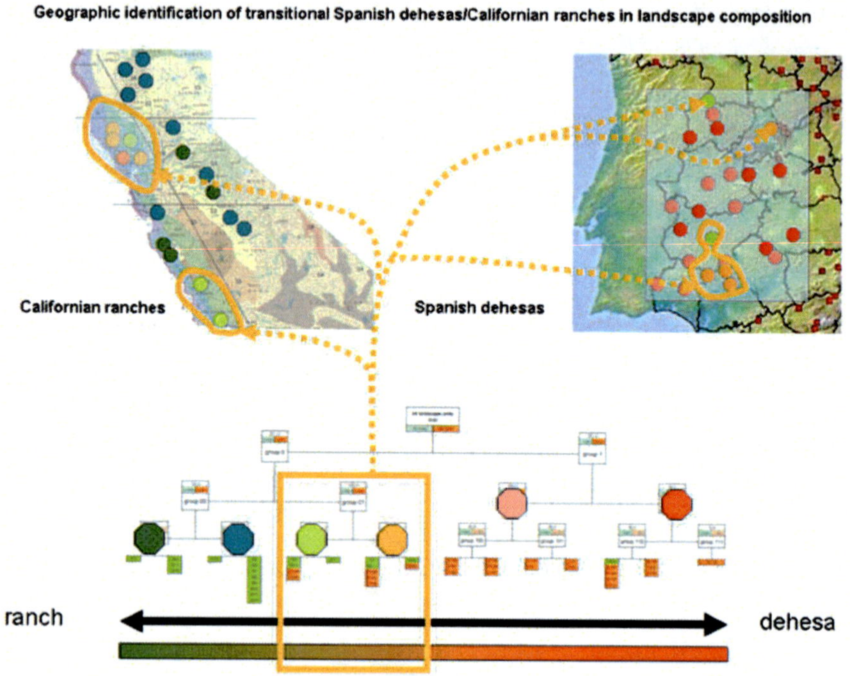

Fig. 14.12 TWINSPAN classification of Spanish and Californian oak woodland landscapes. **a** Dendrogram resulting from the composition data analysis. Line lengths (along the horizontal and vertical axes) are not proportional to distances between groups. The greener boxes are Californian landscapes and redder boxes are Spanish landscapes. **b** The geographical position of the different landscape compositional taxa in California and Spain. Transitional landscapes are distinguished by orange contour lines

other three groups, reaching nearly 0 in California), edge density, and main patch size. Both configuration indices were discriminative for the two main groups in edge density and main patch size reaching extreme values for the mixed group, with greater edge density and lesser mean patch size denoting that the mixed group corresponded to more fragmented landscapes. As the figure shows, the time of sampling did not result in significant differences.

Finally, it is in the mixed group 01 where Spanish and Californian landscapes showed the most similarities (Fig. 14.14). These woodlands tend to be dominated by evergreen oaks in both California and Spain. Here we can explore which physio-climatic variables have caused the convergence of these savanna-like landscapes. No discriminative variables were detected in an analysis of variance but slope is slightly greater in the mixed group than in the rest of the samples (8.8 % vs. 6.9 %, $p = 0.17$). This could mean that Spanish landscapes with rugged relief and more extensive use, that support shrubland and forested patches, are more similar to Californian landscapes where more intensive uses like cultivation and development of improved pastures are residual or absent.

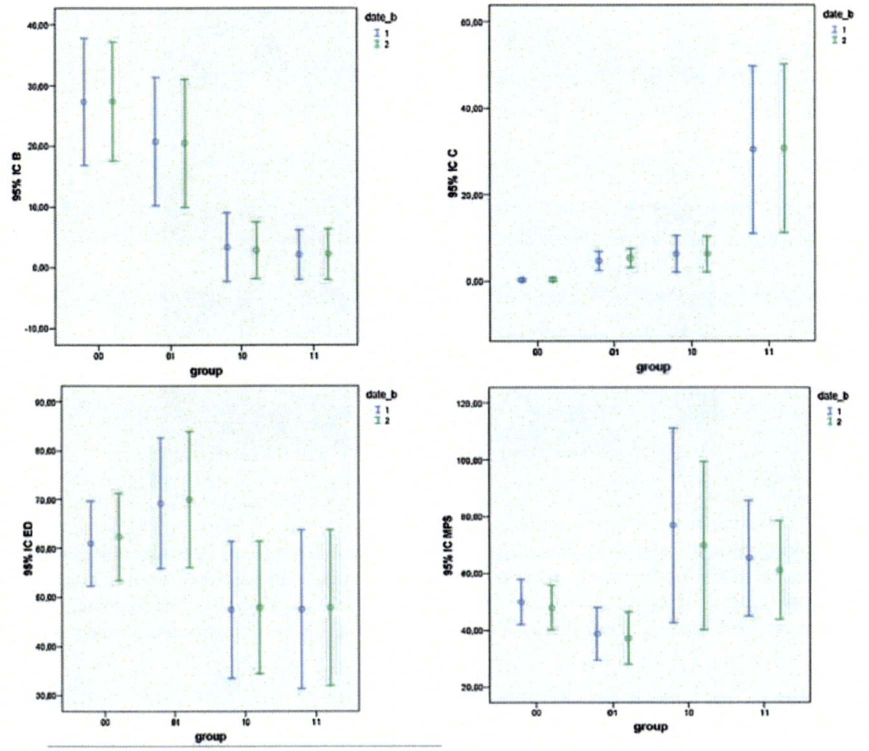

Fig. 14.13 Analysis of variance among TWINSPAN main groups for two dates, as related to composition (3a percentage of forest, 3b percentage of crops) and configuration (3c edge density, 3d main patch size)

14.7 Landscape Drivers: Factors Responsible for Spanish and Californian Oak Woodland Landscape Structure and Function

Analysis of variance using physiographic and climatic variables for 44 Spanish and California oak woodland landscapes showed significant results. Three of six variables were discriminative: annual and summer rainfall, and slope, as is shown in Fig. 14.15, while the other three, annual temperature, the mean of the maximum temperature of the warmest month, and elevation were not.

The topography of the central and southwestern Spanish peneplain is characterized by flats and gentle hill slopes. This made possible the conversion of Mediterranean forest areas first to dense dehesas, but then eventually to dehesas with reduced oak cover, or even to treeless pastures and irrigated croplands. These dynamics are different than in California where complete removal of oaks for conversion to croplands in flat valley bottoms is the historical trend, leaving oak

Fig. 14.14 Using the TWINSPAN classification, some California oak woodland landscapes are grouped together with Spanish landscapes as having similar spatial patterns and composition. This Sonoma county landscape is an example. Although the dense forest intermixed with oaks is most typical of California, oak woodlands mixed with agricultural landscapes, such as vineyards, are more typical of Spain. This landscape is a member of group 01, where Spanish and Californian landscapes are both included because they have similar components. (Photograph by 4Ullas)

woodlands confined to less fertile soils and rougher topography, as is characteristic of the oak woodlands in the Sierra Nevada foothills. Topography favors the persistence of the forest and shrub patches that are more widely distributed in California than in Spain.

Wildfire is an important modifier of Mediterranean landscapes. Fire is also a powerful tool used by humans to control stand structure and consequently to manage and maintain cultural landscapes. However, wildfire regimes were not evaluated in our study. When maps (Fig. 14.16) showing wildfire frequency and hazard were reviewed, we found that dehesa landscapes are among the least-often burned (Ortega et al. 2012). California oak woodlands have a high wildfire occurrence and had large burned areas in the 1950–2010 period, though some were intentionally burned for management purposes (Fig 2.1.1 in CDF-FRAP 2010a). We discuss two components of wildfire vulnerability: wildfire risk and wildfire hazard. Risk is assessed by the proportion of a landscape considered vulnerable, and hazard by the human population density in the landscape and distance to the nearest town. Our oak woodland landscape samples have the following averages for wildfire vulnerability indicators:

- Spain:

  14.1 % vulnerable land cover (rate from 24 samples)
  108.3 km average distance to the nearest town (>250,000 inhabitants)

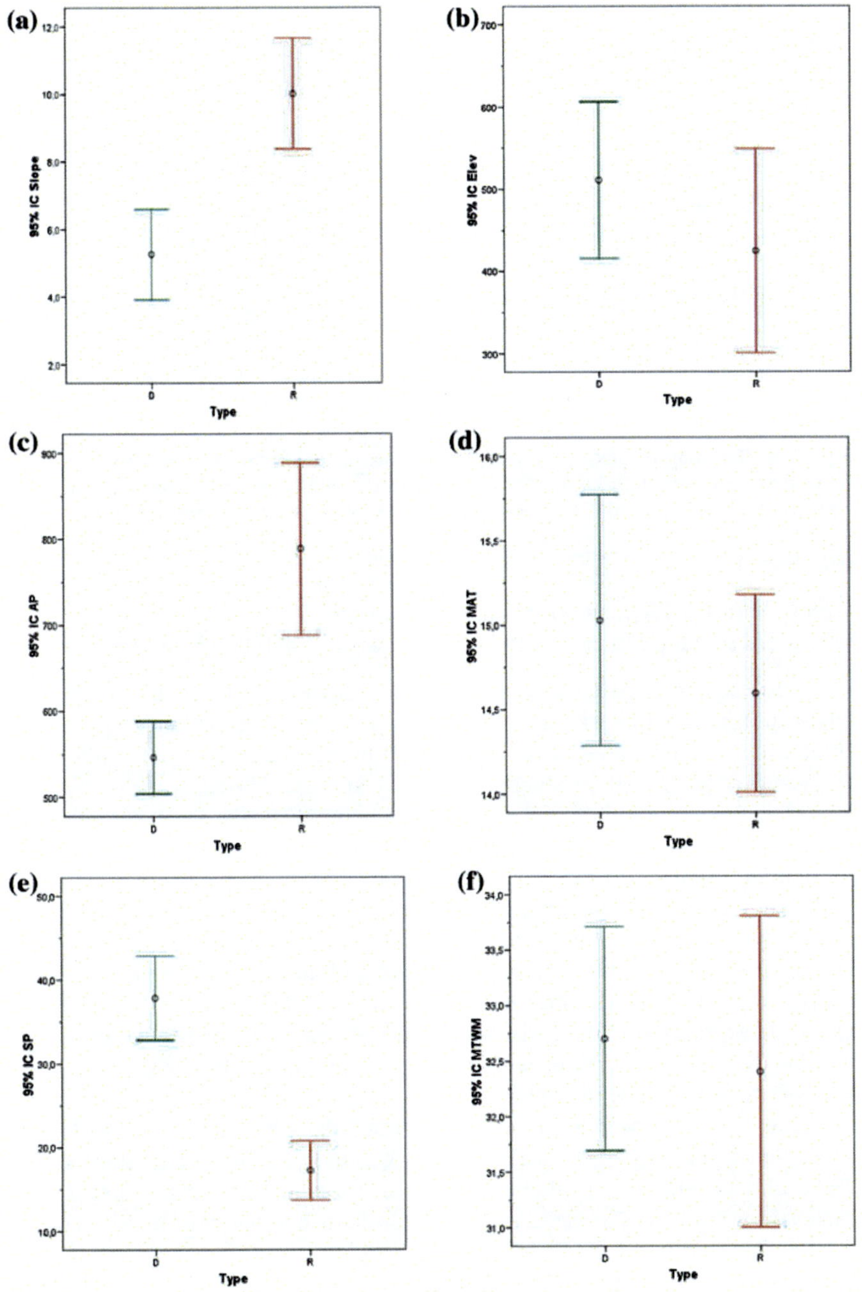

Fig. 14.15 Analysis of variance of Californian (red) and Spanish (green) oak woodland landscapes in relation to physiographic and climatic variables. Slope in percentage (**a**), annual rainfall in mm (**c**), and summer rainfall in mm (**e**) were significantly different between California and Spain ($p < 0.001$). The other three variables, elevation in meters (**b**), annual temperature in degree Celsius (**d**) and the mean of the maximum temperature in the warmest month in degree Celsius (**f**) were not different

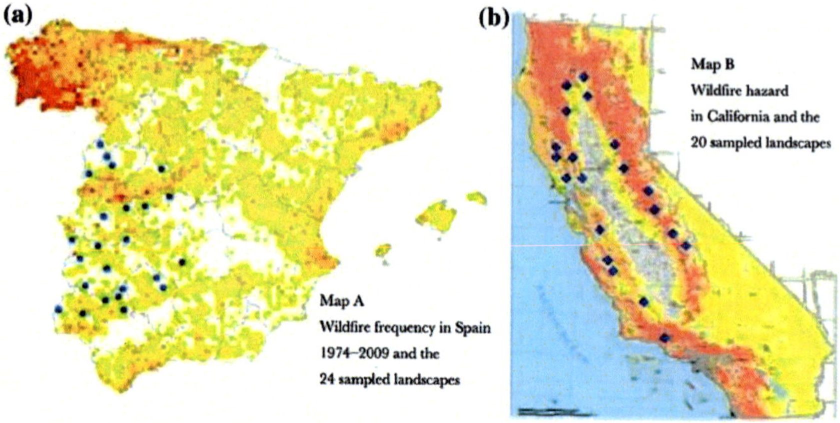

Fig. 14.16 Location of plots (blue dots) in relationship to wildfire in Spain and California. (**a**) wildfire frequency in Spain (EGIF database), and (**b**) wildfire hazard in California mapped in 2007 (http://www.fire.ca.gov/fire_prevention/fire_prevention_wildland_statewide.php)

- California:

  39.6 % vulnerable land cover (rate from 20 samples)
  26.8 km average distance to the nearest town (>250,000 inhabitants)

These data show significant differences between Spanish and Californian landscapes in terms of wildfire vulnerability because both the landscape composition and the human population density are greater in California than in Spain as we would expect based on wildfire frequency and hazard data.

Although these results clarify differences between Spain and California, data can be used to further explain factors driving the two patterns and their relative impacts. Anthropogenic factors and biophysical factors may each play a role.

The analysis of vegetation patterns in both landscapes showed that the most distinctly Californian oak woodland landscapes (group 00) are located in the Sierra Nevada foothills, intermixed with dense forest and shrubland, where the woodland tends to be dominated by deciduous blue oak. When looking at the altitudinal series, dense forests, often coniferous, tend to border the upper edge of the oak woodlands. At the same time, well-developed shrublands often border the woodlands at the lower edge (Stephens 1997).

These open oak woodland landscapes intermix with both upper and lower vegetation types. This is particularly true when they are located on steep slopes: The ecological elevation gradient generates very narrow horizontal projections for each vegetation type. The strips delineating suitable biophysical conditions can be even narrower than 4 km in width, creating highly intermixed landscapes.

These factors can help explain landscape patterns in the Sierra Nevada foothills, but not for the landscapes in Spain or even other areas within California.

14.8 Past and Future Californian and Spanish Oak Woodland Landscapes

Historical land use dynamics are important, as are trends and driving forces that shape the future evolution of landscapes. These are the context of sound management principles for oak woodland landscapes.

In California, differences in the time of ranch settlement are distinguishable in the configuration of oak woodland ranch landscapes. The Californian oak woodland landscapes most different from the Spanish ones are located mainly in the foothills of Sierra Nevada, in general not ranched until the 1860s. Considering the historical data, those regions were not colonized by Spanish pioneer ranchers in the eighteenth century, but about 120 years later during the Gold Rush period in the middle of the nineteenth century. These lands were by and large homesteaded by settlers coming from the East. Instead of the expansive properties bordered by natural features typical of Spanish and Mexican land grants near the coast, the rectangular Township and Range land allocations based on the American rectangular land survey system are more common, with straight property lines remaining as a landscape pattern characteristic of the eastern part of the San Joaquin and Sacramento Valleys. Such a footprint cannot be seen in the oldest ranch landscapes, located along the Camino Real, from San Diego to Santa Rosa.

In considering the future dynamics of these landscapes, the trends detected during the last two decades will continue, absent major macroeconomic changes. Californian and Spanish oak woodland landscapes probably will suffer a slow but constant abandonment of the most traditional rural uses, and will be home to new uses based on services to metropolitan areas including recreation, hunting, and exurban residential development. Consequently, the main changes in California are expected to be an increase in development for primary and secondary residences, including tourist resorts such as golf courses. In contrast, the main change in Spain is expected to be an increase in agro-tourism with hunting-related activities open to Central European markets.

14.9 Final Comments and Conclusions

This chapter utilized information provided by SISPARES in Spain and a replicated sampling design for California. The comparison between oak woodland landscapes was made according to the methodological requirements of stratified sampling. Results from the 24 + 20 sample plots are summarized in Fig. 14.17.

Using an ecological rationale, we have compared the structure, function, main influential factors and recent landscape dynamics in two oak woodland landscapes.

From those comparisons can be drawn the following conclusions:

Composition and configuration:

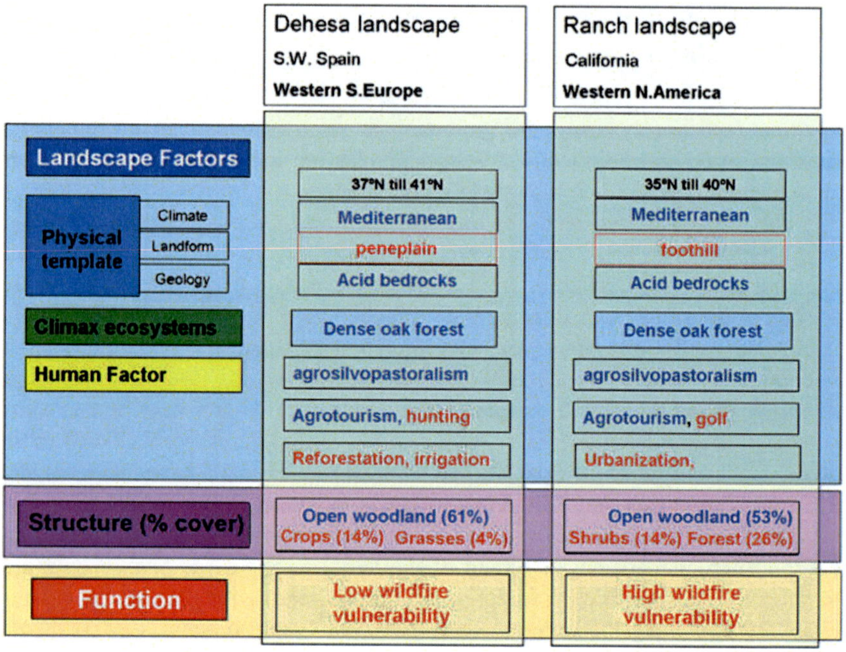

Fig. 14.17 Summary diagram of the main landscape factors and the resulting structural and functional features of Spanish and Californian oak forest landscapes. Red text indicates strong differences and blue text indicates high similarities. Major differences are in macro-relief, recreation activities and urban pressure. The main structural difference is landscape composition. The main functional difference is the wildfire vulnerability. According to our data and results, the main prevailing driver is macro-landform, and the subordinate driver is human land use. Peneplains generate less vulnerable landscapes than foothills

- Typical Spanish oak woodland structure has larger oak woodland patches intermixed with grasses and crops. The woodlands are most frequently located in peneplains with Mediterranean climate and acid bedrock. Spatial configuration is intermixed.
- Typical Californian oak woodland patches are smaller and intermixed with shrublands and dense forest and most frequently located in the foothills of the Sierra Nevada and Coast Ranges with Mediterranean climate and acid bedrock. The configuration is rather artificial with square shaped patterns.
- The oak woodlands that are most similar in Spain and California, according to the TWINSPAN classification, are those in Spain close to the mountains, and in California located near the coast.
- The shape analysis shows very different distribution areas in Spain and California, clearly determined by the macro topography.
- Spanish open oak woodland distribution is highly concentrated in peneplains, resulting in a vast region with a uniform pattern of contiguous oak woodlands most likely to be dehesa farms.

Fig. 14.18 Exurban development is an important influence on California oak woodlands. This woodland is inland from the Central Coast near Santa Barbara. (Photograph by P.F. Starrs)

- California open oak woodland distribution is highly dispersed along the foothills of the Sierra Nevada and the coast range. That results in a higher degree of interspersion with the chaparral and dense forest ecosystems.

Landscape function:

- Wildfire occurrence and starts are an important ecosystem function and there are stark differences between the two study areas. Californian oak woodlands are located in areas with pronounced wildfire hazard and Spanish oak woodlands are located in areas with low wildfire frequency. The apparent contradiction is fully explained by the consideration of two important factors. First the inclusion of very flammable ecosystems, such as shrubland and dense forest, in a high proportion of the California woodland landscape, and secondly, the much shorter distance to metropolitan areas in California than in Spain. This has increased the size of fire-prone wildland-urban interface zones.
- Although open oak woodlands are located in areas with low human population, urban pressure for outdoor recreation and secondary housing is much higher in California than in Spain, due the proximity of metropolitan areas and comparatively flexible land use zoning and controls (Chap. 12) (Fig. 14.18).
- Using Urban's (2002) classification of landscape factors, we have identified the following main drivers for open oak woodland landscapes:

Fig. 14.19 An intermix of urban and suburban development and oak woodlands is more common in California, but is found in the vicinity of the largest Spanish metropolitan region of Madrid (Photographs by P.F. Starrs and 4Ullas)

- Physical template: Landform at the macro scale is the main driver for the mid and long-term, because it determines climate zonation, soil development, woodland distribution, landscape composition, wildfire vulnerability, and landscape patterns.
- Human factors: Agro-silvopastoral practices are the main human factor, and currently they appear to play a larger role in Spain than in California. In Spain, an absence of active and intensive human management results in degradation and eventually destruction of many oak woodland biodiversity and production

values within a time frame of two decades (Chaps. 6, 8, 9, and 10). In California, open oak woodlands are relatively stable, and without direct oak removal or catastrophic wildfire, change is slow on most ecological sites. However, the role of Native Californians is not fully understood in creating these landscapes, and with it, how the landscapes might change without the frequent anthropogenic and natural fire that were ubiquitous in California woodlands until the late nineteenth century. The introduction of livestock grazing in 1769 brought in large numbers of alien species that flourish in today's oak woodlands, changing woodland characteristics and their future trajectories in ways as yet unknown. About two-thirds of the Califonia oak woodlands are grazed by livestock.

– New land uses are increasing in both study areas, including activities such as agro-tourism, game hunting, golf courses, and suburban development, though they affect mainly Californian oak woodlands (Fig. 14.19).

To conclude, the oak woodlands of California and Spain look alike in many ways, and are quite similar in some places in their configuration and composition. However, looking closer at spatial pattern and functional processes reveals that the challenges facing conservationists are very different. Policies affecting land use and development, and traditional management practices, are having a major impact on landscape structure and processes in oak woodlands in each country. Landscape outcomes constrain and create opportunities for long term conservation of the woodlands.

Acknowledgments The present work was partially financed by the Spanish project DECOFOR (AGR2009-07140). Many thanks to the book editors, as well as to the anonymous reviewers who improved earlier drafts of the manuscript. We are thankful to Assistant Professor Cesar López Leiva for providing GIS data and maps about dehesa and ranch landscape potential areas. EGIF Wildfire statistics for Spain were kindly supplied by the Spanish Ministry of the Environment (Dirección General del Medio Natural y Política Forestal).

References

Allen-Diaz BH, Bartolome JW, McClaran MP (1999) California oak savanna. Chapter 20. In: Anderson RC, Fralish JS, Baskin JM (eds) Savannas, barrens, and rock outcrop plant communities of North America. Cambridge University Press, Cambridge

Angelstam P (2004) Habitat thresholds and effect of forest landscape change on the distribution and abundance of black grouse and capercaillie. EcolF210 Bull 51:173–187

Bartolome JW, Muick PC, McClaran MP (1988) Natural regeneration of California hardwoods. In: Plumb TR, Pillsbury NH (Tech. Coord.) (eds). Proceedings of symposium multiple use management of California's hardwood resources. USFS Gen Tech Rep PSW-100

Bunce RGH, Barr CJ, Clarke RT, Howard DC, Lane AMJ (1996) The ITE Meriewood land classification of Great Britain. J Biogeogr 23:625–634

CDF-FRAP [California Department of Forestry and Fire Protection–Fire and Resource Assessment Program] (2002) Bioregions of California (map), Forest and Rangeland

Resources Assessment Program, Calfire, State of California, Sacramento. http://frap.cdf.ca.gov/webdata/maps/statewide/inacregmap.pdf. Accessed 15 Oct 2012

CDF-FRAP [California Department of Forestry and Fire Protection–Fire and Resource Assessment Program] (2010a) Fire frequency (number of times burned) over the period 1950–2008 (map), Fig. 2.1.1 In: FRAP, California's Forests and Rangeland: The 2010 Assessment, p 98. State of California, Sacramento. http://frap.fire.ca.gov/assessment2010/pdfs/california_forest_assessment_nov22.pdf.Accessed 15 Oct 2012

CDF-FRAP [California Department of Forestry and Fire Protection–Fire and Resource Assessment Program] (2010b) Forests and rangelands of California (map), Fig. 1.5 In: FRAP, California's Forests and Rangeland: The 2010 Assessment, p 39. CalFire, State of California, Sacramento. http://frap.fire.ca.gov/assessment2010/pdfs/california_forest_assessment_nov22.pdf. Accessed 15 Oct 2012

CDF-FRAP [California Department of Forestry and Fire Protection–Fire and Resource Assessment Program] (2010c) Priority landscape for population growth and development impacts (map) In: FRAP, California's Forests and Rangeland: The 2010 Assessment. State of California, Sacramento. http://frap.fire.ca.gov/assessment2010/maps/PL_11_1.pdf. Accessed 15 Oct 2012

Díaz M, Campos P, Pulido FJ (1997) The Spanish dehesas: a diversity of land use and wildlife. In: Pain D, Pienkowski M (eds) Farming and birds in Europe: the common agricultural policy and its implications for bird conservation. Academic Press, London, pp 178–209

Díaz M, Pulido FJ (2009) Dehesas perennifolias de Quercus spp. In: Bases ecológicas preliminares para la conservación de los tipos de hábitat de interés comunitario presentes en España. Dirección General de Medio Natural y Política Forestal, Ministerio de Medio Ambiente, y Medio Rural y Marino, Madrid

Elena-Rosselló R, Tella G, Castejón M (1997) Clasificación Biogeoclimática de España peninsular y balear. Ministerio de Agricultura, Pesca y Alimentación. Madrid, Spain

ESRI (2009) ArcGIS Desktop: release 9.3. Environmental Systems Research Institute, Redlands CA

Ewing RA, Tosta N, Tuazon R, Huntsinger L, Marose R, Nielson K, Montroni R, Turan S (1988) The California forest and rangeland resources assessment: growing conflict over changing uses. California Department of Forestry and Fire Protection. Anchor Press, Sacramento, CA

Forman RTT, Godron M (1986) Landscape ecology. Wiley, New York

García del Barrio JM, Bolaños F, Ortega M, Elena-Rosselló R (2004) Dynamics of land use and land cover change in dehesa landscapes of the "REDPARES" network between 1956 and 1998. In: Schnabel S, Ferreira A (eds) Sustainability of agrosilvopastoral systems -dehesas, montados-. CATENA VERLAG GMBH, Reiskirchen, Germany, p 47–54

González Bernáldez F, Morey M, Velasco F (1969) Influences of Quercus ilex rotundifolia on the herb layer at El Pardo woodland. Bol Soc Esp Hist Nat (Biol) 67:265–284

Haynes-Young R (2005) Landscape pattern: context and process. In: Wiens JA, Moss MR (eds) Issues and perspectives in landscape ecology. Cambridge University, Cambridge, pp 103–111

Hijmans RJ, Cameron SE, Parra JL, Jones PG, Jarvis A (2004) The WorldClim interpolated global terrestrial climate surfaces, version 1.3.http://biogeo.berkeley.edu/

Hill MO (1979) TWINSPAN: a FORTRAN program for arranging multivariate data in an ordered two-way table by classification of the individuals and attributes. Cornell University Press, Ithaca

Honnay O, Piessens K, Van Landuyt W, Hermy M, Gulinck H (2003) Satellite based land use and landscape complexity indices as predictors for regional plant species diversity. Landsc and Urb Plan 63:241–250

Huntsinger L, Bartolome JW (1992) Ecological dynamics of Quercus dominated woodlands in California and southern Spain: a state-transition model. Vegetation 99–100:299–305

Huntsinger L, Johnson M, Stafford M, Fried J (2010) California Hardwood Rangeland landowners 1985–2004: ecosystem services, production, and permanence. Rangel Ecol Mgmt 63:325–334

IBM-SPSS (2008) SPSS Inc: release 2008. SPSS Statistics for Windows Version 17.0, Chicago

Joffre R, Lacaze B (1993) Estimating tree density in oak savanna-like "dehesa" of southern Spain from SPOT data. Int J Rem Sens 14:685–697
Joffre R, Rambal S, Ratte JP (1999) The dehesa system of southern Spain and Portugal as a natural ecosystem mimic. Agrofor Sys 45:57–79
Linares-Lujan AM, Zapata-Blanco S (2003) Una visión panorámica de ocho siglos de la dehesa. In: Pulido F, Campos PS, Montero G (eds) La gestión forestal de las dehesas. Junta de Extremadura, IPROCOR. Badajoz
Marañon T (1986) Plant species richness and canopy effect in the savanna like dehesa of S.W. Spain. Ecol Medit XII: 131–139
Marañon T, Pugnaire FI, Callaway RM (2009) Mediterranean-climate oak savannas: the interplay between abiotic environment and species interactions. Web Ecol 9:30–43
McCollin D (1993) Avian distribution patterns in a fragmented wooded landscape (North Humberside, U.K.): the role of between-patch and within-patch structure. Glob Ecol Biogeog Let 3:48–62
Moreno G, Pulido FJ (2009) The functioning, management, and persistence of dehesa. In: Rigueiro A et al. (eds) Agroforestry in Europe: Current status and future prospects. Springer Science+ Business Media, Chapter 7, pp 127–160
Naveh Z, Whittaker RH (1979) Structural and floristic diversity of shrublands and woodlands in northern Israel and other Mediterranean areas. Vegetation 41:171–190
Ortega M, Saura S, González-Ávila S, Gómez-Sanz V, Elena-Rosselló R (2012) Landscape vulnerability to wildfires at the forest-agriculture interface: half-century patterns in Spain assessed through the SISPARES monitoring framework. Agrofor Sys 85:331–349
Pérez-Soba M, San Miguel A, Elena-Roselló R (2007) Complexity in the simplicity: the Spanish dehesas. In: Pedroli B, Van Doorn A, De Blust G, Paracchini ML, Wascher D, Bunce F (eds) Europe's living landscapes. Essays on exploring our identity in the countryside. LANDSCAPE EUROPE/KNNV
Regato-Pajares P, Jimenez-Caballero S, Castejón M, Elena-Rosselló R (2004) Recent landscape evolution in Dehesa woodlands of western Spain. In: Mazzoleni S, di Pasquale G, Mulligan M, di Martino P, Rego F (eds) Recent dynamics of the Mediterranean vegetation and landscape. Wiley, Chichester, pp 57–72
Ruiz de la Torre (2002) Mapa Forestal de España. Escala 1:1000000. Ministerio de Medio Ambiente, Madrid
Sánchez D, Elena Roselló R, Roig S, García del Barrio JM (2007) Los paisajes de dehesa en España y su relación con el ambiente geoclimático. Cuader de la SECF 22:171–176
Standiford RB, Howitt RE (1992) Solving empirical bioeconomic models: a rangeland management application. Am J Ag Econ 74:421–433
Stephens SL (1997) Fire history of a mixed oak-pine forest in the foothills of the Sierra Nevada, El Dorado county, California. USDA Forest Services Gen Tech Rep, pp 191–198
Sulak A, Huntsinger L (2007) Public lands grazing in California: untapped conservation potential for private lands? Rangel 23(3):9–13
Turner MG, Ruscher L (1988) Changes in landscape patterns in Georgia USA. Landsc Ecol 1(4):241–251
Urban DL (2002) Tactical monitoring of landscapes. In: Liu J, Taylor WW (eds) Integrating landscape ecology into natural resource management. Cambridge University Press, Cambridge, pp 294–311
Valladares F (2004) El bosque mediterráneo, un sistema antropizado y cambiante. In: Valladares F (ed) Ecología del bosque mediterráneo en un mundo cambiante. Organismo Autónomo Parques Nacionales. Ministerio de Medio Ambiente, Madrid
Van Doorn AM, Pinto Correia T (2007) Differences in land cover interpretation in landscapes rich in cover gradients: reflections based on the montado of South Portugal. Agrofor Sys 70:169–183

Part VI
Conclusions

Chapter 15
Whither Working Oak Woodlands?

Paul F. Starrs, José L. Oviedo, Pablo Campos, Lynn Huntsinger, Mario Díaz, Richard B. Standiford and Gregorio Montero

Frontispiece Chapter 15. Commodity production and wildlife conservation can be compatible outputs of dehesas and oak woodland ranches as shown by foraging fighting bulls and white storks (*Ciconia ciconia*) sharing a dehesa pasture in the Montes de Toledo region, central Spain. (Photograph by M. Díaz)

P. F. Starrs (✉)
Department of Geography, University of Nevada, Reno, NV, MS 0154 89557–0048, USA
e-mail: starrs@unr.edu

P. Campos et al. (eds.), *Mediterranean Oak Woodland Working Landscapes*, Landscape Series 16, DOI: 10.1007/978-94-007-6707-2_15,
© Springer Science+Business Media Dordrecht 2013

Abstract Comparative research into the human-maintained economic and ecological systems referred to as working landscapes is a rarity in the literature. Nonetheless, developing such comparisons is the end goal of this book. Altogether, 44 field scientists are contributing authors, with some appearing in multiple contributions, but others spelling out the specific knowledge crucial to just one part of a single chapter. In this final commentary, the book's editors lay out the conclusions attained in this extended inquiry, suggest research needs and lessons learned, and raise the issue of policies that are needed, some urgently, to support oak woodland working landscapes. We recognize a number of takeaway lessons from this long-term project, including advances in economic analysis that make it possible to assess the total economic value of a landscape. We have come to, as we journeyed through the production of this volume, an overall conclusion that seems to us important: Just because two places appear similar hardly means that they are alike; oftentimes the variations are far more than skin deep. But with that as an initial concession, it pays to acknowledge how much can be learned from comparative research that matches physical, cultural, historical, economic, and geographical features, and then carefully places likenesses and departures side-by-side, in a deliberate attempt to learn across oceans, landscapes, economies, and societies.

J. L. Oviedo
Institute of Public Goods and Policies (IPP), Spanish National Research Council (CSIC), Albasanz 26–28 28037 Madrid, Spain
e-mail: jose.oviedo@csic.es

P. Campos
Institute of Public Goods and Policies (IPP), Spanish National Research Council (CSIC), Albasanz 26–28 28037 Madrid, Spain
e-mail: pablo.campos@csic.es

L. Huntsinger
Department of Environmental Science, Policy, and Management, University of California, 130 Mulford Hall , Berkeley, CA, MC 3110 94720, USA
e-mail: huntsinger@berkeley.edu

M. Díaz
Department of Biogeography and Global Change, Museo Nacional de Ciencias Naturales (BGC-MNCN), Spanish National Research Council (CSIC), Serrano 115bis E-28006 Madrid, Spain
e-mail: Mario.Diaz@ccma.csic.es

R. B. Standiford
Department of Environmental Science, Policy and Management, University of California, 130 Mulford Hall , Berkeley, CA, MC 3110 94720, USA
e-mail: standifo@berkeley.edu

G. Montero
Forest Research Centre, National Institute for Agriculture and Food Research and Technology, Carretera de la Coruña km 7.5 28040 Madrid, Spain
e-mail: montero@inia.es

Keywords Comparative summary · Complex economic and ecological systems · Long-term conservation · Multiple benefits · Cultural geography

15.1 Challenges in Conservation and Management

The dehesas of Spain and the oak woodland ranches of California are ecologically and economically significant environments. Slow growing and long-lived, oak trees record an enduring history of use and human interaction. These landscapes support economies that feed people, livestock, and wildlife through an ecologically diverse agriculture, and are home to prodigious biological productivity and a world-valued habitat shaped by long-time management. We have poked and prodded many aspects of these working ecosystems in this book, but there remains much to do to understand their multiple benefits, how they work, and how they are likely to change. Just what will happen to these landscapes in the future remains a question.

In significant measure this book is about what the oak woodlands of California and Spain are now, and how they have come to be the way they are. But as an analytical and deliberative study, this is intended to serve as a source of questions, with one first among equals: If these oak woodlands are so valuable, how then do we make sure they will endure as productive environments into another epoch of human history? In a world of 7 billion people the answer will require careful policy and rational planning. In this concluding chapter, we suggest tools that are needed to ensure that biodiverse, working oak woodlands survive. Information is a key factor, and we contribute to that in this book. But a farther-reaching goal is to foster oak woodland environments that will continue to be productive places in the future and sustainably support economies that can satisfy local, regional, national, and hemispheric needs into a distant and unpredictable future.

15.2 The Bio-Political Setting

The settings for conservation in California and Spain show some commonalities but with important differences. Commercial operating income in both locales is low, and returns from non-market ecosystem services are comparatively high. These non-market values are minimally prioritized in the policymaking arena, making oaks vulnerable to loss. The processes of oak regeneration and recruitment, and how best to ensure them, are subjects of scientific and popular debate. What is not in doubt is that costs to maintain oaks over the long run tend to outstrip the capacity of landowners to fund them.

15.2.1 California

For 100 years in the United States, the principal strategy for land conservation involved the government setting aside land in public ownership, along with incentives, property tax policy, zoning, and environmental regulations. But in the familiar story of the occupation of the North American continent, arriving European settlers took up the most productive and best-watered areas, which often went on to become cropland. Oak woodlands in California were acquired early on because they constituted an environment that settlers of Mediterranean origin knew how to use for grazing livestock and forage production.

A modern-day consequence of that history is that game and wildlife—including endangered species—often depend on resources and on land that is privately owned. These selfsame private landowners are, as a rule, profoundly skeptical about government involvement in land-use questions, and express particular concern about what they see as over-regulation (Sulak et al. 2004). Ranchers recognize ecological and oak recruitment issues as likely to affect the sustainability of ranching, yet they also see land use planning as a threat to their independence (Liffmann et al. 2000). Somewhat paradoxically, most are nonetheless willing to take advantage of government incentive programs and compensation for drought, floods, or other disasters (Chap. 12). This sets up a complicated relationship between all levels of government (and increasingly non-governmental organizations, or NGOs) that may want to encourage landowners to undertake certain conservation and management practices, and landowners who are chary of management requirements being imposed on them and are reluctant to work with what they regard as an intrusive regulatory bureaucracy.

California, with its Integrated Hardwood Range Management Program (IHRMP), made an investment in research, outreach, and public education for oak woodland landowners starting in 1985 (Chap. 2), conveying what was learned about landowner response to policymakers. Landowners now mostly recognize the value of moderate oak tree cover not only for forage productivity and livestock shelter, but also for protecting watershed values, wildlife habitat, ecosystem services, and for adding market value to land (Standiford and Bartolome 1997; Huntsinger et al. 2010). Studies show that most ranchers no longer thin oak woodlands unless canopy cover exceeds 50 % (Huntsinger et al. 1997). This is an important development because oak woodlands cleared during the post-1945 era regenerated poorly or not at all (Brooks and Merenlender 2001). An increasing number of landowners plant oaks, but the extent and success of these afforestation plantings remains unknown (Huntsinger et al. 2010). The residential real-estate market crash of 2008 and the subsequent financial crisis and economic recession have reduced short-term development pressures throughout most of California. Yet in the future oak woodland ranches—particularly those near large urban centers—will regain their attraction to developers and urban out-migrants. California's land use controls and planning are notoriously weak and subject to the influence of development pressure (Saving and Greenwood 2002). Further, planners may treat oak woodlands as a place to channel development, because ranching produces lower dollar returns per unit area than crop production or forestry.

Ranchers bank on a rising value of land for development as a major capital asset. Historically, the only way to realize this increase in value was to sell all or part of a property. Many property owners have assumed that when they sell their land, income from the appreciation will make up for poor year-to-year returns of livestock production (Hargreave 1993; Brownsey et al. 2012; Chap. 13). If zoning designations are altered, those urban-derived changes can eradicate this asset without compensation, which consequently encourages rural ranchers to liquidate, selling off land before such re-zoning happens. Traditionally, cash-short ranchers have sold small parcels of their land to gain capital when needed. Surveys show that as much as 1 % of ranch land is sold each year for this reason (Huntsinger et al. 1997; Liffman et al. 2000). The subdivision of ranches brings growth in local population, habitat fragmentation, land use conflicts, and higher taxes—all of which encourage further subdivision, even for ranchers who might have preferred not to break up their holdings. The area of oak woodlands is on the decline, with California losing nearly one-third of its oak woodlands from 1930 to 2002 [Fig. 1 in Sulak et al. (2004)]. Studies by Workman (1986) and recent literature (Chap. 13) show that commercial operating profits from ranching are declining and do not seem to be a promising factor to encourage ranch ownership among the commercially-minded in the future.

For all the development pressure, the California woodland is ecologically a relatively stable landscape. Most woodlands will remain open and suitable for livestock production or wildlife habitat with irregular, limited regeneration and little shrub invasion over a long period without human intervention. Eighty-five percent of woodland understory persists as grassland in the absence of management (Allen-Díaz et al. 1999). In addition, the regeneration and recruitment of oaks is highly variable, seemingly dependent on a host of factors including wildlife and insect populations, grazing, soils, precipitation, aspect, and so forth, with variable outcomes including increases in oak woodlands in some areas (Russell and McBride 2003; Tyler et al. 2006).

15.2.2 Spain

The Spanish dehesa constitutes a less stable oak woodland, subject to rapid change without human intervention (Chap. 14). Property development and urbanization is held in check by relatively strong land use controls. But for traditional agricultural operations to persist in these woodlands, owners must make a consistent investment in maintaining the dehesa yield of grass for fodder and of other products (Chaps. 3, 5, 6) (Fig. 15.1). Absent that, the volatile woodlands become nearly impenetrable brush within a couple of decades and lose many of the characteristics so valued for habitat, amenities, wildlife, and livestock production. Woodland

Fig. 15.1 A fixture in the dehesa, the pig is a feature of ancient interest in Spain, sufficiently so that granite monuments are erected to *Sus scrofa* in Spanish dehesa communities such as La Alberca, at the southern edge of Salamanca province near the Sierra de Francia. Among the perceived charms of the Iberian pig, historically, is the longevity of cured pork products, a dehesa fixture that traditionally has fed rural (and urban) Spaniards through long winter and spring seasons, before other meat sources are available. (Photograph by P.F. Starrs)

instability makes management of competing vegetation and attention to reforestation more important.

The pressures in Spain tend to be implacable, and in some ways the reverse of those faced by landowners in California. Since the early 1900s, rural depopulation has been the norm in Spain, especially after 1950 as young people left to find work opportunities in cities (Hoggart 1997; Sánchez-Alonso 2000). The economic historian Ricardo Robledo succinctly notes that parts of Salamanca province, with ample areas in oak woodland, are a demographic Siberia, with scarcely a handful of residents per hundred hectares [quoted in Elola 2008]. The results of this are complicated and sometimes untested, as when remaining landowners welcome understory shrubs that are then managed as habitat for game species and a hoped-for but not always realized income from fee-hunting.

In places the results are costly, with fire incidence rising on dehesas, particularly when they include sizable remnant plantings of conifers, as rural populations shrink across the southern half of Spain (Campos 2012b; Pausas and Fernández-Muñoz 2012). Absent the once-resident landowners, ongoing care and cultivation of the dehesa slows or ceases. Concern is so high about Spanish rural depopulation that far-reaching proposals are at times developed and advanced, including in perhaps the most radical form an initiative for "Rewilding Europe" that suggests using municipal lands in western Spain and Portugal as the basis for one of ten

European zones set aside for "wildlife and wilderness" (RE 2012). Yet the model is strikingly reminiscent of the so-called Buffalo Commons proposed in the 1980s for the northwestern Great Plains of the United States, where a similar depopulation has occurred (Popper and Popper 1999; Manning 2011; GPRC 2012). Rewilding Europe promoters claim that "The regional government is already giving shape to a new economy that relies on culture, nature and landscape. And since West Iberia has plenty of wilderness to offer, it is attracting more and more nature tourists each year" (RE 2012). This would be a solution that many would not support, but even its voicing out loud—on a sophisticated website targeting the European Union—suggests a significant difference between the dilemmas facing Spain's dehesas and the oak woodland ranches of California.

While abandonment of cultivation and use is a threat to long-sustained dehesa working landscapes, within managed woodlands the oaks are not recruiting at a rate sufficient to replace senescent or dying trees. The long-term maintenance of the dehesa requires attention to oak regeneration, which in the best solution available so far in Spain involves an abandonment of livestock grazing for up to 20 years (BOJA 2009; Campos and Mariscal 2003) to allow shrub encroachment and oak regeneration (Chaps. 5, 6; Ramírez and Díaz 2008 Plieninger et al. 2010). There are advantages, of course, to good oak recruitment: the resulting mosaic of open productive and encroached areas within dehesas maximizes levels of biodiversity and species richness at the individual property and landscape scales (Chap. 8; Martín-Queller et al. 2011). Long-term forestry-style rotations with everything from seedlings to mature trees in evidence over large areas are as essential to traditional dehesa management as is the typical savanna-like landscape with scattered trees, shrub patches, and other landscape elements, a fact fully recognized only quite recently (Chaps. 5, 6, 8; Manning et al. 2009).

15.3 Policy and Research Considerations

Reviewing the comparisons in this book, some particularly important factors that must be considered in the development of policy and research agendas for oak woodland conservation merit highlighting.

15.3.1 Size Matters

Partly as a result of historical processes, and partly due to the nature of the dehesa and oak woodland ranch enterprise, oak woodland properties tend to be sizable private ownerships. Such properties are at times criticized because they put large areas of land in the custody of relatively few owner-users (Artola et al. 1978; Simpson 1995; Starrs 1998). The term latifundia comes from early Roman history (2nd century BC), and in Spain and California alike, large landholdings were distributed as grants to nobility, church, military orders, and soldiers—in California, without regard for Native American traditional territories or land rights. The legacy effects of those sizable (sometimes vast) land bequests has diminished somewhat as later generations of offspring divide holdings through inheritance and pieces are sold off to provide operating capital, but what remain are still large properties: bigger than homesites, and more extensive than all but the very largest farms or commercial forest plantations (Chap. 1).

There are distinct advantages to the retention of extensive oak woodland properties in private hands. Unfragmented properties that extend over hundreds and sometimes thousands of hectares yield significant ecosystem services and benefits to society that fragmented or small parcels of land are unable to offer, a benefit well-understood by conservation biologists who see greater value for wildlife and watershed in contiguous areas of habitat (Huntsinger 2009). In dehesas and ranchlands, large properties ensure landscape-scale management practices aimed at maintaining the mixtures of habitats and elements responsible for high biodiversity (Chap. 8).

But what, it has to be asked, do such landholdings offer the citizenry in terms of environmental quality, ecological function, and quality of life? And, assuming that the public chooses to continue and even increase investment in these woodlands, how can the public be assured of continual benefits from them? The pattern of ownership has been a success in producing large quantities of items that the public values—with oak woodlands home to some of the highest biodiversity of any broad landscape type. In this is an answer to a question often voiced when economic policymakers look at the continued existence of large properties that appear, in terms of economic reasoning, to be underutilized and not producing commercial operating profits at an expected rate. Large properties can and do produce benefits, sometimes for an individual or family, often for society in the form of ecosystem services, and such benefits have until recently been hidden from calculations of economic productivity. For oak woodlands in California and Spain, those economic realities are laid out throughout this book.

15.3.2 People Matter

Brian Fagan, reviewing in 2001 a remarkable study of Mediterranean Europe, argues that "The greatest changes have come since World War II; most rural populations throughout the Mediterranean region have abandoned traditional subsistence economies based on fields and flocks.... Only continued occupation by people gaining their livelihoods locally can maintain the man-made diversity typical of Mediterranean Europe" (2002). Little wonder that Spanish and California landowners express a prevailing worry, when surveyed regarding attitudes and management practices, about watershed health, wildlife populations, endangered species, urban encroachment, failures in the regeneration of oak woodlands, and even hiccups in their quality of life. These are precisely the concerns that policy makers are given to voicing, except they are newcomers to such worries, unlike the landowners and resource stewards who have attempted to manage large properties to the best of their ability for decades if not generations (Standiford and Bartolome 1997).

Dehesas and oak woodland ranches are not landscapes that can thrive forever on their own; they are products of a low-intensity historical management which offers specific products that have changed over time. In dollars and euros earned, the commercial operating income (*renta de explotación comercial* in Spanish) of the lands we have examined is relatively low. Nonetheless, landowners persist in retaining their lands, and, in both Spain and California, often work off-property and sometimes at a considerable distance in part to support their dehesa or ranch (Chaps. 1, 2, 10, 12, 13). They glean a sense of well-being, of social and ecological capital production, from their land, and live on it or visit it as they can, investing in land management and what they envision as a legacy of stewardship. Landowners are willing to give up cash income and alternative investments in order to benefit from non-market ecosystem services, many of which are shared with the public. The costs of conservation of lands to the public are effectively leveraged.

15.3.3 Diversity and Ecosystem Services Matter

The most important ongoing attribute of oak woodland ranches is, without a doubt, what twenty-first century policymakers describe as ecosystem services production (Chap. 12). Oak woodland ranchlands of California have an economy of limited commercial products. The cash-product or commercial economy of the dehesa in Spain traditionally counted on remarkably varied outputs—to a degree that offers some unusual economic stability and an enduring defense against risk, based on a high ecological diversity (Chap. 8) that has largely been lost in the recent economic trend toward specialization. But arguably equal to those products in either site are environmental goods and benefits, captured generally by the term ecosystem services, which are a boon to local, regional, and global populations (Chaps. 12, 13). As the ecological economist Robert Costanza writes, "The ecosystems-services concept makes it abundantly clear that the choice of "the environment versus the economy" is a false choice. If nature contributes significantly to human well-being, then it is a major contributor to the real economy… and the choice becomes how to manage all our assets, including our natural and human-made capital, more effectively and sustainably" (2006) (Fig. 15.2).

The oak woodlands of Spain offer expansive habitat for wildlife, and harbor raptors and predators that utilize the dehesa understory along with hundreds of species of animals, plants, and other organisms that depend on oaks, grasslands, and the intimate mixtures of trees and grass that characterize dehesas. Wildlife populations inhabiting dehesas and California oak woodland ranches are particularly valued by biodiversity researchers who reckon Spanish-Portuguese and

Fig. 15.2 Concern about agricultural land loss, including ranch land being converted to small parcel ranchettes, runs strong in California. The conservation easement is one approach that precludes development "in perpetuity" by allowing landowners to sell development rights (Chaps. 12, 13). (Photograph by P.F. Starrs)

Californian oak woodlands among the crucial ecological habitats for species in North America and Europe.

15.3.4 Market and Non-market Opportunities Matter

One way to encourage oak woodland conservation in Spain and California is to develop markets for products that can be produced without reducing the flow of multiple ecosystem benefits from the woodlands. Another is to offer payments for environmental services to influence landowner decisions and management, or to compensate landowners for the current and past provision of societal benefits derived from these woodlands. But ecosystem services and non-market products are decidedly not all about money; they are also about improvements to society, daily life, and the human prospect with gifts from nature the most important contribution to life on Earth.

15.3.4.1 Hunting

Hunting, and especially of big game species, is seen by landowners in the dehesa as a potentially important source of income—a hope sometimes correct, sometimes wrong. Hunting revenue can actually be two-fold, as direct income and averted cost. Fees are paid by hunt organizers in Spain, who in effect rent a property for the one- or two-day duration of a hunt (some landowners do the organizing themselves, but most not), and on a second front, landowners can save on management costs, since the ongoing expenses of clearing brush and planting cereal crops in the understory are reduced—at times to nothing. Hunting income can mitigate a declining cash flow in the dehesa economy for landowners who need cash to cover family livelihood needs. In California, hunting income can enhance profits on properties long managed for livestock. Because fee-hunting occurs on private land, the return on investment for prime big game habitat where hunters are quite likely to meet with success—and perhaps obtain a trophy-quality animal—is difficult to calculate, in either California or the Spanish dehesa.

Nonetheless, selling the opportunity to hunt (and in Spain, selling the meat obtained in a successful *montería*) can increase landowner income in marginal dehesa areas, which can boost property value. And there are strong suggestions from developing research that a mosaic of invading brush or shrubs can function as a nursery for oak regeneration, protecting seedlings from grazing and rooting by livestock or wildlife (Chaps. 5, 8). Hiring a guard or caretaker to prevent poaching and limit access can be an added expense, however, a fact that may encourage intensive land uses to favor game (feeding grain, in Spain) with negative side effects on other wildlife (Díaz et al. 2009). Published data on Spanish and California hunting in oak woodland ranches and dehesas is limited, lacking official price reporting statistics. It is notable that, other things being equal, prime big game habitat has higher land values than properties with lesser quality game habitat (Chaps. 11, 13) (Fig. 15.3).

Hunting, especially of deer and wild boar in Spain and California, attracts a devoted following (Chap. 11). Some of that is recent history: Spain saw an enormous upswing in foreign hunter interest from 1970 to the early 1990s as borders opened and EU residents grew more financially able to pay. Those numbers dropped after the Berlin Wall was brought down, Germany reunited, and Eastern Europe was opened to hunting. But hunting on a dehesa with first-class accommodations and catered dining brings a high premium from an international elite. California has an active guest-ranch and guided hunting industry on oak woodland ranches. It bears mention, however, that hunting is an activity enjoyed by an ever-diminishing share of the Spanish and California population, and particularly in California, a rising anti-hunting attitude poses a potential barrier to landowners profiting from effective game management.

Fig. 15.3 California has lower profit streams for wildlife and game than properties in Spain. Nonetheless, duck clubs in the Sacramento Valley where oaks once prevailed, and hunting clubs along the Coast Ranges (here near San Luis Obispo) and in the western Sierra Nevada foothills offer income to landowners willing to welcome hunters as guests. (Photograph by P.F. Starrs)

15.3.4.2 Diversification

Shared or societal products in California and Spain can run a gamut from the specific, even the unique, to the general. A particular product might include a historically important site that can be visited, a petroglyph of fame, harvestable medicinal or aromatic plants, a hot springs frequented by visitors, leased space for a cell-phone tower, or opportunities for property owners to hunt with friends and invited guests. The more general products are equally varied, such as genetic resources valued by the pharmaceutical industry, animals kept as pets, areas that can be trekked on foot or traversed on horseback, viewsheds and vistas of note, or distinctive wildlife associated with habitat that can be seen in recreational hikes. These are goods of recognizable economic value that are difficult to price, although ongoing research efforts are attempting to track even the marginal products of oak woodlands (Fig. 15.4).

Historically, landowners have often made properties accessible to neighbors or to visitors for gathering various products or hunting. In California, custom and trespass law, make it necessary to ask permission of the landowner to come on to a property, seems something apparently not known by many urban visitors today, and access is further constrained by landowner concerns about the costs of liability should a user come to harm. In Spain, the right to use the property of others may be an element of national or customary law. Traditionally, there were implicit

Fig. 15.4 Oaks, truffles, mushrooms, and fungi—all collectible, and on occasion with high commercial value—are sometimes sold, sometimes collected and consumed by landowners, and sometimes freely-picked by recreationists. Either way, in California and Spain they represent an important ancillary product of an oak woodland environment not always easily inserted into the roster of commodities produced. Those evident here are on display and identified on a table in the Plaza Mayor of Salamanca, where mycology is much appreciated. (Photograph by P.F. Starrs)

agreements that meshed a landowner's use with deference to the needs of community residents. Hunting small game was widely allowed, as was gathering plant material for broom-making along *cañadas* (stock driveways) that crossed a dehesa; rockroses were uprooted and converted into essential oils for perfume-making; fodder and harvested products included honey, mushrooms, truffles, and other fungi. But now, landowners can legally bar all access to a property, especially when land is registered as a protected reserve, and there is a possibility of charging for such access. Mushroom collection, for example, is a prized social, edible, and cultural undertaking, and anecdotal evidence suggests that some landowners can earn significant income by selling access to only a few collectors. Landowners do not typically enforce the law if they see recreationists and mushroom gatherers on their properties, since these rarely impact owner privacy. In California, allowing ranch access for birdwatching is new to some ranches, and wildlife or landscape photography classes, or painting and art courses, are sometimes offered in cooperation with landowners—and on occasion taught by them (Fig. 15.5).

The commercialization of oak woodland recreational services, meals, and accommodation is in its infancy in California, with formal *turismo rural* and agro-tourism programs considerably more advanced in rural Spain—many government supported. Guest stays offer an opportunity to increase income and encourage the maintenance of the oak woodland and dehesa landscape. If a site degrades through time—loses oaks, becomes overgrown by shrubs, has less wildlife, or is crowded

Fig. 15.5 While it may seem a stretch to attribute religious values to ecosystem services, in this church in Sonoma County, California, the "Church of the Oaks" is arranged around a phalanx of stout oaks that add character and identity to the site. (Photograph by P.F. Starrs)

by new homes—then the value of a property for recreation declines. In Spain, the diversified and somewhat unique nature of dehesa production contributes to opportunities in product labeling, as opposed to California production which has been single-product and homogenous for some time. Guijuelo, in Salamanca (Elola 2008) and Jabugo, in Huelva, are experiencing relatively rapid population growth that is linked to the region's notable denomination of origin hams from Iberian pigs. By contrast, California ranchers believe they have fewer opportunities to significantly increase income with diversified production (Sulak and Huntsinger 2001), especially where land use change is rapid.

15.3.4.3 Carbon Sequestration

New developments for markets for credits from carbon sequestration (Chap. 12) may also offer income opportunities. However, oaks grow very slowly and have low rates of carbon capture as compared to other forest species. On the other hand, the shrubs that readily invade dehesas in Spain (and much less vigorously in California oak woodlands) do also accumulate carbon. The net balance is difficult to estimate, but a stand of oaks offers a long-term carbon sink that will exist for 250 years, if managed in a silvicultural rotation (Chap. 9), and provides benefits so long as the trees are standing. Newly planted native oaks can be significant accumulators, since they grow relatively quickly at a young age. "Avoided deforestation" programs, if applied to oak woodlands and agroforestry systems, may hold considerable benefit for oak woodland ranches and dehesa, given the difficulty of demonstrating consistent "additionality" on arid rangelands, the multiple benefits of management for diverse goods and services, and the desirability of a cleared understory for long periods of time along with matured oaks. Exact estimates of carbon sequestration markets for carbon storage and markets for avoided deforestation are works in progress.

15.3.4.4 Mitigation

United States and California laws protecting endangered species and their habitats have created considerable demand for sites for mitigation. To develop a property with a resident endangered species, or to address significant water quality and yield issues, mitigation of environmental impacts may be required. The developer must purchase and set aside for perpetuity habitat that substitutes for the loss on a developed site. Sometimes the habitat to be set aside is required to be much larger than the habitat lost. These strict environmental policies have created a market for mitigation properties and easements. A mitigation easement means that the habitat is on private ground, sometimes part of a working ranch, but in exchange for payment from the developer, the landowner agrees to protect the property, and all future landowners are compelled to protect the habitat as a condition of the property title. Speculation in what are termed "conservation banks" occurs in areas where development is presumed to be on the horizon: private investors buy lands that they believe will provide needed mitigation opportunities, and hence income, to the owners.

15.3.5 Partnerships Matter

Some costs, like expenses involved in regenerating an oak stand, or resisting an outrageously high offer from a real-estate developer, are so high that a partnership with the public may be required to maintain a healthy and intact woodland. Otherwise, too often the sale, subdivision, neglect, or abandonment of land is a result.

Regulation will always be part of the conservation scene, but incentive-based methods may be much more palatable to land managers. Some of the efforts to reduce development of working woodlands in California have largely been led by the private sector through the efforts of non-governmental entities such as The Nature Conservancy, The California Rangeland Trust, and local land trusts as they broker conservation easements on private lands. Although far from perfect as a conservation strategy, conservation easements allow ranchers to preserve a way of life they are overwhelmingly fond of, while continuing to manage their land (Chap. 12; Merenlender et al. 2004). At the same time they represent a vernacular response to what are perceived as ineffectual government land-use controls.

Government and non-governmental organizations can encourage landowners to follow specific practices that are deemed environmentally helpful. These programs take many forms: conservation easements, direct payments for ecological services, assistance with product marketing, or shared-governance agreements that tailor land-use decisions to something more sought-after by a property owner. Each is a form of income, either one-time or ongoing, that makes conservation practices and management that supports ecological products a more desirable path for landowners to follow (Spash and Hanley 1995; Nunes and Nijkamp 2010). With the significant advantage accrued in spreading costs across a society, governments are finding ways of doing this in Europe and Spain, in the U.S. and California. The ongoing global economic crisis that began in 2008 makes this less easy now than during more economically flush times, even as understanding of such principles grows more sophisticated (EEA 2010, 2011; Haines-Young and Potschin 2010).

The actions of oak woodland landowners and managers can increase, or decrease, net water yield. Dehesa and ranch runoff stored in reservoirs may also have an income value (Berbel and Mesa 2007; Berbel et al. 2011). How and whether improved water flow across oak woodlands is a service to be paid by government and/or downstream users is a topic addressed in part in Chap. 12. There are differences between California and Spain, since large percentages of upland areas in California are publically owned. Much of the California water supply comes from mountain snowpack melts and travels down streams and rivers that frequently flow across private land—and landowners often have irrigation or diversion rights to some of those flows. The public-private relationship is a complicated one, and much negotiated, since without rights to water diversion there is no irrigation, and that impedes irrigated crop agriculture. An increasing demand for stored water may result in opportunities to market water to downstream.

15.4 Research Approaches and Needs

A variety research topics and approaches are needed to learn more about the oak woodlands that constitute significant parts of ranches and dehesas, and to make the case for conservation.

15.4.1 Field-Based Interviews

In an age of office cubicles, powerful laptop computers, online bibliographic databases of unlimited range, agglomerated datasets, and research institutions that make it difficult for staff to conduct field research, the value of going to the source is at times forgotten. The research included in this volume is based in overwhelming measure on field-obtained information gathered from locally

Fig. 15.6 Landowners in California and Spain (as here) are a tremendous knowledge pool for researchers interested in understanding the stewardship of oak woodland properties. (Photograph by P.F. Starrs)

knowledgeable people who are willing to contribute to our larger knowledge as investigators and field scientists. That makes studies such as this one unusual, and also complicated, since working in Spanish and American English at once is something all of us authors can do, though with varying levels of comfort and skill. Having the ability to work across the boundaries of language and literature was crucial to our efforts (Fig. 15.6).

Field-based research makes four things possible, with each of them absolutely essential to the formulation of sound studies. First, there is no substitute for being there. As the American anthropologist Clifford Geertz noted in a 1988 essay of that title, "being there" is not only important in establishing the credibility of a researcher with peers, it is often the only way to ground truth hypotheses, surmises, and to refine tentative conclusions. Second, places matter, and while some research sites can be treated normatively—as part of a larger whole—sometimes-subtle distinctions can have deep significance, and discussions on-site make those differences evident. They cannot be captured by satellite imagery or aerial photography (although each of those has good uses), and have to be unearthed by researchers able to figure out the right questions, gauging the reactions of those being interviewed, and iteratively recognizing where a following question should go. To acquire such skills requires experience, a degree of patience, and access. Third, variations from place to place are at times subtle, at other times massive, and finding those lines of division and commonality is necessary to understanding the data. Institutional fault lines—regional or county borders, divergent political regimes, census tract edges, boundaries of public-private land—grow more obvious in the field, and explain much. Finally, field-based interviews are as important for the interviewee as for the interviewer. Information flows in interviews go both ways, and in many respects, the researcher-interviewer is a mendicant, coming with notebook (or digital camera or scanner or digital sound recorder) to gain knowledge from local sources. The local residents are teaching us, and that two way flow of information restores urban–rural connections that are too often lost in the scholarly world where the researcher's self-image may be as "the expert" and the local landowner or caretaker or policymaker is regarded only as a "source." Routinely, in their particular world, they are the experts, and the researcher the student.

15.4.2 Attitudinal Surveys: Owners, Managers, Policymakers

Formal surveys employing rigorous methodologies with data analyzed by tested protocols are a remarkable font of information. While methodologies followed in the studies cited in these chapters vary somewhat, the results are carefully balanced to make sure they are representative and accurate within acceptable standards of probability. Questionnaire surveys must be prepared carefully, pretested to make sure that the questions are understandable and answerable, and when that is the case, quite significant conclusions can be reached. Much of the data on owner, manager, and public attitudes toward oak woodlands in California has evolved through studies that began more than 25 years ago (Huntsinger and Fortmann 1990; Huntsinger et al. 1997, 2004, 2010; Liffman et al. 2000; Oviedo et al. 2012), with the studies re-tasked and recalibrated as attitudes change and new areas of interest emerge. Significantly, this research can be considered representative of actual active beliefs, and many of the data on owner attitudes toward economic choices develops from similar methodologies, sometimes pursued in mail or on-line survey, at other times by one-on-one interviews (Fig. 15.7).

Too often, policy decisions are made without a clear understanding of what landowners, managers, and local populations actively care about and want to see happen. When scientific survey data is used to determine landowner resources and needs, management reccommendations can be developed that are in concert with the capacities and expectations of communities and landowners. Sometimes the discoveries made in such inquiries are startling, and lead to significant shifts in policymaker decisions, to better accord with what is considered acceptable.

15.4.3 Economic Survey Research

The "million dollar view" attributed to a given ranch or dehesa may not in truth be worth that exact dollar figure, but it certainly has value. Economic surveys, done in the field with an intensive inquiry not only into landowner and community-asserted beliefs, but by examining ledger books and enterprise accounts, can go

Fig. 15.7 Whenever possible, it is important to get information right from the source. Survey-based research can explore local attitudes and beliefs, and inform policymakers, education experts, and technical advisory services what is needed on the ground. Needless to say, a survey needs willing respondents. These Iberian pigs are too busy eating acorns. (Photograph by P.F. Starrs)

beyond assertion to capture economic fact. These are not always what, for example, a municipality that owns a dehesa would like to believe is true. An acute scrutiny of financial accounts can reveal other truths that landowners may or may not themselves realize to be the case. Many landowners in California, for example, will admit to keeping an oak woodland ranch intact because she or he likes living there, enjoys solitude and stewardship of an evolving landscape, and wants to see a ranch kept intact; that this comes at substantial cost is clear from a detailed examination of household or enterprise accounts. Such a conclusion should hardly be startling; in the early 1970s, two New Mexico researchers coined the term "ranch fundamentalism" to describe the affinity of ranchers in that state for their extensive properties, and their willingness to work off-ranch to support them (Smith and Martin 1972). Economic techniques have evolved in substantial ways, and the economic surveys shared in this work by Spanish and California researchers (Chaps. 11, 12, 13) demonstrate how case studies can provide the scope, dimensions, and dollar-euro value of landowner, manager, and policymaker choices (Figs. 15.8 and 15.9).

The techniques of economic valuation are gaining sophistication in the scholarly literature and in publically-supported undertakings such as the Andalusian

Fig. 15.8 Near Santa Olalla del Cala (Huelva), the Fundación Monte Mediterráneo has created a shrine of sorts to the holm oak, complete with a text-etched granite monument. The management clearly favors oaks. If evidence of ecosystem value attributed by landowners—in this case, the Dehesa San Francisco supported by German-based EU members—were ever wanted, this tree qualifies. (Photograph by P.F. Starrs)

Fig. 15.9 Presenting a nearly reverential attitude toward a wooded oak landscape is by no means a solely Spanish practice. Oaks in California are often highly respected elements of the countryside, and meet with a respectful treatment. (Photograph by L. Huntsinger)

RECAMAN [1]project. These results are incorporated into the calculation of market and non-market ecosystem service value. Only now is work being completed in a significant area of the dehesa that makes it possible to assess total economic value. Those considerations, this book suggests, should address oak woodland and dehesa fires, oak conservation (recruitment and regeneration), the preservation of threatened biodiversity, hunting policy and economics, the effects of abandonment of extensive multi-species livestock grazing in the dehesa, and official statistics that incorporate ecosystem services as direct dehesa products. Governments will need to extend official accounting systems to include oaks, but metrics already exist that would allow estimating the values of tree growth, costs of dehesa oak disappearance, the production and provision of firewood and cork, and the value of products including cork, grass (forage), acorn, sequestered carbon, water, and other ecosystem services. These would yield the total capital value of the dehesas of Spain in a system easily extended to other areas and even other countries. The language of the estimates involves terminology such as hedonic pricing, contingent valuation, stated choice, and production functions that may prove difficult for the faint of heart, but the end result is a viable and accurate total value, as demonstrated in Chaps. 12 and 13.

Just how important the existence value of oak woodland dehesa and ranchland is can be ascertained by careful economic survey research. This offers another means of determining the value of ecosystem services, along with willingness to pay, which establishes a market price that visitors would pay to see a specific site or to preserve specific ecosystem services. The name economists and policymakers attach to programs that seek to preserve an environment so future generations may make use of it is "option value" and the later has been recognized as "a significant part of the real income of many individuals" (Krutilla 1967, 779). Oak woodlands are rich in preserved options (Fig. 15.10).

15.4.4 Archival Investigation

The interplay between a human-created ecosystem like the oak woodland and the dehesa can be analyzed in part through painstaking study of ecological data and time series analysis. But the human element in historical landscape creation and change is often best captured through archival research. Routinely that incorporates printed books and maps, newspapers, and manuscript sources, but also folkloric materials: narratives, spoken accounts, literary fiction, poetry, nonfiction, even songs, and through artwork, which at the least can include landscape paintings and portraits, mosaics (Chap. 2), block-prints, and sketches. And a

[1] The RECAMAN project (*Valoración de la Renta y el Capital de los Montes de Andalucía*) of the Junta de Andalucía, initiated in 2008, is ongoing and applies the Agroforestry Accounting System at the regional scale to measure total income and capital from the montes of Andalucía in Spain.

Fig. 15.10 Livestock feeders, painstakingly hewn from stone, speak to long labors on behalf of livestock generations past, in the Valle de los Pedroches (Sierra north of Córdoba). These are, themselves, significant historical archives, especially when accompanied by knowledgeable "captions" of explanation that are provided by local caretakers and residents. (Photograph by S. García)

cautionary note sounds here: reaching back through time to understand human influence and shaping of past environments is no easy matter, but has to be done. Assertions and speculations are one thing, but data and evidence-based conclusions require painstaking work, sometimes in the archives, sometimes in cooperation with earth scientists, linguists, and anthropologists, sometimes with cultural-historical geographers, botanists, zoogeographers, or plant geneticists.

In California, the earliest sources are necessarily archaeological; the material-culture records of Native American use go back millennia, and only in the last 30 years has a serious scientific understanding of pre-European land management been developed. In Spain, some remarkable use is being made of convent, monastery, and other ecclesiastical records, and of the commercial records of guilds and associations such as the Honorable Council of the Mesta (founded in 1273 by the King Alfonso Tenth the Wise), the Merino sheep-growers association that survived well into the nineteenth century (Chap. 2).

These information sources are an essential element of ecological and economic history. Mastery of the subtleties of written sources requires skills that extend to ancient scripts and paleography, a broad understanding of past social relations, and the ability to visualize geographical relationships as evolved through time and space. Without drawing comprehensively on such skills, a study such as this risks offering only a snapshot, a quick view of a moment; we have attempted to go well beyond that.

Ecological sources, including analysis of pollen records, dendro-chronological dating, and well-aged human records, suggest that human tinkering with what would become the dehesa began four- to six-thousand years ago. Gaining an understanding of what was done, what the original vegetation might have been, and how changes were wrought is nothing simple, but definitely worth continued study. Paleoecological studies in California have advanced well in recent decades; comparable research in Spain is proceeding more slowly, though with developing enthusiasm.

15.4.5 Ecological Data Acquisition

Comparative research is particularly difficult, and in some ways this book-length study is an ultimate example of cross-cultural and comparative ecological study. While field research is sometimes encouraged for social scientists (economists, geographers), and humanists (historians, artists, literary scholars), time in the field is not only expected of most ecologists, biologists, soil scientists, and climatologists, it is unavoidable. And, to be fair, the gathering of field data is what many environmental scientists—including veterinarians, animal behaviorists, and wildlife biologists—relish most in their professional work (Fig. 15.11).

There is already data available that captures key ecological traits of working oak landscapes. That data includes records of outstanding biological diversity at most, if not all, spatial and temporal scales; documentation of poor and variable tree recruitment; cataloguing of long to very long life cycles; and the key role of drought. From these data, several hypotheses about the biological causes of these traits have been put forward, as developed in Chaps. 3, 4, 5, 6, 7, 8. Some management proposals are derived from these hypotheses, such as landscape-scale rather than field-scale management approaches and temporal rotations.

However, it is clear that much more data is needed to fully test the hypotheses proposed and to refine management prescriptions. Needs for more data are linked mostly to two main sources of ecological variation: spatial, rooted in the huge areas over which working oak woodlands are established, and temporal, grounded in the high variability of Mediterranean climates and the long life cycles of oaks. For this reason, the set-up and the maintenance of networks of ecological long-term monitoring for dehesas and California oak woodland ranches is essential. A start on monitoring acorn crops is the project of a handful of Spanish and California researchers, as described in Chap. 7, but such monitoring programs should be expanded to include the key ecological interactions influencing tree recruitment (Díaz et al. 2011).

Fig. 15.11 A profoundly human-inflected landscape is evident in this view of the Montes de Toledo, an area as well known for its small- and big-game habitat as for its expanses of oaks and underlying shrubs. A diversity of ecological niches is in evidence, ranging from a narrow gallery forest notching into the hillside, to an extensive area of shrubs on some intriguing soil substrates. Evident are two areas of cleared understory where cereal crops were recently harvested. Many a dehesa is a landscape mix of natural and owner-initiated processes, and this site reflects the cycle of clearing and regrowth. (Photograph by M. Díaz)

15.4.6 Local Knowledge

A perhaps predictable divide exists between traditional beliefs and management practices and those that are developed following standard scientific methodologies—and there is debate about which is more reliable (Fuentes Sánchez 1994; Montoya Oliver 1996; all chapters, this book). Evaluating the quality and potential of local and "expert" knowledge usually involves a carefully set up study, or several. Both types of knowledge may turn out to be misguided, too generally prescribed, or simply incorrect. Someone who is brought in for consultation may know everything about one kind of tree, or be the reigning intelligence about a specific silvicultural product; he may be great on theory yet hopeless on practice; or she can be highly skilled in management of cattle or Iberian pigs or soil health or mushrooms and fungi (Fig. 15.12).

As research discussed in this book establishes, sometimes it is scientists, consultants, or resource managers who draw incorrect conclusions and embed them in

Fig. 15.12 Local breeds of cattle (handsome horned retintas, in this case) approach warily, with a Sierra Norte de Sevilla dehesa woodland in the background. This site is managed in part for cattle, in part for Iberian pigs, and to a significant degree for fee-hunting. Managing such a mix of uses demands considerable time-tested local knowledge and the ability of a manager to assess resource utilization and management needs quickly and accurately. (Photograph by A. Caparrós)

the literature—and in practice. Sorting fact from fiction or surmise entails collaboration, transparency, and communication within the community of those involved with oak woodland and dehesa. But landscape-level management demands a broader understanding of synergies between different types of resources, detailed comprehension of soils and parent materials, climate and weather, annual variation, and long-term land use. It also requires the ability to adapt to change and to draw on multiple sources of information. If recommendations are not always successful it is important to try, to experiment, and to see what works. To make that happen effectively, cooperation (and confidence) among landowners, local inhabitants and workers, managers, and scientific experts is essential. From that—and only that—can emerge a plausible economy and land use practice (Fig. 15.13).

15.5 Conserving Working Landscapes: Fostering Policy Development

15.5.1 Spain

A rise in official EU and Spanish government interest in the dehesa as a productive landscape began 30 or more years ago, and recognition of the value of the

Fig. 15.13 The cork oak (*Quercus suber*) is, along with the Iberian pig and possibly hunting, the most reliable commercial income source in the dehesa, which makes it a linchpin in the working landscapes where such trees are found. In general, Portugal offers better cork oak habitat than south-west Spain, although there are exceptions to this generalization. (Photograph by P.F. Starrs)

characteristic multiple use management of Spanish oak woodlands was initially satisfied with an emphasis on the traditional production of grass and understory forage, acorns, cork, firewood, and grains (Parsons 1962a, b; Martín Bolaños 1943). Dehesa owners concurred, though recently and with a slightly different emphasis. Oaks today offer an aesthetic accent to the land, but the market for natural resources in the last 50 years has trended toward valuing acorns and grass more than wood products such as firewood or charcoal (Bermejo 1994). The production of grain and cereal crops, which in the mid-twentieth-century helped maintain a shrub-free substrate under and beyond the oak canopy, is today proving worth ever less on the market, especially with concern within the EU's Common Agricultural Policy (CAP) about surplus production of agricultural crops including staple grains. On the other hand, fees gained from selling the rights to hunt a property are a change (Chap. 11) that in the most favorable circumstances can offer sizable added income to landowners.

Edicts, directives, and legislation issuing from the Common Agricultural Policy (CAP) tend to favor high added-value products with strong constituencies and established local markets—cheeses, meats, wines—that thanks to denomination of origin labeling carry the flavor of home and find a ready-made local and sometimes international demand. This fits well with the dehesa. Other changes are somewhat more startling and unexpected. Legal changes in the 1970s (the *Ley de Caza*, or Hunting Law, of 1970) made it easier for landowners to profit from the sales of hunting and game rearing. Re-colonization of the oak understory by shrubs provides game habitat that favors deer and wild boars that can yield a significant income while minimizing cultivation expenses. This has produced significant changes in dehesa management, especially in game-rich areas, where understory cultivation can be relaxed to allow a mosaic of shrubs, and maintenance costs for clearing the understory are reduced. Biodiversity can be a beneficiary if encroached areas are interspersed with open areas and are also re-opened afterwards in accordance with the traditional rotation of trees and shrubs necessary for the long-term functioning of dehesa farming (Chap. 8). In the absence of managed rotations, local or regional reductions in biodiversity will follow due to crop set-asides, grazing depletion, tree population decline, or landscape homogenization (Díaz 2009; Díaz et al. 2009). As Guy Beaufoy noted in 1995, "by far the largest concentration and the greatest diversity of low-intensity farming systems are to be found in Spain" and he concludes, "Low-intensity livestock raising also has a significant social role, in helping to maintain rural communities in remote areas" (Beaufoy 1995).

California oak woodland and Spanish dehesa managers recognize a government responsibility in conservation of natural resources, but do not want this to impinge on them—in interviews most say they are doing a good job taking care of their property. California ranchers typically comment that they are good stewards of their land, they understand how the land works better than others, and that an increasingly urban American population does not understand them or the ecosystem. Although they recognize a "problem" with oaks, they do not necessarily see the problem on their own property. In California and in Spain owners report that government agencies do a poor job of understanding or working with them. California ranchers often believe government is more responsive to urban needs than to smaller, rural constituencies (Huntsinger et al. 2004) (Fig. 15.14).

In Spain, only an active encouragement of oak regeneration, and rotation through shrub incursions (Chap. 6), will sustain wooded dehesa (Pulido et al. 2010; Senado 2010; Campos 2012a). In California, numerous species depend on grazing and the water developments that accompany ranching (Chaps. 8 and 12). Active support for regeneration may be needed in some areas. Work on the economics of conventional silviculture in oak woodlands shows that private landowners are unlikely to invest sizable funds in oak recruitment and regeneration on their own in Spain or California because of high costs and very slow financial returns. But future generations will benefit from investment in oak woodlands now, whether such investment fosters further research, encourages ecosystem services, subsidizes woodland regeneration by direct public investment, or rewards landowners for not subdividing their properties and keeping them instead in extensive tracts of oak ranchlands.

There are glimmers of government and policymaker awareness that a regeneration problem exists in the Spanish dehesa, as in the 2010 "Report on the Protection of the Dehesa Ecosystem" issued by the Spanish Senate, which concluded that the crisis of natural regeneration in the dehesa is a pressing environmental problem. A variety of corrective measures have been attempted, generally supported by the EU or by autonomous regional governments (Ovando et al. 2007). But widespread recognition of the problem of a lack of new trees has not motivated the Spanish government to cope with or even address the economic effects of diminishing stands of aging oaks. In Spain, the national government

Fig. 15.14 Tree mortality, either through the natural death of old trees or increased mortality driven by recent disease spread or climate changes, should be compensated for by the recruitment of young trees to ensure the long-term persistence of dehesas. The picture shows just the opposite, a decline in tree numbers, as is happening in the National Park of Cabañeros and in most Spanish dehesas. (Photograph by M. Díaz)

from the 1950s through the 1960s imposed an afforestation regime on some landowners that required planting of non-native conifers and eucalypts on land that previously held dehesa and oaks. That established a plantation system of pines such as Aleppo (*Pinus halepensis*), maritime (*P. pinaster*), Monterrey (*P. radiata*), Scotch (*P. sylvestris*), European black (*P. nigra*), and eucalyptus (500,000 ha; mainly *Eucalyptus globulus*), to satisfy government interest in afforestation following the German model. Only in the last 20 years, with direct investment from the European Union (EU) rather than support from Spanish oak woodland policy, has a return to native species brought subsidies back to encourage landowners to remove exotic trees and reintroduce (or reemphasize) holm and cork oaks in the dehesa (Beaufoy et al. 1994; Ovando et al. 2007).

Desire is widespread to preserve dehesa landscapes, but a continuing controversy about costs and benefits and the most socially just ways to work with land-owners, along with particularly tight current budgets, hampers willingness to intervene and preserve. To say that further study is needed is almost a tired cliché. But an important fact about dehesas and ranchlands, as this book should by now have made clear, is that expert knowledge exists and can be put into play by policymakers who inform themselves about the needs of the oak woodland enterprises that constitute such a large portion of the land in Spain and California (Fig. 15.15).

Fig. 15.15 So integrated into village and rural community life is the wild boar that, in some sites such as this town in the Sierra Norte de Sevilla, statues are erected to the boar honoring its place as a feature in the local economy and culture. Survey-based research can further explore local attitudes and beliefs, but a statue is a statue, and this one its own way speaks volumes. (Photograph by P.F. Starrs)

One of the most promising new EU programs is a social contract between dehesa landowners and the public administration that is termed a *land contract*. This contract is in fact a government purchase of dehesa ecosystem services for the public, including both ecological and cultural services. This new policy is based on a compromise between policy demands that follow on political, ecological, and economic goals: to avoid incentives for surplus production, which are forbidden by the World Trade Organization (WTO); to maintain dehesa land operations to help address unemployment in marginal rural areas; and to preserve endangered wildlife and plants and culturally significant animal breeds. The EU is now in a deep budget crisis, yet European citizens show a continuing willingness to pay for protection of natural resources while asking for increased government regulations to improve nature conservation (European Commission 2010a, b). However, in spite of this global economic context, dehesa conservation will continue to be based on promoting livestock rearing (Iberian pigs excluded, as they are economically profitable and easily adapted, thanks to their short production cycle), since today's main policy trend is to support low-intensity agriculture, habitat improvement for wild species, encourage ecosystem services, and discourage production of products that already are in surfeit in the official view of the European Union's CAP. However, a broader conception is needed to preserve dehesa in the long term, as current policy does not directly address protecting the biodiversity that lies in the core of the current and future benefits from dehesa management.

15.5.2 California

Ranching in California is very much derived in inspiration and technique from the experience of Spanish-Mexican colonists and Native American labor that came to be involved in raising cattle, sheep, goats, and horses starting in the late eighteenth century (Chaps. 2, 10). The livestock economy has proved relatively reliable for ranchers. Those who wish to stay have practiced ingenuity and adaptability in finding means to remain on the land. In terms of products, forage production has always been a dominant focus in California, accompanied by some thinning of oaks designed to stimulate forage growth. Fuelwood and hunting are two other sources of income that have a role in some places and in some circumstances. Concern about the recruitment of new oaks is the subject of years of California-based study, science, and experimentation. Nearly 30 years ago, range scientists and economists understood that an active program, deriving from academic researchers and extension foresters, should undertake research to address regeneration and conservation issues in oak woodlands (Chaps. 5, 6, 7). Conservation easements and government expenditures on cost sharing, advisory, and technical assistance programs aid in these efforts, sometimes forming complex public-private-NGO collaborations. The rapid very recent rise in the prices of supplemental feedstuffs will impact oak woodland ranchers in ways that may be difficult to predict.

The control of oaks in California was consciously relinquished to landowners by the State Board of Forestry, which otherwise controls forest harvest (Chap. 2; Giusti et al. 2004). The University of California's Integrated Hardwood Range Management Program and its legacy of research and extension programs made oak woodland policy, management, economics, and product expansion a significant object of study over more than 25 years, and work continues through University of Caifornia auspices and initiatives, with direct outreach to landowners. California in 2004 passed the Oak Woodlands Conservation and Environmental Quality bill, which requires mitigation if a project threatens to have a significant effect on oak woodlands (Giusti et al. 2004).

California has examples of significant ecological and agricultural preservation programs that have simply been ended because of costs, requiring lower-level governments (counties, regions, cities) or non-governmental organizations to take on the debt of farm- and ranch-preservation efforts. The now-orphaned California Land Conservation Act or Williamson Act is one such case, and the long-term effects of its dissolution are likely to be disruptive and encouraging of land development that is not in the interest of oak woodland preservation (Wetzel et al. 2012) (Chap. 12).

15.6 Conclusions

An important policy goal is intergenerational equity: How are we to protect the interests of future generations? All the benefits discussed and analyzed in this book speak to the desirability of preserving oak woodland benefits and services for future generations. "Sustainability" is easy to define—it is the sustained flow of goods and services into the future—but how to measure and accomplish sustainability is not so easy, especially when society and ecosystems are undergoing such accelerated change.

An even larger and more challenging issue involves changes in climate and the collective effect of climate change on sought-after landscapes, species richness, and biodiversity. In that vein, a group of Australian researchers, looking at the superior resistance of scattered trees to the effects of climate change, concluded that "ensuring appropriate levels of tree regeneration, and preventing premature mortality of mature trees, is critical for maintaining scattered trees in landscapes…. The encouragement of scattered trees would require the protection and restoration of traditional cultural landscapes, or the establishment of new ones" (Manning et al. 2009). Avoiding irreversible losses in natural and human capital is included as a goal in wealth calculations by the United Nations (UNU-IHDP and UNEP 2012). Oak woodland dehesas and ranchlands certainly can be considered and fitted within this global conservation framework.

The central question of the new paradigm for conserving nature to enhance human well-being—now and in the future—is just how much judicious and forward-looking management is likely to cost. An intolerance for public spending, particularly in the United States, and a shortage of funds in California and Spain, do limit current options. On the other hand, the societal costs of failing to take action immediately could be far higher than investing right now in dehesa and ranch sustainability. But the public has demonstrated a persistent inability to weigh long-term costs against the advantages of spending money right now, which is the time when problems can be dealt with most effectively. Is it possible to find "win–win" solutions that minimize costs? Can we view public expenditures in this realm as a bridge to a future when such systems can stand on their own with their plethora of public benefits? Or is investing in oak woodlands just a first step down a path of continual public dependency? These questions need answering. In terms of market-based solutions, a dedication to buying wines where a real cork is used, or purchasing certified sustainable woodland products, can contribute in ways that span international borders if consumers are attentive.

There is, in this study, a significant conclusion: Just because two places appear similar hardly means that they are alike; oftentimes the variations are far more than skin deep (Aschmann 1973). But with that as an initial concession, it pays to acknowledge how much can be learned from comparative research that matches physical, cultural, historical, economic, and geographical features, and then carefully places likenesses and departures side-by-side, in a deliberate attempt to learn across oceans, landscapes, economies, and societies.

References

Allen-Diaz BH, Bartolome JW, McClaran MP (1999) California oak savanna. In: Anderson RC, Fralish JS, Baskin JM (eds) Savannas, barrens, and rock outcrop plant communities of North America. Cambridge University Press, Cambridge, pp 322–341

Artola M, Bernal AM, Contreras J (1978) El latifundio: propiedad y explotación, ss. XVIII–XX Ministerio de Agricultura, Madrid

Aschmann H (1973) Man's impact on the several regions with mediterranean climates. In: di Castri F, Mooney H (eds) Mediterranean type ecosystems: origin and structure. Springer, Berlin, pp 363–372

Beaufoy G (1995) Distribution of extensive farming systems in Spain. In: McCracken DI, Bignal EM, Wenlock SE (eds) Farming on the edge: the nature of traditional farmland in Europe. Joint Nature Conservation Committee, Peterborough, pp 46–49

Beaufoy G, Baldock D, Clark J (eds) (1994) The nature of farming: low-intensity farming systems in nine European countries. IEEP (Institute for European Environmental Policy), WWF (World Wildlife Fund), and the Joint Nature Conservation Committee, London

Berbel J, Mesa P (2007) Valoración del agua de riego por el método de precios quasi-hedónicos: aplicación al Guadalquivir. Econ Agrar y Recurs Natu 7:127–144

Berbel J, Mesa-Jurado MA, Pistón JM (2011) Value of irrigation water in Guadalquivir Basin (Spain) by residual value method. Water Res Mgmt 25:1565–1579

Bermejo I (1994) The conservation of dehesa systems in Extremadura. In: Bignal EM, McCracken DI, Curtis DJ (eds) Nature conservation and pastoralism in Europe: proceedings of the 3rd European forum on nature conservation and pastoralism, University of Pau, France, 21–24 July 1992, Joint Nature Conservation Committee, Peterborough, UK, 14–19 July 1992, pp 14–18

BOJA (2009) ORDEN de 26 de marzo de 2009, por la que se regula el régimen de ayudas para el fomento de la primera forestación de tierras agrícolas en el marco del Programa de Desarrollo Rural de Andalucía 2007–2013, y se efectúa su convocatoria para el año 2009, BOJA, Núm 66, Consejería de Agricultura y Pesca, Sevilla, 6 Apr 2009, pp 15–39

Brooks CN, Merenlender AM (2001) Determining the pattern of oak woodland regeneration for a cleared watershed in Northwest California: a necessary first step for restoration. Rest Ecol 9:1–12

Brownsey P, Oviedo J, Huntsinger L, Allen-Diaz B (2012) (forthcoming) Historic forage productivity and cost of capital for cow-calf production in California. Rangel Ecol Mgmt (in press) http://dx.doi.org/10.2111/REM-D-11-00059.1

Campos P (2012a) Economías privada y pública de la dehesa. Seminario paisajes culturales agroforestales. Consorcio Plasencia, Trujillo, Parque Nacional de Monfragüe y Biodiversidad Territorial. Univ de Extremadura, Plasencia, 18–21 Apr 2012. http://www.eweb.unex.es/eweb/accionporladehesa/opinion/valoreseconomicos3may_pablocampos.pdf. Accessed Sept 2012

Campos P (2012b) ¿Qué cambio de política de incendios forestales? Hoy (Cáceres, Extremadura, opinion essay), 24 Sept 2012, p 14

Campos P, Mariscal P (2003) Preferencias de los propietarios e intervención pública: El caso de las dehesas de la comarca de Monfragüe. Invest Agrar: Sist y Recur For 12:87–102

Costanza R (2006) Nature: ecosystems without commodifying them. Nature 443:749

GPRC [Great Plains Restoration Council] (2012) Buffalo commons. http://gprc.org/research/buffalo-commons/. Accessed Sept 2012

Díaz M (2009) Biodiversity in the dehesa. In: Mosquera MR, Rigueiro A (eds) Agroforestry systems as a technique for sustainable land management. Programme Azahar. AECID, Madrid, pp 209–225

Díaz M, Alonso CL, Beamonte E, Fernández M, Smit C (2011) Desarrollo de un protocolo de seguimiento a largo plazo de los organismos clave para el funcionamiento de los bosques mediterráneos. In: Ramírez L, Asensio B (eds) Proyectos de investigación en parques nacionales: 2007–2010. Organismo Autónomo Parques Nacionales, Madrid, pp 47–75

Díaz M, Campos P, Pulido FJ (2009) Importancia de la caza en el desarrollo sustentable y en la conservación de la biodiversidad. In: Sáez de Buruaga M, Carranza J (coords) Gestión cinegética en ecosistemas mediterráneos. Consejería de Medio Ambiente de la Junta de Andalucía, Sevilla, pp 21–33

EEA [European Environment Agency] (2010) Scaling up ecosystems benefits: a contribution to the economics of ecosystems and biodiversity (TEEB) study. European Environment Agency, Copenhagen

EEA [European Environment Agency] (2011) An experimental framework for ecosystem capital accounting in Europe. EEA. Copenhagen. http://unstats.un.org/unsd/envaccounting/seeaLES/egm/EEA_bk1.pdf. Accessed Oct 2012

Elola J (2008) La España que aún se desangra. El Pais (Madrid, National Ed.), 20 Jan 2012. http://elpais.com/diario/2008/01/20/sociedad/1200783601_850215.html. Accessed Sept 2012

European Commission (2010a) Attitudes of Europeans towards the issue of biodiversity. Summary Wave 2. Flash Eurobarometer No 290. http://ec.europa.eu/public_opinion/flash/fl_219_en.pdf. Accessed Oct 2012

European Commission (2010b) Farm accounting data network: an A to Z of methodology. http://ec.europa.eu/agriculture/rica/pdf/site_en.pdf. Accessed Oct 2012

Fagan B (2002) Book review of Grove AT, Rackham O, The Nature of Mediterranean Europe: An Ecological History (Yale University Press, New Haven, Conn., 2001), In: J Interdisc Hist, 32:454–455

Fuentes Sánchez C (1994) La encina en el centro y suroeste de España (Su aprovechamiento y el de su contorno). Consejería de Medio Ambiente y OT, Junta de Castilla y León, Salamanca

Geertz C (1988) Being There: Anthropology and the Scene of Writing, In: Geertz C, Works and Lives: The Anthropologist as Author, 1–25 (Stanford University Press, Stanford, California)

Giusti GA, Standiford RB, McCreary DD, Merelender A, Scott T (2004) Oak woodland conservation in California's changing landscape. October issue. http://ucanr.org/sites/oak_range/files/60502.pdf. Accessed Aug 2012

Haines-Young R, Potschin M (2010) Proposal for a common international classification of ecosystem goods and services (CICES) for integrated environmental and economic accounting (V1). In: 5th meeting of the UN Committee of experts on environmental-economic accounting, Department of Economic and Social Affairs, Statistics Division, United Nations, New York, 23–25 June 2010

Hargreave T (1993) The impact of a federal grazing fee increase on land use in El Dorado County, California. Energy and Resources Group, Thesis, UC Berkeley

Hoggart K (1997) Rural migration and the counter-urbanization in the European periphery: the case of Andalucía. Sociolog Ruralis 37:134–153

Huntsinger L (2009) Into the wild: vegetation, alien plants, and familiar fire at the exurban frontier. In: Esparza A, McPherson G (eds) The planner's guide to natural resource conservation: the science of land development beyond the metropolitan fringe. Springer, Berlin, pp 133–156

Huntsinger L, Fortmann LP (1990) California's privately owned oak woodlands: owners, use, and management. J Range Mgmt 42:147–152

Huntsinger L, Buttoloph L, Hopkinson P (1997) Ownership and management changes on California hardwood rangelands: 1985 to 1992. J Range Mgmt 50:423–430

Huntsinger L, Sulak L, Gwin L, Plieninger T (2004) Oak woodland ranchers in California and Spain: conservation and diversification. Adv Geoecol 37:309–326

Huntsinger L, Johnson M, Stafford M, Fried J (2010) California hardwood rangeland landowners 1985 to 2004: ecosystem services, production, and permanence. Rangel Ecol Mgmt 63:325–334

Krutilla JV (1967) Conservation reconsidered. Amer Econ Rev 57:777–786

Liffmann RH, Huntsinger L, Forero LC (2000) To ranch or not to ranch: home on the urban range? J Range Mgmt 53:362–370

Manning R (2011) Rewilding the West: restoration in a prairie landscape. University of California Press, Berkeley

Manning AD, Gibbons P, Lindenmayer DB (2009) Scattered trees: a complementary strategy for facilitating adaptive responses to climate change in modified landscapes? J Appl Ecol 46:915–919

Martín Bolaños M (1943) Consideraciones sobre los encinares de España. Año 14, Núm 27. Instituto Forestal de Investigaciones y Experiencias, Madrid

Martín-Queller E, Gil-Tena A, Saura S (2011) Species richness of woody plants in the landscapes of central Spain: the role of management disturbances, environment, and non-stationarity. J Veg Sci 22:238–250

Merenlender AM, Huntsinger L, Guthey G, Fairfax SK (2004) Land trusts and conservation easements: who is conserving what for whom? Conserv Biol 18:65–76

Montoya Oliver JM (1996) La poda de los arboles forestales. Agroguías mundi-prensa. 3a edn, Ediciones Mundi-Prensa, Madrid, Barcelona, México

Nunes P, Nijkamp P (2010) Sustainable biodiversity: evaluation lessons from past economic. Reg Sci Inquiry J 2:13–46

Ovando P, Campos P, Montero G (2007) Forestaciones con encina y alcornoque en el área de la dehesa en el marco del Reglamento (CE) 2080/92 (1993–2000). Rev Esp Estud Agrosoc y Pesq 214:173–186

Oviedo JL, Huntsinger L, Campos P, Caparrós A (2012) Income value of private amenities assessed in California. Cal Agric 66:91–96

Parsons JJ (1962a) The Acorn-Hog economy of the oak woodlands of Southwestern Spain. Geogr Rev 52:211–235

Parsons JJ (1962b) The cork oaks forests and the evolution of the cork industry in southern Spain and Portugal. Econ Geog 38:195–214

Pausas JG, Fernández-Muñoz S (2012) Fire regime changes in the Western mediterranean Basin: from fuel-limited to drought-driven fire regime. Climatic Change 110:215–226

RE [Rewilding Europe: Making Europe a Wilder Place] (2012) Rewilding Europe: projects/western Iberia/economic perspective, webpage consulted at: http://www.rewildingeurope.com/projects/western-iberia/economic-perspective/. Accessed Sept 2012

Plieninger T, Rolo V, Moreno G (2010) Large-scale patterns of Quercus ilex, Quercus suber, and Quercus pyrenaica regeneration in Central-Western Spain. Ecosystems 13:644–660

Popper DE, Popper FJ (1999) The buffalo commons: metaphor as method. Geog Rev 89:491–510

Pulido F, Picardo A, Campos P, Carranza J, Coleto JM, Díaz M, Diéguez E, Escudero A, Ezquerra FJ, Fernández P, López L, Montero G, Moreno G, Olea L, Roig S, Sánchez E, Solla A, Vargas JD, Vidiella A (2010) Libro Verde de la Dehesa. Documento para el debate hacia un Estrategia Ibérica de gestión. Pulido F, Picardo Á (coords), Consejería de Medio Ambiente, Junta de Castilla y León, Sociedad Española de Ciencias Forestales (SECF), Sociedad Española para el Estudio de los Pastos (SEEP), Asociación Española de Ecología Terrestre (AEET) y Sociedad Española de Ornitología (SEO). Salamanca. http://www.uco.es/integraldehesa/images/stories/doc/Jornadas/libro_verde_dehesa.pdf. Accessed Sept 2012

Ramírez JA, Díaz M (2008) The role of temporal shrub encroachment for the maintenance of Spanish holm oak *Quercus ilex* dehesas. Forest Ecol Manag 255:1976–1983

Russell WH, McBride JR (2003) Vegetation succession and fire in California's bay area. Landsc Urb Plan 64:201–208

Sánchez-Alonso B (2000) Those who left and those who stayed behind: explaining emigration from the regions of Spain, 1880–1914. J Econ Hist 60:730–755

Saving SC, Greenwood G (2002) The potential impacts of development on wildlands in El Dorado County, California. In: Standiford RB, McCreary D, Purcell KL (tech cords) Proceedings of the 5th symposium on oak woodlands: oaks in California's changing landscape, San Diego, CA, 22–25 Oct 2001, pp 443–461. Gen Tech Rep PSW–GTR–184. Pacific Southwest Research Station, Forest Service, US Department of Agriculture: Albany, CA

Senado (2010) Ponencia de estudio sobre la protección del ecosistema de la dehesa http://www.uco.es/integraldehesa/images/stories/doc/bilioteca/Normativa/nacional/informe_23_noviembre_ponencia_del_senado.pdf. Accessed Sept 2012
Simpson J (1995) Spanish agriculture: the long siesta, 1765–1965. Cambridge University Press, Cambridge
Smith AH, Martin WE (1972) Socioeconomic behavior of cattle ranchers, with implications for rural community development in the west. Amer J Agric Econ 54:217–225
Spash CL, Hanley N (1995) Preferences, information, and biodiversity preservation. Ecol Econ 12:191–208
Standiford RB, Bartolome J (1997) The integrated hardwood range management program: education and research as a conservation strategy. USDA Forest Service Gen Tech Rep PSW–GTR–160, PSW, Albany, CA, pp 569–581
Starrs PF (1998) Let the cowboy ride: cattle ranching in the American West. Johns Hopkins University Press, Baltimore
Sulak A, Huntsinger L (2001) The importance of federal allotments to central Sierran oak woodland permittees: a first approximation. In: Standiford RB, McCreary D, Purcell KL (tech cords) Proceedings of the 5th symposium on oak woodlands: oaks in California's changing landscape, San Diego, CA, pp 43–52. Gen Tech Rep PSW–GTR–184. Pacific Southwest Research Station, Forest Service, US Department of Agriculture: Albany, CA
Sulak A, Huntsinger L, Standiford R, Merelender A, Fairfax SK (2004) A strategy for oak woodland conservation: the conservation easement in California. In: Schabel S, Gonçalves A (eds) Sustainability of agrosilvopastoral systems: dehesas and montados. Advances in geoecology, 37:353–364
Tyler C, Kuhn B, Davis FW (2006) Demography and recruitment limitations of three oak species in California. Q Rev Biol 81:127–152
UNU-IHDP and UNEP [United Nations University-International Human Dimensions Programme and United Nations Environment Programme] (2012) Inclusive wealth report 2012: measuring progress toward sustainability. Cambridge University Press, Cambridge
Wetzel WC, Lacher IL, Swezey DS, Moffitt SE, Manning DT (2012) Analysis reveals potential rangeland impacts if Williamson Act eliminated. California Agriculture 66(4):131–6
Workman JP (1986) Range economics. Macmillan Publ. Co., New York

Index

A

AAS. *See* Agroforestry Accounting System (AAS)

Accipiter cooperii, 215

Acorn
- dispersal, 167
- flour, 28
- forage, 277
- gathering, 27, 42
- harvesting, 27
- predation, 168
- size, 183, 194
- trade-offs, 195

Acorn crop, 140, 202–204, 486
- climatic influence, 185
- comparison, 184
- factors, 184
- quantifying, 192
- recruitment, 140
- size, 183, 184

Acorn production, 18, 19, 183, 184, 186, 253, 278, 279
- asynchrony, 199
- biotic factors, 188
- communities, 199
- comparison, 182, 185
- components, 191
- environmental factors, 185
- estimation methods, 201
- management and habitat factors, 190
- pollen limitation, 196
- spatial synchrony, 197, 198
- trade-offs, 195

Aaptive traits, 165

Adenostoma fasciculatum, 337

Aegypius monachus, 232

Afforestation, 137, 138, 259, 260, 281, 362, 377, 466, 490

Aforador, 292

Age determination, 128

Agricultural extension, 35

Agro-environmental schemes, 376

Agroforestry Accounting System (AAS), 20, 363, 391–393, 419, 421

Air Resources Board (ARB)

Alcornocal, 8

Alectoris rufa, 227, 313–315

Alternative management, 254

Ambystoma californiense, 219, 375

Amenity value, 262, 380, 419

Ammotragus lervia, 314, 343

Andalucía, 8, 10, 11, 13, 14, 19, 40, 99, 137, 247, 313, 333, 334, 336, 337, 362, 365, 397, 413

Annual grasses, 10, 108, 111, 132, 135, 146, 152, 166, 275, 276, 297, 435

Antlerless harvest, 332

Aphelocoma californica, 199, 223

Apodemus sylvaticus, 234

Aquila
- *adalberti*, 227, 344
- *chrysaetos*, 215, 218

Arab, 6, 15, 41

ARB. *See* Air Resources Board (ARB)

Arboreal predators, 189

Arbutus spp., 337

Arctostaphylos spp., 337

Area shape index (ASI), 434

Arthropod herbivores
- borer insects, 189

Athene cunicularia, 375

Autonomous regions, 10, 11

P. Campos et al. (eds.), *Mediterranean Oak Woodland Working Landscapes*, Landscape Series 16, DOI: 10.1007/978-94-007-6707-2,

© Springer Science+Business Media Dordrecht 2013

Avena, 147
fatua, 53
spp., 297

B

Baccharis spp., 337
Baeolophus inornatus, 221
Bag limits, 332
Batrachoceps stebbinsi, 215
Big conservation, 36
Bioclimatic
analysis, 65
modeling, 66, 72, 73
Biodiversity, 19, 214, 224, 225, 233, 235, 360, 470, 472, 488
biological diversity patterns, 214, 229
breeding bird diversity, 220
hotspots, 214
threatened, 483
Bioenergy Interagency Working Group (BIWG)
BIWG. *See* Bioenergy Interagency Working Group (BIWG)
Boars (wild), 40, 326, 329, 343, 344, 473, 495. *See also* Sus scrofa, pig
Boletín Oficial del Estado (BOE)
Bottom-up management practices, 329, 336
Bovine tuberculosis, 286
Brenneria, 189
Bromus, 147
Burning
early Spain, 39
native Californians, 127
prescribed, 224, 338

C

Cabrillo, Juan Rodríguez, 26
Cáceres, 39
Arroyo de la Luz, 400, 417
CalFire, 434
California Chapter, American Society of Farm Managers and Rural Appraisers (CASFMRA)
California Department of Fish and Game, 332
California Department of Forestry and Fire Protection—Fire and Resource Assessment Program, CalFire (CDF-FRAP)
California Interagency Wildlife Task Group (CIWTG)
California Land Conservation Act of 1965. *See* Williamson Act
California Legislative Analyst's Office (CALAO)
California Oak Woodland Conservation Working Group (CA-OWCW)
California Rangeland Trust, The, 477
Californios, 29
California's Global Warming Solutions Act, 267
Callipepla
californica, 313, 315
spp., 314
Canis
lupus signatus, 336
lupus, 232, 344
Canopy effects, 147, 149, 153, 155, 156, 168, 217, 277
wildlife, 229
CA-OWCW. *See* California Oak Woodland Conservation Working Group
Capital, 392, 393, 412, 467, 471, 483
depreciation, 328
fixed, 256, 394, 414
gains, 395, 396, 416
immobilized, 397
income, 254, 257, 395, 397
income gains/losses, 257
indicators, 407, 410, 419
manufactured, 395, 414, 419
natural, 20, 382, 419
operating, 470
revaluation, 396, 413
Capra pyrenaica, 314, 344
Capreolus capreolus, 314
Carbon markets, 369
Carbon sequestration, 267, 268, 362, 369, 375, 381, 476
Carbon sink, 288
Case studies, 410, 411
comparative analysis, 410, 417
dehesa, 397, 402, 404, 408, 412
land use and vegetation, 398
economic analysis, 407
ranches, 399, 402, 405, 409, 410, 415
CASFMRA. *See* California Chapter, American Society of Farm Managers and Rural Appraisers
Castilla-La Mancha, 10, 11, 16, 79, 137, 247, 283
Castilla-León, 10, 11, 79, 392
Castizos, 279
Catchment hydrology
California, 112
Spain, 102, 103

Catholic Church, 29, 40, 42, 53
Cattle, 285, 298, 487
  production, 299
Cave art, 319
CDF-FRAP. *See* California Department of Forestry and Fire Protection—Fire and Resource Assessment Program, CalFire
CdV. *See* Equipo Pluridiciplinar de la Casa deVelázquez (CdV)
Ceanothus spp., 337
Centaurea solstitialis, 343
Cervus
  canadensis, 311, 321
  elaphus hispanicus, 313
  elaphus, 227
Chicago Climate Exchange (CCX), 362, 369
Chondestes grammacus, 223
CICES. *See* Common International Classification of Ecosystem Goods and Services (CICES)
Ciconia, 463
  nigra, 232
Cistus, 132, 147, 154, 337
  ladanifer, 132, 337
CIWTG. *See* California Interagency Wildlife Task Group (CIWTG)
Classification and regression tree, 72
Climate, 17
  California, 63, 65
  climatic niche model, 65
  climatic requirements for California oak species, 65
  climatic zones, 77, 78, 82, 83, 85, 86
  forage production, 276
  parameters, California, 65
  parameters, Spain, 79
  Spain, 75, 80
  zones, 431
Climate change, 73, 74, 82, 86, 103, 132, 140, 172, 493
Coccothraustes vespertinus, 224
Cogeneration, 266
Columba
  oenas, 227
  palumbus, 227, 232
Columbidae spp., 316
Commercial returns, 254
Common International Classification of Ecosystem Goods and Services (CICES), 355
Conservation, 171, 214, 267, 369
  California, 466
  challenges, 465
  policy, 487
  Spain, 467
  status, 215, 227, 228
  wildlife, 222
Conservation banks, 477
Contopus sordidulus, 224
Cork
  harvesting, 12, 43, 130
  production, 47
  stripping, 393
  wine, 373
Cotos, 315
Coturnix coturnix, 314
Cowboy, 8
Curculio spp., 189
Current value, 357
Cydia spp., 189

D

Dama dama, 313, 314
Dehesa, 7, 8, 12, 13, 39, 41, 51
  consolidation, 42, 43
  conversion, 43, 125
  decay, 48
  improvement, 46
  privatization, 44
Dehesa boyal, 8, 42, 292
Delichon urbica, 232
Dendrochronological studies, 82
Diario Oficial de la Unión Europea (DOUE)
Dipodomys stephensi, 222, 375
Distribution
  oak species, 66
  oak woodlands, 62
  predicting, 70
Distribution factors, 85, 86
Disturbance regime, 221
  fire, 224, 451
Dogs, 330, 331
Donkeys, 288
DOUE. *See* Diario Oficial de la Unión Europea (DOUE)
Drippy nut disease, 189
Dromaius novaehollandiae, 297

E

Easements, 38
  conservation, 369, 371, 381, 472, 477, 492
  mitigation, 369, 477
Ecohydrological equilibrium, 160

Ecological comparison
  composition, 444
  conclusions, 453
  differences, 442
  landscape similarity, 447
  methods, 438, 439
  questions, 438
Ecological functions, 214
Economic
  analysis, 247
  sustainability, 225, 235
Ecosystem productivity, 148
Ecosystem services, vi, 10, 11, 20, 39, 43, 47, 54, 93, 95, 97, 99, 125, 126, 128, 135, 137, 153, 157, 247, 256, 282, 283, 286, 289, 303, 313, 336, 372, 389, 413, 432
  conservation, 365
  cultural, 355
  existence values, 360
  incentivizing management practices, 374
  landowner decisions, 377
  landscape scale, 360
  marketing, 371
  payment, 375
  payment for ecosystem services (PES), 366
  policies and programs, 366
  provisioning, 355
  regulating, 355
  trade-offs, 362, 365
  value, 356, 358
Ecotone, 225
Empidonax difficilis, 224
Encinar, 8
Ensatina eschscholtzii croceater, 215
Environmental Quality Incentives Program (EQIP), 366, 377, 387
EQIP. *See* Environmental Quality Incentives Program (EQIP)
Equipo Pluridiciplinar de la Casa de Velázquez (CdV)
Erica, 147, 337
Erosion processes
  Spain, 97, 100
Erosion rates
  California, 116
Erythrobalanus, 199
Eucalyptus globulus, 490
Euphydryas editha bayensis, 222
European birds and habitats directives, 227
European habitats directive, 224
European regulations 2080/92 and 1257/99, 259
European Union Common Agricultural Policy (CAP), 48, 50, 261, 367, 376, 414, 488
Existence value, 357
Expropriation, 34
Extremadura, 10, 11, 39, 40, 43, 47, 54, 93–95, 98, 99, 125, 126, 128, 135, 137, 153, 157, 246, 247, 256, 283, 286, 289, 290, 303, 313, 336, 366, 372, 389, 413, 432

F

Fagus sylvatica, 76
Falco tinnunculus, 345
Felis silvestris, 345
Festuca, 147
Fighting bulls, 287
Fire, 224, 468
  impacts on regeneration, 130, 133
  suppression, 32, 132, 133
  wildfire, 445
Firewood, 191, 224, 248, 251, 260, 262, 263, 266
Forage production, 112, 127, 148, 154–156, 170, 191, 261, 275, 278, 375, 492
Forage unit (FU), 249
Forest Practice Act of 1974, 38
Forest products, 16, 50, 363, 474
Forest Reserve Act, 33
Forestry, 46, 419. *See also* Silvopastoral management
  activities, 412, 415
Fragmentation, 223, 341, 379, 382, 420, 467
Franco, 47
Fraxinus angustifolia, 76, 431
Fungi, 475
FU. *See* Forage unit (FU)
Future dynamics, 453

G

Galerida cristata, 232
Game
  fencing, 317, 323, 333, 334, 341
  restocking and farming, 334
  small, 314
  species, 313, 315, 323, 466
  supplemental feeding, 336, 338
  water provisioning, 338
Garrulus glandarius, 199, 234
General Disentitlement Act, 44
Genetta genetta, 345
Geographic Information System (GIS), 22, 65, 73, 429, 439
Global Warming Solutions Act, 369
Goats, 284, 300
Gold Rush, 21, 32, 128, 294, 321, 453

Gómez Mendoza, 40
GPRC. *See* Great Plains Restoration Council (GRPC)
Grassland Reserve Program (GRP), 377
Grazing, 436
  behavior, 281
  exclusion, 249, 260
  leases, 296
Great Plains Restoration Council (GPRC)
Green revolution, 48
Growth function
  holm oak, 248
GRP. *See* Grassland Reserve Program (GRP)
Grus grus, 232
Gyps *fulvs*, 213

H

Habitat elements, 218, 221, 227
Hedonic value, 263, 357, 362, 483
Hirundo
  daurica, 232
  rustica, 232
History, 15
  California, 27, 429
  Spain, 38
Homoclime matching, 65
Horses, 288, 300
  purebred spanish horse, 288
  supplemental feeding, 406
Hounds. *See* dogs
Hunters, 319, 320
Hunting, 50, 52, 254, 262, 266, 285, 468, 473
  commercialization, 325
  costs, 326, 328, 337
  dehesa development, 47
  demographic shift, 318
  dogs, 321
  environmental effects, 339, 340
  harvest methods, 329
  historical Spain, 40
  history, 319
  lands, 313
  net profit, 327
  regulations, 333, 343
  revenue, 326, 345
  seasons, 316
  wildlife management, 227, 329
Hunting Law of 1970 (Spain), 315, 317, 325, 488
Hunting operations, 317, 327
  coto privado, 317
  coto social, 317
  day hunts, 318
  year-long leases, 318
Hydrological cycle, 92

I

Icterus bullockii, 224
IHRMP. *See* Integrated Hardwood Range Management Program (IHRMP)
Immobilized capital (IMC), 397
Income, 38, 49, 50, 137, 214, 225, 294, 325, 326, 345, 377, 380, 396, 399, 406, 419, 473, 474, 476, 488
  commercial operating, 396, 465, 471
  indicators, 407, 410, 419
  maximization, 378
  measurement, 421
  mixed, 414
  results, 412, 415
  total, 359, 363, 391–393, 395, 396, 416
Infinite productivity, 35
Insect herbivores, 188, 190
  acorn weevils, 189
Integrated Hardwood Range Management Program (IHRMP), 37, 377, 378, 466, 492
Integrated management, 37
Intensive agriculture, 32, 33, 43, 48, 49, 54, 125, 222
Interactions
  environmental, 169
  seedling-understory, 165
  shrub-seedling, 169
  shrub-tree, 162, 171
  tree-herb, 163
  tree-tree, 158
Intergenerational equity, 493
Introduced species, 147
Invasive species, 343
Irrigated pasture, 406
Islamic conquest, 40

J

Jamón, 20, 273, 277, 289, 294, 301, 372
  de bellota, 289, 292, 293
Judges of the Plains, 29
Juniperus
  oxycedrus, 76
  thurifera, 76, 79

K

Keystone
  dispersers, 234

species, 225
structures, 220
Kyoto Protocol, 369

L
Lacerta lepida, 232
Lama glama, 297
Land
contract, 490
grants, 31
ownership, 319, 466, 469
price, 247
trust, 370, 477
use, 444
use change, 304
Landowner
characteristics, 13, 303
income, 225
Landscape
ecology, 429, 430
heterogeneity, 216, 220, 225, 227, 337, 469, 488
patterns, 429
processes, 429
Landscape-scale management, 470, 486
Laterallus jamaicensis coturniculus, 361
Lepus
californicus, 314
granatensis, 227, 314
Light availability, 149
Livestock
breeds, 280, 304, 406, 418
exclusion, 134, 136, 260
production, 20
Livestock grazing, 222, 260
impacts on regeneration, 132
impacts on wildlife, 231, 232, 236
Livestock production, 20, 146, 170, 182, 247, 262, 266, 274, 278, 294, 297, 367, 374, 379, 419, 492
comparison, 302
overview, 274
Logistic regression, 66, 69, 72, 73
Lupinus luteus, 191
Lynx
pardina, 232, 333
pardinus, 229
rufus, 335

M
Madrid, 10, 11, 79, 289, 381, 392, 443, 456
Management
objectives, 443
regime, 430, 432
MARM. *See* Ministerio de Medio Ambiente Rural y Marino (MARM)
Martes foina, 345
Masting, 183, 192, 197
Matrix, 215
Measure, 37, 368
Melanaplus spp., 136
Melanerpes formicivorus, 199–221
Meleagris gallopavo, 313, 315, 316
Meles meles, 345
Melojal, 8
Merino, 43, 53, 282, 283, 389, 392, 400, 413
Mesta, 29, 40, 43, 44, 389, 413, 484
Mexican–American War, 31
Microclimate, 103, 168
Microtus californicus, 132, 135
Miliaria calandra, 232
Millennium Ecosystem Assessment (MEA), 355, 358
Mimus polyglottos, 223
Ministerio de Medio Ambiente Rural y Marino (MARM)
Missions, 29, 30, 321
Mitigation, 477
Monitoring, 486
Montado, 10
Montanera, 50, 249, 254, 291, 401
Montería, 331, 343
Montes, 8, 320
Mules, 288
Municipios, 40
Murcia, 40, 291
Mus
domesticus, 232
spretus, 234
Myrtus spp., 337

N
National forests, 296, 301
National parks, 32, 33
National Parks Law, 367
Native Americans, 27, 28, 35, 215, 221, 321, 470, 484, 492
Native Californians, 9, 16, 18, 27, 29, 32, 53, 85, 127, 133, 455
management, 27
Natural Patrimony Law, 367
Natural Resources Conservation Service (NRCS), 377
Nature Conservancy, The, 138, 477
Neolithic caves, 39

Net operating margin, 328, 395, 397, 412, 415, 418, 419
Net operating subsidies, 395
Net Present Value (NPV), 256–259, 265
Net value added, 396
NFI. *See* Spanish National Forest Inventory (NFI)
Non-market products, 472
Non-target species, 344
Normalized Difference Vegetation Index (NDVI), 185
North American Model of Wildlife Conservation, 322
Nurse plant effect, 168, 172
Nutrient dynamics, 150
  California, 106, 107, 114
  Spain, 96

O
Oak forest, 7, 160
Oak life cycle, 131
Oak removal, 111, 115, 127, 148, 161
Oak savanna, 7, 63
Oak woodland, 6, 431
Oak woodland loss, 127
Oak Woodlands Conservation and Environmental Quality Bill, 492
Odocoileus hemionus, 313, 315, 316
Olea europaea, 76
Option value, 357, 483
Oreortyx spp., 314
Organic Act, 33
Organizations, 477
Oryctolagus cuniculus, 227, 229, 313, 314
Output, 246, 254, 267, 355, 364, 392, 394, 397, 408, 409, 463, 471
  final, 357
  indicators, 401, 404, 405
  intermediate, 357
  market, 257
  non-market, 258
  valuation, 356
Overgrazing, 222, 227, 340, 367
Ovis
  musimon, 314
  orientalis musimon, 313

P
Parametric autoecological studies, 76
Passer domesticus, 232
Patches, 215
  proporties, 216
Pathogens, 188
  bacterial, 189
Perennial native grasses, 343
  bunchgrasses, 132, 147, 152
Perognathus inornatus, 215
PES. *See* ecosystem services
Phainopepla nitens, 224
Phasianus colchicus, 314
Phoradendron villosum, 223
Phyllirea spp., 337
Phytophthora
  cinnamomii, 82
  ramorum, 360
Phytophthora ramorum, 37, 266
Pica pica, 199, 335
Picoides nuttallii, 221
Pig, 5, 20, 27, 43, 45, 46, 49, 50, 148, 205, 208, 255, 273, 289, 290, 292, 300, 303, 313, 332, 412, 468, 476, 488. *See also* Sus scrofa, boars (wild)
  breeds, 289
  production, 291
  supplemental feeding, 401
Pinchot, Gifford, 33, 35
Pinus
  halepensis, 490
  nigra, 490
  pinaster, 490
  pinea, 76
  ponderosa, 220
  radiata, 490
  sabiana, 401
  sabiniana, 159, 221
  sylvestris, 490
Pistacia lentiscus, 337
Poaching, 40, 47, 319, 320, 333, 473
Poecile rufescens, 221
Pollination, 196
Post-war management, 37
Predator control, 335, 344
Private amenities, 254, 255, 259
  benefits, 391
Private Lands Management Program (PLM), 325
Product diversification, 474
Profitability, 247, 256, 317, 327, 391, 397, 407, 411, 417
  measure, 413, 416
  operating, 396, 413, 414
Profitability rates. *See* profitability
Progne subis arboricola, 220
Protobalanus, 199
Pruning, 43, 45, 125, 130, 191, 251, 279
  studies, 191

Psaltriparus minimus, 221
Psammodromus hispanicus, 232
Public partnerships, 477
Puma concolor, 335

Q

Quejigal, 8
Quejigos, 287
Quercus, 27, 199, 337
  agrifolia, 12, 64, 135, 147, 152, 185, 186, 190, 218, 266
  canariensis, 8, 147
  chrysolepis, 64, 185, 186, 435
  coccifera, 19, 39, 200
  douglasii, 7, 13, 64, 106, 127, 136, 139, 159, 185, 186, 198, 218, 266, 399, 435
  engelmannii, 12, 64, 128, 147, 190, 192, 435
  faginea, 8, 147, 200, 287, 431
  faginea broteroi, 76
  faginea faginea, 76
  garryana, 64, 75, 192, 435
  humilis, 185, 186
  ilex, 9, 13, 39, 76, 147, 185, 186, 203, 227, 246, 405, 431
  ilex ballota, 192, 278
  ilex ilex, 192
  kelloggii, 64, 185, 186, 220, 435
  lobata, 12, 64, 127, 185, 186, 197, 218, 220, 279, 434, 435
  pyrenaica, 8, 76, 147, 200, 392, 431
  robur, 189
  rotundifolia, 278
  rubra, 185
  suber, 7, 76, 147, 185, 186, 226, 278, 404, 405, 431, 488
  velutina, 185
  wislizenii, 12, 64, 147, 159, 192, 266, 435

R

Ranch, 7, 12, 13
Ranchers, 294
Ranching techniques, 295
Ranchos, 29, 31, 51
Rangeland improvement, 37
RE. *See* Rewilding Europe (RE)
RECAMAN. *See* Valoración de la Renta y el Capital de los Montes de Andalucía (RECAMAN)
Reconquest, 39
Recruitment, 132, 492
  failure, 130, 132, 138
  limitations, 131, 139, 165
  understory effects, 165, 167
Regeneration, 18, 125, 128, 130, 132, 133, 135, 137, 138, 171, 234, 247, 256, 266, 466, 467, 469, 489
  trade-offs, 196
Regeneration felling, 249
Regulation, 477
Rehaleros, 321
Remote Sensing (RS), 429–431
Report on the Protection of the Dehesa Ecosystem, 490
Reproductive stages, 130
Res nullius, 322
Research approaches, 478
  archival, 483
  comparative, 485
  field research, 478
  interviews, 479
  surveys, 480
Residual dry matter, 297
Restoration, 134
Retama, 154
Rewilding Europe (RE), 469
Riparian habitat, 219
Rock walls, 280
Roman, 6, 15, 39, 54, 41, 470
Root profiles, 162, 163
RS. *See* Remote Sensing (RS)
Rupicabra rupicabra, 314
Rural depopulation, 468

S

Sampson, 35
Scientific range management, 35, 44
Sciurus griseus, 316
Scolopax rusticola, 314
Seca disease, 82
SHARE. *See* Shared Habitat Alliance for Recreational Enhancement (SHARE)
Shared Habitat Alliance for Recreational Enhancement (SHARE), 325
Sheep, 282, 299
  production, 300
  products, 283
Shoot defoliation, 189
Shrub
  encroachment, 18, 39, 97, 125, 132, 160, 162, 167, 170, 171, 234, 236, 336, 468
  management, 337

Sialia mexicana, 223
Silvopastoral
  management, 19, 246, 247, 261, 266
  model, 246, 249, 258, 265, 267
  system, 280, 433
Simulated Exchange Value (SEV) method, 363
SISPARES. *See* Spanish Rural Landscape Monitoring System (SISPARES)
Soil, 17, 434
  California, 104, 105
  degradation, 97, 101
  dynamics comparison, 116
  effects on regeneration, 169
  erosion, 341
  properties, 92, 151
  properties, California, 104, 106, 107, 113
  properties, Spain, 94, 95
  Spain, 93, 95
  water dynamics, 103, 151
Soil types
  Acrisol, 93
  Alfisol, 105
  California, 105, 109
  cambisol, 93
  inceptisol, 105
  leptosol, 93
  luvisol, 93
  mollisol, 105
  Spain, 93
Spanish Civil War, 332
Spanish Constitution, 44
Spanish Ministry of Agriculture, 41
Spanish National Forest Inventory (NFI), 77, 247
Spanish Rural Landscape Monitoring System (SISPARES), 436
Spanish War of Independence, 44, 53
Spatial synchrony, 197
  Moran effect, 198
  pollen coupling, 198
Species densities, 225, 226, 229, 231
Species-environment interactions, 148
Species richness, 215, 216, 225, 236
State and transition models, 266
State Board of Forestry, 492
Stockponds, 219
Streptopelia turtur, 314
Strix occidentalis caurina, 361
Struthio camelus, 297
Study of Critical Environmental Problems (SCEP), 355
Subsidies, 38, 254, 286, 366, 367, 394
  planting, 137
Sudden Oak Death (SOD). *See Phytophthora ramorum*
Sus scrofa, 227, 313, 314, 316, 468. *See also* pig, boars (wild)
Sustainability, 493
Sylvilagus spp., 316
System of National Accounts (SNA), 256, 391, 392

T

Thomomys spp., 136
Thryomanes bewickii, 223
Top-down population regulation, 329
Topography, 451
Toro de lidia, 282
Toros bravos. *See* fighting bulls
Total Economic Value (TEV), 357
Transhumance, 32, 40, 46, 276, 280, 283, 296, 299, 381
Translation, 6, 479
Tree
  density, 158, 172, 253
  distribution, 161, 183, 184, 197, 279
  removal, 170
Turdus
  merula, 232
  spp., 314
TWINSPAN. *See* Two-Way Indicator Species Analysis (TWINSPAN)
Two-Way Indicator Species Analysis (TWINSPAN), 439, 447, 449

U

Ulex, 147
Understory, 153
  effects on wildlife, 231
  production, 149, 151, 155
  species composition, 148, 152
United Nations University—International Human Dimensions Programme and United Nations Environment Programme (UNU-IHDP and UNEP)
United States Forest Service, 32, 35, 223, 266, 301
Urban development, 223, 467
Urocyon cinereoargenteus, 335
Ursus arctos horribilis, 344
Useful Agricultural Land (UAL), 247

V

Valoración de la Renta y el Capital de los Montes de Andalucía (RECAMAN), 13, 365, 396
Vareador, 292
Vegetation conversion, 115, 116
Vegetation patterns, 452
Vegetation Type Map (VTM), 65, 67, 68, 70, 72, 73
Vicugna pacos, 297
Vineyards, 222
Vireo
  bellii pusillus, 219
  cassini, 223
  huttoni, 224
VTM. *See* Vegetation Type Map (VTM)
Vulpes
  macrotis mutica, 222
  vulpes, 335

W

Water dynamics, 17, 92, 162, 163
  California, 112
  comparison, 116
  Spain, 101
Water quality, 341, 375
  California, 114
Water yield, 115, 478
Weed control, 135, 136, 165
WHIP. *See* Wildlife Habitat Incentives Program (WHIP)
Wildlife, 281, 466
  California oak woodland, 215
  conservation status, California, 217
  grazing, 260
  habitat, 375, 471
  impacts on regeneration, 278
  landscape effects, 232
  practices comparison, 322
  Spain, 224
Wildlife Habitat Incentives Program (WHIP), 377
Wildlife management
  animal densities, 340
  genetic impacts, 341
  population structure, 343
  scale, 217, 233, 234
Williamson Act, 38, 368, 492
Woodland conversion, 223
Working landscape, 10, 236, 380, 391
World Trade Organization (WTO), 491
World War II, 6, 32, 35, 36, 54, 470

Z

Zoning, 368, 467